AF594963

Technological Advances in Seismic Data Processing and Imaging

Technological Advances in Seismic Data Processing and Imaging

Editors

Guofeng Liu
Zhifu Zhang
Xiaohong Meng

Basel • Beijing • Wuhan • Barcelona • Belgrade • Novi Sad • Cluj • Manchester

Editors
Guofeng Liu
China University of
Geosciences (Beijing)
Beijing, China

Zhifu Zhang
China University of
Geosciences (Beijing)
Beijing, China

Xiaohong Meng
China University of
Geosciences (Beijing)
Beijing, China

Editorial Office
MDPI
St. Alban-Anlage 66
4052 Basel, Switzerland

This is a reprint of articles from the Special Issue published online in the open access journal *Applied Sciences* (ISSN 2076-3417) (available at: https://www.mdpi.com/si/95670).

For citation purposes, cite each article independently as indicated on the article page online and as indicated below:

Lastname, A.A.; Lastname, B.B. Article Title. *Journal Name* **Year**, *Volume Number*, Page Range.

ISBN 978-3-0365-9788-1 (Hbk)
ISBN 978-3-0365-9789-8 (PDF)
doi.org/10.3390/books978-3-0365-9789-8

Contents

About the Editors

Guofeng Liu

Guofeng Liu graduated with a degree in Applied Geophysics from the China University of Geosciences (Beijing) in 2004, an M.Sc. degree in Earth Exploration and Information Technology in 2007, and a Ph.D. degree in Reservoir Geophysics in 2010 from the Chinese Academy of Sciences. Guofeng Liu is a professor at the School of Geophysics and Information Technology, China University of Geosciencesn (Beijing). Guofeng Liu has participated in more than 40 scientific projects and has published more than 50 papers. Guofeng Liu has wide research interests, including seismic imaging, high-performance computing, and passive seismic imaging.

Zhifu Zhang

Zhifu Zhang graduated with a degree in Applied Geophysics from the China University of Mining and Technology in 2000, an M.Sc. degree in Earth Exploration and Information Technology in 2003 in Guilin University of Technology, and a Ph.D. degree in Solid Geophysics in 2006 from the Chinese Academy of Sciences. Zhifu Zhang is a vice professor at the School of Geophysics and Information Technology, China University of Geosciences (Beijing). Zhifu Zhang has participated in more than 20 scientific projects and has published more than 30 papers. Zhifu Zhang mainly engaged in research on surface wave imaging.

Xiaohong Meng

Xiaohong Meng is a professor and doctoral supervisor at the School of Geophysics and Information Technology, China University of Geosciences (Beijing). She has devoted himself to front-line teaching for 42 years and has won honorary titles such as Beijing's Outstanding Young Teacher, and the Ministry of Education's IET Education Award. Xiaohong Meng has participated in more than 80 scientific projects and has published more than 150 papers. She has made great contributions to the processing and inversion of gravity and magnetic data.

Preface

Seismic exploration is a method with deeper exploration capabilities and higher resolution than potential geophysical methods, and has been widely used in the fields of oil and gas exploration, mineral exploration, engineering, and environmental exploration. Seismic data processing imvolves many conventional steps such as static correction, denoising, deconvolution, migration, and multiple eliminations, all of which play an important role in improving the final imaging quality. This Special Issue focuses on the development of advanced seismic data processing and imaging technologies and their successful application, and it provides a platform for scientific researchers to display their latest research results, which will provide a reference for practitioners engaged in oil and gas exploration, mineral exploration, and engineering exploration to solve complex problems. We are grateful to all authors who contributed articles to the Special Issue and to reviewers and editors who contributed to the publication of these papers.

Guofeng Liu, Zhifu Zhang, and Xiaohong Meng
Editors

Editorial

Special Issue on Technological Advances in Seismic Data Processing and Imaging

Guofeng Liu

School of Geophysics and Information Technology, China University of Geosciences (Beijing), Beijing 10083, China; liugf@cugb.edu.cn

Seismic exploration is a geophysical method with deeper exploration capabilities and higher resolution than potential geophysical methods [1], and has been widely used in the fields of oil and gas exploration, mineral exploration, engineering, and environmental exploration [2–4]. This Special Issue will focus on the development of advanced seismic data processing and imaging technologies and their successful application.

There are a total of 23 papers presented in this Special Issue, and we divide them into the following directions. In the seismic noise elimination section, five papers introduce their new methods to deal with multiples, periodic noise, coherent noise, and ground roll. Xiao et al. developed an iterative virtual event internal multiple suppression method for post-stack data and used it with the Tarim data successfully. Wu et al. presented another alternative based on sparse parabolic radon transform for multiple attenuation. Sun et al. proposed a new method to attenuate periodic noise based on sparse representation. Based on the multi-scale and multi-direction properties of the non-subsampled Shearlet transform, Yu et al. proposed a seismic coherent noise removal method. Liang et al. applied multivariate empirical mode decomposition to multi-component seismic data for the purposes of attenuating ground roll, and this method is more robust than traditional empirical mode decomposition. In the migration and inversion section, there are a total of seven papers about new improvements and applications. Zhou et al. presented a modified inverse scattering imaging condition in the reverse time migration to separate the effect of the impedance and velocity perturbations from the reflectivity. Zhou et al. discussed the reverse time migration imaging of compressional waves (P waves) and horizontally polarized shear waves (SH waves) of seismic data, together with P- and SH shear wave-constrained velocity model building, and the imaging quality of the Sanhu area in Qaidam Basin was greatly improved. Zhou et al. used the combined compact difference scheme and the combined super-compact difference scheme for the numerical simulation of the shear-wave reverse time migration, which proved to be practical and effective. Song et al. reported a method to improve insufficient illumination in reverse time migration in the presence of a shielding effect on seismic waves caused by overburden rock. Li et al. proposed a new least square migration method in the imaging domain to overcome unstable convergence and excessive computational cost problems. Sun et al. proposed a method for establishing the initial inversion model based on the velocity spectrum and Siamese network for the velocity inversion. Rachman presented a case study that performed travel time tomography in order to image the subsurface and approximate the Molucca Sea Plate subduction angle beneath Sulawesi's north arm. Gouda et al. reported a study that aims at estimating zero-offset acoustic and shear impedances based on partial-stack inversion. There are also two papers in this section that introduce the progress in passive seismic exploration. Fang et al. used synchro-squeezed continuous wavelet transform to eliminate the automatic noise in the Sichuan project. Liu et al. proposed a 3D method for passive reflection exploration, which can reduce the computing intensity. Seismic wave propagation is always a hot topic for research, and in this section, there are four papers about this field. Liu et al. discussed the simulation of elastic wave propagations, and the Poisson disk node generation algorithm

Citation: Liu, G. Special Issue on Technological Advances in Seismic Data Processing and Imaging. *Appl. Sci.* **2023**, *13*, 11789. https://doi.org/10.3390/app132111789

Received: 18 October 2023
Accepted: 23 October 2023
Published: 28 October 2023

and the centroid Voronoi node adjustment algorithm were combined to obtain an even and random node distribution to improve the accuracy of the simulation. Jin et al. studied seismic attenuation, focusing on accurately predicting the velocity dispersion at low porosity and permeability. Yang et al. applied the multi-axis perfectly matched layer to the wave-field simulations in anisotropic viscoelastic media to overcome instability. Sun et al. simulated the seismic wave propagation of the translational and rotational motions in two-dimensional isotropic and transverse isotropic media with a vertical axis of symmetry media under different source mechanisms with the staggered-grid finite-difference method, and also analyzed the characteristics. Artificial intelligence technology is also widely used in seismic data processing and interpretation, and there are three papers on this field in this Special Issue. Li et al. reported on velocity analysis, in which a modified U-Net framework is proposed and applied directly to the seismic shot gathersto identify anomalies in the early stage of velocity analysis and provide a suitable initial model. Wu et al. proposed a modified U-Net architecture in fault interpretation. Zheng et al. reported a multi-task deep learning solution for the impendence inversion when the number of logging curves is limited. In this Special Issue, we also present a paper about near-surface structure imaging. In order to realize a rapid, accurate, robust near-surface seismic imaging, a minimum variance spatial smoothing beamforming method is proposed.

The research field on seismic data processing and imaging is very extensive. Although we have published 23 articles in this Special Issue, there are still many interesting topics that have not been discussed. We hope to have more opportunities to publish further studies in the future, to foster further communication on this topic.

Funding: This research received no external funding.

Acknowledgments: Thanks to all the authors and peer reviewers for their valuable contributions to this Special Issue 'Technological Advances in Seismic Data Processing and Imaging'. I would also like to express my gratitude to all the staff and people involved in this Special Issue.

Conflicts of Interest: The author declares no conflict of interest.

List of Contribuitions

1. Xiao, M.; Xie, J.; Wang, W.; Liu, W.; Sun, J.; Jin, B.; Zhang, T.; Zhao, Y.; Wang, Y. Application of Iterative Virtual Events Internal Multiple Suppression Technique: A Case of Southwest Depression Area of Tarim, China. *Appl. Sci.* **2023**, *13*, 8832. https://doi.org/10.3390/app13158832.
2. Wu, Q.; Hu, B.; Liu, C.; Zhang, J. Sparse Parabolic Radon Transform with Nonconvex Mixed Regularization for Multiple Attenuation. *Appl. Sci.* **2023**, *13*, 2550. https://doi.org/10.3390/app13042550.
3. Sun, L.; Qiu, X.; Wang, Y.; Wang, C. Seismic Periodic Noise Attenuation Based on Sparse Representation Using a Noise Dictionary. *Appl. Sci.* **2023**, *13*, 2835. https://doi.org/10.3390/app13052835.
4. Yu, M.; Gong, X.; Wan, X. Seismic Coherent Noise Removal of Source Array in the NSST Domain. *Appl. Sci.* **2022**, *12*, 10846. https://doi.org/10.3390/app122110846.
5. Xiao, L.; Zhang, Z.; Gao, J. Ground Roll Attenuation of Multicomponent Seismic Data with the Noise-Assisted Multivariate Empirical Mode Decomposition (NA-MEMD) Method. *Appl. Sci.* **2022**, *12*, 2429. https://doi.org/10.3390/app12052429.
6. Liang, H.; Zhang, H.; Liu, H. Estimation of Relative Acoustic Impedance Perturbation from Reverse Time Migration Using a Modified Inverse Scattering Imaging Condition. *Appl. Sci.* **2023**, *13*, 5291. https://doi.org/10.3390/app13095291.
7. Zhou, C.; Yin, W.; Yang, J.; Nie, H.; Li, X. Reverse Time Migration Imaging Using SH Shear Wave Data. *Appl. Sci.* **2022**, *12*, 9944. https://doi.org/10.3390/app12199944.
8. Zhou, C.; Wu, W.; Sun, P.; Yin, W.; Li, X. The Combined Compact Difference Scheme Applied to Shear-Wave Reverse-Time Migration. *Appl. Sci.* **2022**, *12*, 7047. https://doi.org/10.3390/app12147047.
9. Song, L.; Shi, Y.; Liu, W.; Zhao, Q. Elastic Reverse Time Migration for Weakly Illuminated Structure. *Appl. Sci.* **2022**, *12*, 5264. https://doi.org/10.3390/app12105264.

10. Li, B.; Sun, M.; Xiang, C.; Bai, Y. Least-Squares Reverse Time Migration in Imaging Domain Based on Global Space-Varying Deconvolution. *Appl. Sci.* **2022**, *12*, 2361. https://doi.org/10.3390/app12052361.
11. Sun, L.; Ding, L.; Wang, X. Research on Initial Model Construction of Seismic Inversion Based on Velocity Spectrum and Siamese Network. *Appl. Sci.* **2022**, *12*, 10593. https://doi.org/10.3390/app122010593.
12. Rachman, G.; Santosa, B.J.; Nugraha, A.D.; Rohadi, S.; Rosalia, S.; Zulfakriza, Z.; Sungkono, S.; Sahara, D.P.; Muttaqy, F.; Supendi, P.; et al. Seismic Structure Beneath the Molucca Sea Collision Zone from Travel Time Tomography Based on Local and Regional BMKG Networks. *Appl. Sci.* **2022**, *12*, 10520. https://doi.org/10.3390/app122010520.
13. Gouda, M.F.; Abdul Latiff, A.H.; Moussavi Alashloo, S.Y. Estimation of Litho-Fluid Facies Distribution from Zero-Offset Acoustic and Shear Impedances. *Appl. Sci.* **2022**, *12*, 7754. https://doi.org/10.3390/app12157754.
14. Fang, J.; Liu, G.; Liu, Y. Application of an Automatic Noise or Signal Removal Algorithm Based on Synchrosqueezed Continuous Wavelet Transform of Passive Surface Wave Imaging: A Case Study in Sichuan, China. *Appl. Sci.* **2021**, *11*, 11718. https://doi.org/10.3390/app112411718.
15. Liu, Y.; Liu, G. Three-Dimensional Processing of Reflections for Passive-Source Seismology Based on Geometric Design. *Appl. Sci.* **2023**, *13*, 6126. https://doi.org/10.3390/app13106126.
16. Liu, S.; Zhou, Z.; Zeng, W. Simulation of Elastic Wave Propagation Based on Meshless Generalized Finite Difference Method with Uniform Random Nodes and Damping Boundary Condition. *Appl. Sci.* **2023**, *13*, 1312. https://doi.org/10.3390/app13031312.
17. Jin, Z.; Zheng, X.; Shi, Y.; Wang, W. Study on Seismic Attenuation Based on Wave-Induced Pore Fluid Dissolution and Its Application. *Appl. Sci.* **2023**, *13*, 74. https://doi.org/10.3390/app13010074.
18. Yang, J.; He, X.; Chen, H. Processing the Artificial Edge-Effects for Finite-Difference Frequency-Domain in Viscoelastic Anisotropic Formations. *Appl. Sci.* **2022**, *12*, 4719. https://doi.org/10.3390/app12094719.
19. Sun, L.; Wang, Y.; Li, W.; Wei, Y. The Characteristics of Seismic Rotations in VTI Medium. *Appl. Sci.* **2021**, *11*, 10845. https://doi.org/10.3390/app112210845.
20. Li, Z.; Jia, J.; Lu, Z.; Jiao, J.; Yu, P. Seismic Velocity Anomalies Detection Based on a Modified U-Net Framework. *Appl. Sci.* **2022**, *12*, 7225. https://doi.org/10.3390/app12147225.
21. Wu, J.; Shi, Y.; Wang, W. Fault Imaging of Seismic Data Based on a Modified U-Net with Dilated Convolution. *Appl. Sci.* **2022**, *12*, 2451. https://doi.org/10.3390/app12052451.
22. Zheng, X.; Wu, B.; Zhu, X.; Zhu, X. Multi-Task Deep Learning Seismic Impedance Inversion Optimization Based on Homoscedastic Uncertainty. *Appl. Sci.* **2022**, *12*, 1200. https://doi.org/10.3390/app12031200.
23. Peng, M.; Wang, D.; Liu, L.; Liu, C.; Shi, Z.; Ma, F.; Shen, J. Near-Surface Geological Structure Seismic Wave Imaging Using the Minimum Variance Spatial Smoothing Beamforming Method. *Appl. Sci.* **2021**, *11*, 10827. https://doi.org/10.3390/app112210827.

References

1. Lü, Q.; Qi, G.; Yan, J. 3D geologic model of Shizishan ore field constrained by gravity and magnetic interactive modeling: A case history. *Geophysics* **2013**, *78*, B25–B35. [CrossRef]
2. Malehmir, A.; Durrheim, R.; Bellefleu, G.; Urosevic, M.; Juhlin, C.; White., D. Seismic methods in mineral exploration and mine planning: A general overview of past and present case histories and a look into the future. *Geophysics* **2012**, *77*, WC173–WC190. [CrossRef]
3. Liu, G.; Meng, X.; Chen, Z.; Ni, J.; Zhang, D. Evaluation of the 2D reflection seismic method toward the exploration of thrust-controlled mineral deposits in southwestern Fujian Province, China. *Geophysics* **2018**, *83*, B209–B220. [CrossRef]
4. Sloan, S.D.; Peterie, S.L.; Miller, R.D.; Ivanov, J.; Schwenk, J.T.; McKenna, J.R. Detecting clandestine tunnels using near-surface seismic techniques. *Geophysics* **2015**, *80*, EN127–EN135. [CrossRef]

Article

Application of Iterative Virtual Events Internal Multiple Suppression Technique: A Case of Southwest Depression Area of Tarim, China

Mingtu Xiao [1], Junfa Xie [1], Weihong Wang [2,3,4,*], Wenqing Liu [1], Jiaqing Sun [1], Baozhong Jin [1], Tao Zhang [1], Yuhe Zhao [1] and Yihui Wang [1]

1 Northwest Branch, Research Institute of Petroleum Exploration & Development of Petro China, Lanzhou 730020, China; xiaomt@petrochina.com.cn (M.X.); xiejunfa@petrochina.com.cn (J.X.); liuwq@petrochina.com.cn (W.L.); sunjiaq@petrochina.com.cn (J.S.); jinbz@petrochina.com.cn (B.J.); zhangtao_xb@petrochina.com.cn (T.Z.); zhaoyuhe@petrochina.com.cn (Y.Z.); wangyh0801@petrochina.com.cn (Y.W.)

2 School of Earth Sciences, Northeast Petroleum University, Daqing 163318, China

3 National Engineering Research Center of Offshore Oil and Gas Exploration, Beijing 100027, China

4 Heilongjiang Provincial Key Laboratory of Oil and Gas Geophysical Exploration, Daqing 163318, China

* Correspondence: wangweihong@nepu.edu.cn

Abstract: The seismic records in the Cambrian southwest depression of the Tarim Basin exhibit discrepancies when compared to the actual geological setting, which is caused by the presence of multiples. Despite the application of the Radon transform, multiple interferences persist beneath the Cambrian salt in the pre-stack data, with significant variations in energy and frequency across the horizontal direction. In addition, other multiple suppression methods are also difficult to handle this problem. To address this issue, we have developed an iterative virtual event internal multiple suppression method for post-stack data. This novel algorithm extends the traditional virtual event internal multiple suppression approach, eliminating the need for data regularization and avoiding the problem of the traditional virtual events method requiring sequential extraction of primaries from relevant layers, which greatly improves computational efficiency and simplifies the implementation steps of the traditional method. Numerical experiments demonstrate the efficacy of our method in suppressing internal multiples in both synthetic and field data while preserving primary signals. When applied to real seismic data profiles, the iterative method yields structural characteristics that align more closely with sedimentary laws and reduces disparities in energy and frequency of multiples along the horizontal axis. Consequently, our method provides a robust foundation for subsequent hydrocarbon source rock prediction.

Keywords: internal multiple suppression; virtual events; iterative; post-stack data; field data

Citation: Xiao, M.; Xie, J.; Wang, W.; Liu, W.; Sun, J.; Jin, B.; Zhang, T.; Zhao, Y.; Wang, Y. Application of Iterative Virtual Events Internal Multiple Suppression Technique: A Case of Southwest Depression Area of Tarim, China. *Appl. Sci.* **2023**, *13*, 8832. https://doi.org/10.3390/app13158832

Academic Editors: Guofeng Liu, Zhifu Zhang and Xiaohong Meng

Received: 11 March 2023
Revised: 22 May 2023
Accepted: 31 May 2023
Published: 31 July 2023

1. Introduction

The increasing complexity of geological structures necessitates advancements in oil and gas exploration and development technologies, making multiple suppression an essential and challenging problem to address. The presence of multiples interferes with the identification of primary reflections, reducing the signal-to-noise ratio of seismic data, compromising the accuracy and reliability of seismic imaging, and potentially leading to erroneous interpretations and source rock investigations. Consequently, effective suppression of multiples is crucial for enhancing seismic data quality and facilitating oil and gas exploration and development. In northwest China, there are abundant resources of oil and natural gas. However, the complexity of its geological features and strong reflective interfaces in underground media, such as rock mounds, coal seams, basalt, etc., can generate strong internal multiples. The understanding of internal geological structures is constrained. Therefore, the demand for effective suppression of seismic multiples is very urgent.

Seismic multiples are generally classified into two categories based on the location of reflective layers: surface-related multiples and internal multiples. Surface-related multiples exhibit relatively stronger energy, particularly in offshore seismic data. Consequently, numerous researchers have focused on predicting and suppressing surface-related multiples, resulting in the maturation of suppression techniques for this category. In 1992, Verschuur et al. employed the concept of series expansion to decompose the seismic wavefield into primaries and multiples of all orders, effectively suppressing surface-related multiples [1]. This method obviates the need for subsurface geological model information, is a data-driven method, delivers superior results in field data processing, and exhibits high computational efficiency. Many geophysicists have expressed interest in and conducted research on this approach, leading to its widespread adoption in the exploration and development field. In 2009, Van Groenestijn and Verschuur proposed using sparse inversion to estimate primaries, eliminating the requirement for adaptive matched subtraction and near offset extrapolation [2,3]. This advancement significantly optimized the traditional surface-related multiple elimination (SRME) methods. In 2015, Ma et al. developed a suppression method for 3D surface-related multiples [4]. This technique assumes that the multiple contribution trace set corresponding to each seismic record in 3D seismic data is hyperbolic, enabling multiple predictions and suppression in 3D seismic data through sparse inversion. In 2022, Zhu et al. introduced a least squares datum continuation method for suppressing surface-related multiples within the least squares inversion theoretical framework [5]. This approach targets seafloor-separated up- and down-wave records, and, without multiple separation or fractional extraction, iterative inversion yields pre-stack seismic data from virtual observations on the seafloor, enhancing the resolution of effective signals and improving seismic imaging quality.

In comparison to surface-related multiples, internal multiples exhibit smaller differences in frequency, stacking velocity, and normal moveout (NMO) correction, making their suppression more challenging. In 1998, Jakubowicz introduced a data-driven method for internal multiple suppression [6]. This approach directly employs surface observed seismic data to construct internal multiples, enhancing computational efficiency. However, the method necessitates the accurate selection of primary reflection events from seismic data, which is difficult to implement in complex field data. In 2005, Berkhout and Verschuur proposed an internal multiple prediction method based on common focus technology, which is grounded in wavefield extrapolation [7,8]. This method is better suited to complex geological conditions but relies on a velocity model, limiting its application in field data. In 2006, Weglein et al. presented the inverse scattering series method based on the point scattering model [9]. As a data-driven, wave equation-based method, it does not require prior information such as velocity models. However, the method involves substantial computation and is only effective for near offsets, making it insufficient for current data processing demands. In 2013, Ypma and Verschuur introduced an internal multiple suppression method based on sparse inversion [10]. This approach minimizes the damage caused by adaptive matched subtraction to primary signals and effectively protects effective waves. Nevertheless, the method still entails considerable computation and yields unstable wavelets, complicating large-scale field data processing. In 2015, inspired by Marchenko's imaging method, Meles et al. proposed a self-focusing-based Marchenko internal multiple suppression method [11]. This technique does not require accurate velocity models, only an inaccurate macro velocity model to forward direct waves. Subsequently, direct waves and original data are used to construct the up- and down-going Green functions for relevant virtual source points, generating the associated internal multiples. However, this method demands significant computation due to the need for virtual source points at various depths, making it difficult to apply widely to large scale, complex structured seismic data. In 2018, Zhang and Staring proposed a one-step internal multiple suppression method based on Marchenko's self-focusing approach, also known as the data domain internal multiple suppression method [12]. This technique does not require macro velocity models for direct wave estimation and directly outputs primaries without internal multiples. However, this method imposes high requirements on the observation system, and its large-scale application in field data remains limited. In 2020, Zhang et al. applied the Marchenko internal multiple suppression method to field seismic data processing, achieving favorable results [13]. Nonetheless, in low signal-to-noise ratio scenarios, energy may not be sufficiently focused, rendering the Marchenko internal multiple suppression method ineffective. In 2022, Zhang et al. compared the self-focusing-based Marchenko method for internal multiple suppression with the data domain method [14]. In the same year, Zhang et al. introduced a multiple suppression method based on self-attention convolutional auto encoders [15]. This method can reduce artificial cost, reduce dependence on unknown prior information, and improve data processing efficiency.

In 2006 and 2009, Ikelle proposed a virtual seismic event construction method to predict internal multiples [16,17]. This approach does not require prior information, such as velocity models or subsurface structures, enabling accurate internal multiple predictions. However, this method requires sequential extraction of primaries to construct internal multiples generated by relevant layers, which increase the computational complexity. In 2013, Wu et al. utilized a multi-channel L_1 norm adaptive matching algorithm to suppress multiples, considering the differences in amplitude and phase between virtual event method-predicted internal multiples and actual internal multiples, achieving favorable results in model data [18]. In 2018, Liu et al. introduced an adaptive virtual event method to address the excessive dependency on the matching algorithm in the traditional virtual event method, enhancing internal multiple suppression capabilities [19]. Although researchers have improved and modified the virtual event method to address various issues, the traditional virtual event method still requires sequential internal multiple constructions on subsurface interfaces, incurring high computational costs for seismic data derived from complex structures and making primary extraction difficult. To overcome this challenge, we propose an iterative virtual event method. Compared to the traditional virtual event method, the iterative virtual event technique incorporates an iterative calculation process, mitigates the influence of false frequencies generated by data space convolution on predicted multiple models, and significantly improves computational efficiency. When applied to actual onshore post-stack seismic data from the southwest depression of the Tarim Basin, our method effectively suppresses internal multiples that are challenging to eliminate in pre-stack traces, emphasizes the energy of deep effective waves, and further enhances the quality of seismic profiles.

The paper is organized as follows. After the introduction, we provide an overview of the study region. Subsequently, we present the theory of virtual events for internal multiple eliminations, followed by synthetic and field data experiments that validate the effectiveness of our approach. Finally, we draw conclusions based on detailed analyses and discussions.

2. The Overview of the Study Region

The Southwest Depression of the Tarim Basin, situated in the southwestern part of the basin, is bordered by the South Tianshan orogenic belt to the north and the West Kunlun orogenic belt to the south. Encompassing an area of approximately 100×100 km^2, it is the largest sub-basin within the Tarim Basin and a favorable area for oil and gas exploration in Cambrian Ordovician carbonate rocks. The traps formed during the Himalayan movement predominantly consist of effective hydrocarbon source rocks distributed in the Bachu fault uplift and Maigaiti slope area (Qu et al., 2000) [20]. After extensive research, a consensus has emerged concerning the Cambrian subsalt of the Maigaiti slope: the Maigaiti slope contains Ordovician Carboniferous oil and gas sources, while drilling through the Cambrian south of Bachu revealed no Cambrian source rock, indicating the development of Cambrian subsalt source rock in the slope area. The Middle-Lower Cambrian serves as an important reservoir and regional cap rock for the area's deep carbonate rocks and is a significant horizon for hydrocarbon source rock development (Cui et al., 2017; Zhang et al., 2015) [21,22]. Its seismic sequence features three strong reflective interfaces within the Middle Cambrian. The energy at the bottom boundary of the Lower Cambrian is not strong, whereas the bottom of the upper salt section displays a continuous and stable strong wave crest. A noticeable phase change occurs at the bottom of the lower salt section, and the sedimentary pattern from the Early Cambrian to Middle Cambrian exhibits significant changes.

The three strong reflective interfaces in the Middle Cambrian provide favorable conditions for internal multiple formations. However, internal multiples can distort seismic waveforms, decrease the signal-to-noise ratio, and affect effective wave identification, thereby complicating seismic processing, diminishing seismic imaging authenticity and reliability, influencing hydrocarbon source rock studies, and increasing the difficulty of oil and gas exploration and development in the northwest region. To better understand and identify the internal multiples in the study area, well-seismic calibration was performed using existing seismic and logging information, as illustrated in Figure 1. The well-seismic calibration results reveal that the bottom interface of the upper salt section at the Middle Cambrian is stable. The lithology of the lower salt section of the Middle Cambrian comprises pure gypsum salt rock with a strong wave crest at the bottom, while the lower salt section contains gypsum-bearing dolomite and exhibits a trough at the bottom boundary. The bottom boundary of the Lower Cambrian primarily appears as a trough, and the time thickness of the Lower Cambrian varies significantly across different drilling wells, as does the mismatch between well synthetic records and seismic records. Subsequently, velocity spectra were generated for multiple identification and confirmation, as depicted in Figure 2. Figure 2 displays the velocity spectra of two distinct CMP gathers, revealing numerous low-velocity energy clusters. Following NMO correction of CMP gathers using primary velocity, downward-bending events indicate the presence of a large number of multiples in the original seismic data, necessitating multiple suppression processing.

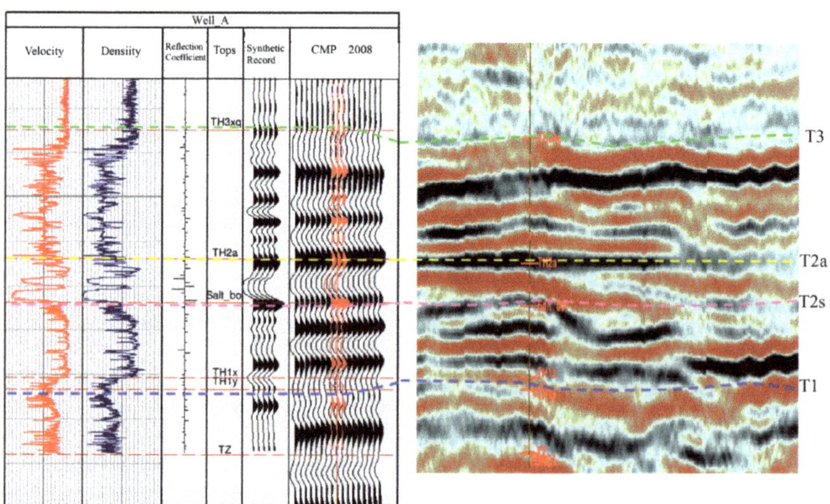

Figure 1. Well A synthetic record calibration and well seismic profile.

Considering the characteristics of multiples in the study area, we initially employed the Radon transform to suppress internal multiples in the pre-stack data. However, the Radon transform cannot entirely eliminate all internal multiples present in the data, and this method can only suppress some long-period multiples. Multiples interference still exists under the Cambrian salt, as shown by the red circle in the velocity spectrum in Figure 3. Additionally, there are significant differences in energy and frequency of multiples in the horizontal direction. It is challenging to suppress short-path internal multiples in the pre-stack data, necessitating the selection of an appropriate multiple suppression method from the post-stack data for this purpose. The iterative method for suppressing internal multiples based on virtual events is capable of accurately predicting internal multiples without requiring prior information, such as velocity models and underground structures, while maintaining high computational efficiency. Consequently, we opt for the iterative method to suppress internal multiples in the seismic data for the study area.

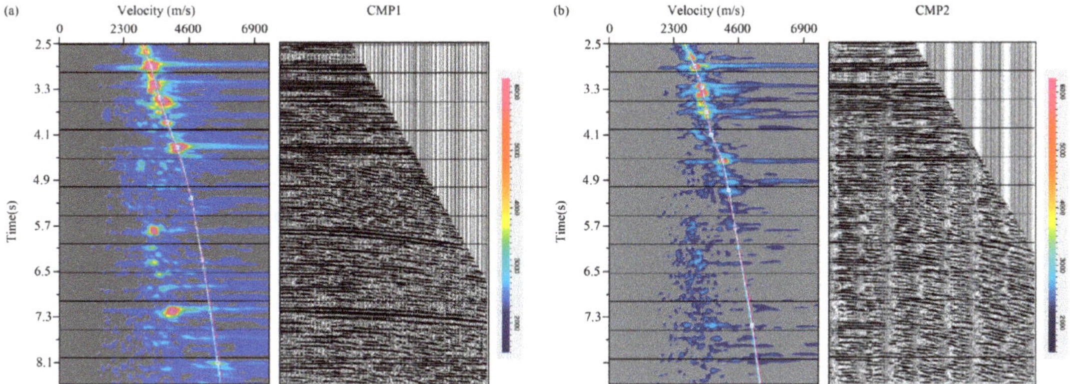

Figure 2. Velocity spectrum of the different original CMP trace sets; (**a**) velocity spectrum of CMP1; (**b**) velocity spectrum of CMP2.

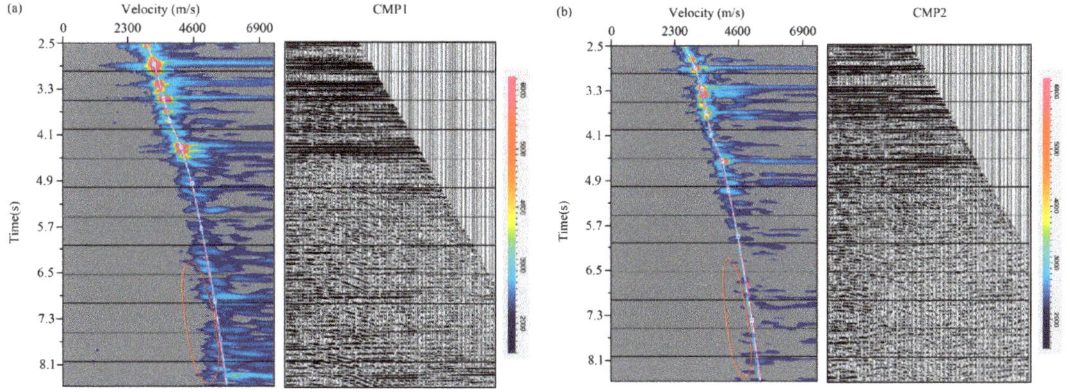

Figure 3. Velocity spectrum of the different CMP gathers after multiple suppression by the Radon transform (corresponds to Figure 2); (**a**) velocity spectrum of CMP1; (**b**) velocity spectrum of CMP2.

3. Internal Multiple Elimination Using the Theory of Virtual Events

The iterative virtual event internal multiple suppression approach extends the traditional virtual event internal multiple suppression method and avoids constructing internal multiple from every layer, which greatly improves computational efficiency [23,24]. Seismic virtual events are not directly recorded in standard seismic data acquisition, but their existence can help us construct internal multiples with scattering points at the sea surface [16,17]. Therefore, the total seismic wave field data $P(x_R, x_S, \omega)$ received from the surface can be used to successfully construct internal multiples. The process of predicting internal multiples mainly includes two parts. For the internal multiples related to the first underground interface, it is necessary to first extract the primary event $P_i(x_R, x_S, \omega)$ of ith interface and the primary $\overline{P_i}(x_R, x_S, \omega)$ generated by the interface below this interface. Then, correlating the primary $\bar{P_i}(x_R, x_S, \omega)$ generated by other interfaces below this interface with the time reversal $P_i^H(x_R, x_S, \omega)$ of the primary event $P_i(x_R, x_S, \omega)$ to build the virtual event $P_{iv}(x_R, x_S, \omega)$ of this interface (Figure 4a). Finally, the convolution operation is carried out by using the constructed virtual event $P_{iv}(x_R, x_S, \omega)$ and the primary $\bar{P_i}(x_R, x_S, \omega)$ generated by the interface below this interface to construct the internal multiples about this interface (Figure 4b). Namely,

$$P_{iv}(x_R, x_S, \omega) = \sum_x \bar{P_i}(x, x_S, \omega) P_i^H(x_R, x, \omega) \quad (1)$$

$$M_i(x_R, x_S, \omega) = \sum_x P_{iv}(x_R, x, \omega) \bar{P_i}(x, x_S, \omega) \quad (2)$$

where x represents any point on the surface, R is the receiver point, S is the source, H represents complex conjugation, and ω represents angular frequency. The subscripts represent the number of layers, namely, the lower subscript i in Formulas (1) and (2) represents the ith layer. $P_{iv}(x_R, x_S, \omega)$ represents the constructed post-stack virtual events. $M_i(x_R, x_S, \omega)$ is the internal multiple of the constructed post-stack data.

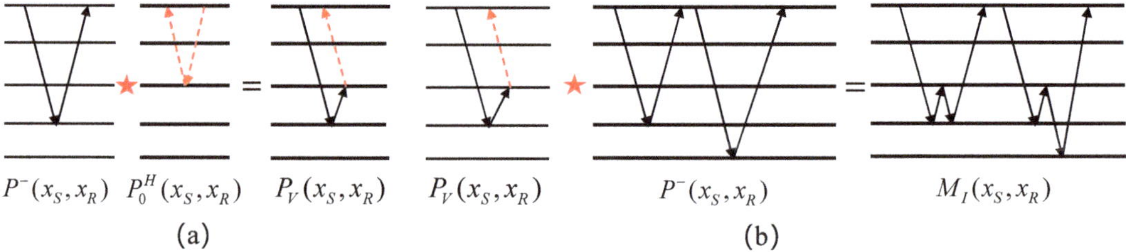

Figure 4. Construction of the virtual events and the internal multiples; (**a**) construction of the virtual events; (**b**) construction of the internal multiples. There pentagram represents convolution, the black line and black arrow represent the actual detectable seismic wave paths, red line and red arrow is used to assist in constructing virtual events and internal multiples.

Figure 4 shows the construction process of virtual events and related internal multiples; constructing post-stack virtual events $P_v(x_R, x_S, \omega)$ for the correlation of delay wave $\bar{P}(x, x_S, \omega)$ and lead wave $P^H(x_R, x, \omega)$. Then, the constructed virtual events and primary convolution is used to obtain the internal multiples of the post-stack-related layers $M(x_R, x_S, \omega)$, where the red star represents the convolution operation.

Assuming that the median of seismic data does not include surface-related multiples but only internal multiples, the effective wave after internal multiples suppression is as follows:

$$P_0(x_R, x_S, \omega) = P(x_R, x_S, \omega) - M(x_R, x_S, \omega), \quad (3)$$

where $P_0(x_R, x_S, \omega)$ is the data of internal multiple suppression and $P(x_R, x_S, \omega)$ is the original seismic data. According to the estimation of multiple scattering by iterative inversion [25,26], the multiple can be written as follows:

$$M(x_R, x_S, \omega) = P_0(x_R, x_S, \omega)\}_0 AP(x_S, x_R, \omega) \quad (4)$$

where A is the surface factor. Substituting Formula (4) into Formula (3):

$$P_0(x_R, x_S, \omega) = P(x_R, x_S, \omega) - P_0(x_R, x_S, \omega)\}_0 AP_0(x_S, x_R, \omega) \quad (5)$$

In terms of the Neumann series, Formula (5) can be written as follows:

$$\{P_0(x_R,x_S,\omega)\}^{(n)} = P(x_R,x_S,\omega) - \{P_0(x_R,x_S,\omega)\}_0^{(n-1)}A^{(n)}P(x_R,x_S,\omega), \tag{6}$$

where n represents the number of iterations and the expression of surface factor A is as follows:

$$\begin{aligned} A^{(n)} = {} & \left(\{P_0(x_R,x_S,\omega)\}_0^{(n-1)}\right)^{-1}(P(x_R,x_S,\omega) \\ & - \{P_0(x_R,x_S,\omega)\}^{n})(P(x_R,x_S,\omega))^{-1} \end{aligned} \tag{7}$$

In order to obtain more abundant internal multiple information, the multiple iteration theory can be used for multiple models. The multiples after iteration n + 1th can be expressed as follows:

$$\{M(x_R,x_S,\omega)\}^{(n+1)} = \{P_0(x_R,x_S,\omega)\}_0^{(n)}A^{(n)}P(x_R,x_S,\omega) \tag{8}$$

The multiples after n iterations are expressed as follows:

$$\begin{aligned} \{M(x_R,x_S,\omega)\}^{(n)} & = \{P_0(x_R,x_S,\omega)\}_0^{(n-1)}[\{P_0(x_R,x_S,\omega)\}_0^{(n-2)}]^{-1} \\ & \times(P(x_R,x_S,\omega) - \{P_0(x_R,x_S,\omega)\}_0^{(n-1)}) \end{aligned} \tag{9}$$

Bringing Formula (7) into Formula (9), the internal multiple model after n iterations is as follows:

$$\begin{aligned} \{M(x_R,x_S,\omega)\}^{(n)} & = \{P_0(x_R,x_S,\omega)\}_0^{(n-1)}\left(\{P_0(x_R,x_S,\omega)\}_0^{(n-1)}\right)^{T} \\ & \times[\{P_0(x_R,x_S,\omega)\}_0^{(n-2)}(\{P_0(x_R,x_S,\omega)\}_0^{(n-2)})^T]^{-1} \\ & \times(P(x_R,x_S,\omega) - \{P_0(x_R,x_S,\omega)\}_0^{(n-1)}) \end{aligned} \tag{10}$$

where T represents matrix transpose.

The primary focus of this article is on the suppression of internal multiples in post-stack profiles. In the post-stack profile, firstly, we can employ the tracing of horizons method to obtain the primary events. Secondly, by shifting the window up and down, we can acquire the primary data volume and other primary data volume below this primary used to construct the virtual events. Subsequently, we input the seismic data obtained, the iterative virtual event internal multiple suppression technique initially utilizes the virtual events internal multiple methods, such as Formulas (1) and (2), to construct the internal multiple models for the corresponding layer. Finally, through iterative updates, such as Formula (10), a more comprehensive multiple model is obtained, ultimately leading to the suppression of the multiple results. This approach not only enhances the computational efficiency, accuracy, and applicability but also reduces the requirements for the original input data, thereby improving imaging accuracy, which can make it widely applicable in the processing of field data.

4. Numerical Examples

To verify the effectiveness of the iterative virtual events internal multiple suppression method, first of all, we use a simple horizontal layered model as a test case, with the velocity model displayed in Figure 5. The velocity model size is 4000 m × 1600 m, and forward modeling is carried out using the higher order finite difference method based on the acoustic equation. An absorption boundary is applied at the top of the model. At the acquisition surface, seismic data are modeled with 501 sources and 501 receivers on a fixed spread with a spacing of 20 m, and the grid size is 2.5 m × 2.5 m. The source emits a Ricker wavelet with a 25 Hz center frequency. The number of time sampling points is 1024, and the time sampling interval is 4 ms. Figure 6 shows the seismic record of one shot, including primaries and internal multiples in forward modeling, with the direct wave removed. Since the primary focus of this paper is the suppression of internal multiples in post-stack data, zero offset gathers are extracted from the forward shot records for internal multiple suppression. The extracted zero offset gathers are shown in Figure 7a.

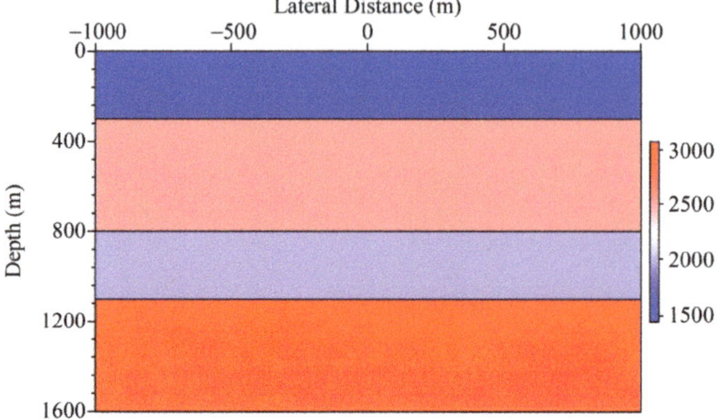

Figure 5. Velocity model.

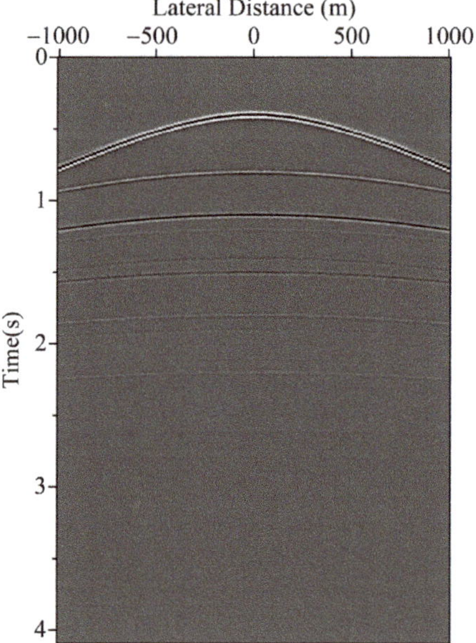

Figure 6. Original seismic data with the internal multiple.

The iterative virtual event multiple suppression method is applied to suppress the internal multiples, and the resulting zero offset gathers are displayed in Figure 7b. The figure reveals that all internal multiples are essentially suppressed without damaging the primaries, indicating that the iterative virtual event method can effectively suppress stacked multiples in model data. Figure 8a displays the predicted initial internal multiples model, while Figure 8b shows the internal multiples after iteration. The internal multiple information after iteration is more abundant and matches the actual multiples better. The suppression effect is sufficiently ideal, demonstrating that the iterative virtual event multiple suppression method can effectively suppress the internal multiples in synthetic data.

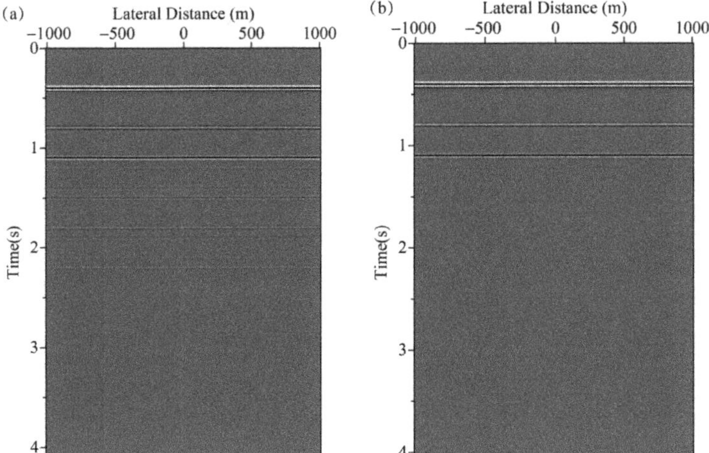

Figure 7. Zero-offset data with the internal multiples suppression. (**a**) Original zero-offset data with the internal multiples; (**b**) result after the internal multiples are suppressed.

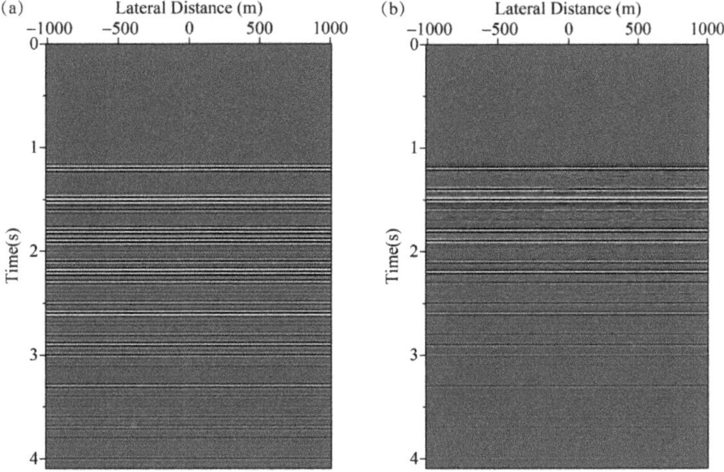

Figure 8. The predicted internal multiple. (**a**) The predicted internal multiples using the traditional virtual events method; (**b**) the predicted internal multiples after iterating.

5. Field Example

Per the interpreter's request, the primary focus is on suppressing internal multiples below 6 s to observe the original seismic profile more clearly. The seismic profile below 4 s is shown in Figure 9. From Figure 9a, it can be seen that the events below 6 s are similar in wave shape to the strong energy primaries, and the energy and frequency of these events are significantly different in the horizontal direction. Based on the relevant characteristics of multiples and multiples identification methods, these events are determined as internal multiples generated by the three sets of strong reflection interfaces of Middle Cambrian. Proof by facts, these internal multiples are challenging to suppress in pre-stack data using the filtering methods, such as the Radon transform method. Therefore, on top of the pre-stack suppression of multiples, we suppressed those internal multiples that are difficult to suppress in post-stack data using the iterative virtual events internal multiples suppression method.

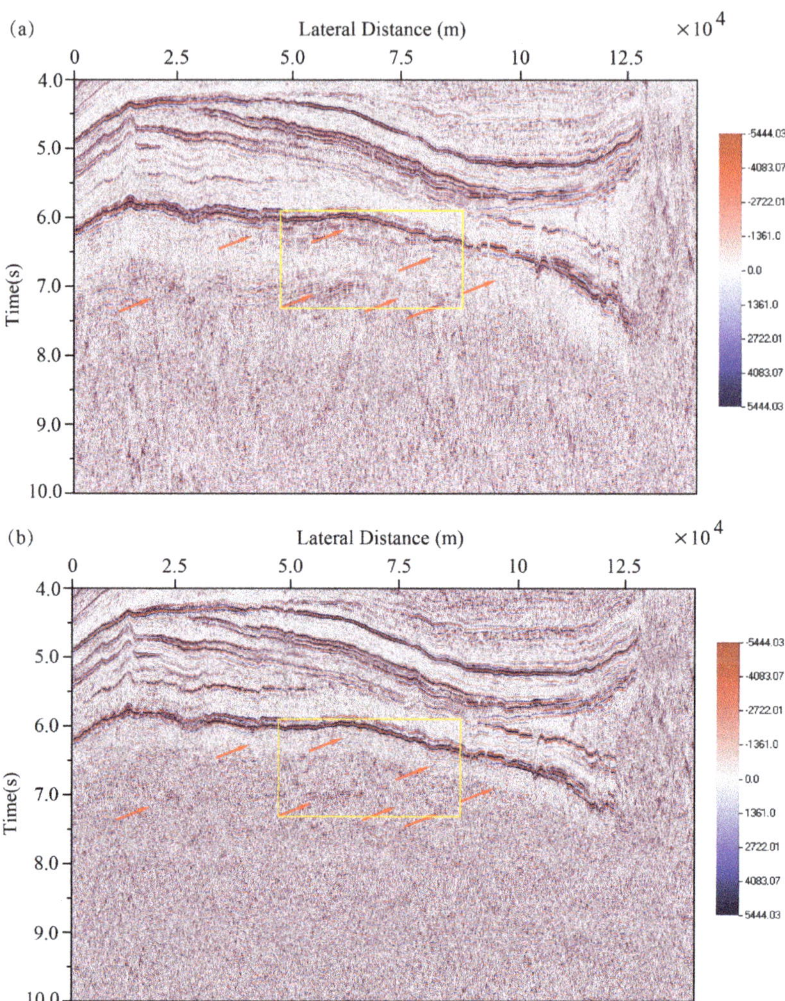

Figure 9. Profile comparison before and after the internal multiple suppression. (**a**) Original stacked profile; (**b**) stacked section after the internal multiple suppression.

Figure 9 shows the profile comparison before and after the internal multiple suppression. Figure 9a displays the partial stack profile before internal multiple suppression, while Figure 9b shows the corresponding stack profile after the internal multiple suppression. As seen in Figure 9b, the internal multiples below 6 s are mostly suppressed, and the horizontal difference in energy and frequency of the events below 6 s is smaller after the internal multiples are suppressed, particularly at the position indicated by the red arrow. To more clearly observe the effect of internal multiple suppression before and after, the part in the yellow box in Figure 9 is enlarged, as shown in Figure 10. In Figure 10, the event, which is similar to the upper strong reflection waveform with the opposite phase, periodic appearance, and significant lateral variation, is essentially suppressed. This is consistent with the analysis and identification results of multiples discussed earlier (Figures 2 and 3). The results demonstrate that the iterative virtual event multiple suppression method can effectively suppress the internal multiples in field data.

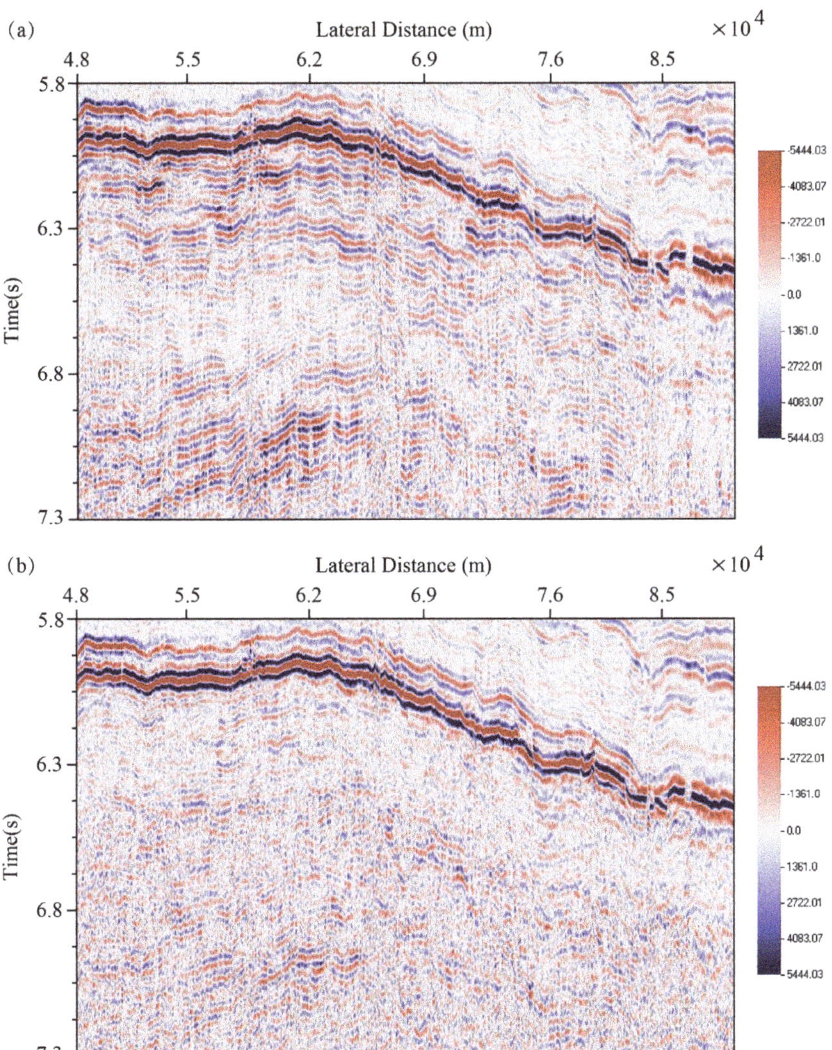

Figure 10. Part of profile comparison before and after the internal multiple suppression. (**a**) Original stacked profile; (**b**) stacked section after the internal multiple suppression.

To further analyze the suppression effect and accuracy of internal multiples, some sections before and after internal multiple suppression are selected for spectrum analysis, as shown in Figure 11. Figure 11a is the partial superimposed section before internal multiple suppression, Figure 11b is the partial superimposed section after internal multiple suppression, and Figure 11c shows the frequency distribution curve before and after internal multiple suppression. It can be seen from the figure that after suppressing internal multiples, the frequency spectrum is significantly broadened, and the spectral energy of the seismic profile is increased. This indicates that while using the iterative virtual event technology to suppress the internal multiples, the effective wave is protected, and the resolution of seismic data is improved. The method can provide reliable seismic data for subsequent studies of hydrocarbon source rocks.

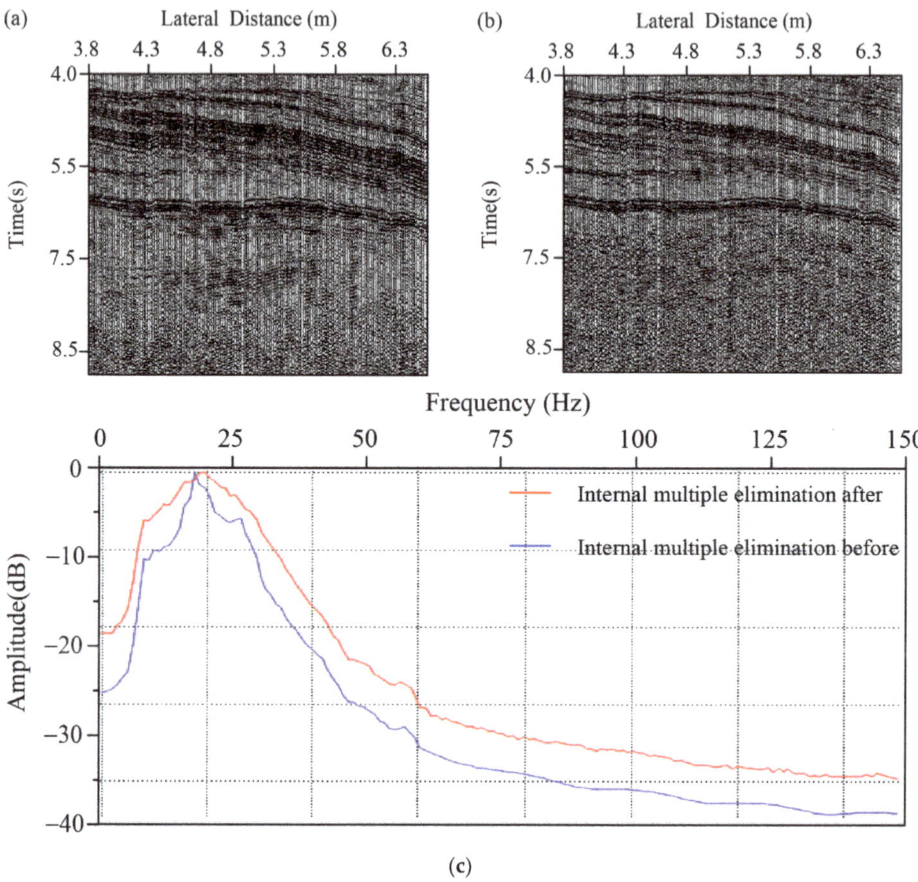

Figure 11. Spectrum analysis before and after the internal multiple suppression. (**a**) Partial stacked sections before the internal multiple suppression; (**b**) partial stack section after the internal multiple suppression; (**c**) frequency distribution curve before and after the internal multiple suppression.

6. Conclusions

The iterative virtual events method, when used to suppress internal multiples, involves first using the traditional virtual events method to build the initial multiple models, and then iterating the initial multiple models to complete the suppression of internal multiples. This approach not only simplifies the process of multiple suppression and improves computational efficiency and accuracy, but also obtains more abundant multiple information. From the model test, field data processing, and profile and spectrum analysis before and after processing, we can conclude the following:

(1) The internal multiples in the post-stack data are suppressed using the iterative virtual event internal multiples suppression method, significantly reducing the number of events close to the primary shape.
(2) After suppressing the internal multiples, the spectrum is significantly broadened, the overall energy is raised, and the resolution is improved.
(3) The phenomenon of multiples' energy and frequency varying greatly in the horizontal direction is weakened. The obtained seismic profile is more conforming to the actual geologic structure.

These results indicate that this method can effectively suppress the internal multiples in seismic data and can be widely used in the internal multiples suppression of field data.

Author Contributions: Conceptualization, M.X. and W.W.; methodology, J.X. and W.L.; software, J.S.; validation, W.L., J.S. and B.J.; formal analysis, Y.Z. and W.L.; investigation, T.Z. and Y.Z.; resources, Y.W. and B.J.; data curation, J.X., Y.W. and T.Z.; writing—original draft preparation, M.X. and Y.W.; writing—review and editing, W.W. and J.X.; visualization, B.J. and M.X.; supervision, W.W. and W.L.; project administration, Y.Z.; funding acquisition, T.Z. and W.W. All authors have read and agreed to the published version of the manuscript.

Funding: We thank the prospective basic projects of CNPC under grant no. 2021DJ-0304 and the National Natural Science Foundation of China under grant nos. 41974116 and 4227040468 for supporting this work.

Institutional Review Board Statement: Not applicable.

Informed Consent Statement: Not applicable.

Data Availability Statement: Not applicable.

Conflicts of Interest: The authors declare no conflict of interest.

References

1. Verschuur, D.J.; Berkhout, A.J.; Wapenaar, C.P.A. Adaptive surface-related multiple elimination. *Geophysics* **1992**, *57*, 1166–1177. [CrossRef]
2. Van Groenestijn, G.J.A.; Verschuur, D.J. Estimation primaries by sparse inversion and application to near-offset data reconstruction. *Geophysics* **2009**, *74*, A23–A28. [CrossRef]
3. Van Groenestjin, G.J.A.; Verschuur, D.J. Estimation primaries by sparse inversion near-offset reconstruction: Marine data applications. *Geophysics* **2009**, *74*, R119–R128. [CrossRef]
4. Ma, J.; Chen, X.; Xue, Y. 3D surface-related multiple elimination. *Oil Geophys. Prospect.* **2015**, *50*, 14+33–40. [CrossRef]
5. Zhu, F.; Cheng, J.; Wang, T.; Geng, J. Elimination of free-surface multiples using least-squares redatuming II: Marine seismic data acquired by streamers. *Chin. J. Geophys. Chin.* **2022**, *65*, 3123–3138. [CrossRef]
6. Jakubowicz, H. Wave equation prediction and suppression of interbed multiples. In *SEG Technical Program Expanded Abstracts*; Society of Exploration Geophysicists: New Orleans, LA, USA, 1998; pp. 1527–1530. [CrossRef]
7. Berkhout, A.J.; Verschuur, D.J. Removal of internal multiples with the common-focus-point (CFP) approach: Part 1—Explanation of the theor. *Geophysics* **2005**, *70*, V45–V60. [CrossRef]
8. Verschuur, D.J.; Berkhout, A.J. Removal of internal multiples with the common-focus-point (CFP) approach: Part 2—Application strategies and data examples. *Geophysics* **2005**, *70*, V61–V72. [CrossRef]
9. Weglein, A.B.; Nita, B.; Innanen, K.; Otnes, E.; Shaw, S.A.; Liu, F.; Zhang, H.; Ramírez, A.C.; Zhang, J.; Pavlis, G.L.; et al. Using the inverse scattering series to predict the wave-field at depth and the transmitted wave-field without an assumption about the phase of the measured reflection data or back propagation in the overburden. *Geophysics* **2006**, *71*, 1125–1137. [CrossRef]
10. Ypma, F.H.C.; Verschuur, D.J. Estimating primaries by sparse inversion, a generalized approach. *Geophys. Prospect.* **2013**, *61*, 94–108. [CrossRef]
11. Meles, G.A.; Löer, K.; Ravasi, M.; Curtis, A.; Filho, C.A.D.C. Internal multiple prediction and removal using Marchenko autofocusing and seismic interferometry. *Geophysics* **2015**, *80*, A7–A11. [CrossRef]
12. Zhang, L.; Staring, M. Marchenko scheme based internal multiple reflection elimination in acoustic wavefield. *J. Appl. Geophys.* **2018**, *159*, 429–433. [CrossRef]
13. Zhang, L.; Slob, E. A field data example of Marchenko multiple elimination. *Geophysics* **2020**, *82*, S65–S70. [CrossRef]
14. Zhang, L.; Shao, J.; Zheng, Y.; Wang, Y.; Slob, E. A comparison of imaging domain and data domain Marchenko multiple elimination schemes. *Chin. J. Geophys. Chin.* **2022**, *65*, 3123–3138. [CrossRef]
15. Zhang, M. A multiple suppression method based on self attention convolutional auto encoder. *Geophys. Prospect. Pet.* **2022**, *61*, 454–462.
16. Ikelle, L.T. A construct of internal multiples from surface data only: The concept of virtual seismic events. *Geophys. J. Int.* **2006**, *164*, 383–393. [CrossRef]
17. Ikelle, L.T.; Erez, I.; Yang, X. Scattering diagrams in seismic imaging: More insights into the construction of virtual events and internal multiples. *J. Appl. Geophys.* **2009**, *67*, 150–170. [CrossRef]
18. Wu, J.; Wu, Z.; Hu, T.; He, Y.; Wang, P.; Yan, G.; Li, L. Seismic internal multiple attenuation based on constructing virtual events. *Chin. J. Geophys. Chin.* **2013**, *56*, 985–994. [CrossRef]
19. Liu, J.; Hu, T.; Peng, G. Suppressing seismic inter-bed multiples with the adaptive virtual events method. *Chin. J. Geophys. Chin.* **2018**, *61*, 1196–1210. [CrossRef]

20. Qu, Q.; Tang, Y.; Wang, H.; Lang, W. The petroleum system and objective interval for exploration of cambrian in southwest de-pression of Tarim Basin. *Xing Jiang Pet. Geol.* **2000**, *21*, 101–104+168–169.
21. Cui, H.; Tian, L.; Liu, J.; Zhang, N. Ordovician tectonic activities and oil and gas accu-mulation in the Southwest Depression of Tarim Basin. *Oil Geophys. Prospect.* **2017**, *52*, 834–840+850.
22. Zhang, L.; Han, E.; Zhu, L.; Zeng, C.; Fan, Q.; Wu, K.; Cao, Y.; Jiao, P. Characteristics of Evaporites Sedimentary Cycles and Its Controlling Factors of Paleocene Aertashi Formation in the Southwestern Tarim Depression. *Acta Geol. Sin.* **2015**, *89*, 2161–2170.
23. Bao, P.; Wang, X.; Xie, J.; Wang, W.; Shi, Y.; Xu, J.; Yang, Y. Internal multiple suppression method based on iterative inversion. *Chin. J. Geophys. Chin.* **2021**, *64*, 2061–2072. [CrossRef]
24. Bao, P.; Shi, Y.; Wang, W.; Zhang, W.; Pan, Z. An improved method for internal multiple elimination using the theory of virtu-al events. *Pet. Sci.* **2022**, *19*, 2663–2674. [CrossRef]
25. Berkhout, A.J.; Verschuur, D.J. Estimation of multiple scattering by iterative inversion, Part I: Theoretical considerations. *Geophysics* **1997**, *62*, 1586–1595. [CrossRef]
26. Verschuur, D.J.; Berkhout, A.J. Estimation of multiple scattering by iterative inversion, Part II: Practical aspects and examples. *Geophysics* **1997**, *62*, 1596–1611. [CrossRef]

Article

Sparse Parabolic Radon Transform with Nonconvex Mixed Regularization for Multiple Attenuation

Qiuying Wu, Bin Hu *, Cai Liu * and Junming Zhang

College of GeoExploration Science and Technology, Jilin University, Changchun 130026, China
* Correspondence: binhu@jlu.edu.cn (B.H.); liucai@jlu.edu.cn (C.L.)

Abstract: The existence of multiple reflections brings difficulty to seismic data processing and interpretation in seismic reflection exploration. Parabolic Radon transform is widely used in multiple attenuation because it is easily implemented, highly robust and efficient. However, finite seismic acquisition aperture of seismic data causes energy diffusion in the Radon domain, which leads to multiple residuals. In this paper, we propose a sparse parabolic Radon transform with the nonconvex $L_{q_1} - L_{q_2} (0 < q_1, q_2 < 1)$ mixed regularization (SPRT$L_{q_1} - L_{q_2}$) that constrains the sparsity of primary and multiple reflections to overcome the energy diffusion and improve the effect of multiple attenuation, respectively. This nonconvex mixed regularization problem is solved approximately by the alternating direction method of multipliers (ADMM) algorithm, and we give the convergence conditions of the ADMM algorithm. The proposed method is compared with least squares parabolic Radon transform (LSPRT) and sparse parabolic Radon transform based on L_1 regularization (SPRTL_1) for multiple attenuation in the synthetic data and field data. We demonstrate that it improves the sparsity and resolution of the Radon domain data, and better results are obtained.

Keywords: multiple attenuation; parabolic Radon transform; L_q regularization; sparse inversion

Citation: Wu, Q.; Hu, B.; Liu, C.; Zhang, J. Sparse Parabolic Radon Transform with Nonconvex Mixed Regularization for Multiple Attenuation. *Appl. Sci.* **2023**, *13*, 2550. https://doi.org/10.3390/app13042550

Academic Editor: Roberto Zivieri

Received: 12 January 2023
Revised: 10 February 2023
Accepted: 14 February 2023
Published: 16 February 2023

1. Introduction

Multiple reflections are usually regarded as coherent noise in reflection seismic data especially in marine exploration. The existence of multiple reflections brings difficulty to seismic data processing [1]. It makes the structural illusion appear in the migrated-stacked section, which can affect the accuracy of subsequent data interpretation [2]. The Radon transform is a common method for multiple attenuation because of its high efficiency and due that it is easily implemented [3,4]. Due to the kinematic differences between primary and multiple reflections, most multiples are considered as coherent noise with low velocity [5]. Therefore, the primary reflections are upturned or flat, and multiple reflections are parabolic after normal movement correction with certain velocities in the CMP gather, which provides a theoretical basis for multiple attenuation using Radon transform. In 1986, Hampson [6] used the parabolic Radon transform to suppress multiples. The parabolic Radon transform makes their seismic events of different curvatures projected in different regions of the Radon domain, so as to separate primary and multiple reflections. However, finite seismic acquisition aperture of seismic data causes energy diffusion in the Radon domain, and it causes the smearing shown by the red arrows in Figure 1b. The energy diffusion limits the effect of multiple attenuation. To improve the above problem, the conception of optimization inversion is introduced into Radon transform [7,8].

When solving the inverse problems of Radon transform, the solution of inversion Radon transform is not unique. That is to say, there will be multiple sets of different solutions in the Radon domain, and these solutions do not describe the model function well.

Hampson [6] proposed the least squares inversion method, and the original data are reconstructed by optimizing the data in the Radon domain. This method reduces the smearing in the radon domain, but there are still some artefacts and data outside the offset that are not restricted. After Radon inverse transform, the data within a restricted offset still consist of the original data, but they are different when the data are outside the restricted range. In addition, the solution of the Radon domain is not optimal. Therefore, to solve this problem, it is hoped that the solution of the Radon domain will be sparser [2]. In 1985, the Radon transform was considered as a sparse inversion problem by Thorson et al. [9]. However, a lot of computation is disadvantageous. Sacchi et al. [10] adopted the method of Thorson to implement the sparse inversion Radon transform, and it is in the frequency domain. A high-resolution Radon transform was proposed by Herrmann to distinguish primary and multiple reflections, and it tackled the aliasing and resolution issues of the Radon transform [11]. Trad et al. [12] discussed fast implementations for the Radon transform in the time and frequency domains. With this sparse constraint, the seismic events have better localization characteristics in the Radon domain and the separation effect of primaries and multiples is guaranteed to a certain extent. Lu [13] proposed iterative 2D model shrinkage to solve the sparse inverse problem of RT and obtained the analogously sparse Radon model. Xiong et al. [14] established a mathematical model to combine L_2-norm and the adaptive multiple subtraction method on L_1-norm by weighted combination. This method can suppress the energy of multiples relatively better than non-combination.

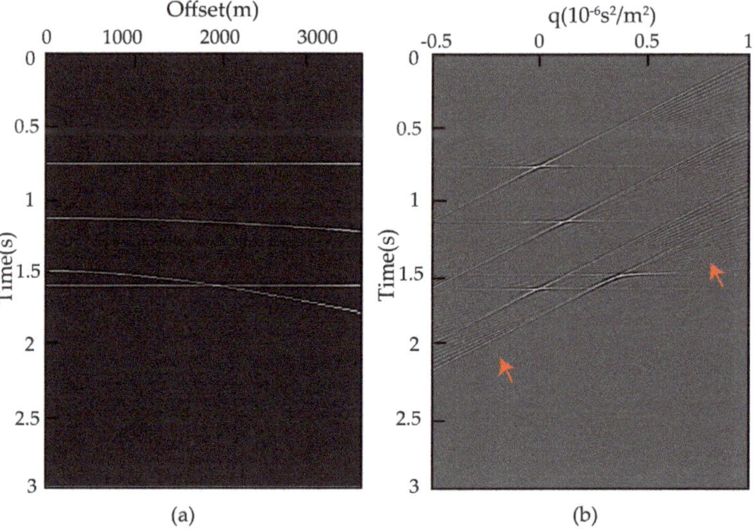

Figure 1. (**a**) CMP gather after NMO; (**b**) Radon domain data after parabolic Radon transform. The red arrows indicate the smearing.

In order to further improve the separation effect, researchers use the regularization methods of L-norm as the sparsity constraint condition that fully reflects the basic characteristics of the seismic data [15–17]. We can obtain the high-resolution Radon transform, and it is constrained sparsely, such as in L_0 regularization, but it is generally not used because it is hard to solve. Tisbshirani [18] proposed L_1 regularization, and it provides an alternative. L_1 regularization [19] is a convex optimization problem. This means that L_1-norm regularization is easy to solve. Donoho et al. [20] proved that sometimes the solution of the L_0 regularization is equivalent to that of the L_1 regularization for the sparsity problem. L_2 regularization can avoid the overfitting of the model, but its solutions do not have the sparse property. Moreover, some numerical experiments showed that sparse signals are recovered from fewer linear measurements by L_q $(0 < q < 1)$ regularization [21–24]. $L_{1/2}$ regularization can be taken as a representative of the L_q $(0 < q < 1)$ regularization, and it has been proved that the solution of $L_{1/2}$ regularization is sparser and more stable than the solution of L_1 regularization [25]. However, the $L_{1/2}$ regularization leads to a nonconvex, nonsmooth and non-Lipschitz optimization problem that is difficult to solve quickly and efficiently. Some scholars proposed corresponding solving methods. The reweighted iteration algorithm [25], the iterative reweighted least squares method [26], the iterative half thresholding algorithm [27] and the generalized iterated shrinkage algorithm [28,29] are efficient methods for solving the $L_{1/2}$ regularization.

Generally, sparse regularization constrains the sparsity of the whole seismic wavefield, which means a unique regularization parameter for different wavefields. However, the different wavefields have different amplitudes, and the same regularization parameter setting easily causes signal damage or redundant residual information. Sparse regularization for multi-tasking exploits differences in features of the data to demix the two distinct components for constraining the sparsity. For the multiple attenuation case in the Radon domain, primary and multiple reflections are distributed in different zones of the Radon domain and have different amplitudes. The unique regularization parameter may limit the performance of multiple attenuation.

In this paper, we propose a sparse parabolic Radon transform in the frequency domain with the nonconvex $L_{q_1} - L_{q_2}(0 < q_1, q_2 < 1)$ mixed regularization (SPRT$L_{q_1} - L_{q_2}$) that constrains the sparsity of primary and multiple reflections, respectively. In addition, we use the alternating direction method of multipliers algorithm (ADMM) [30–32] to approximately solve the multi-task regularization problem. Furthermore, conditions for convergence [33–36] are indicated. The rationality and effectiveness of the proposed method in multiple attenuation are verified by synthetic and real data. The method in this paper is compared with least squares parabolic Radon transform (LSPRT) and sparse parabolic Radon transform based on L_1 regularization (SPRTL_1) for multiple attenuation. This method improves the precision and focusing ability of Radon transform, obtains a better result of multiple suppression, reduces the loss of primary reflections, and improves the interpretability of seismic data.

2. Methodology

2.1. Parabolic Radon Transform

The fundamental strategy of multiple suppression based on Radon transform is to use the difference in kinematic features between primary and multiple reflections. The velocity between primary and multiple reflections is used to perform normal movement correction of the seismic data of CMP gather, and then the seismic events are more like parabolas [6]. Therefore, after the parabolic Radon transform, the seismic events with different curvatures can be mapped to different regions of the Radon domain to realize the separation of primary and multiple reflections.

The sum trajectories of parabolic Radon transform follow a parabola, and its definition is as follows:

$$m(\tau, q_j) = \sum_{j=0}^{N_x} d\left(t = \tau + q_j x^2, x\right) \tag{1}$$

$$d(t, x) = \sum_{j=0}^{N_q} m\left(\tau = t - q_j x^2, q_j\right) \tag{2}$$

where $d(t,x)$ is seismic data, $m(\tau, q_j)$ is Radon domain data after parabolic Radon transform, x is offset distance, t is time, q is curvature parameter and τ is intercept time. The meaning of parabolic Radon transform is that the seismic events having a parabolic shape in the time domain are mapped to a point in the Radon domain.

For the sake of efficiency, Radon transform can be transformed via the time variable to the frequency domain, as shown in Formulations (3) and (4):

$$M(q_j, f) = \sum_{k=1}^{N_x} D(x_k, f) e^{i2\pi f q_j x_k^2} \tag{3}$$

$$D(x_k, f) = \sum_{j=1}^{N_q} M(q_j, f) e^{-i2\pi f q_j x_k^2} \tag{4}$$

where f is frequency, $k = 1, 2, \cdots, N_x$ and $j = 1, 2, \cdots, N_q$

The matrix vector form of Formulations (3) and (4) can be represented by

$$\boldsymbol{m} = \boldsymbol{A}^H \boldsymbol{d} \tag{5}$$

$$\boldsymbol{d} = \boldsymbol{A}\boldsymbol{m} \tag{6}$$

where $\boldsymbol{d}$ is CMP gather and $\boldsymbol{m}$ is the matrix form of Radon domain data. The Radon operators $\boldsymbol{A}$ and $\boldsymbol{A}^H$ of forward and inversion transform are defined as

$$\boldsymbol{A}^H = e^{i2\pi f q_j x_k^2} \tag{7}$$

$$\boldsymbol{A} = e^{-i2\pi f q_j x_k^2} \tag{8}$$

Due to the limited aperture and discretization leading to low resolution and aliasing, Hampson [6] proposed the least squares inversion method, and the solution is presented as

$$\boldsymbol{m} = \left(\boldsymbol{A}^H \boldsymbol{A}\right)^{-1} \boldsymbol{A}^H \boldsymbol{d} \tag{9}$$

The solution of Tikhonov regularization [37] is

$$\boldsymbol{m} = \left(\boldsymbol{A}^H \boldsymbol{A} + \sigma \boldsymbol{I}\right)^{-1} \boldsymbol{A}^H \boldsymbol{d} \tag{10}$$

where σ is a stable factor and $\boldsymbol{I}$ is the identity matrix.

The least squares parabolic Radon transform improves the resolution of the Radon domain data as well as the accuracy and focusing ability of the transformation. This method limits partial energy diffusion and ensures the consistency of the reconstructed data with the original data. However, for multiples and primaries with little difference in the time difference of normal moveout, this method cannot separate them without distortion, and there is still some energy overlap in the Radon domain. It is necessary to improve the parabolic Radon transform to make the solution of the Radon domain more sparse. Radon transform is regarded as a nonlinear inversion problem. The data transformation error is constrained by L_2-norm, and the model sparsity is constrained by L-norm to improve the sparsity and resolution of the model. Therefore, L_1 regularization [20] and L_q regularization [25,27] are proposed to constrain data.

2.2. High-Resolution Sparse Parabolic Radon Transform

The regularization method of using L-norm as a penalty term can make the result sparse and obtain a high-resolution parabolic Radon transform with sparse constraints. Common regularization methods include L_0 regularization, L_1 regularization, L_2 regularization, L_q regularization and the unit sphere geometry diagram of L-norm, shown in Figure 2. Since there are no corners in the constraint region of L_0 regularization, and it is difficult to have zero solutions, this method has sparsity. The constraint region of L_2 regularization also has no corners and it can avoid the overfitting of the model, but its solutions do not have the sparsity property. The constraint region of L_1 regularization is a square. The L_1 regularization is convex optimization problem, and it is sparse. It can be solved using the iterative soft threshold algorithm (ISTA) [38] and the fast iterative shrinkage-thresholding algorithm (FISTA) [39]. However, some numerical experiments [21] showed that L_q $(0 < q < 1)$ regularization has much better signal recovery capability than L_1 regularization minimization. As shown in Figure 2, the solution of L_q $(0 < q < 1)$ regularization is more likely to be at a corner, which proves its sparsity.

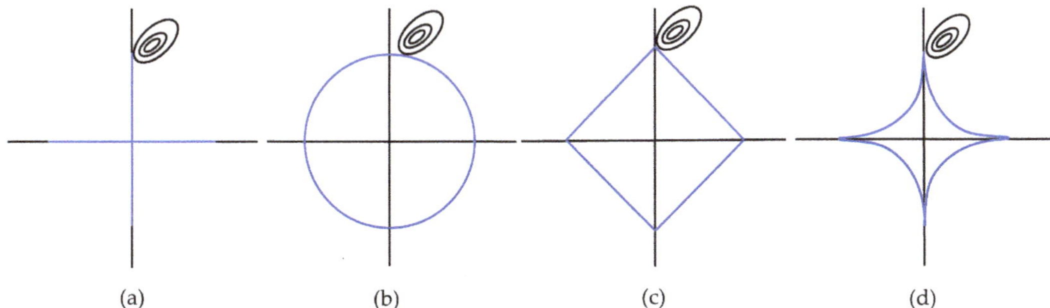

Figure 2. Unit ball pictures for **(a)** L_0, **(b)** L_2, **(c)** L_1 and **(d)** $L_{1/2}$ regularization.

The multiple suppression based on the parabolic Radon transform problem can be formulated as following minimization problem of L_q regularization:

$$\min_{m}\{\frac{1}{2}||d - Am||_2^2 + \alpha||m||_q^q\} \tag{11}$$

where $||m||_q = (\sum_{i=1}^{N}|m_i|^q)^{1/q}$ and $\alpha > 0$ is the regularization parameter.

L_q $(0 < q < 1)$ norm constrains whole seismic wavefield data to improve the inversion accuracy. However, in order not to cause signal damage and noise residue, sparse constraints of primary and multiple reflections are considered, respectively, according to the differences between them to achieve a better multiple suppression.

2.3. High-Resolution Parabolic Radon Transform with $L_{q_1} - L_{q_2}$ Mixed Regularization

To suppress multiple m_2 from d, we use $L_{q_1} - L_{q_2}$ mixed regularization with $0 < q_1, q_2 < 1$ for sparsity promotion and use the following formulation:

$$\min_{m_1 m_2}\left\{\frac{1}{\beta}||A_1 m_1 + A_2 m_2 - d||_2^2 + \mu||m_1||_{q_1}^{q_1} + ||m_2||_{q_2}^{q_2}\right\} \tag{12}$$

where μ is a positive parameter, $\beta > 0$ is a penalty parameter, $A = A_1 = A_2$ is the Radon operator, m_1 is primary and m_2 is multiple. To determine q_1 and q_2, we explore the $L_q(0 < q < 1)$ regularization. Xu et al. [25] introduced $L_q(0 < q < 1)$ regularization and proved that the L_q regularization can obtain much sparser solutions than L_1 regularization. Meanwhile, they proposed $L_{1/2}$ regularization, and it is the most representative regularization method for $L_q(0 < q < 1)$ regularization. The study shows that the $L_{1/2}$ regularization is the sparsest and the most robust among the $L_q(1/2 \leq q < 1)$ regularization, and when $0 < q < 1/2$, the L_q regularization has similar properties to the $L_{1/2}$ regularization. Therefore, we select $q_1 = q_2 = 1/2$ in this paper.

$L_{q_1} - L_{q_2}$ mixed regularization is the nonconvex problem, and it is a very complicated process to solve. In this paper, we use an improved alternative direction method of multipliers algorithm (ADMM) to approximate the solution [34]. The ADMM algorithm can be used to solve many high-dimensional problems and uses a decomposition-coordination procedure to decouple the variables [34–36,40]. This algorithm enables the difficult global problem to be properly solved. The Problem (12) can be formulated as

$$\min_{m_1 m_2}\left\{||A_1 m_1 + A_2 m_2 - d||_2^2 + \beta\mu||z_1||_{q_1}^{q_1} + \beta||z_2||_{q_2}^{q_2}\right\} \tag{13}$$

where the auxiliary variables are $z_1 = m_1$, $z_2 = m_2$.

The augmented Lagrangian function is

$$L(\boldsymbol{m}_1, \boldsymbol{m}_2, z_1, z_2, \boldsymbol{w}_1, \boldsymbol{w}_2) = \|\boldsymbol{A}_1\boldsymbol{m}_1 + \boldsymbol{A}_2\boldsymbol{m}_2 - \boldsymbol{d}\|_2^2 + \beta\mu\|z_1\|_{q_1}^{q_1} + \beta\|z_2\|_{q_2}^{q_2} + \langle \boldsymbol{w}_1, \boldsymbol{m}_1 - z_1\rangle + \langle \boldsymbol{w}_2, \boldsymbol{m}_2 - z_2\rangle + \frac{\rho_1}{2}\|\boldsymbol{m}_1 - z_1\|_2^2 + \frac{\rho_2}{2}\|\boldsymbol{m}_2 - z_2\|_2^2$$

where ρ_1 and ρ_2 are positive penalty parameters and $\boldsymbol{w}_1$ and $\boldsymbol{w}_2$ are the dual variables. The dual variables and $z_1, z_2, \boldsymbol{m}_1, \boldsymbol{m}_2$ are alternatively updated as follows:

$$z_1^{k+1} = \underset{z_1}{argmin}(\beta\mu\|z_1\|_{q_1}^{q_1} + \frac{\rho_1}{2}\left\|\boldsymbol{m}_1^k - z_1 + \frac{\boldsymbol{w}_1^k}{\rho_1}\right\|_2^2) \tag{14}$$

$$z_2^{k+1} = \underset{z_2}{argmin}(\beta\|z_2\|_{q_2}^{q_2} + \frac{\rho_2}{2}\left\|\boldsymbol{m}_2^k - z_2 + \frac{\boldsymbol{w}_2^k}{\rho_2}\right\|_2^2) \tag{15}$$

$$\boldsymbol{m}_1^{k+1} = \underset{\boldsymbol{m}_1}{argmin}(\left\|\boldsymbol{A}_1\boldsymbol{m}_1 + \boldsymbol{A}_2\boldsymbol{m}_2^k - \boldsymbol{d}\right\|_2^2 + \frac{\rho_1}{2}\left\|\boldsymbol{m}_1 - z_1^{k+1} + \frac{\boldsymbol{w}_1^k}{\rho_1}\right\|_2^2) \tag{16}$$

$$\boldsymbol{m}_2^{k+1} = \underset{\boldsymbol{m}_2}{argmin}(\left\|\boldsymbol{A}_1\boldsymbol{m}_1^{k+1} + \boldsymbol{A}_2\boldsymbol{m}_2 - \boldsymbol{d}\right\|_2^2 + \frac{\rho_1}{2}\left\|\boldsymbol{m}_2 - z_2^{k+1} + \frac{\boldsymbol{w}_2^k}{\rho_2}\right\|_2^2) \tag{17}$$

$$\boldsymbol{w}_1^{k+1} = \boldsymbol{w}_1^k + \rho_1(\boldsymbol{m}_1^{k+1} - z_1^{k+1}) \tag{18}$$

$$\boldsymbol{w}_2^{k+1} = \boldsymbol{w}_2^k + \rho_2(\boldsymbol{m}_2^{k+1} - z_2^{k+1}) \tag{19}$$

Due to the proximity operator of L_q regularization, $prox_{q,\eta}(\boldsymbol{t})$ is defined as

$$prox_{q,\eta}(\boldsymbol{t}) = \underset{\boldsymbol{m}}{argmin}\left\{\|\boldsymbol{m}\|_q^q + \frac{\eta}{2}\|\boldsymbol{m} - \boldsymbol{t}\|_2^2\right\} \tag{20}$$

where $\eta > 0$ is a penalty parameter.

When $0 < q < 1$, it can be updated as [41]

$$prox_{q,\eta}(\boldsymbol{t})_i = \begin{cases} 0, & |\boldsymbol{t}_i| < \tau \\ \{0, sign(t_i)\beta\}, & |\boldsymbol{t}_i| = \tau \\ sign(t_i)z_i, & |\boldsymbol{t}_i| > \tau \end{cases} \tag{21}$$

for $i = 1, \cdots, n$, where $\beta = [2(1-q)/\eta]^{1/(2-q)}$ and $\tau = \beta + q\beta^{q-1}/\eta, z_i$ is the result of $h(z) = qz^{q-1} + \eta z - \eta|t_i| = 0$. The exact solutions of $\boldsymbol{m}_1$ and $\boldsymbol{m}_2$ denote

$$\boldsymbol{m}_1^{k+1} = (2\boldsymbol{A}_1^T\boldsymbol{A}_1 + \rho_1\boldsymbol{I})^{-1}[2\boldsymbol{A}_1^T\left(\boldsymbol{d} - \boldsymbol{A}_2\boldsymbol{m}_2^k\right) + \rho_1 z_1^{k+1} - \boldsymbol{w}_1^k] \tag{22}$$

$$\boldsymbol{m}_2^{k+1} = (2\boldsymbol{A}_2^T\boldsymbol{A}_2 + \rho_2\boldsymbol{I})^{-1}[2\boldsymbol{A}_2^T\left(\boldsymbol{d} - \boldsymbol{A}_1\boldsymbol{m}_1^{k+1}\right) + \rho_2 z_2^{k+1} - \boldsymbol{w}_2^k] \tag{23}$$

The standard ADMM algorithm often fails to converge and converges under some conditions. Therefore, the sufficient condition for the ADMM convergence is [34]

$$\rho_1 > \frac{16\lambda_1^2}{\rho_1} + \frac{16\lambda_1\lambda_2}{\rho_2} - 2\varphi_1 \tag{24}$$

$$\rho_2 > \frac{16\lambda_2^2}{\rho_2} + \frac{16\lambda_1\lambda_2}{\rho_1} - 2\varphi_2 \tag{25}$$

where $\lambda_i = \lambda_{max}(A_i^T A_i)$ and $\varphi_i = \lambda_{min}(A_i^T A_i)$, i = 1,2. The convergence condition of (24) and (25) causes the sequence $\left\{(z_1^k, z_2^k, \boldsymbol{m}_1^k, \boldsymbol{m}_2^k, \boldsymbol{w}_1^k, \boldsymbol{w}_2^k)\right\}$ generated by (14)–(19) to converge to a critical point of the Problem (13).

3. Synthetic Data Application

To validate the effectiveness of the proposed method (SPRT$L_{q_1} - L_{q_2}$), we used least squares parabolic Radon transform (LSPRT) and sparse parabolic Radon transform based on L_1 regularization (SPRTL_1) as the control group to perform a multiple attenuation test on the noisy synthetic data. The velocity model of the synthetic seismic data is shown in Figure 3a. Figure 3b is the CMP gather of a full wavefield. There are 750 sampling points in the time direction, the sampling interval is 4 ms and the offset range is 0 to 2000 m with an interval dx = 25 m.

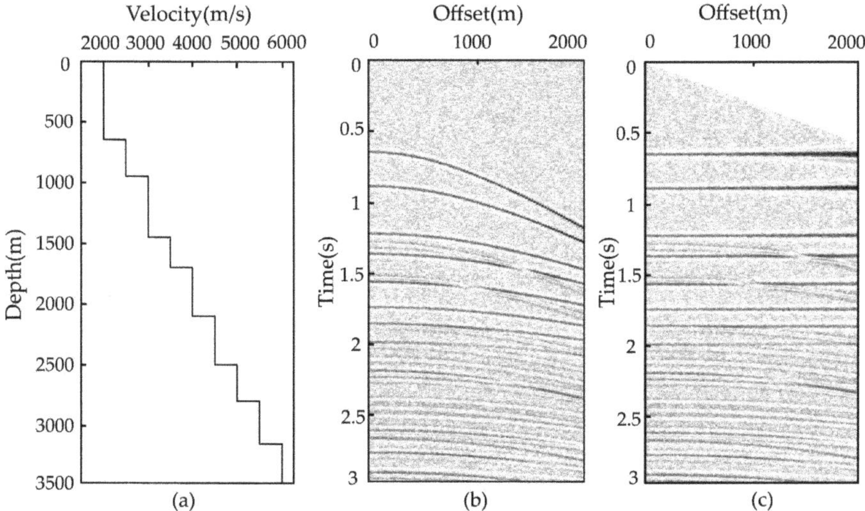

Figure 3. (**a**) Velocity model; (**b**) CMP gather; (**c**) CMP gather after NMO.

In order to make the kinematic characteristics of seismic events closer to parabolas, the CMP gather should be normal movement corrected first. The primary reflections are flat and multiple reflections, which are parabolic because of inadequate correction after normal movement correction with velocities of the primary reflections in the CMP gather, as shown in Figure 3c.

After the Radon transform, the seismic events of primaries are mapped in the region with a negative q value and q = 0, and the seismic events of multiples are mapped in the region with a positive q value, as shown in Figure 4. In Figure 4a, due to the influence of the noise, the seismic events do not converge to a point well in the Radon domain, and finite seismic acquisition aperture causes severe smearing. In Figure 4b, although the smearing is alleviated, the result fell short. Figure 4c shows the result of the SPRT$L_{q_1} - L_{q_2}$ method. Because the primary and multiple reflections are sparsely constrained, respectively, only the results of the primary seismic events are left in the Radon domain, and the mapping of multiples is removed directly. There is almost no smearing in Figure 4c, and it has higher resolution. The transformation accuracy and focusing ability are improved.

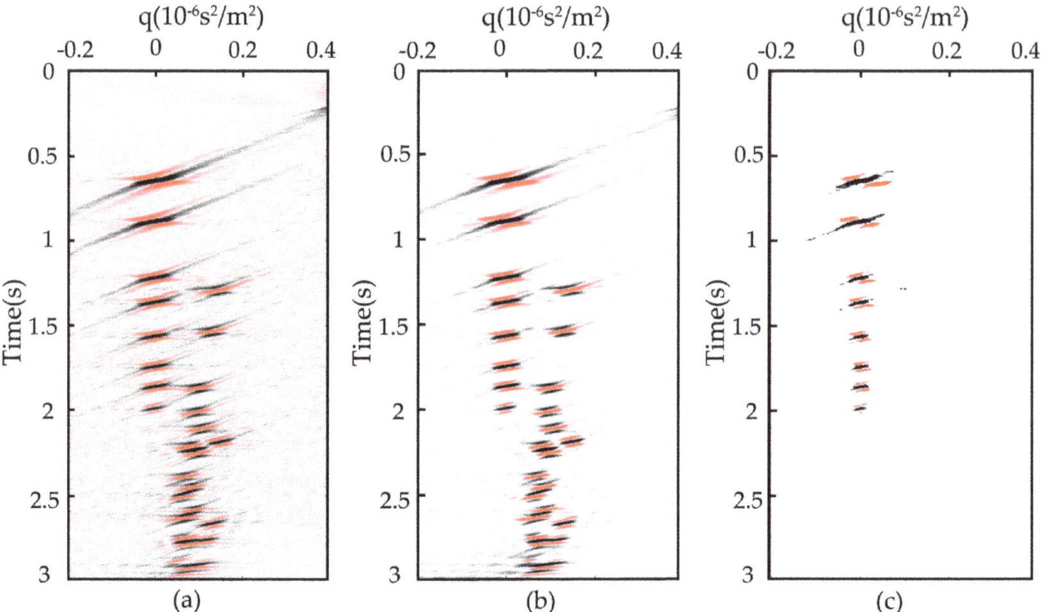

Figure 4. Radon domain results: (**a**) LSPRT method; (**b**) SPRTL_1 method; (**c**) SPRT$L_{q1} - L_{q2}$ method.

Figure 5b is the multiple attenuation result obtained by LSPRT; Figure 5c is the result obtained by SPRTL_1. Due to the overlapping energy of multiple and primary reflections in the Radon domain, there are many artifacts at near offset. They still have some residual multiple energy, especially at near offset. Figure 5d is the result obtained by SPRT$L_{q_1} - L_{q_2}$. There are almost no residual multiples at the arrow, and the reconstructed data have great consistency with the original data. Figure 6 shows the difference between the suppression results and those without multiple wavefield data. It is proved that the SPRT$L_{q_1} - L_{q_2}$ method is effective in suppressing multiples, especially at the near-offset position. In order to quantitatively analyze the consistency between the reconstructed data and the original data, the following formula is applied to calculate the reconstruction error [42]:

$$s = \frac{\|m' - \hat{m}\|_2^2}{\|m'\|_2^2} \times 100\% \tag{26}$$

where m' is no multiple wavefield data, $\hat{m}$ is data after multiple attenuation and s is reconstruction error.

Based on the examples of synthetic data, Table 1 lists the reconstruction error of three methods. It is obvious that compared with the LSPRT and SPRTL_1 methods, the SPRT$L_{q1} - L_{q2}$ method is superior in reconstruction capability.

Table 1. Reconstruction error comparison.

Method	**LSPRT**	**SPRTL_1**	**SPRT$L_{q1} - L_{q2}$**
Reconstruction error	11.2%	8.3%	7.6%

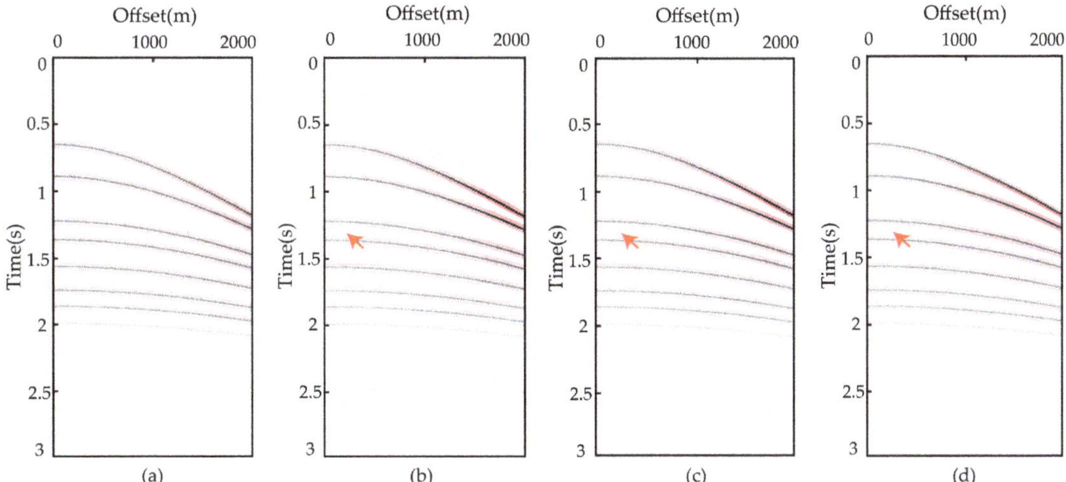

Figure 5. Multiple attenuation result of the theoretical model data: (**a**) without multiple wavefield data; (**b**) LSPRT method; (**c**) SPRTL_1 method; (**d**) SPRT$L_{q1} - L_{q2}$ method. The red arrows indicate the multiples suppression effect of the three algorithms, and there is no residual multiples in Figure 5d.

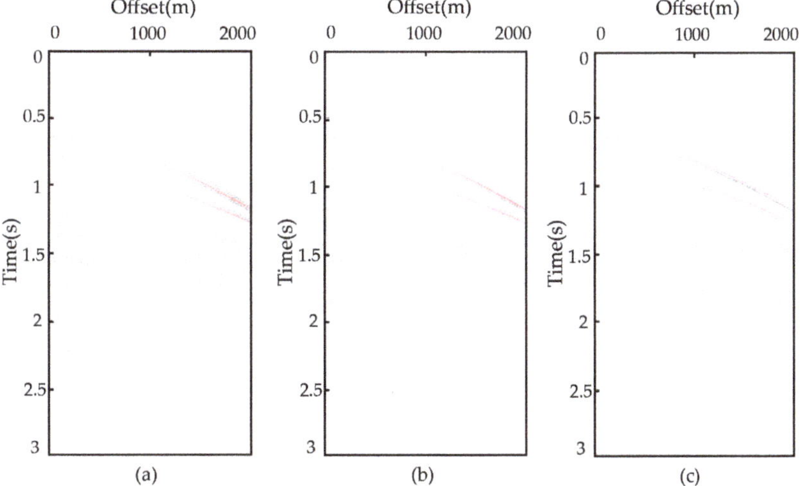

Figure 6. Difference image map: (**a**) by LSPRT; (**b**) by SPRTL_1; (**c**) by SPRT$L_{q1} - L_{q2}$.

In order to verify whether the proposed method is suitable for seismic data with missing traces, we randomly selected 30%, 50% and 70% seismic trace from the synthetic seismic record containing noise and filled them with zero. We applied the proposed method to suppress multiples of missing seismic trace data. As can be seen from Figure 7a,b, the result of suppressing multiples is not affected, and there are also no residual multiples at the near offset. With the increase in the percentage of missing traces, the reconstruction effect becomes worse. When the percentage of missing traces reaches 70%, the phase distortion appears in the suppression result. Based on the reconstruction results, it can be proved that the SPRT$L_{q_1} - L_{q_2}$ method is also suitable for missing trace data.

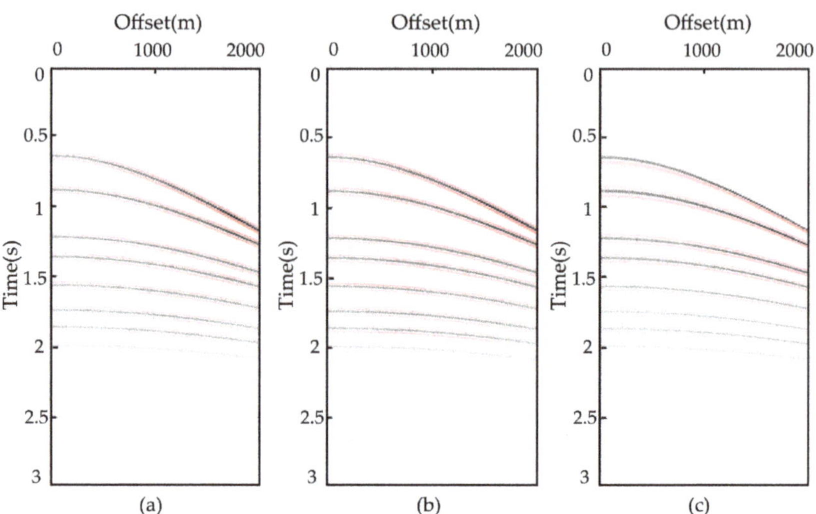

Figure 7. Multiple attenuation results of missing trace data by SPRT$L_{q1} - L_{q2}$: (**a**) 30% missing traces; (**b**) 50% missing traces; (**c**) 70% missing traces.

4. Real Data Application

4.1. CMP Gather of Real Data Application

A marine CMP gather after normal movement correction was applied to further examine the effectiveness of the SPRT$L_{q_1} - L_{q_2}$ method. We chose the time window of 3.2 s–4.8 s in the field data to process because there exists a large number of multiple reflections, as shown in Figure 8a. The reason is that this part of the data has more developed multiple waves, and the effect of suppression can be clearly seen in the final result. There are 400 sampling points in the time direction and 92 offset traces. The sampling interval is 4 ms. In addition, those multiple reflections cover the primary data and affect the imaging effect of the primary reflections.

Figure 8 is the multiple attenuation results. Figure 8b,c show the multiple attenuation results of the LSPRT and SPRTL_1, and there are still some residual multiple reflections, as indicated by the rectangle. The information of the primaries is masked, and it makes the seismic events discontinuous. Shown in Figure 8d is the multiple attenuation result obtained by SPRT$L_{q1} - L_{q2}$, and there are no obvious residual multiples. Because of the sparsity of SPRT$L_{q1} - L_{q2}$, the multiples are suppressed effectively at both near and far offsets. The primaries are highlighted, and the black arrows indicate that the continuity of the seismic events are improved. In addition, we can obviously see that the CMP gather is clear and clean.

In order to further compare the advantages of the proposed algorithm with respect to the LSPRT and SPRTL_1, we extracted three single trace amplitudes after multiple attenuation, and they were compared with single-channel amplitudes before suppression. As shown in Figure 9, there is almost no disturbance in the original location of multiples after multiple attenuation by SPRT$L_{q1} - L_{q2}$, and it is obvious at 4.4 s–4.6 s, and the red rectangle marks the difference in this location. The amplitude of the primary reflection is intact and not attenuated, as at 3.4 s and 3.6 s. The proposed method is effective in suppressing multiples and achieves a better performance.

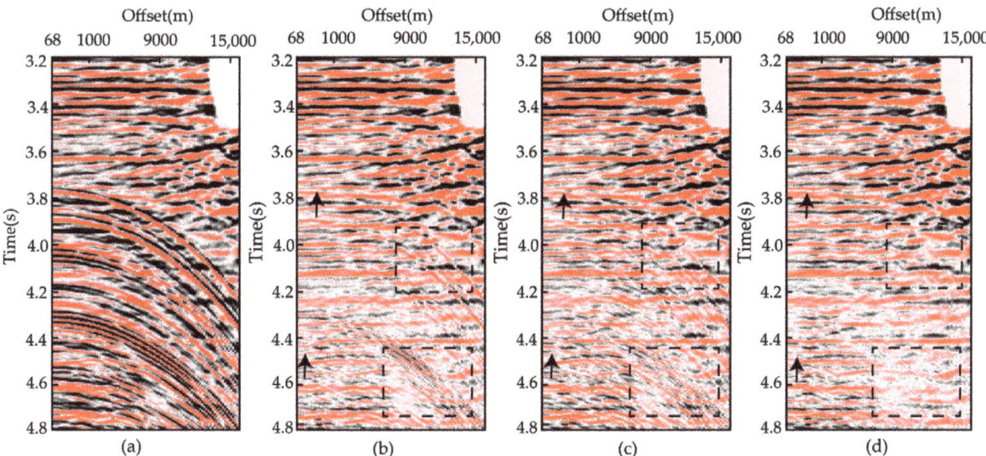

Figure 8. Multiple attenuation result: (**a**) CMP gather of marine real data; (**b**) by LSPRT method; (**c**) by SPRTL_1 method; (**d**) by SPRT$L_{q1} - L_{q2}$ method. The black arrows indicate the continuity of the seismic events are improved in the Figure 8d.

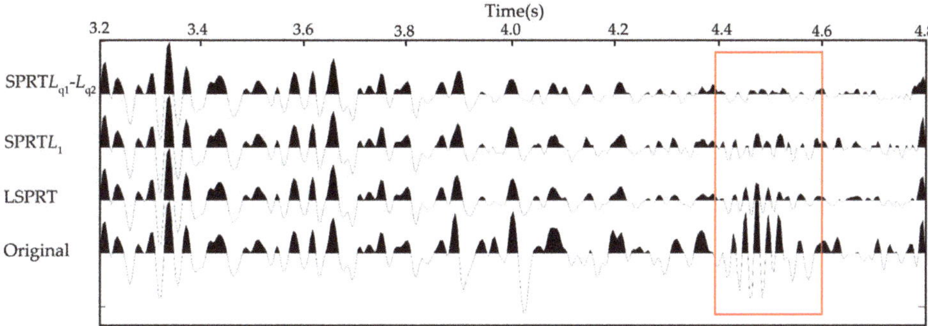

Figure 9. Comparison of single trace amplitude after multiple attenuation by three algorithms. The red rectangle indicate that the amplitude of multiple decays more significantly in the proposed method.

4.2. Prestack Field Data

The SPRT$L_{q1} - L_{q2}$ method was further tested with multi-shot prestack field data. In order to comprehensively evaluate the multiple suppression effect, we used a 3D method to display the multi-shot data volume, then analyzed the effectiveness of the method from the time slices, the common middle point gathers and the common offset gathers.

Figure 10a is original data, and multiples can be clearly seen at the arrow position in the figure. Figure 10b,c are the multiple attenuation results obtained by the LSPRT and the SPRTL_1 methods; Figure 10d is the result obtained by the SPRT$L_{q1} - L_{q2}$. Overall, all three methods suppress multiples, but the effects are different. In Figure 10b,c, it can be obviously observed from the time slice that a certain amount of the primary energy is lost when multiples are suppressed. In the CMP gathers the multiples of far offset are sufficiently suppressed by SPRT$L_{q1} - L_{q2}$, as indicated by the arrow. In the common offset gathers, the effectiveness of the SPRT$L_{q1} - L_{q2}$ method is demonstrated by the continuity of the seismic events. This method has a sparser representation of the seismic data and fully retains the characteristics and information of the primaries, so as to improve the continuity of the seismic events.

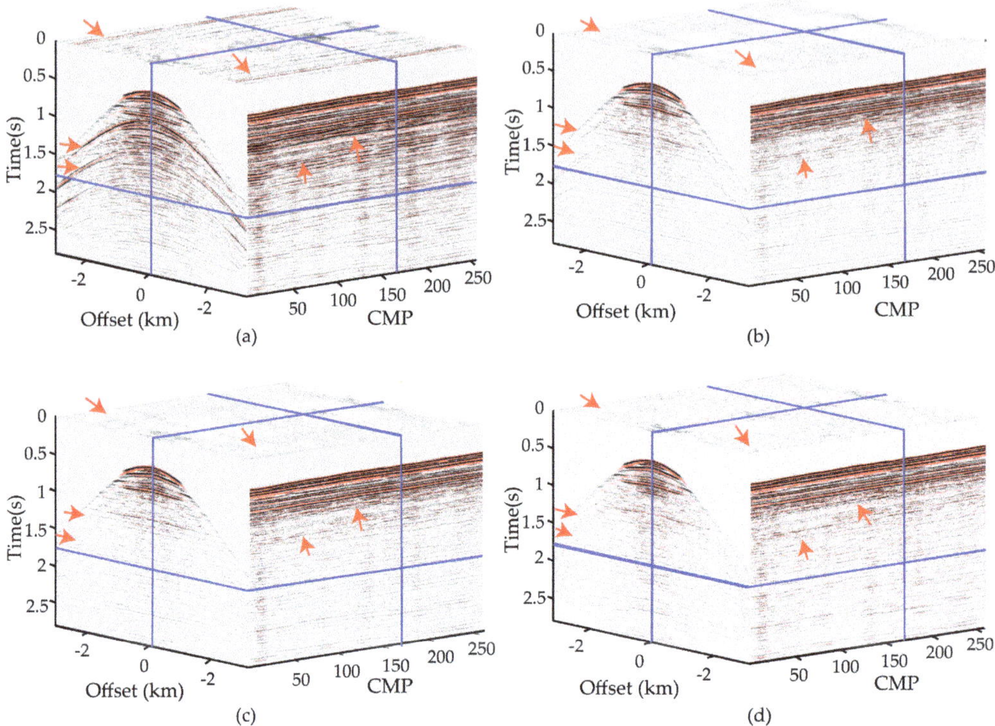

Figure 10. Multiple attenuation result: (**a**) pre-stack real data; (**b**) LSPRT method; (**c**) SPRTL_1 method; (**d**) SPRT$L_{q1} - L_{q2}$ method. The red arrows indicate the cleaner multiple attenuation in the Figure 10d.

Figure 11 shows the stack section after multiple attenuation by three methods. The residual of multiples leads to the discontinuity of the events of the primaries (black rectangle), which increases the difficulty of subsequent interpretation. The continuity of the seismic events is obviously enhanced in the stack sections using the method presented in this paper, and the effect of multiple attenuation between 0.5 s and 1 s has a remarkable effect. In addition, as can be seen by the black arrow region of Figure 11d, there is almost no multiple energy compared to Figure 11b,c.

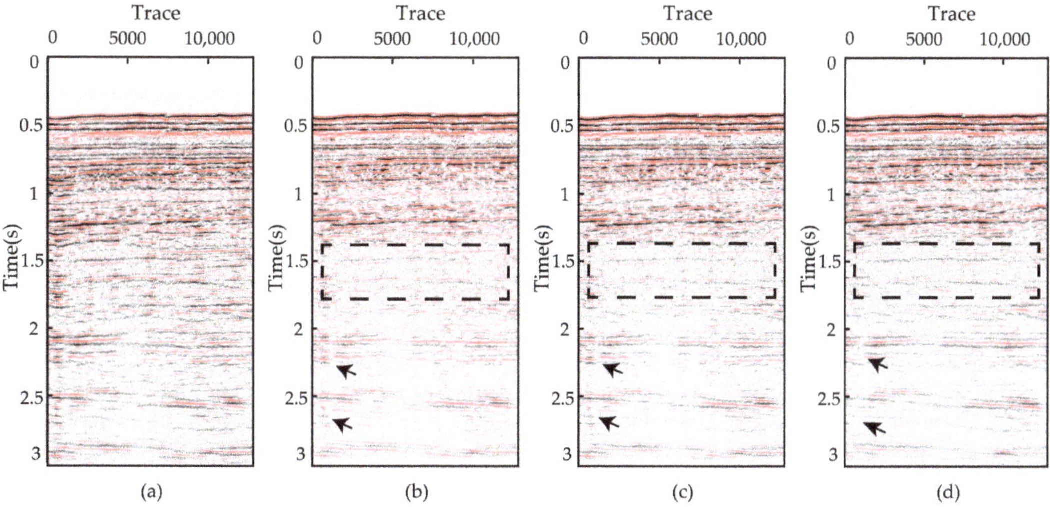

Figure 11. Multiple attenuation result: (**a**) stacked real data; (**b**) LSPRT method; (**c**) SPRTL_1 method; (**d**) SPRT$L_{q1} - L_{q2}$ method. The black arrows indicate the continuity of the seismic events are improved in the Figure 11d.

Therefore, the SPRT$L_{q1} - L_{q2}$ method has some advantages in suppressing multiple reflections and retaining effective reflections.

5. Discussion

Multiple attenuation can provide a good foundation for subsequent data processing and data interpretation. The traditional parabolic Radon transform is unable to separate primaries and multiples without distortion. Sparse parabolic Radon transform based on L_1 regularization also does not achieve the desired effect of suppressing multiples. We introduce the nonconvex $L_{q_1} - L_{q_2}(0 < q_1, q_2 < 1)$ mixed regularization sparse inversion method. This method obviously improves the accuracy of inversion, suppresses multiples and obtains clean seismic data. Moreover, the reconstructed primary data have higher accuracy, which was verified by experimental data.

In this paper, $L_{1/2}$ regularization is used to constrain primary and multiple waves because it has many properties such as sparsity and unbiasedness and oracle properties [25]. Meanwhile, it is more stable and sparse than the L_1 regularization, and easier to solve than L_0 regularization. In other practical applications, the values of q_1 and q_2 can be selected according to the actual problem. For example, if the coefficients are not strictly sparse, a moderate to large value of q can yield better results [34].

In complex geological conditions, the weak effective signals in the wavefield are easily covered by noise. Therefore, the parameter setting needs to be more careful, otherwise it may lose an effective signal because of noise suppression.

Since the proposed method introduced the $L_{1/2}$-norm, the computation time is increased when the sparsity of seismic signals is improved. Therefore, the computational efficiency of this method has no advantage over the L_1-norm sparse inversion method.

In future research, we propose three research directions that can be improved:

1. To solve the problem of the destruction of amplitude versus offset (AVO) signature in seismic data, the proposed method is combined with the orthogonal polynomial transform.
2. The algorithm for solving nonconvex regularization is further improved to improve the computational efficiency.
3. The proposed method is combined with other multi-wave suppression methods to process seismic data with high efficiency and high quality under complex geological conditions.

4. In addition, with the rapid development of deep learning, fields such as mechanics, medicine and geophysics [43–45] have been actively combined with deep learning, and more possibilities have been developed. Therefore, in future studies, we will also combine multiple suppression with deep learning to solve problems such as computational efficiency.

6. Conclusions

Multiple attenuation is an important problem in seismic data processing. The suppression results directly affect the quality of stacked seismic data. Radon transform is an efficient and low-cost method. In order to solve the problem of low resolution of Radon transform, a sparse parabolic Radon transform in the frequency domain with the nonconvex $L_{q_1} - L_{q_2}(0 < q_1, q_2 < 1)$ mixed regularization is proposed. This method has higher sparsity, and it can restrict primaries and multiples, respectively. The mappings of multiple seismic events are muted directly in the Radon domain. The theoretical data and field data results show that the resolution of the parabolic Radon transform and reconstruction capability are improved when using SPRT$L_{q_1} - L_{q_2}(q_1 = q_2 = 1/2)$, especially at near offset. The effectiveness of the sparse constraint method is demonstrated in the aspect of seismic event continuity. This method greatly improves the effect of multiple attenuation and reduces the unnecessary energy loss of useful signals, which provides high-quality data for the subsequent primary imaging.

The proposed method can also be used in other processing methods, such as data reconstruction, interpolation and dispersion curve extraction.

Author Contributions: Conceptualization, Q.W. and B.H.; methodology, Q.W. and B.H.; software, B.H.; validation, Q.W., B.H. and J.Z.; formal analysis, Q.W. and B.H.; investigation, Q.W.; resources, C.L.; data curation, Q.W. and B.H.; writing—original draft preparation, Q.W.; writing—review and editing, Q.W., B.H. and J.Z.; visualization, Q.W.; project administration, C.L.; funding acquisition, C.L. All authors have read and agreed to the published version of the manuscript.

Funding: This research was funded by the National Natural Science Foundation of China, grant number 41874125.

Data Availability Statement: Not applicable.

Acknowledgments: The authors give thanks for the open-source program of Matlab.

Conflicts of Interest: The authors declare no conflict of interest.

References

1. Weglein, A.B. Multiple attenuation: An overview of recent advances and the road ahead. *Lead. Edge* **1999**, *18*, 40–44. [CrossRef]
2. Verschuuur, D.J. *Seismic Multiple Removal Techniques: Past, Present and Future*, 1st ed.; Chen, H., Zhang, B., Liu, J., Eds.; Petroleum Industry Press: Beijing, China, 2010; pp. 20–28.
3. Wang, X.; Wang, H. A research of high-resolution plane-wave decomposition based on compressed sensing. *Chin. J. Geophys.* **2014**, *52*, 1068–1077. [CrossRef]
4. Li, Z.N.; Li, Z.C.; Wang, P. Wavefield separation by a modified linear Radon transform in borehole seismic. *Chin. J. Geophys.* **2014**, *57*, 2269–2277.
5. Zhang, J.; Wang, D.; Hu, B.; Gong, X. An Automatic Velocity Analysis Method for Seismic Data-Containing Multiples. *Remote Sens.* **2022**, *14*, 5428. [CrossRef]
6. Hampson, D. Inverse velocity stacking for multiple elimination. In *SEG Technical Program Expanded Abstracts*; Society of Exploration Geophysicists: Houston, TX, USA, 1986. [CrossRef]
7. Goncharov, F.O. An Iterative Inversion of Weighted Radon Transforms along Hyperplanes. *Inverse Probl.* **2017**, *33*, 124005. [CrossRef]
8. Ambartsoumian, G.; Goula-Zarrad, R.; Lewis, M.A. Inversion of the Circular Radon Transform on an Annulus. *Inverse Probl.* **2010**, *26*, 11. [CrossRef]
9. Thorson, J.R.; Claerbout, J.F. Velocity-stack and slant-stack stochastic inversion. *Geophysics* **1985**, *50*, 2727–2741. [CrossRef]
10. Sacchi, M.D.; Ulrych, T.J. High-resolution velocity gather and offset space reconstruction. *Geophysics* **1995**, *60*, 1169–1177. [CrossRef]
11. Herrmann, P.; Mojesky, T.; Magesan, M.; Hugonnet, P. De-aliased, high-resolution Radon transforms. In Proceedings of the 70th Annual International Meeting, SEG, Calgary, AB, Canada, 6–11 August 2000; pp. 1953–1956. [CrossRef]

12. Trad, D.; Ulrych, T.; Sacchi, M. Latest views of the sparse Radon transform. *Geophysics* **2003**, *68*, 386–399. [CrossRef]
13. Lu, W. An accelerated sparse time-invariant Radon trans-form in the mixed frequency-time domain based on iterative 2D model shrinkage. *Geophysics* **2013**, *78*, V147–V155. [CrossRef]
14. Xiong, F.S.; Huang, X.W.; Zhang, D.; Li, R.; Wang, P. Adaptive multiple subtraction method based on hybrid L1/L2 norm. *Geophys. Geochem. Explor.* **2014**, *38*, 996–1002. [CrossRef]
15. Chen, S.S.; Donoho, D.L.; Saunders, M.A. Atomic decomposition by basis pursuit. *SIAM Rev.* **2001**, *20*, 33–61. [CrossRef]
16. Figueiredo, M.A.; Nowak, R.D.; Wright, S.J. Gradient projection for sparse reconstruction: Application to compressed sensing and other inverse problems. *IEEE J. Sel. Top. Signal Process.* **2007**, *1*, 586–597. [CrossRef]
17. Xue, Y.R.; Wang, M.; Chen, X.H. High resolution Radon transform based on SL0 and its application in data reconstruction. *Oil Geophys. Prospect.* **2018**, *53*, 1–7. [CrossRef]
18. Tibshirani, R. Regression shrinkage and selection via the Lasso: A retrospective. *J. R. Statist. Soc. B* **2011**, *73*, 273–282. [CrossRef]
19. Han, J.; Zhang, S.; Zheng, S.; Wang, M.; Ding, H.; Yan, Q. Bias Analysis and Correction for III-Posed Inversion Problem with Sparsity Regularization Based on L_1 Norm for Azimuth Super-Resolution of Radar Forward-Looking Imaging. *Remote Sens.* **2022**, *14*, 5792. [CrossRef]
20. Donoho, D.L. Compressed sensing. *IEEE Trans. Inf. Theory* **2006**, *52*, 1289–1306. [CrossRef]
21. Chartrand, R.; Staneva, V. Restricted isometry properties and nonconvex compressive sensing. *Inverse Probl.* **2008**, *24*, 657–682. [CrossRef]
22. Foucart, S.; Lai, M.J. Sparsest solutions of underdetermined linear systems via L_p-minimization for $0 < q \leq 1$. *Appl. Comput. Harmon. Anal.* **2009**, *26*, 395–407. [CrossRef]
23. Lai, M.J.; Xu, Y.; Yin, W. Improved iteratively reweighted least squares for unconstrained smoothed L_q minimization. *SIAM J. Numer. Anal.* **2013**, *51*, 927–957. [CrossRef]
24. Marjanovic, G.; Solo, V. L_p Sparsity penalized linear regression with cyclic descent. *IEEE Trans. Signal Process.* **2014**, *62*, 1464–1475. [CrossRef]
25. Xu, Z.; Zhang, H.; Wang, Y.; Chang, X.; Liang, Y. $L_{1/2}$ regularization. *Sci. China Inf. Sci.* **2010**, *53*, 1159–1169. [CrossRef]
26. Chartrand, R.; Yin, W. Iteratively reweighted algorithms for compressive sensing. In Proceedings of the 2008 IEEE International Conference on Acoustics, Speech and Signal Processing, Las Vegas, NV, USA, 31 March–4 April 2008; pp. 3869C–3872C. [CrossRef]
27. Xu, Z.; Chang, X.; Xu, F.; Zhang, H. $L_{1/2}$ Regularization: A Thresholding Representation Theory and a Fast Solver. *IEEE Trans. Neural Netw. Learn. Syst.* **2012**, *23*, 1013–1027. [CrossRef] [PubMed]
28. Zuo, W.; Meng, D.; Zhang, L.; Feng, X.; Zhang, D. A Generalized Iterated Shrinkage Algorithm for Non-convex Sparse Coding. In Proceedings of the IEEE International Conference on Computer Vision, Sydney, Australia, 1–8 December 2013; pp. 217–224. [CrossRef]
29. Li, Y. Research on image recovery technology via nonconvex regularization method. Ph.D. Thesis, Nanjing University of Posta and Telecommunications, Nanjing, China, 2020. [CrossRef]
30. Huang, S.; Xu, Y.; Ren, M.; Yang, Y.; Wan, W. Rain Removal of Single Image Based on Directional Gradient Priors. *Appl. Sci.* **2022**, *12*, 1628. [CrossRef]
31. Wang, R.; Wang, D.; Zhang, W.; Liu, Y.; Hu, B.; Wang, L. Pseudo-3D Receiver Deghosting of Seismic Streamer Data Based on l_1 Norm Sparse Inversion. *Appl. Sci.* **2022**, *12*, 556. [CrossRef]
32. Zhang, H.; Chen, J. Robust L_p-Norm Inversion for High-Resolution Fluid Contents from Nuclear Magnetic Resonance Measurements. *Appl. Sci.* **2021**, *11*, 1298. [CrossRef]
33. Chen, J.; Zhang, Z.; Wen, X. Target Identification via Multi-View Multi-Task Joint Sparse Representation. *Appl. Sci.* **2022**, *12*, 955. [CrossRef]
34. Wen, F.; Lasith, A.; Pei, L.; Roummel, F.M.; Liu, P.; Robert, C.Q. Nonconvex Regularization-Based Sparse Recovery and Demixing with Application to Color Image Inpainting. *IEEE Access* **2017**, *5*, 11513–11527. [CrossRef]
35. Wen, F.; Pei, L.; Yang, Y.; Yu, W.; Liu, P. Efficient and Robust Recovery of Sparse Signal and Image using Generalized Nonconvex Regularization. *IEEE Trans. Comput. Imaging* **2017**, *3*, 566–579. [CrossRef]
36. Mei, J.; Xi, J. Efficient Sparse Recovery and Demixing using Nonconvex Regularization. *IEEE Access* **2019**, *7*, 59771–59779. [CrossRef]
37. Dykes, L.; Huang, G.X.; Silvia Noschese, S.; Reichel, L. Regularization Matrices for Discrete ill-posed Problems in several Space Dimensions. *Numer. Linear Algebr. Appl.* **2018**, *25*, 1. [CrossRef]
38. Daubechies, I.; Defrise, M.; De Mol, C. An iterative thresholding algorithm for linear inverse problems with a sparsity constraint. *Commun. Pure Appl. Math.* **2004**, *57*, 1413–1457. [CrossRef]
39. Beck, A.; Teboulle, M.A. fast iterative shrinkage-thresholding algorithm for linear inverse problems. *SIAM J. Imaging Sci.* **2009**, *2*, 183–220. [CrossRef]
40. Boyd, S.; Parikh, N.; Chu, E.; Peleato, B.; Eckstein, J. Distributed optimization and statistical learning via the alternating direction method of multipliers. *Found. Trends Mach. Learn.* **2011**, *3*, 1–122. [CrossRef]
41. Marjanovic, G.; Solo, V. On l_q optimization and matrix completion. *IEEE Trans. Signal Process.* **2012**, *60*, 5714–5724. [CrossRef]
42. Duan, S.Y.; Yang, B.T.; Wang, F.; Liu, G.R. Determination of Singular Value Truncation Threshold for Regularization in Ill-Posed Problems. *Inverse Probl. Sci. Eng.* **2021**, *29*, 1127–1157. [CrossRef]

43. Biousse, V.; Danesh-Meyer, H.V.; Saindane, A.M.; Lamirel, C.; Newman, N.J. Imaging of the Optic Nerve: Technological Advances and Future Prospects. *Lancet Neurol.* **2022**, *21*, 1135–1150. [CrossRef]
44. Soleimani-Babakamali, M.H.; Sepasdar, R.; Nasrollahzadeh, K.; Sarlo, R. A system reliability approach to real-time unsupervised structural health monitoring without prior information. *Mech. Syst. Signal Proc.* **2022**, *171*, 108913. [CrossRef]
45. Shotaro, N.; Gerrit, B. Machine-Learning-Based Data Recovery and Its Contribution to Seismic Acquisition: Simultaneous Application of Deblending, Trace Reconstruction, and Low-Frequency Extrapolation. *Geophysics* **2021**, *86*, 13–24. [CrossRef]

Communication

Seismic Periodic Noise Attenuation Based on Sparse Representation Using a Noise Dictionary

Lixia Sun [1], Xinming Qiu [2] Yun Wang [2,*] and Chao Wang [3]

[1] SINOPEC Research Institute of Petroleum Engineering Co., Ltd., Beijing 102206, China
[2] School of Geophysics and Information Technology, China University of Geosciences, Beijing 100083, China
[3] The State Key Laboratory of Ore Deposit Geochemistry, Institute of Geochemistry, Chinese Academy of Sciences, Guiyang 550081, China
* Correspondence: yunwang@mail.iggcas.ac.cn

Abstract: Periodic noise is a well-known problem in seismic exploration, caused by power lines, pump jacks, engine operation, or other interferences. It contaminates seismic data and affects subsequent processing and interpretation. The conventional methods to attenuate periodic noise are notch filtering and some model-based methods. However, these methods either simultaneously attenuate noise and seismic events around the same frequencies, or need expensive computation time. In this work, a new method is proposed to attenuate periodic noise based on sparse representation. We use a noise dictionary to sparsely represent periodic noise. The noise dictionary is constructed based on ambient noise. An advantage of our method is that it can automatically suppress monochromatic periodic noise, multitoned periodic noise and even periodic noise with complex waveforms without pre-known noise frequencies. In addition, the method does not result in any notches in the spectrum. Synthetic and field examples demonstrate that our method can effectively subtract periodic noise from raw seismic data without damaging the useful seismic signal.

Keywords: periodic noise; notch; dictionary; sparse representation

Citation: Sun, L.; Qiu, X.; Wang, Y.; Wang, C. Seismic Periodic Noise Attenuation Based on Sparse Representation Using a Noise Dictionary. *Appl. Sci.* **2023**, *13*, 2835. https://doi.org/10.3390/app13052835

Academic Editor: Lamberto Tronchin

Received: 15 January 2023
Revised: 18 February 2023
Accepted: 19 February 2023
Published: 22 February 2023

1. Introduction

Noise decreases the signal-to-noise ratio (SNR) of seismic data and affects the quality of subsequent processes [1,2]. There are six types of noise recorded using geophones [3], in which periodic noise is a kind of noise caused by power lines, pump jacks [4], engine operation [5], or other interferences, shown as monochromatic noise or multitoned noise. Sometimes, periodic noise is so strong that seismic records are severely contaminated. However, it is not easy to attenuate periodic noise, since it overlaps with seismic waves in the time domain and the frequency domain.

For periodically monochromatic noise, the conventional method of attenuation is notch filtering, which requires exact knowledge of the frequency of monochromatic noise [6]. Obviously, the notch filtering method can attenuate seismic waves at the cutoff frequency. Model-based approaches [7–10] such as sinusoidal approximation are used to remove power line noise, though this method requires accurate estimation of noise frequency and needs significant computation time [11]. Henley [12] presented a spectral clipping method to detect monochromatic noise automatically; however, it is not applicable to weak periodic noise. Karsli and Dondurur [13] used an improved mean filtering method to attenuate power line harmonic noise without noise frequency estimation; however, this requires knowledge of the rough frequency band.

In recent years, following the application of wide-azimuth, broadband, high-density seismic acquisition technology—including micro-seismic observation [14] and time-lapse monitoring based on fiber sensing—the size of 3D seismic data is increasing, and the frequency band of reflections is becoming wider. Besides monochromatic noises suppression, recognition of multitoned noises and their automatic suppression are urgently needed.

Methods based on sparsity representation, such as S-transform, singular spectrum analysis and empirical mode decomposition, extend realization of automatic denoising [15–17]. Xu et al. [4] proposed a method based on morphological diversity of monochromatic noise and seismic waves, which assumed that the raw data are only composed of two kinds of signals, the monochromatic noise and seismic waves. It is not applicable to seismic data with strong white Gaussian noise or multitoned noise.

In this paper, a new method is proposed to attenuate periodic noise. The proposed method is based on sparse representation using a noise dictionary. The novelty is the construction of a noise dictionary, which can represent periodic noise sparsely. First, the noise dictionary is constructed using ambient noise. Next, the noise dictionary is used to sparsely represent periodic noise. Then, the de-noised data are obtained by subtracting the periodic noise from the raw seismic data. The method is applied to synthetic and field seismic data. The effectiveness of the proposed method is that it can subtract periodic noise from raw seismic data without any notches in the spectrum compared with the notch filtering method.

2. Method

Periodic noise is recurring in raw seismic data, and its amplitude is constant within the recording time [6]. Based on these characteristics, we can differentiate and estimate periodic noise from ambient noise. Because ambient noise, such as periodic noise, white Gaussian noise and other non-stationary [1,18] random noise, is ubiquitous in seismic records before the first arrivals, we can extract noise features and construct the noise dictionary from it. The algorithm is outlined in the flow skeleton (Figure 1).

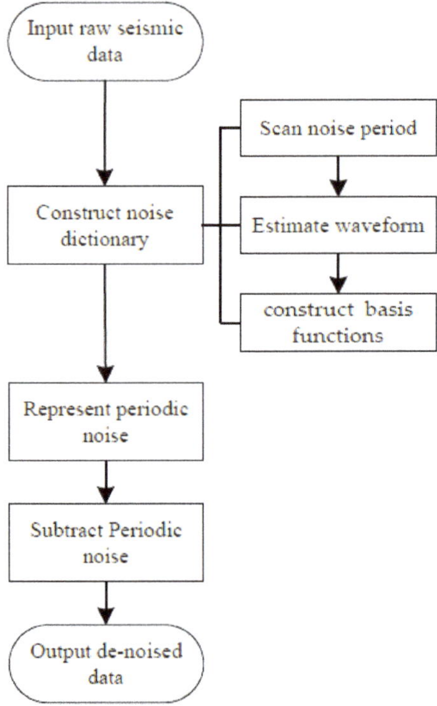

Figure 1. Flow skeleton of the proposed method.

2.1. Noise Period Scanning

The waveform of periodic noise is similar in a single trace. Based on the similarity, we scan the noise period from ambient noise, which exists in the records before the first arrivals. First, given the scanned period interval $[T_{\min}, T_{\max}]$, the noise period T is changed from the initial scanned period $T_{\min}$ to the final scanned period $T_{\max}$. The data S_j on the single trace $h_j(j = 1, 2, \ldots, m)$ are split into many time windows $s_i(i = 1, 2, \ldots, n)$, denoted as

$$S_j = [s_1, s_2, \ldots, s_n] \tag{1}$$

where the length of time window s_i is equal to the scanned period T. The size of S_j is $N \times 1$ $(N \geq nT)$. Next, the correlation coefficients for the adjacent two time windows s_i and s_{i+1} are calculated:

$$\text{Corr}(s_i, s_{i+1}) = \frac{\text{Cov}(s_i, s_{i+1})}{\sqrt{\text{Var}(s_i)}\sqrt{\text{Var}(s_{i+1})}} \tag{2}$$

where $\text{Cov}(s_i, s_{i+1})$ is the covariance of the two time windows and $\text{Var}(s)$ is the variance of s. For accuracy, we average the correlation coefficients to obtain a coefficient to evaluate the similarity of all time windows.

$$C(T) = \frac{1}{n-1}\sum_{i=1}^{n-1} \text{Corr}(s_i, s_{i+1}) \tag{3}$$

Then, the scanned period T is increased and the former procedures are repeated until all periods in the interval $[T_{\min}, T_{\max}]$ are scanned. The period which matches the maximum value of the correlation coefficients is our estimated period:

$$\widetilde{T} = \underset{T}{\text{argmax}}\{C(T)\} \tag{4}$$

where $\widetilde{T}$ is the period of periodic noise on the trace h_j.

2.2. Waveform Estimation by Stacking

The time window, whose length is $\widetilde{T}$, is chosen and is represented as $\widetilde{s}_i$. The chosen time window $\widetilde{s}_i$ is approximated as the noise waveform; however, it is affected by white Gaussian noise. A stacking method is widely used to improve the SNR of the seismic profile [19], because it weakens white Gaussian noise and emphasizes seismic waves. We stack the approximate waveforms for an accurate noise waveform:

$$w_j = \sum_{i=1}^{n} \widetilde{s}_i \tag{5}$$

where w_j is the noise waveform on the trace h_j and its length $\widetilde{T}$ is equal to that of $\widetilde{s}_i$. The waveforms of different traces are similar to the near traces $h_1, h_2, \ldots, h_m$ when the traces are interfered with by the same noise source. Therefore, we can stack the similar waveforms $w_1, w_2, \ldots, w_m$ along different traces:

$$w = \sum_{j=1}^{m} w_j \tag{6}$$

where w is the noise waveform estimated from the traces $h_1, h_2, \ldots, h_m$. We note that waveforms $w_1, w_2, \ldots, w_m$ obtained by Equation (5) are not in phase. Before stacking, waveforms $w_j(j \neq 1)$ need to be cyclically shifted to the same phase as w_1 by scanning the shift length in an interval $\left[0, 1, \ldots, \widetilde{T} - 1\right]$. The suitable shift length corresponds to the maximum correlation coefficient between $w_j(j \neq 1)$ and w_1.

2.3. Periodic Noise Representation

Based on the stationarity of periodic noise, the basis function φ_0 of periodic noise is obtained by extending the waveform w iteratively to the length of seismic data on a single trace and then energy-normalized:

$$\varphi = [w, w, \ldots, w] \tag{7}$$

$$\varphi_0 = \frac{\varphi}{\|\varphi\|_2} \tag{8}$$

where $\|\bullet\|_2$ is the l_2 norm. The size of φ is $N \times 1$, which is equal to that of S_j. The noise dictionary is constructed by basis functions of different phases:

$$\mathbf{D} = \begin{bmatrix} \varphi_0 & \varphi_1 & \varphi_2 & \cdots & \varphi_{T_j-1} \end{bmatrix} \tag{9}$$

where $\varphi_1, \varphi_2, \ldots, \varphi_{\widetilde{T}-1}$ are obtained by cyclically shifting $1, 2, \ldots, \widetilde{T}-1$ time samples. The size of $\mathbf{D}$ is $N \times \widetilde{T}$. The dictionary provides a sparse representation of periodic noise. In sparse representation theory [20], the signals can be efficiently explained as linear combinations of prespecified basis functions, where the linear coefficients are sparse. Based on sparse representation theory, the mathematical model of our noise representation is

$$S_j = \mathbf{D}x \tag{10}$$

where $S_j (j = 1, 2, \ldots, m)$ is the actual periodic noise and x is its coefficient represented by the dictionary $\mathbf{D}$. The size of x is $\widetilde{T} \times 1$. An approach to solve Equation (10) is using the sparsity constraint via an L_0 regularization term. Then, the periodic noise of multi-trace seismic data is obtained by solving the following optimization problem:

$$\widetilde{S}_j = \underset{S_j}{\operatorname{argmin}} \|S_j - \mathbf{D}x\|_2^2 \ \ s.t. \|x\|_0 \leq 1 \tag{11}$$

where $\widetilde{S}_j (j = 1, 2, \ldots, m)$ is the approximate periodic noise and x is its coefficient represented by the dictionary $\mathbf{D}$. In the optimization problem, the sparsity of this representation is 1. Because only one basis function corresponds to a single trace, the coefficient x has only one non-zero component. This is the reason for the condition on the right in Equation (11) $\|x\|_0 \leq 1$. Equation (11) is solved by the matching pursuit algorithm, which entails computing the inner products between the residual and the dictionary elements, updating the coefficient, and updating the residual iteratively [20,21]. Finally, the de-noised data are obtained by subtracting the periodic noise from the raw seismic data.

3. Synthetic Example

A synthetic dataset consisting of 21 traces is shown in Figure 2 with a 1 ms sampling rate and a 10 m geophone interval. Multitoned noises of 40 Hz and 50 Hz are incorporated to the noise-free data. In addition, a small amount of white Gaussian noise is added. Figure 3 shows the contaminated seismic dataset and its frequency–wavenumber (f-k) spectrum. The reflections are hardly identifiable owing to the strong periodic noise. The contaminations in the seismic data appear as two horizontal bands at 40 and 50 Hz in the f-k domain. The signal on the 11th trace (Figure 4a) shows the influence of multitoned noise clearly. Its amplitude spectrum (Figure 4b) also shows high spikes at frequencies of 40 Hz and 50 Hz.

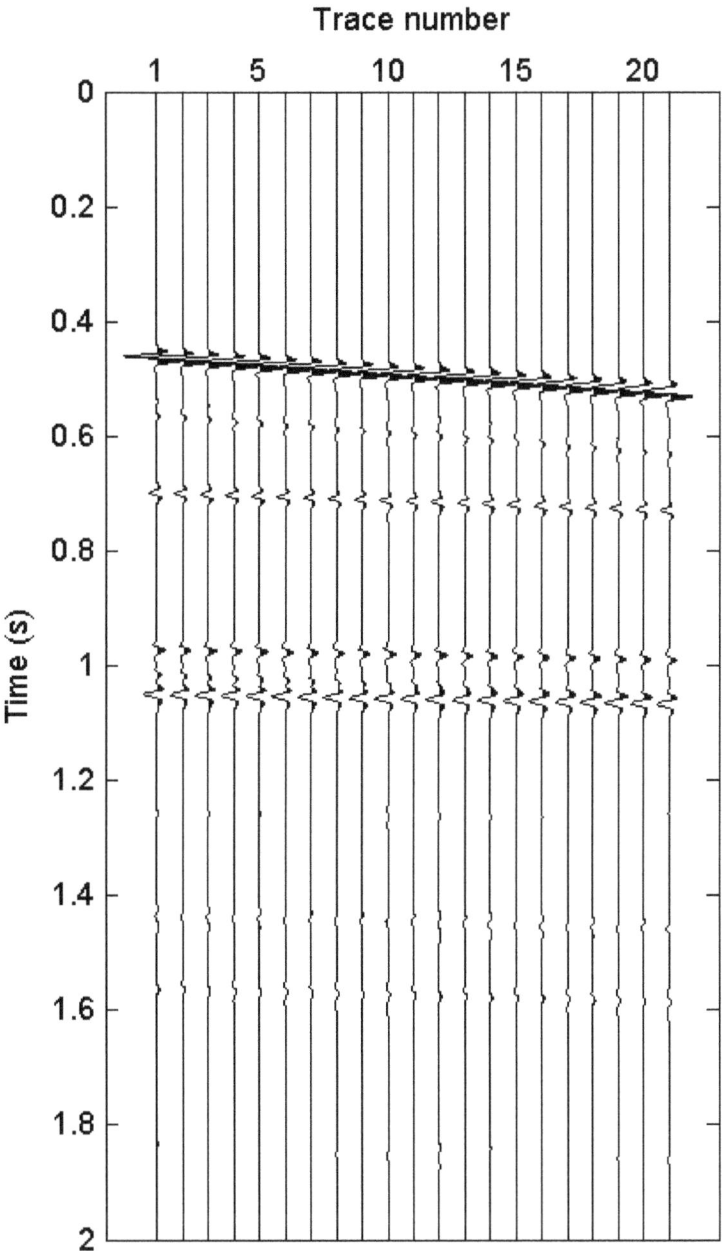

Figure 2. Synthetic seismic data.

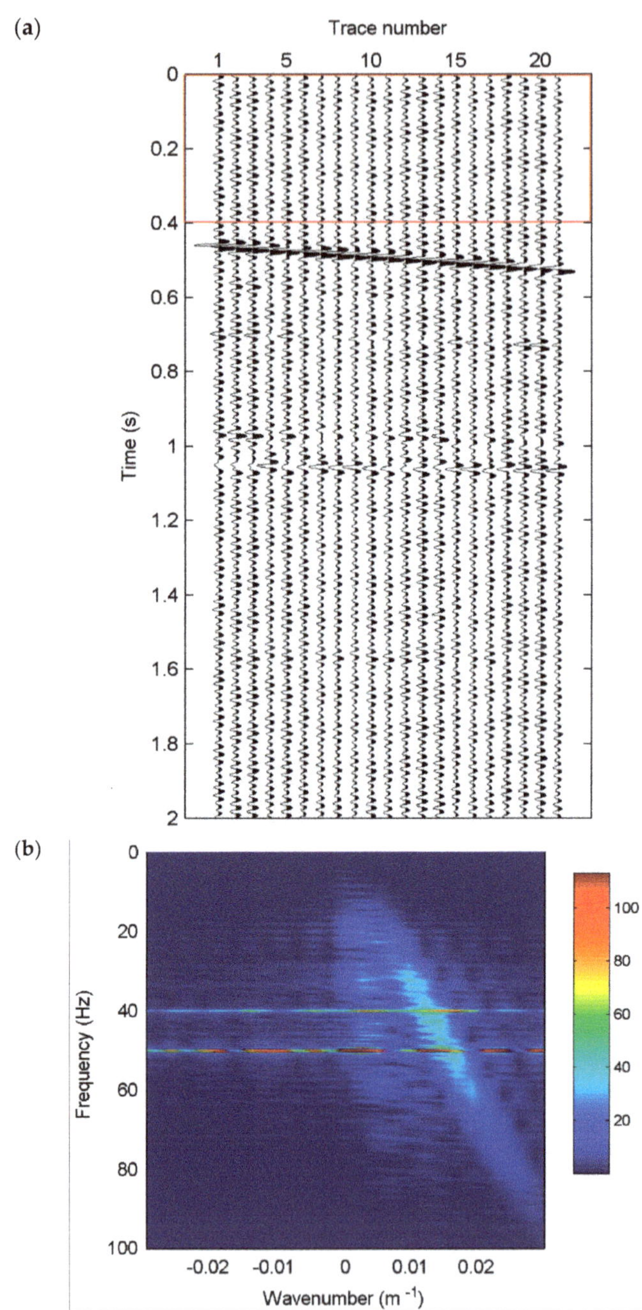

Figure 3. Synthetic contaminated seismic data. (**a**) Contaminated seismic data in offset-time domain and (**b**) its frequency-wavenumber (f-k) spectrum where the red rectangle marks the ambient noise we used.

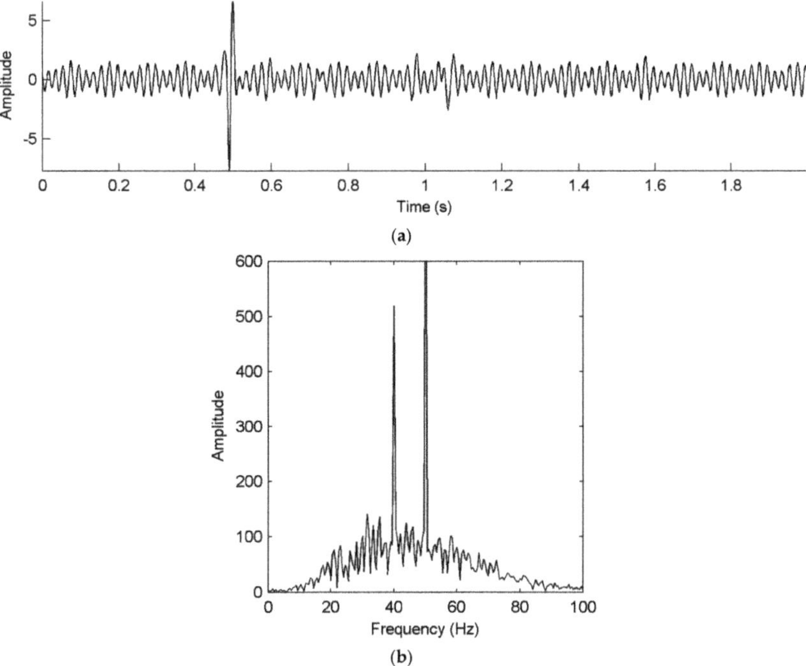

Figure 4. Seismic signal on 11th trace. (**a**) Signal on the 11th trace in time domain and (**b**) its amplitude spectrum.

The first 0.4 s of data is dominated by periodic noise marked by a red rectangle in Figure 3a. In addition, it does not consist of seismic waves. Therefore, it is chosen to construct the noise dictionary. The waveform estimated from the chosen data is shown in Figure 5. After sparse representation, the result of noise attenuation is shown in Figure 6. The main noise is attenuated and reflections can be clearly seen in Figure 6a. The eliminated noise is periodic noise as shown in Figure 6b. Their spectra show that the multitoned noise is attenuated and no seismic waves are eliminated. The de-noising result on the 11th trace and the corresponding amplitude spectra are shown in Figure 7. The de-noised seismic signal is close to the theoretical signal in both time sequence and amplitude spectrum, except for weak white Gaussian noise.

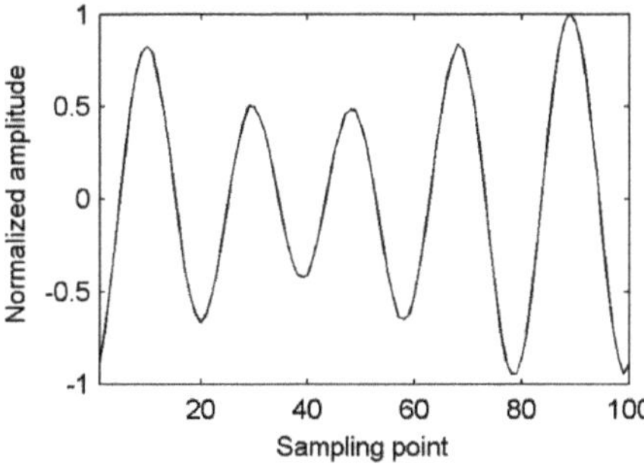

Figure 5. Noise waveform estimated by the proposed method.

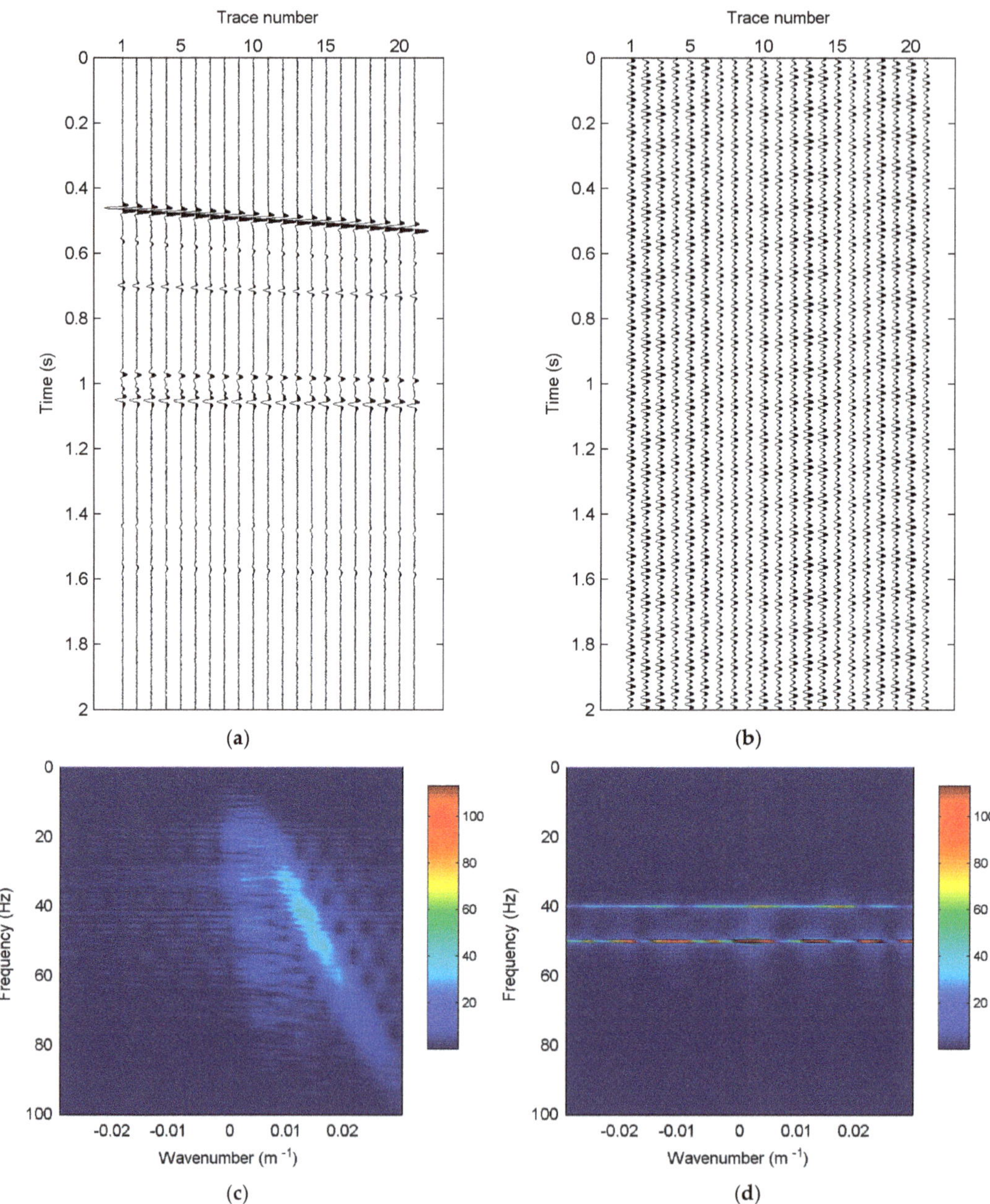

Figure 6. De-noising result of the synthetic seismic data by the proposed method. (**a**) De-noised data and (**b**) eliminated noise by the proposed method; and (**c**,**d**) their corresponding f-k spectra.

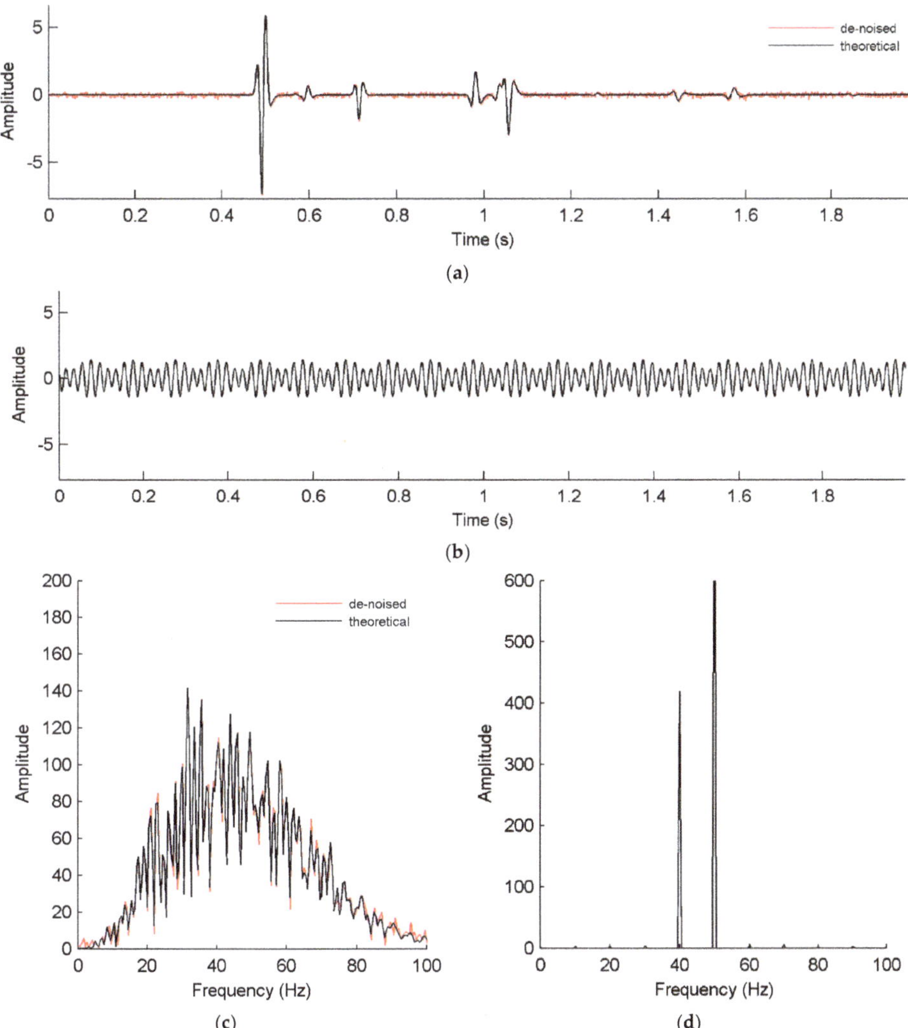

Figure 7. De-noising result on the 11th trace. (**a**) De-noised signal and (**b**) eliminated noise on the 11th trace; and (**c**,**d**) their amplitude spectra. The red line corresponds to the de-noised signal and the black line corresponds to the theoretical signal in (**a**,**c**).

A synthetic signal is shown in Figure 8a, where the main noise is stationary and the waveform is recurring except for weak white Gaussian noise. The eliminated noise and de-noised signal are shown in Figure 8b,c, respectively. Their amplitude spectra are shown in Figure 9. Both the time sequence and amplitude spectrum of the de-noised signal are consistent with the theoretical signal. Because the amplitude spectrum is complex, it is hard to estimate its amplitudes, frequencies and phases, which are important for model-based approaches [8,9]. Therefore, the proposed method can effectively attenuate the multitoned noise and even periodic noise with a complex waveform but not influence the seismic events.

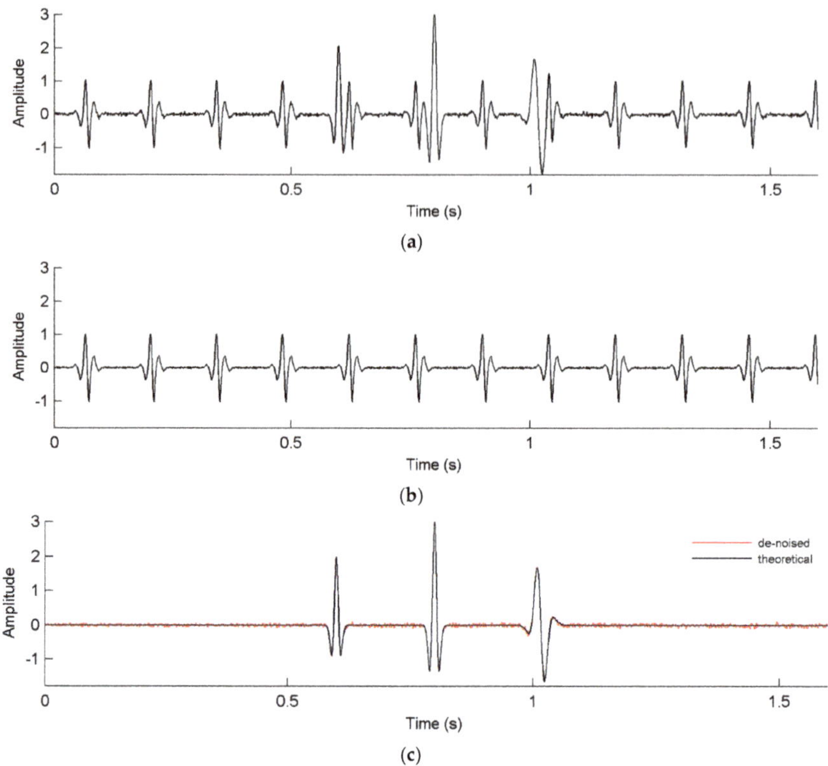

Figure 8. De-noising result of the synthetic noisy signal. (**a**) Noisy signal, (**b**) eliminated noise and (**c**) de-noised signal. The de-noised signal is shown as a red line and the theoretical signal as a black line in (**c**).

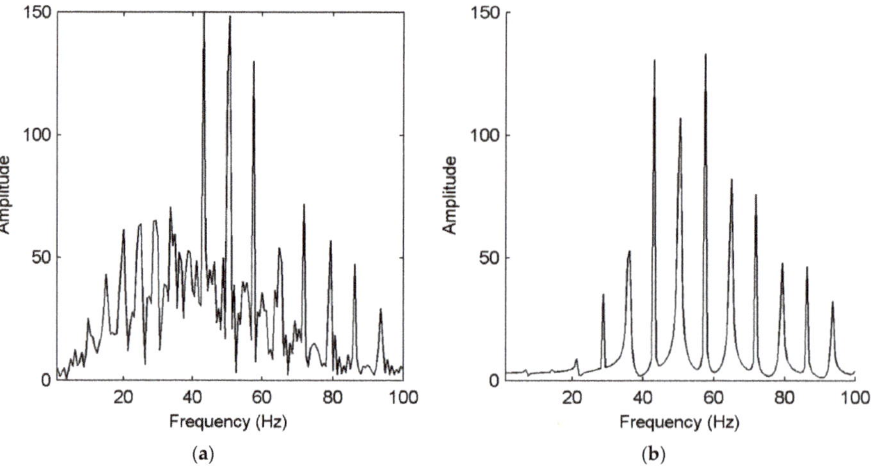

Figure 9. *Cont.*

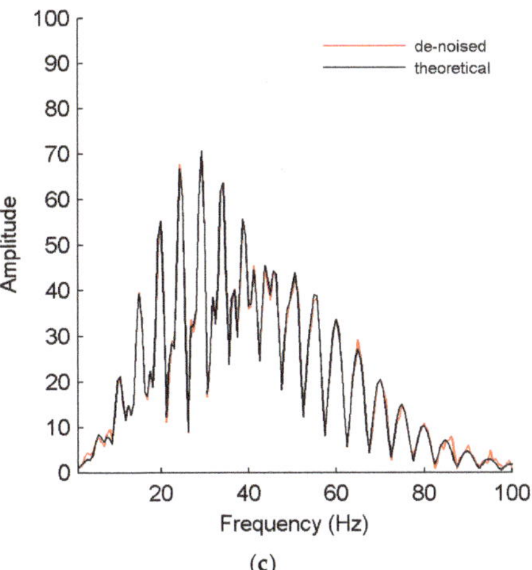

Figure 9. Amplitude spectra of the (**a**) noisy signal, (**b**) eliminated noise and (**c**) de-noised signal. The de-noised signal is shown as a red line and the theoretical signal as a black line in (**c**).

4. Field Example

The proposed method was tested using field data. The field data are a land seismic shot recorded with a 2 ms sample rate and a 40 m geophone interval, as shown in Figure 10a. The data contain periodic noise which may be caused by power lines, engine operation and other interferences. The noise is so strong that it affects the quality of the subsequent processes, especially for the first 12 traces. As shown in Figure 10c, we cannot see any waves after the time 1.5 s. There are three horizontal bands around the frequencies of 7 Hz, 14 Hz and 50 Hz in the f-k domain (Figure 10b). The periodic noise at the frequency of approximately 14 Hz, highlighted in Figure 10b, is the weakest, and the periodic noise at the frequency of 50 Hz is the strongest.

We use the ambient noise marked by a red rectangle in Figure 10a to construct the noise dictionary. The estimated waveform is shown in Figure 11. The result of noise attenuation by the proposed method is shown in Figure 12. The periodic noise is attenuated to a large degree. The horizontal bands of the f-k spectrum are almost eliminated, except for weak residual noise of about 7 Hz frequency. In addition, there are no spectral notches in the f-k spectrum (Figure 12b).

The notch filtering method is applied to the field data for a comparison. The narrow stop band of the notch filter is the frequency range $[f_0 - \Delta f, f_0 + \Delta f]$, where f_0 is the noise frequency and $2 \times \Delta f$ is the noise bandwidth. The values of f_0 for the three noise bands are set to 7 Hz, 14 Hz and 50 Hz, respectively. The noise bandwidth is estimated to be 2 Hz and $\Delta f = 1$Hz is set. The f-k spectra of the filtering result are shown in Figure 13. The horizontal bands around the frequencies of 7 Hz, 14 Hz and 50 Hz are separated from the seismic data. However, the seismic waves are also eliminated at those frequencies. This causes spectral notches of seismic waves (Figure 13a). To further show the effectiveness of our method, the spectra of the single-trace de-noised signals by the two methods are compared in Figure 14. For the notch filtering method, the spectral notch causes amplitude loss of seismic events around the frequencies of 7 Hz, 14 Hz and 50 Hz. However, amplitude loss does not occur using our proposed method. The eliminated noise (Figure 15a) is stationary. This is consistent with the characteristics of periodic noise caused by the power lines or engine operation. For comparison, the noise eliminated by the notch filtering method is not stationary because it contains seismic signals.

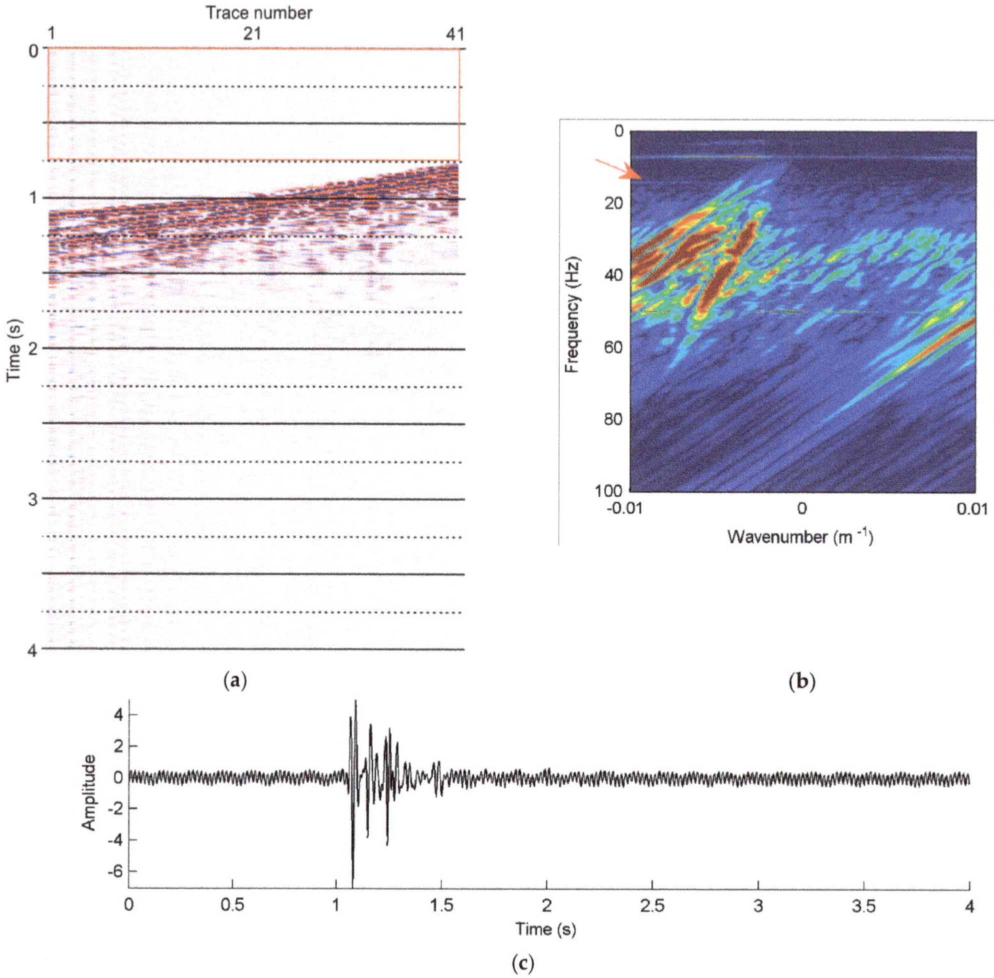

(a) (b)

(c)

Figure 10. Field data. (**a**) Field data in offset-time domain, (**b**) its f-k spectrum and (**c**) the signal on the 8th trace, where the red rectangle marks the ambient noise and the red arrow points to the 14 Hz weak noise.

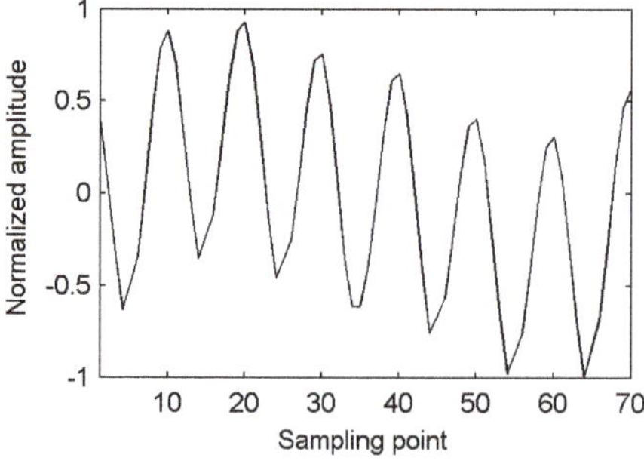

Figure 11. Waveform estimated from the ambient noise.

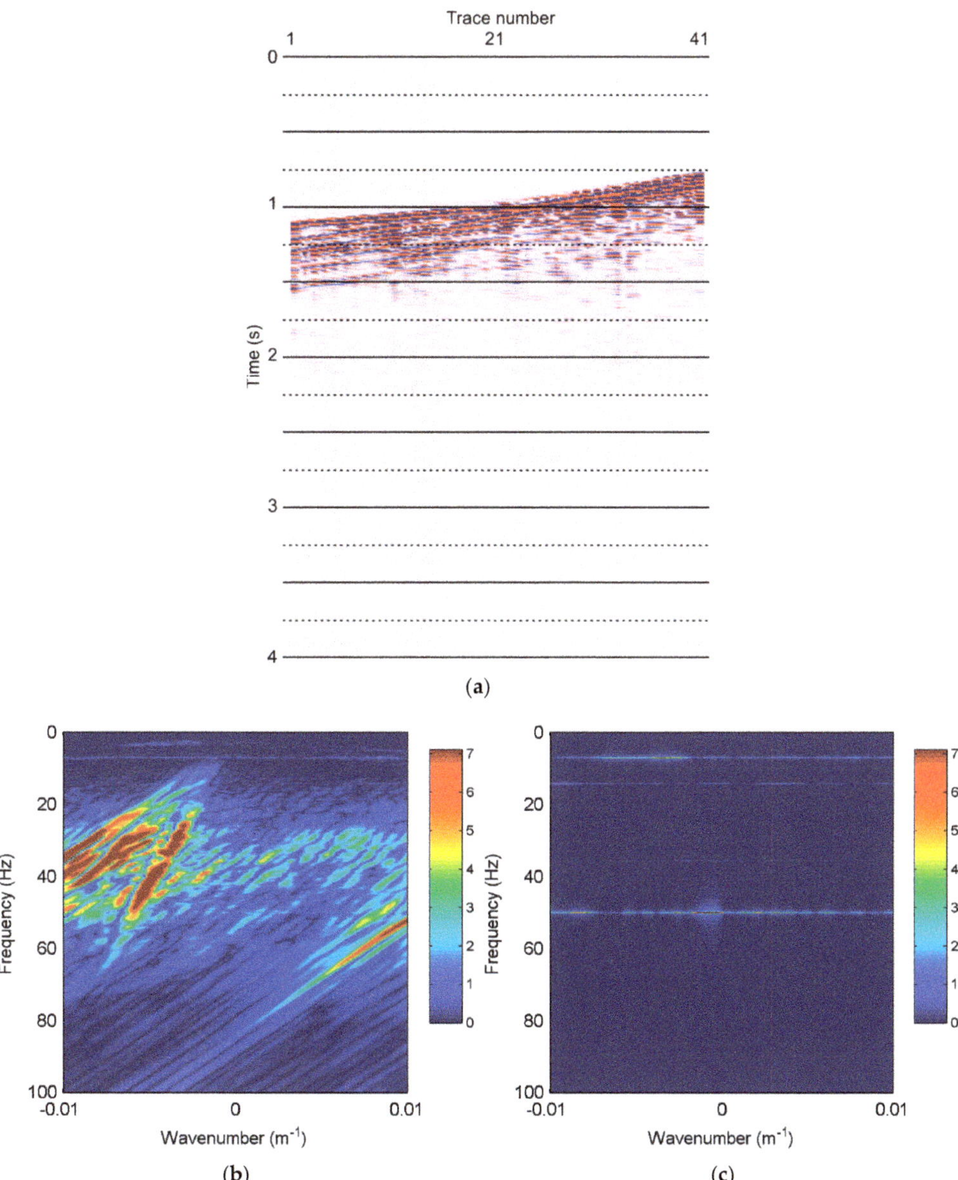

Figure 12. Result of noise attenuation by our method: (**a**) de-noised data, (**b**) its f-k spectrum and (**c**) the f-k spectrum of eliminated noise.

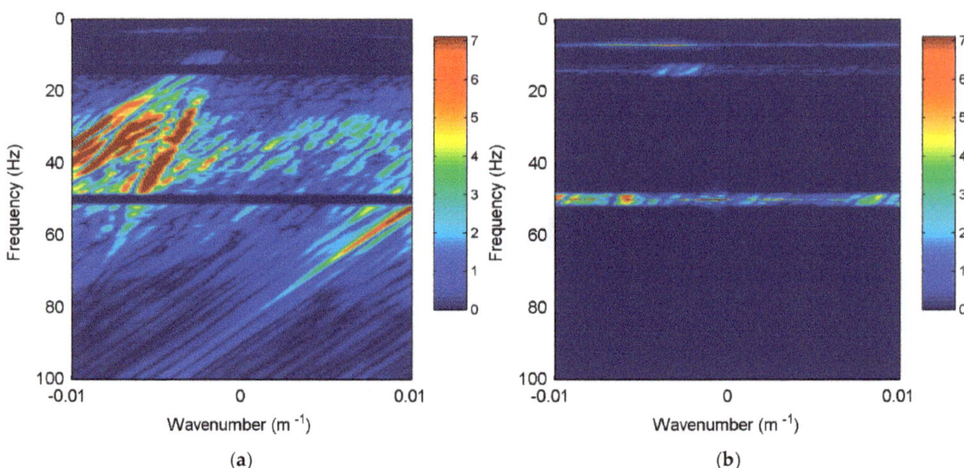

Figure 13. Spectra of (**a**) the de-noising data and (**b**) the eliminated noise by the notch filtering method.

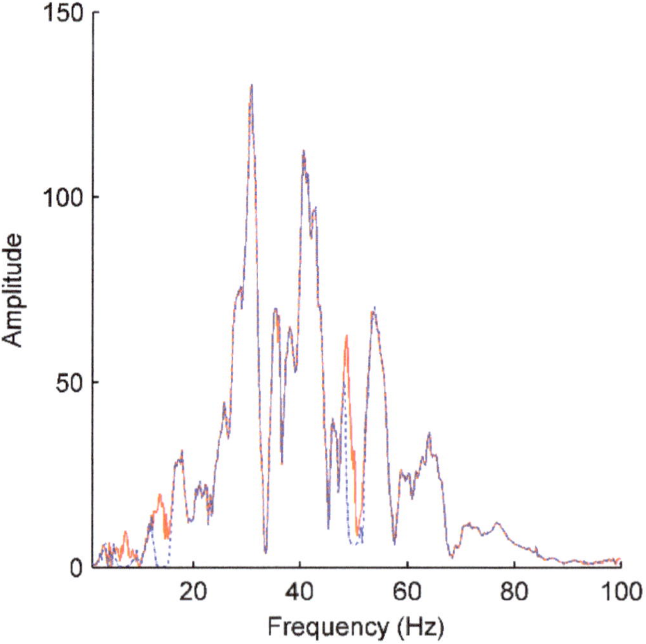

Figure 14. Amplitude spectra of the de-noised signals on the 8th trace by the proposed method (the red solid line) and the notch filtering method (the blue dash-dotted line).

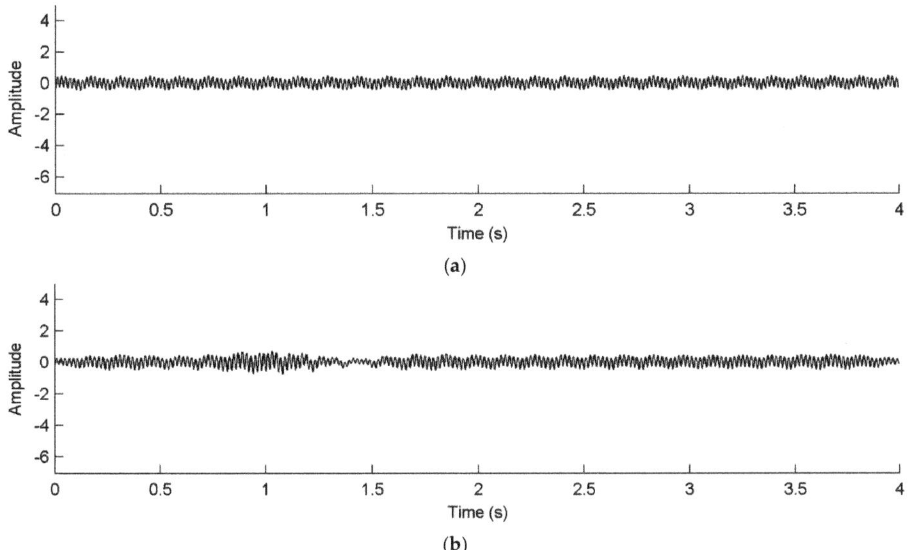

Figure 15. Eliminated noise by (**a**) the proposed method and (**b**) the notch filtering method.

5. Discussions

The proposed de-noising method is based on sparse representation of periodic noise. The key to our method is the construction of the noise dictionary. Because ambient noise contains no seismic waves but dominant periodic noise and other random noise, periodic noise can be estimated without the influence of seismic reflections. Therefore, our method is useful regardless of whether the seismic waves are strong or weak. A scanning method is used to estimate the noise period. The accuracy of the noise period largely influences our de-noising result. To obtain an accurate noise waveform, the waveforms in the time domain and the space domain are stacked shown as the Equations (5) and (6), respectively. It must be emphasized that our de-noising method is only applicable to stationary noise with a constant period, waveform and amplitude. It can be used to attenuate power line harmonic noise, pump jack noise and engine operation noise in land or oceanic seismic exploration.

Based on the proposed method, ambient noise detection is urged for noise estimation. However, ambient noise has not specifically been detected in oil exploration. Therefore, we suggest that ambient noise should be acquired for one second before source excitation. Long-term ambient noise will be helpful for building a perfect noise dictionary.

Wind- or water-induced noise is not strictly stationary, and the stationarity becomes weak with increasing recording time [22,23]. Our method cannot be used to attenuate this kind of noise. To broaden the method for non-stationary noise, higher-order statistics [24] need to be considered further. In addition, methods of noise feature extraction from ambient noise based on machine learning [25] will draw increased attention in the future.

6. Conclusions

A new method is proposed to attenuate periodic noise based on sparse representation. The novelty is the construction of a noise dictionary based on ambient noise. Our method can attenuate monochromatic or multitoned periodic noise automatically without pre-known noise frequencies. The noise is assumed to be stationary noise with a constant period, waveform and amplitude. Synthetic and field tests show the effectiveness of the proposed method. Compared with the conventional notch filtering method, the proposed method can obtain de-noised data with no distortion in the time and frequency domains. Therefore, our method can attenuate periodic noise without damaging the seismic events.

Author Contributions: Conceptualization, Y.W.; investigation, C.W.; writing—original draft preparation, L.S.; writing—review and editing, L.S. and X.Q.; visualization, X.Q. All authors have read and agreed to the published version of the manuscript.

Funding: This work was supported by the National Natural Science Foundation of China (62127815) and Guizhou Science and Technology Cooperation Platform Talents Program: [2021] 5629.

Institutional Review Board Statement: Not applicable.

Informed Consent Statement: Not applicable.

Data Availability Statement: Not applicable.

Acknowledgments: We thank Yijun Yuan for providing the field data.

Conflicts of Interest: The authors declare no conflict of interest.

References

1. Li, G.; Li, Y.; Yang, B. Seismic exploration random noise on land: Modeling and application to noise suppression. *IEEE Trans. Geosci. Remote Sens.* **2017**, *55*, 4668–4681. [CrossRef]
2. Zhong, T.; Zhang, S.; Li, Y.; Yang, B. Simulation of seismic-prospecting random noise in the desert by a brownian-motion-based parametric modeling algorithm. *Comptes Rendus Geosci.* **2019**, *351*, 10–16. [CrossRef]
3. Groos, J.; Ritter, J. Time domain classification and quantification of seismic noise in an urban environment. *Geophys. J. Int.* **2009**, *179*, 1213–1231. [CrossRef]
4. Xu, J.; Wang, W.; Gao, J.; Chen, W. Monochromatic noise removal via sparsity-enabled signal decomposition method. *IEEE Geosci. Remote Sens. Lett.* **2013**, *10*, 533–537. [CrossRef]
5. Meunier, J.; Bianchi, T. Harmonic noise reduction opens the way for array size reduction in vibroseis™ operations. *In Seg Tech. Program Expand. Abstr.* **2002**, *21*, 70–73. [CrossRef]
6. Karsli, H.; Dondurur, D.; Güney, R. A comparison of post-stack results after filtering of harmonic noise using two filter methods. In Proceedings of the Near Surface Geoscience 2016-Second Applied Shallow Marine Geophysics Conference, Barcelona, Spain, 4–8 September 2016.
7. Larsen, J.J.; Dalgaard, E.; Auken, E. Noise cancelling of mrs signals combining model-based removal of powerline harmonics and multichannel wiener filtering. *Geophys. J. Int.* **2013**, *196*, 828–836. [CrossRef]
8. Yao, J.; Di, D.; Jiang, G.; Gao, S.; Yan, H. Real-time acceleration harmonics estimation for an electro-hydraulic servo shaking table using kalman filter with a linear model. *IEEE Trans. Control Syst. Technol.* **2014**, *22*, 794–800. [CrossRef]
9. Olsson, P.-I.; Fiandaca, G.; Larsen, J.J.; Dahlin, T.; Auken, E. Doubling the spectrum of time-domain induced polarization by harmonic de-noising, drift correction, spike removal, tapered gating and data uncertainty estimation. *Geophys. Suppl. Mon. Not. R. Astron. Soc.* **2016**, *207*, 774–784. [CrossRef]
10. Ghanati, R.; Hafizi, M. Statistical de-spiking and harmonic interference cancellation from surface-nmr signals via a state-conditioned filter and modified nyman-gaiser method. *Boll. Di Geofis. Teor. Ed Appl.* **2017**, *58*, 181–204.
11. Saucier, A.; Marchant, M.; Chouteau, M. A fast and accurate frequency estimation method for canceling harmonic noise in geophysical records. *Geophysics* **2006**, *71*, V7–V18. [CrossRef]
12. Henley, D.C. Spectral clipping: A promax module for attenuating strong monochromatic noise. *CREWES Calg. AB Can.* **2001**, *13*, 311–320.
13. Karslı, H.; Dondurur, D. A mean-based filter to remove power line harmonic noise from seismic reflection data. *J. Appl. Geophys.* **2018**, *153*, 90–99. [CrossRef]
14. Shao, J.; Wang, Y.; Yao, Y.; Wu, S.; Xue, Q.; Chang, X. Simultaneous denoising of multicomponent microseismic data by joint sparse representation with dictionary learning. *Geophysics* **2019**, *84*, KS155–KS172. [CrossRef]
15. Huang, C.C.; Liang, S.F.; Young, M.S.; Shaw, F.Z. A novel application of the s-transform in removing powerline interference from biomedical signals. *Physiol. Meas.* **2008**, *30*, 13–27. [CrossRef]
16. Ghanati, R.; Fallahsafari, M.; Hafizi, M.K. Joint application of a statistical optimization process and empirical mode decomposition to magnetic resonance sounding noise cancelation. *J. Appl. Geophys.* **2014**, *111*, 110–120. [CrossRef]
17. Ghanati, R.; Hafizi, M.K.; Mahmoudvand, R.; Fallahsafari, M. Filtering and parameter estimation of surface-nmr data using singular spectrum analysis. *J. Appl. Geophys.* **2016**, *130*, 118–130. [CrossRef]
18. Wang, D.; Li, Y.; Nie, P. A study on the gaussianity and stationarity of the random noise in the seismic exploration. *J. Appl. Geophys.* **2014**, *109*, 210–217. [CrossRef]
19. Yilmaz, O. *Seismic Data Analysis: Processing, Inversion, and Interpretation of Seismic Data*; Society of Exploration Geophysicists: Tulsa, OK, USA, 2001.
20. Mallat, S. *A Wavelet Tour of Signal Processing: The Sparse Way*, 3rd ed.; Elsevier: Amsterdam, The Netherlands, 2008.

21. Mallat, S.G.; Zhang, Z. Matching pursuits with time-frequency dictionaries. *IEEE Trans. Signal Process.* **1993**, *41*, 3397–3415. [CrossRef]
22. Zhong, T.; Li, Y.; Wu, N.; Nie, P.; Yang, B. Statistical properties of the random noise in seismic data. *J. Appl. Geophys.* **2015**, *118*, 84–91. [CrossRef]
23. Zhong, T.; Li, Y.; Wu, N.; Nie, P.; Yang, B. A study on the stationarity and gaussianity of the background noise in land-seismic prospecting. *Geophysics* **2015**, *80*, V67–V82. [CrossRef]
24. Mousavi, S.M.; Langston, C.A. Hybrid seismic denoising using higher-order statistics and improved wavelet block thresholding. *Bull. Seismol. Soc. Am.* **2016**, *106*, 1380–1393. [CrossRef]
25. Jia, Y.; Ma, J. What can machine learning do for seismic data processing? An interpolation application. *Geophysics* **2017**, *82*, V163–V177. [CrossRef]

Article

Seismic Coherent Noise Removal of Source Array in the NSST Domain

Minghao Yu, Xiangbo Gong * and Xiaojie Wan

College of Geo-Exploration Science and Technology, Jilin University, Changchun 130026, China
* Correspondence: gongxb@jlu.edu.cn

Abstract: The technique of the source array based on the vibroseis can provide the strong energy of a seismic wave field, which better meets the need for seismic exploration. The seismic coherent noise reduces the signal-to-noise ratio (SNR) of the source array seismic data and affects the seismic data processing. The traditional coherent noise removal methods often cause some damage to the effective signal while suppressing coherent noise or cannot suppress the interference wave effectively at all. Based on the multi-scale and multi-direction properties of the non-subsampled Shearlet transform (NSST) and its simple mathematical structure, the seismic coherent noise removal method of source array in NSST domain is proposed. The method is applied to both the synthetic seismic data and the filed seismic data. After processing with this method, the coherent noise of the seismic data is greatly removed and the effective signal information is greatly protected. The analysis of the results demonstrates the effectiveness and practicability of the proposed method on coherent noise attenuation.

Keywords: source array; NSST; seismic coherent noise removal; seismic data analysis

Citation: Yu, M.; Gong, X.; Wan, X. Seismic Coherent Noise Removal of Source Array in the NSST Domain. *Appl. Sci.* **2022**, *12*, 10846. https://doi.org/10.3390/app122110846

Academic Editors: Guofeng Liu, Xiaohong Meng and Zhifu Zhang

Received: 7 September 2022
Accepted: 25 October 2022
Published: 26 October 2022

1. Introduction

The seismic exploration method is widely used in resource exploration and other fields, and the seismic source, as an important part of the seismic data acquisition, directly affects the exploration effect of the seismic exploration. With the rapid development of vibroseis technology, vibroseis is gradually replacing the traditional dynamite source for seismic data acquisition due to its advantages of safety, environmental protection, high quality, economy, and efficiency. However, limited by its own structure and exploration environment, the single vibroseis has the problems of small output force and limited exploration depth. In order to meet the demand of seismic exploration, the source array based on vibroseis technology is paid more and more attention. In the seismic data acquisition of source array, there are generally more serious coherent noises, such as multiple refraction waves and surface waves of strong energy, strong acoustic interference, and surface direct waves with wide frequency bands, which greatly affect the quality of the seismic data. The removal of coherent noise is essential for the exploration technique of the source array.

How to better remove the seismic noise and improve the SNR of the seismic data has been a hot research topic in the seismic exploration field [1–3]. For coherent noise, scholars of seismic exploration have presented many seismic denoising approaches to suppress it. The traditional noise removal methods mainly suppress coherent noise according to the difference between effective signal and coherent noise in apparent velocity and frequency, such as f-k filter [4,5], K-L transform [6,7], Radon transform [8,9], radial channel transform [10,11], empirical mode decomposition (EMD) [12,13], and so on. Wavelet transform [14,15] is an effective tool for seismic data processing, which has great advantages in one-dimensional data processing, though this advantage cannot be simply extended to two-dimensional or three-dimensional data. Therefore, multi-scale geometric analysis methods such as Ridgelet transform [16,17], Curvelet transform [18,19], Contourlet transform [20,21] and Shearlet transform [22] have been proposed successively. At present, multi-scale geometric analysis has been widely used in seismic data processing and achieved certain effect.

G. Easley, K. Guo, and D. Labate [23] proposed the Shearlet transform based on the basic theory of wavelet transform and the multi-scale analysis theory. Shearlet transform not only has good localization characteristics, scaling characteristics, direction sensitivity, and near-optimal sparse representation performance, but also has a simple mathematical structure and the transformed coefficients have a one-to-one correspondence with the image points. Therefore, compared with other multi-scale geometric analysis methods, Shearlet transform has a more delicate representation of the scale direction and the physical significance of the transformed coefficients is clearer, which is more conducive to the processing of the coefficients. In the process of the conventional Shearlet transform, the subsampling operation is carried out, which leads to spectrum aliasing in the low frequency and high frequency sub-bands decomposed by Shearlet transform. The spectrum aliasing will cause the information of the same direction to appear in several different sub-bands at the same time, which will weaken the direction selectivity. In addition, the subsampling operation makes the Shearlet coefficient low redundancy, resulting in the lack of the translation invariance. If it is used for image denoising, there will be an obvious ringing effect. In order to overcome the shortcomings mentioned above and enhance the directional selectivity and the translation invariance of Shearlet transform, the NSST came into being. NSST is a non-orthogonal transform. The NSST abandons the subsampling operation and combines the non-subsampled Laplacian pyramid transform (NSLP) with the non-subsampled directional filter. After the NSST, the size of each direction sub-band at each scale is the same as the original image, which greatly improves the image redundancy. The enhancement of the image coefficient redundancy allows the NSST to have the translation invariance. The NSST has been successfully applied in many fields.

In the field of image processing, the NSST has been widely used. Guo, et al. (2013) proposed a new multi-focus image fusion algorithm based on the NSST [24]. Karami et al. (2016) adopted the NSST and the fully constrained least squares unmixing (FCLSU) to the denoising of hyperspectral images [25]. Priya et al. (2017) used the NSST based on the multislice fusion to the edge enhancement of liver CT images [26]. Qu et al. (2018) presented an image enhancement method based on the NSST and the directional information measurement [27]. Li et al. (2019) performed the enhancement of hyperspectral remote sensing images based on improved fuzzy contrast in the NSST domain [28]. Ramakrishnan et al. (2020) applied the NSST to the fusion of multiple exposure images [29]. Shen et al. (2021) made a change detection in SAR images based on the improved NSST and the multi-scale feature fusion CNN [30]. The NSST also has been increasingly applied in the seismic exploration field in recent years. Liang et al. (2018) presented a denoising method of microseismic data noise by the NSST based on singular value decomposition [31]. Sang et al. (2018) applied the NSST to petroleum seismic exploration denoising [32]. Sang et al. (2020) proposed a seismic random noise attenuation based on PCC classification in the NSST Domain [33].

The existing traditional denoising methods have achieved certain effects in the process of attenuating coherent noise but they also have some problems. For example, some methods will produce some new interference and some will lead to uneven energy after attenuating, even damaging the effective signal. Based on the coherent noise characteristics of the vibrator seismic data and the multi-scale and multi-directional properties of Shearlet transform, coherent noise removal of the source array in NSST domain is proposed in this article. The method is applied in both the synthetic seismic data and the filed seismic data. After processing with this method, the coherent noise in the seismic data is greatly removed and the effective signal information is greatly protected. The analysis of the results demonstrates the effectiveness and practicability of the proposed method on seismic coherent noise removal.

2. Theory and Method

2.1. Shearlet Transform

Shearlet transform is an asymptotically optimal sparse representation of multidimensional functions constructed by a special form of affine systems with synthetic expansion, which is composed of the Laplace pyramid transform and a set of directional filters.

When the dimension is two-dimensional, the scale matrix and the shear matrix are:

$$A = \begin{bmatrix} j & 0 \\ 0 & \sqrt{j} \end{bmatrix}, B = \begin{bmatrix} 1 & -l \\ 0 & 1 \end{bmatrix} \tag{1}$$

In Equation (1), A is the scale matrix, which can be divided into scales; $j > 0$, j is the scale parameter; B is the shear matrix, which can be dissected into directions; and l is the shear parameter.

The affine system with synthetic expansion is defined as:

$$\begin{array}{c} \psi_{AB}(\varphi) = \left\{ \varphi_{j,l,k}(x) = \left| det\ A \right|_{j/2} \cdot \right. \\ \left. \varphi\left(B^l A^j x - k \right) : j, l \in R, k \in R^2 \right\} \end{array} \tag{2}$$

In Equation (2), $\varphi \in L^2(R^2)$, φ is the shear wave function, L represents the productable space, x is the data to be processed, $det\,(\cdot)$ is the determinant, and k is the translation parameter.

If $\psi_{AB}(\varphi)$ satisfies the Parseval framework (the tight framework), that is, for any function, it satisfies the following equation:

$$C_{j,l,k} = S\{f\} = \left\langle f, \varphi_{j,l,k} \right\rangle \tag{3}$$

In Equation (3), $C_{j,l,k}$ is the Shearlet domain coefficient, S is the Shearlet forward transform, and $\langle \cdot \rangle$ is the inner product. $\psi_{AB}(\varphi)$ is called the composite wavelet. Parameters $j,\ l,\ k$ can control the division of the direction and the scale. By taking the inner product of the function and the basis function f and the basis function $\varphi_{j,l,k}$ of different scales, directions, and positions, the Shearlet domain coefficients in the corresponding scale, direction, and position can be obtained. This process essentially uses the basis function after translation, expansion, and rotation to approximate the signal.

The frequency domain subdivision of the shear wave and its support interval are shown in Figure 1. As can be seen from Figure 1, the support interval of any shear wave element $\varphi_{j,l,k}$ in the frequency domain is a trapezoidal region symmetric about the origin; the direction is distributed along a straight line with the slope of $l2^{-j}$. The size of the trapezoidal region is about $2^{2j} \times 2^j$.

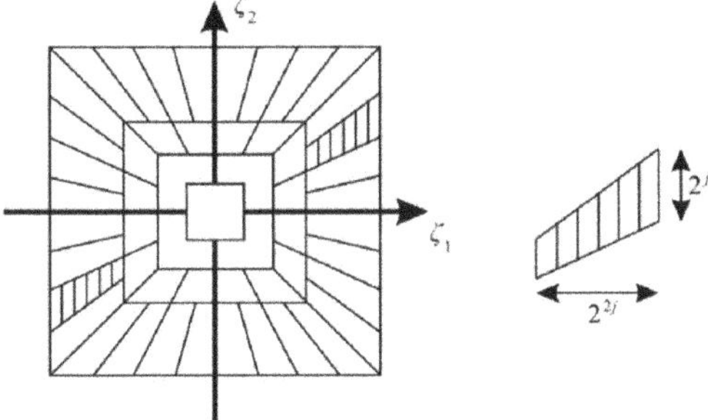

Figure 1. Frequency domain subdivision and its support interval.

2.2. Non-Subsampled Shearlet Transform

Because Shearlet transform does not have the translational invariance, it is easy to generate the pseudo-Gibbs phenomenon near singular points when it is applied to image fusing and denoising. The NSST with the translational invariance can solve this problem. The discretization process of the NSST is composed of the multi-scale subdivision and the direction localization:

(1) The multi-scale subdivision of the NSST is achieved by the non-subsampled pyramid transform (NSP). The image is decomposed by the j-level of the non-subsampling filters to obtain $j + 1$ sub-bands of the same size as the original image, including a low-frequency image and j high-frequency images;
(2) The direction localization of the NSST is achieved by the improved shearing filter (SF). The standard shearing filter is realized by the translation operation of the window function in the pseudo-polarization grid; the subsampling operation is required during the execution, so it does not have the translation invariance. However, the NSST maps the standard shearing filter from the pseudo-polarization grid system to the Cartesian coordinate system and then directly completes its operation through the inverse Fourier transform and the two-dimensional convolution. Therefore, NSST can avoid the subsampling operation and meet the translation invariance. Figure 2 shows the multi-scale and multi-direction decomposition process of the NSST.

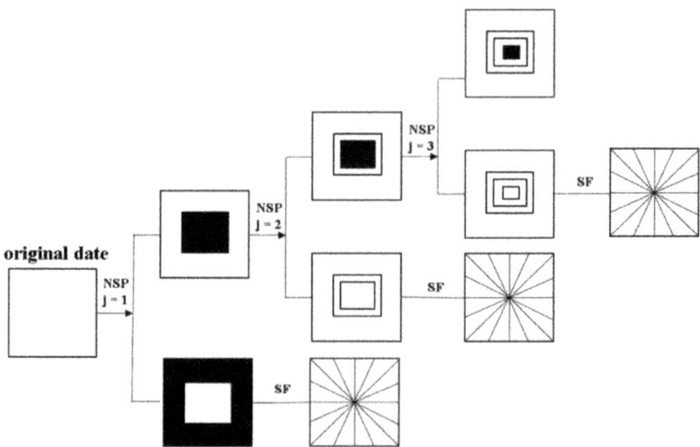

Figure 2. The decomposition process of the NSST.

2.3. The Denoising Method in NSST Domain

The NSST abandons the subsampling operation and combines the non-subsampled Laplacian pyramid transform (NSLP) with the non-subsampled directional filter. After the NSST, the size of each direction sub-band at each scale is the same as the original image, which greatly improves the image redundancy. The enhancement of the image coefficient redundancy makes the NSST have the translation invariance and avoid the spectrum aliasing and the pseudo-Gibbs phenomenon. Therefore, the NSST has a stronger data processing performance. The NSST has a more delicate representation of the scale direction and the physical significance of the transformed coefficients is clearer, which is more conducive to the processing of the coefficients. The NSST can transform the time-space domain signal $f(x)$ into the NSST domain (j, l, k), so the different signals can be easily separated in the NSST domain. Each NSST coefficient has its own specific frequency, direction, and position. The scale parameter j reflects the frequency difference; the direction parameter l reflects the direction difference, that is, mainly reflects the apparent velocity difference; and the position parameter k reflects the spatial location difference and the amplitude difference.

As shown in Figure 3, The specific steps of the coherent noise removal method of the vibrator source array seismic data in the NSST domain in this article are as follows:

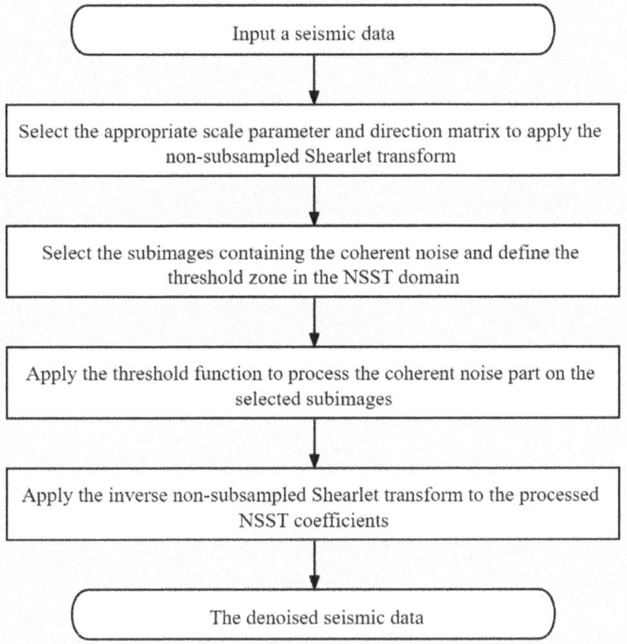

Figure 3. The flowchart of seismic coherent noise removal method in the NSST domain.

In this article, we choose the L_2 norm to construct the threshold function.

The L_2 norm of the NSST domain coefficient is:

$$e_{j,l} = \sqrt{\sum C_{j,l}^2} \tag{4}$$

In Equation (4), $e_{j,l}$ is the L_2 norm of the NSST coefficient in the j scale l direction.

The threshold function is the adaptive threshold changing with scale and direction, which is expressed as follows:

$$T_{j,l} = \lambda_j \frac{\sigma\sqrt{2\ln N}\lg(j+1)}{\lg\left(e_{j,l} - \min e_{j,l} + 10\right)} \tag{5}$$

In Equation (5), $T_{j,l}$ is the threshold function; N is the total number of sampling points for the seismic data; $e_{j,l}$ is the L_2 norm of the NSST coefficient in the j scale l direction, $j = 1, 2, \ldots, J$, J is the total scale of the decomposition; $\sigma = \frac{\text{median}(|C|)}{0.6745}$, σ is the noise standard deviation, median( ) is a median value for all the elements in the data matrix; min( ) is minimum value function; and λ_j is the test constant.

3. Results

3.1. Synthetic Example

In order to verify the validity of the coherent noise removal method based on the NSST, the synthetic data example is tested in this article. The synthetic seismic data is shown in Figure 4a. The seismic data to be tested includes 100 channels, each channel contains 1000 sampling points, and the sampling interval is 1 ms.

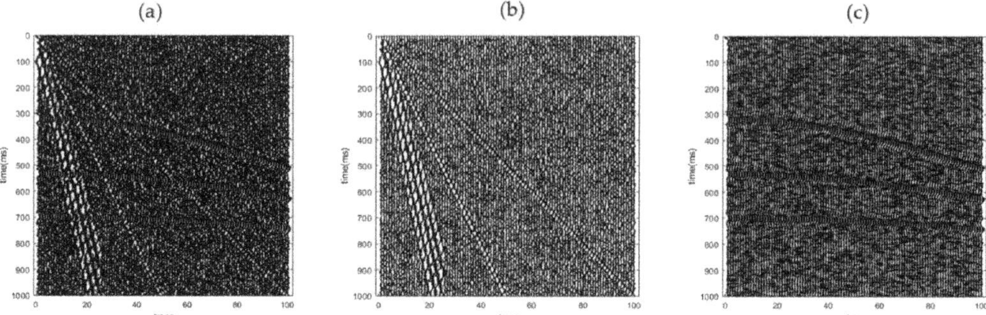

Figure 4. The processing results of the synthetic seismic data by the proposed method. (**a**) The original synthetic seismic data. (**b**) The removed noise. (**c**) The seismic data after the denoising.

As can be seen from Figure 4a, the effective signal with weak energy in the in-phase axis of the seismic record is covered by the coherent noise, which brings the interference to the subsequent processing of the seismic data. We apply the denoising method to the synthetic seismic data. Among the process, the decomposition scale of the NSST is 2, and the direction matrix is [12]. Figure 5 shows the coefficients after the decomposition of the NSST. There is an overlap of the frequency band between the effective signal and the noise. NSST can map the data into several sub-domains, which can represent the data sparsely, so that the noise and the signal can be separated better and the processing effect is better. Figure 4b shows the removed noise. Figure 4c is the result of the denoising method in this article. This method removes most of the coherent noise. This method can not only suppress the coherent noise but can also protect the effective signal to a great extent, restore the in-phase axis of the seismic record, enhance the continuity of the seismic record, and improve the SNR of the seismic record. The denoising results proves the feasibility of the denoising method in this article.

3.2. Field Data Example

The above section has verified the feasibility and validity of the proposed method using the synthetic data example. Now, we further verify the effectiveness and practicality of the proposed method from the field data example.

In this article, the field seismic data of source array in Chaganhua area is selected. The overview of the field exploration area is shown in Figure 6. The field exploration area is located in a grassland near Chaganhua Town, Qianguerroth Mongolian Autonomous County, Songyuan City, Jilin Province, China. The terrain of the exploration area is flat and the diving surface of the exploration area is relatively shallow. Table 1 shows the field seismic data parameters.

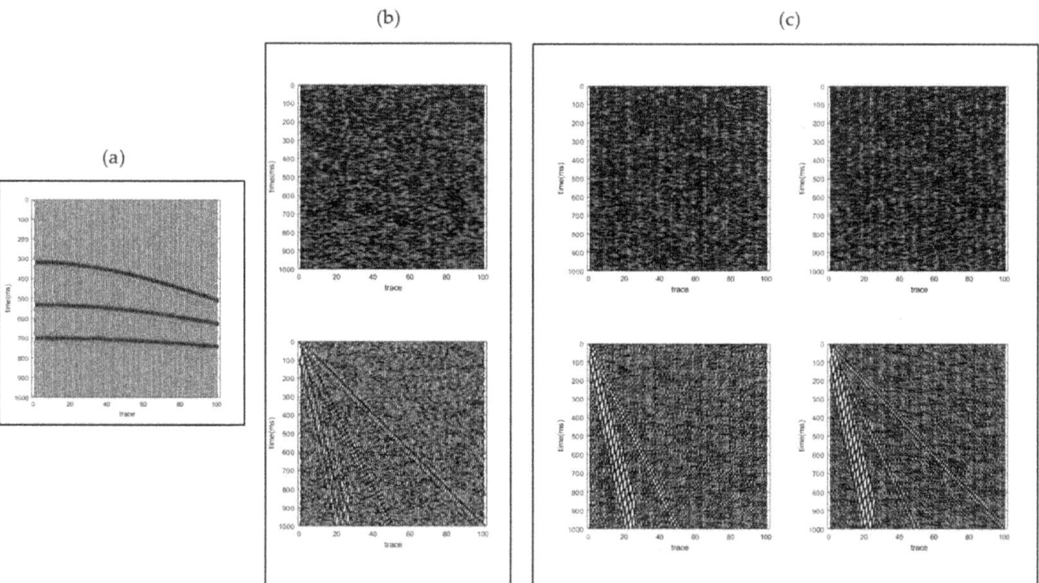

Figure 5. The NSST coefficients after the decomposition of the NSST. (**a**) The approximate NSST coefficients. (**b**) The detail NSST coefficients at scale 2, 2 directions. (**c**) The detail NSST coefficients at scale 1, 4 directions.

Figure 6. The overview of the field exploration area.

Table 1. Field seismic data parameters.

The Source Excitation Mode	The Simultaneous Excitation of the Four Vibrators
the sweep frequency	5–120 Hz
the scanning duration	20 s
the receiving mode	one-sided 120 channels
the sampling interval	1 ms
the trace interval	10 m
the recording length	5 s
the time window	0–3 s

Figure 7a shows the field seismic gather. From Figure 7a, we can see that the seismic gather contains obvious coherent noise, including surface wave, acoustic wave, and direct wave. The coherent noise seriously affects the SNR of the seismic gather and bring certain difficulties to the subsequent seismic processing and interpretation. The SNR of the seismic gather in this work area is low, which is suitable for testing the effectiveness and adaptability of the proposed method in this article. We apply the method in this article to remove the coherent noise from this seismic data. Among the processes, the decomposition scale of the NSST is 3 and the direction matrix is [4 5 5]. Figure 8 shows the main NSST coefficients containing the coherent noise after the decomposition of the NSST. Figure 7b shows the removed noise. Figure 7c shows the seismic data after noise removal using the method in this article. From the denoising results in Figure 7, the coherent noise in Figure 7a is effectively removed by this method, and the effective information of the seismic data is greatly protected. The coherent noise removed in Figure 7b is obvious, including surface wave, acoustic wave, and direct wave. The SNR of the seismic data in Figure 7c is greatly improved, the effective reflection signal is well recovered, and the seismic lineups becomes very clear and continuous, which can provide the better basic data for the subsequent seismic processing and interpretation. In order to show the ability of the proposed method, we use the conventional coherent noise suppression method—FK filtering—to compare with this method. As shown in the Figure 7d,e, FK filtering can cause certain damage to the effective signal while suppressing the coherent noise and have certain noise residue. To sum up, the method proposed in this article has a good effect on the removal of the coherent noise in various kinds of the field seismic data.

We analyzed the average amplitude spectrum, the single-channel amplitude spectrum, and the f-k spectrum of the original data and the denoised data after processing using this method. The frequency spectrum of the original data and the denoised data in Figures 9 and 10 shows that the effective signal is distributed in the frequency range of 25–40 Hz and the coherent noise is mainly distributed in the frequency range of 5–20 Hz, and 70–85 Hz. The spectrum analysis in Figure 9 shows that the noise signal in the frequency range of coherent noise is well suppressed after the denoising by this method. The deviation of the signal waveform outside the coherent noise frequency range is very small compared with the original signal. The spectral analysis of the f-k spectrum in Figure 10 also shows that the coherent noise is significantly suppressed. The spectral analysis of Figures 9 and 10 further illustrate that the proposed method in this article has good removal ability for coherent noise and can protect the effective signal well.

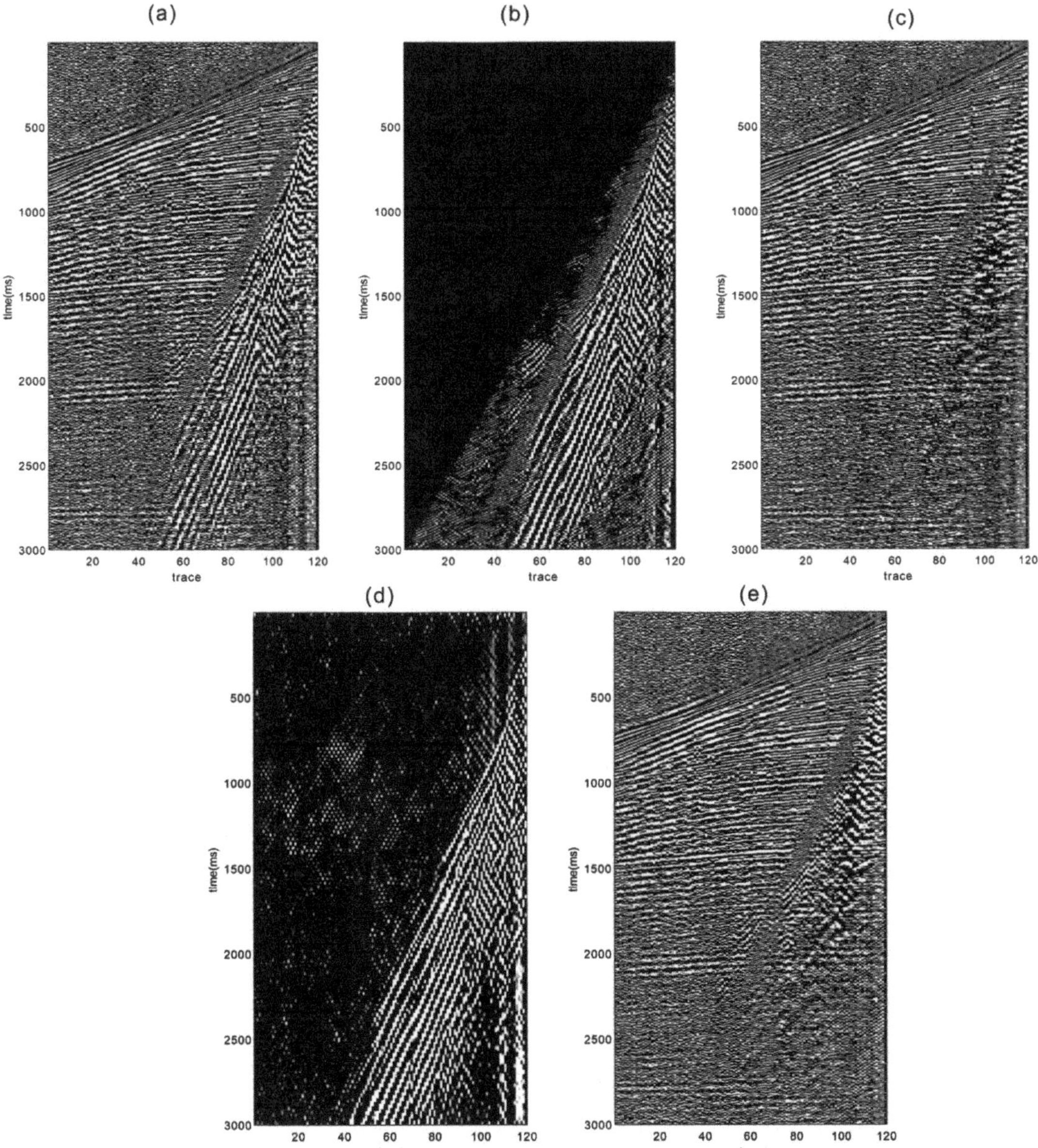

Figure 7. The processing results of the field seismic data using the proposed method. (**a**) The field seismic data. (**b**) The removed noise by NSST. (**c**) The seismic data after the NSST. (**d**) The removed noise by FK filter. (**e**) The seismic data after the FK filter.

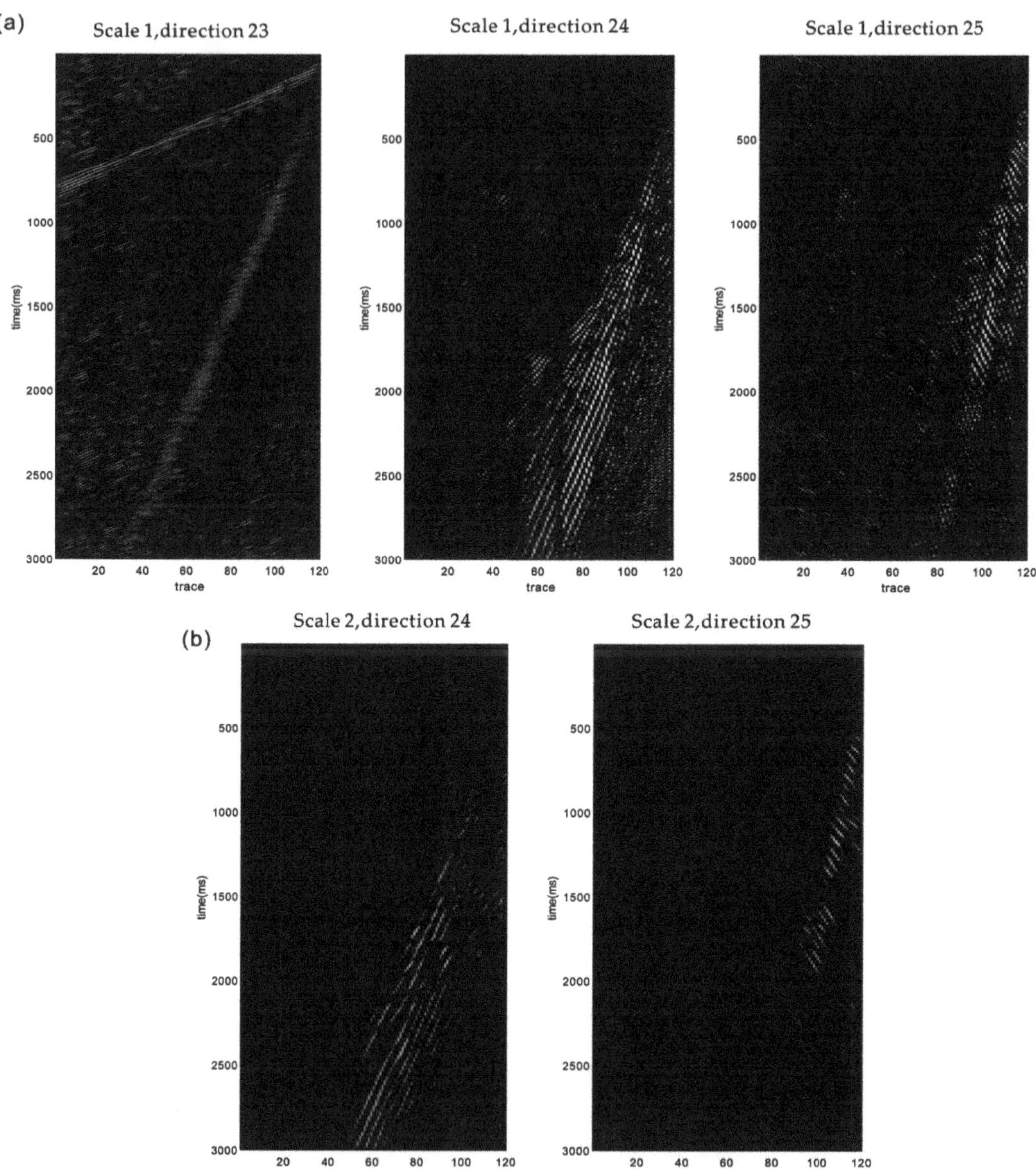

Figure 8. The main NSST coefficients containing the coherent noise. (**a**) The detail NSST coefficients at scale 2, 2 directions. (**b**) The detail NSST coefficients at scale 1, 3 directions.

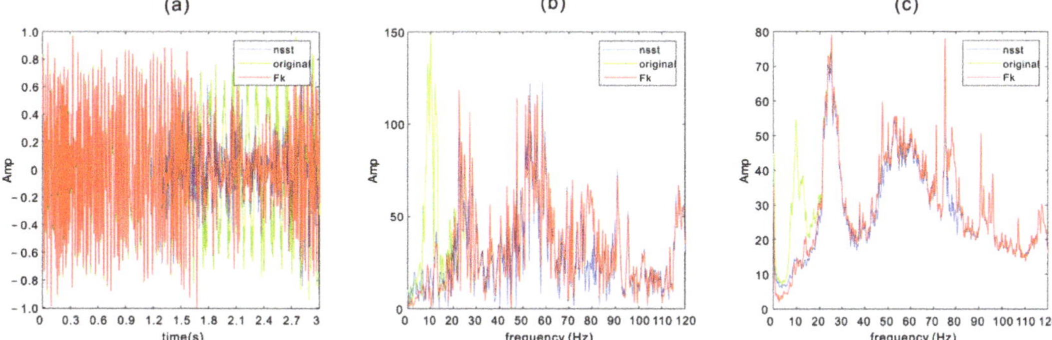

Figure 9. Waveform and amplitude spectra of the seismic data. (**a**) The waveform of trace 80 of the original seismic data (the green curve), the denoised seismic data by the NSST (the blue curve), and the denoised seismic data by the FK filter (the red curve). (**b**) The amplitude spectra of trace 80 of the original seismic data (the green curve), the denoised seismic data by the NSST (the blue curve) and the denoised seismic data by the FK filter (the red curve). (**c**) The average amplitude spectra of the original seismic data (the green curve), the denoised seismic data by the NSST (the blue curve) and the denoised seismic data by the FK filter (the red curve).

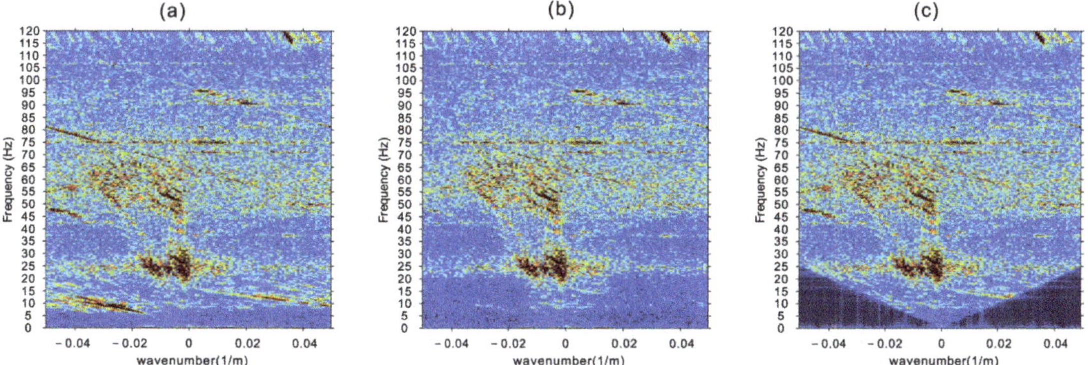

Figure 10. The results of applying f-k transform in the original seismic data and the denoised seismic data. (**a**) The original data. (**b**) The seismic data after the NSST. (**c**) The seismic data after the FK filter.

4. Discussion

As described in Sections 3.1 and 3.2, we used synthetic data examples and field data examples to analyze the denoising performance of the proposed method in this article. We mainly used the difference in the distribution of noise and signal in the scale and direction through NSST, a multi-scale and multi-directional analysis method, to suppress the noise. The denoising results show that this method can effectively remove the coherent noise from the source array seismic data, especially surface wave, and it can greatly protect the effective information of the seismic data from being lost and greatly improve the SNR of the seismic data. The proposed NSST method is not only a method to suppress noise, but also a beneficial tool for signal analysis. The accurate data analysis, targeted strategy, and sparse representation make NSST have more desirable denoising results. There are some limitations to the proposed method: we need to artificially select the scale, direction, and threshold zone according to the seismic data, which is not adaptive enough, and then we will study further to simplify the process.

5. Conclusions

Coherent noise removal is very important for seismic data processing. The traditional coherent noise removal methods often cause some damage to the effective signal while suppressing coherent noise or cannot suppress the interference wave effectively at all. In this removal, based on the multi-scale and multi-direction properties of the NSST and its simple mathematical structure, a coherent noise removal method in NSST domain is proposed which can effectively remove the coherent noise while ensuring the effective signal is not damaged. The effectiveness and practicability of the proposed method on coherent noise attenuation of the source array seismic data are fully proven using synthetic examples and filed data examples. According to the processing results of the synthetic seismic data and the filed seismic data, the denoising method proposed in thixs article can effectively protect the effective signal, enhance the continuity of in-phase axis, and effectively improve the signal-to-noise ratio and resolution of seismic data while removing the coherent noise, which is of great significance to accurately identify the geological information contained in seismic data.

Author Contributions: Conceptualization, M.Y.; methodology, M.Y., X.G. and X.W.; software, M.Y.; validation, M.Y., X.G. and X.W.; formal analysis, M.Y.; investigation, M.Y.; resources, M.Y.; data curation, M.Y.; writing—original draft preparation, M.Y.; writing—review and editing, M.Y.; visualization, M.Y.; supervision, X.G.; project administration, X.G.; funding acquisition, X.G. All authors have read and agreed to the published version of the manuscript.

Funding: This research was supported by the National Natural Science Foundation of China (42074151).

Institutional Review Board Statement: Not applicable.

Informed Consent Statement: Not applicable.

Data Availability Statement: Not applicable.

Conflicts of Interest: The authors declare no conflict of interest.

References

1. Mafakheri, J.; Kahoo, A.R.; Anvari, R.; Mohammadi, M.; Radad, M.; Monfared, M.S. Expand Dimensional of Seismic Data and Random Noise Attenuation Using Low-Rank Estimation. *IEEE J. Sel. Top. Appl. Earth Obs. Remote Sens.* **2022**, *15*, 2773–2781. [CrossRef]
2. Anvari, R.; Kahoo, A.R.; Monfared, M.S.; Mohammadi, M.; Omer, R.M.D.; Mohammed, A.H. Random Noise Attenuation in Seismic Data Using Hankel Sparse Low-Rank Approximation. *Comput. Geosci.* **2021**, *153*, 104802. [CrossRef]
3. Yaghmaei-Sabegh, S. Evaluation of Pulse Effect on Frequency Content of Ground Motions and Definition of a New Characteristic Period. *Earthq. Struct.* **2021**, *20*, 457–471.
4. Linville, A.F.; Meek, R.A. A Procedure for Optimally Removing Localized Coherent Noise. *Geophysics* **1995**, *60*, 191–203. [CrossRef]
5. Duncan, G.; Beresford, G. Slowness Adaptive f-k Filtering of Prestack Seismic Data. *Geophysics* **1994**, *59*, 140–147. [CrossRef]
6. Liu, X. Ground Roll Supression Using the Karhunen-Loeve Transform. *Geophysics* **1999**, *64*, 564–566. [CrossRef]
7. Montagne, R.; Vasconcelos, G.L. Optimized removal of Coherent Noise from Seismic Data Using the Karhunen-Loève Transform. *Phys. Rev. E* **2006**, *74*, 016213. [CrossRef] [PubMed]
8. Turner, G. Aliasing in the Tau- p Transform and the Removal of Spatially Aliased Coherent Noise. *Geophysics* **1990**, *55*, 1496–1503. [CrossRef]
9. Foster, D.J.; Mosher, C.C. removal of Multiple Reflections Using the Radon Transform. *Geophysics* **1992**, *57*, 386–395. [CrossRef]
10. Henley, D.C. The Radial Trace Transform: An Effective Domain for Coherent Noise Attenuation and Wavefield Separation. In *SEG Technical Program Expanded Abstracts 1999*; Society of Exploration Geophysicists: Houston, TX, USA, 1999; pp. 1204–1207.
11. Henley, D.C. Coherent Noise Attenuation in the Radial Trace Domain. *Geophysics* **2003**, *68*, 1408–1416. [CrossRef]
12. Bekara, M.; van der Baan, M. Random and Coherent Noise Attenuation by Empirical Mode Decomposition. *Geophysics* **2009**, *74*, V89–V98. [CrossRef]
13. Xiao, L.; Zhang, Z.; Gao, J. Ground Roll Attenuation of Multicomponent Seismic Data with the Noise-Assisted Multivariate Empirical Mode Decomposition (NA-MEMD) Method. *Appl. Sci.* **2022**, *12*, 2429. [CrossRef]
14. Goudarzi, A.; Riahi, M.A. Seismic Coherent and Random Noise Attenuation Using the Undecimated Discrete Wavelet Transform Method with WDGA Technique. *J. Geophys. Eng.* **2012**, *9*, 619–631. [CrossRef]
15. Sinha, S.; Routh, P.S.; Anno, P.D.; Castagna, J.P. Spectral Decomposition of Seismic Data with Continuous-Wavelet Transform. *Geophysics* **2005**, *70*, P19–P25. [CrossRef]

16. Zhang, H.; Liu, T.; Zhang, Y. Denoising of Seismic Data via Multi-Scale Ridgelet Transform. *Earthq Sci.* **2009**, *22*, 493–498. [CrossRef]
17. Bao, Q.-Z.; Gao, J.-H.; Chen, W.-C. Ridgelet Domain Method of Ground-Roll removal. *Chin. J. Geophys.* **2007**, *50*, 1041–1047. [CrossRef]
18. Zhang, J.; Liang, X.; Fu, J.; Guo, J.; Zheng, X. Curvelet Transform and Its Application in Seismic Data Denoising. In *Proceedings of the Beijing 2009 International Geophysical Conference and Exposition, Beijing, China, 24–27 April 2009*; Society of Exploration Geophysicists: Beijing, China, 2009; p. 110.
19. Neelamani, R.; Baumstein, A.I.; Gillard, D.G.; Hadidi, M.T.; Soroka, W.L. Coherent and Random Noise Attenuation Using the Curvelet Transform. *Lead. Edge* **2008**, *27*, 240–248. [CrossRef]
20. Song, P.; Li, Y.; Ma, H.; Sun, H.; He, X. Attenuation of Random Noise for Seismic Data Based on Nonsubsampled Contourlet Transform. In *Recent Advances in Computer Science and Information Engineering*; Qian, Z., Cao, L., Su, W., Wang, T., Yang, H., Eds.; Lecture Notes in Electrical Engineering; Springer: Berlin/Heidelberg, Germany, 2012; Volume 128, pp. 103–108. ISBN 978-3-642-25791-9.
21. Li, Y.; Yuan, C.; Zhong, Y.; Wang, Y. Non-Subsampled Contourlet Transform Based Seismic Signal De-Noising. In *Proceedings of the 2009 WRI World Congress on Computer Science and Information Engineering, Washington, DC, USA, 31 March–2 April 2009*; IEEE: Los Angeles, CA, USA, 2009; pp. 456–459.
22. Liu, J.; Gu, Y.; Chou, Y.; Gu, J. Seismic Random Noise Reduction Using Adaptive Threshold Combined Scale and Directional Characteristics of Shearlet Transform. *IEEE Geosci. Remote Sens. Lett.* **2020**, *17*, 1637–1641. [CrossRef]
23. Easley, G.; Guo, K.; Labate, D. Analysis of Singularities and Edge Detection Using the Shearlet Transform. In *Proceedings of the SAMPTA'09, Marseille, France, 18–22 May 2009*; Fesquet, L., Torrésani, B., Eds.; Special session on geometric multiscale analysis; HAL: Marseille, France, 2009.
24. Guorong, G.; Luping, X.; Dongzhu, F. Multi-focus Image Fusion Based on Non-subsampled Shearlet Transform. *IET Image Process.* **2013**, *7*, 633–639. [CrossRef]
25. Karami, A.; Heylen, R.; Scheunders, P. Denoising of Hyperspectral Images Using Shearlet Transform and Fully Constrained Least Squares Unmixing. In *Proceedings of the 2016 8th Workshop on Hyperspectral Image and Signal Processing: Evolution in Remote Sensing (WHISPERS), Los Angeles, CA, USA, 21–24 August 2016*; IEEE: Los Angeles, CA, USA, 2016; pp. 1–5.
26. Priya, B.L.; Jayanthi, K. Edge Enhancement of Liver CT Images Using Non Subsampled Shearlet Transform Based Multislice Fusion. In *Proceedings of the 2017 International Conference on Wireless Communications, Signal Processing and Networking (WiSPNET), Chennai, India, 22–24 March 2017*; IEEE: Chennai, India, 2017; pp. 191–195.
27. Qu, Z.; Xing, Y.; Song, Y. An Image Enhancement Method Based on Non-Subsampled Shearlet Transform and Directional Information Measurement. *Information* **2018**, *9*, 308. [CrossRef]
28. Li, L.; Si, Y. Enhancement of Hyperspectral Remote Sensing Images Based on Improved Fuzzy Contrast in Nonsubsampled Shearlet Transform Domain. *Multimed Tools Appl.* **2019**, *78*, 18077–18094. [CrossRef]
29. Ramakrishnan, V.; Pete, D.J. Non Subsampled Shearlet Transform Based Fusion of Multiple Exposure Images. *SN Comput. Sci.* **2020**, *1*, 326. [CrossRef]
30. Shen, F.; Wang, Y.; Liu, C. Change Detection in SAR Images Based on Improved Non-Subsampled Shearlet Transform and Multi-Scale Feature Fusion CNN. *IEEE J. Sel. Top. Appl. Earth Obs. Remote Sens.* **2021**, *14*, 12174–12186. [CrossRef]
31. Liang, X.; Li, Y.; Zhang, C. Noise removal for Microseismic Data by Non-subsampled Shearlet Transform Based on Singular Value Decomposition. *Geophys. Prospect.* **2018**, *66*, 894–903. [CrossRef]
32. Sang, Y.; Guo, P.; Liu, S.; Song, Z.; Gao, D. Denoising of Petroleum Seismic Exploration Based on Non-Subsampled Shearlet Transform. In *Proceedings of the International Geophysical Conference, Beijing, China, 24–27 April 2018*; Society of Exploration Geophysicists and Chinese Petroleum Society: Beijing, China, 2018; pp. 468–471.
33. Sang, Y.; Sun, J.; Meng, X.; Jin, H.; Peng, Y.; Zhang, X. Seismic Random Noise Attenuation Based on PCC Classification in Transform Domain. *IEEE Access* **2020**, *8*, 30368–30377. [CrossRef]

Article

Ground Roll Attenuation of Multicomponent Seismic Data with the Noise-Assisted Multivariate Empirical Mode Decomposition (NA-MEMD) Method

Liying Xiao, Zhifu Zhang * and Jianjun Gao

Institute of Geophysics, China University of Geoscience (Beijing), Beijing 100083, China; xiaoliying7949@163.com (L.X.); gaojianjun@cugb.edu.cn (J.G.)
* Correspondence: zhangzhf@cugb.edu.cn

Abstract: Multicomponent seismic exploration provides more wavefield information for imaging complex subsurface structures and predicting reservoirs. Ground roll is strongly coherent noise in land multicomponent seismic data and exhibits similar features, which are strong energy, low frequency, low velocity and dispersion, in each component. Ground roll attenuation is an important step in seismic data processing. In this study, we utilized multivariate empirical mode decomposition to multicomponent seismic data for attenuating ground roll. By adding extra components containing independent white noise, noise-assisted multivariate empirical mode decomposition is adopted to overcome the mode-mixing effect in standard empirical mode decomposition. This method provides a more robust analysis than the standard empirical mode decomposition EMD method performed separately for each component. Multicomponent seismic data are decomposed into different intrinsic mode functions in frequency scale. According to different frequency scales of seismic reflection wave and ground roll, intrinsic mode functions with low frequency are eliminated to suppress ground roll, and the remaining are reconstructed for seismic reflection waves. Synthetic and field data tests show that the proposed approach performs better than the traditional attenuation method.

Keywords: EMD; MEMD; ground roll; multicomponent seismic data

Citation: Xiao, L.; Zhang, Z.; Gao, J. Ground Roll Attenuation of Multicomponent Seismic Data with the Noise-Assisted Multivariate Empirical Mode Decomposition (NA-MEMD) Method. *Appl. Sci.* **2022**, *12*, 2429. https://doi.org/10.3390/app12052429

Academic Editor: Andrea Paglietti

Received: 14 January 2022
Accepted: 23 February 2022
Published: 25 February 2022

1. Introduction

Recently, multicomponent seismology has gained considerable attention due to providing more seismic wavefield information for imaging complex structures and predicting reservoirs [1]. Generally, three-component (3C) geophones simultaneously record two horizontal components and one vertical component of the incident seismic wave. Ground roll is coherent noise with characteristics of strong energy, low frequency, low velocity and elliptical polarization, which shield the seismic reflection wave from the middle-deep layer in near offset. Ground roll attenuation is a key step in seismic data processing. An adaptive matched filter is conventionally used to process each component separately for attenuating ground roll. This method consists of a prediction step and a robust subtraction step. Ground roll is predicted by data-driven interferometry method [2] or model-driven semi-analytic modeling [3] methods. Then, an optimization subtraction operation is used to adaptively separate the ground roll [2]. Filter methods in transform domain are also commonly used, such as the frequency-wavenumber (F-K) filter, Radon transform [4], Curvelet-domain filter [5] and time-frequency-domain filter [6], to attenuate the ground roll based on the different characteristics of frequency, velocity and polarization between the ground roll and seismic reflection waves. However, when the acquisition geometry is under sampled in space, the aliasing problem can decrease the effect of these methods. In land multicomponent seismic exploration, 3C seismic data are recorded and processed as a vector, which provides not only a robust analysis of each individual component but also valuable information about the coherency components [7,8].

Empirical mode decomposition (EMD) proposed by Huang et al. [9] decomposes a signal into a series of intrinsic mode functions (IMFs). Each IMF has a relatively local constant frequency ranging from high frequency to low frequency. The EMD method does not require predetermined base functions and can adaptively separate nonlinear and non-stationary signals. EMD and its extensions have good performance on random and ground roll attenuation [10–12]. Chen et al. [13] used complete ensemble empirical mode decomposition (CEEMD) to improve the standard EMD method and obtained better results when they were applied to attenuate the ground roll.

The standard EMD method is conventionally conducted on each component separately, which results in misaligned IMFs for multicomponent data. The ground roll attenuation based on the standard EMD method damages the vector coherence of seismic data. Combining multiple components with a higher-dimensional signal could circumvent the mode alignment problem caused by the standard EMD method [14]. Rehman and Mandic [15] proposed a multivariate empirical mode decomposition (MEMD) method to process multicomponent signals and obtained the IMFs of aligned frequency range. These IMFs match well in the frequency scale and number properties for each component of the multicomponent data. In order to solve the mode mixing effect in standard EMD, extended EMD methods [13,16] have been proposed. Rehman and Mandic [17] proposed noise-assisted MEMD (NA-MEMD) to improve MEMD by adding extra components containing independent white noise to the original multivariate signal. This method helps overcome the mode-mixing problem existing in the extracted IMFs.

In this paper, we propose an alternative ground roll attenuation method for multicomponent seismic data in time domain. First, we start this article with a description of the MEMD and NA-MEMD methods. Then, we demonstrate the principle of the proposed method for ground roll attenuation of multicomponent seismic data. Finally, the synthetic and field seismic data tests demonstrate that the NA-MEMD method can effectively attenuate the ground roll. Compared to the F-K filter method, the proposed method preserves more low-frequency content of the seismic reflection wave and coherency information between components.

2. MEMD Method

The EMD adaptively decomposes the original signal into a finite set of signals, called intrinsic mode functions, abbreviated IMFs [9]. Each IMF represents different vibrational modes embedded in the data, and also has a localized frequency content by preventing frequency spreading because of asymmetric waveforms. Obtaining the IMFs with a frequency range from high to low frequency is an iterative sifting procedure, as follows:

1. Compute the mean envelope $m_n = (env_n^{up} + env_n^{down})/2$, where env_n^{up} and env_n^{down} are the upper and lower envelopes that pass through the maxima and minima of the input signal s_n, respectively;
2. Set the residue $r_n = s_n - m_n$ as the new input and repeat step 1.

This two-step iterative process is repeated until the stopping criterion is satisfied. The resulting residue of this iteration can be regarded as the first IMF with the lowest frequency content. In standard EMD, the local mean is computed by taking an average of the upper and lower envelopes, and IMFs in turn are obtained between the local maxima and minima. For multicomponent signals, the local maximum and minimum values are not directly defined by the EMD. Rehman and Mandic [15] proposed to generate multiple n-dimensional envelopes by taking signal projections along different directions in n-dimensional spaces. Cubic spline interpolations in different directions are adopted to form multiple local envelops, and then these values are averaged. Later, the MEMD proposed was used to process the nonlinear n-dimensional time-series signals. Let the sequence $\{V(t)\} = \{v_1(t), v_2(t), \ldots, v_n(t)\}$ represent a multivariate signal with n components, and $X^{\theta_k} = \left\{x_1^k, x_2^k, \ldots, x_n^k\right\}$ denote a set of direction vectors along the directions given by angles

$\theta^k = \left\{\theta_1^k, \theta_2^k, \ldots, \theta_{(n-1)}^k\right\}$ on an $(n-1)$-dimensional sphere. The main steps of the MEMD method are as follows,

1. Take advantage of Hammersley sequence sampling method, choose a suitable pointset for sampling on an $(n-1)$-dimensional sphere;
2. Calculate a projection set, denoted by $\{p^{\theta_k}(t)\}$, of the input signal $\{V(t)\}$ along the direction vector X^{θ_k}, for all k, which is the whole set of direction vectors, giving $\{p^{\theta_k}(t)\}$ as the set of projections;
3. Find the time instants $\left\{t_i^{\theta^k}\right\}$ corresponding to the maxima of the set of projected signals $\{p^{\theta_k}(t)\}_{k=1}^{K}$;
4. Interpolate $\left[t_i^{\theta^k}, V\left(t_i^{\theta^k}\right)\right]$ to perform multiple spline interpolation and obtain multivariate envelope curves $\{e^{\theta_k}(t)\}_{k=1}^{K}$;
5. For a set of K direction vectors, the mean value $m(t)$ of the envelope curves is calculated as $m(t) = \frac{1}{K}\sum_{k=1}^{K} e^{\theta_k}(t)$;
6. Extract the 'detail' $d(t)$ using $d(t) = v(t) - m(t)$. If the 'detail' $d(t)$ fulfills the stopping criterion for a multivariate IMF, apply the above procedure to $v(t) - d(t)$, otherwise apply the above procedure to $d(t)$.

Through a series of MEMD processes, like the EMD decomposition algorithm, the n-dimensional multi-signal $\{V(t)\} = \{v_1(t), v_2(t), \ldots, v_n(t)\}$ is decomposed into a series of IMF components and a residue $r(t) = V(t) - \sum_{i=1}^{q} d_i(t)$, where q denotes the number of IMF components.

The ability of MEMD which aligns to the intrinsic mode functions is demonstrated in Figure 1. We tested the MEMD method on synthetic data using a three-component signal s = [X, Y, Z] whose components are mixtures of a 16 Hz sinusoid common to all data channels, a 64 Hz and 4 Hz tone in the X component, a 32 Hz tone in the Y component and a 4 Hz tone in the Z components. We observed that all IMFs are three-dimensional and scale-aligned; the 16 Hz tone present in all data channels is localized in a single IMF3, while the 64 Hz, 32 Hz, and 4 Hz tones are localized in IMF1, IMF3, and IMF4, respectively. Such a strict mode alignment cannot be achieved when applying the standard EMD channel wise. Figure 2 shows that the different frequency scale modes exhibit the same IMF.

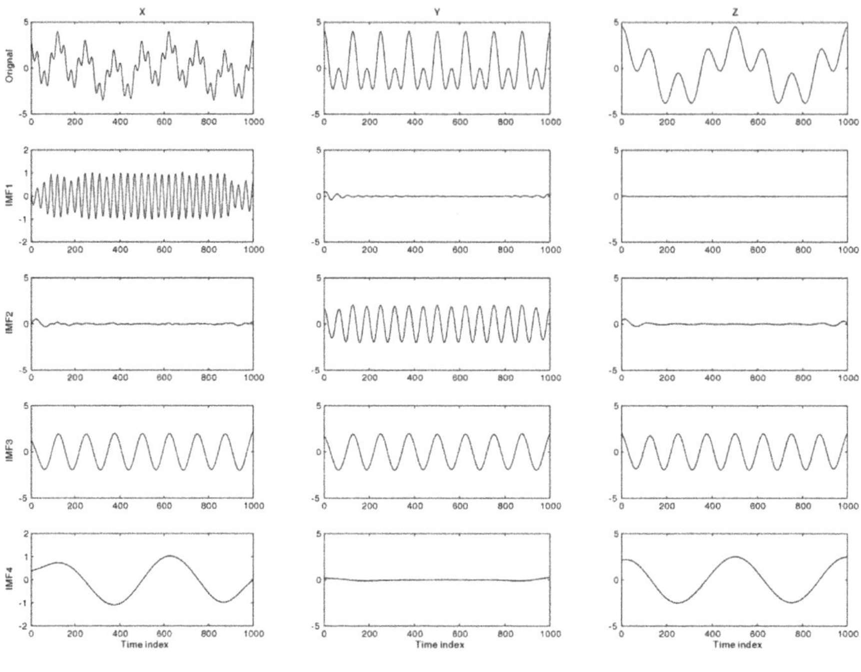

Figure 1. IMFs of a synthetic signal obtained by MEMD.

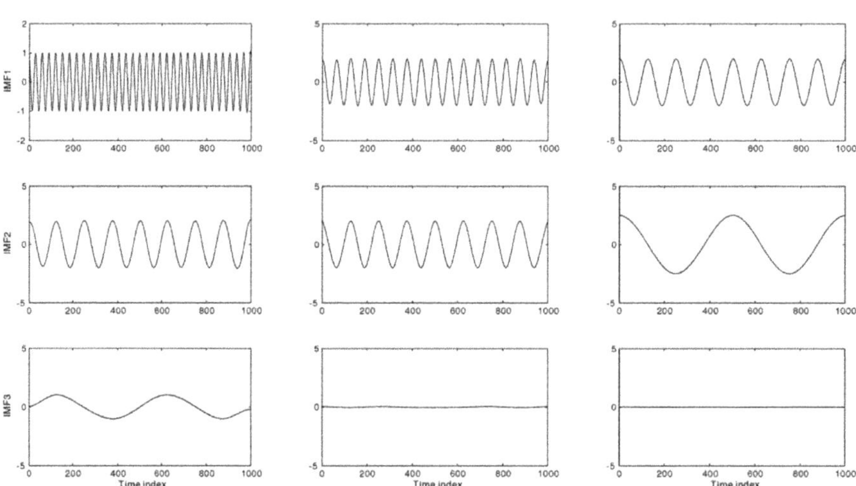

Figure 2. IMFs of the synthetic signal in Figure 1 obtained by standard EMD.

3. NA-MEMD Method

To further reduce the mode-mixing effect in an MEMD, a noise-assisted MEMD method, named NA-MEMD, was introduced by adding white noise. The method is different from the ensemble EMD method in that multiple realizations of white noise are added to the input signal before being decomposed via EMD [13]. The NA-MEMD method adds extra multivariate channels, such as l dimensional independent white noise to the original n dimensional multivariate signal, and then processes such an $n + l$ dimensional composite signal via MEMD. The NA-MEMD algorithm is described as follows:

1. Create an l-dimensional independent Gaussian white noise series with the same length as that of the input signal;
2. Generate an $n + l$-dimensional signal by adding the noise series created in Step 1 to the n-dimensional input multivariate;
3. Process the resulting $n + l$-dimensional multivariate signal via the MEMD algorithm, to obtain the multivariate IMFs;
4. Discard the l-dimensional IMFs corresponding to the noise and obtain a set of n-dimensional IMFs corresponding to the original signal from the resulting $(n + l)$-variate IMFs.

For adding Gaussian white noise, the NA-MEMD method has better binary filtering characteristics than the MEMD method. Both the mode mixing and misalignment problems were significantly reduced. Figure 3 illustrates the benefits of noise-assisted MEMD algorithms for reducing mode mixing via a bivariate signal example. The Z component signal contains a 20 Hz sinusoid and two intermittent signals with frequencies 8 Hz and 80 Hz. The X component was generated by rotating the Z component 90 degrees on the vertical direction. For convenience, we plotted only the IMFs of the Z component. We can see that mode mixing in a single channel is obvious for the standard EMD in Figure 3a. A single-frequency scale mode exists in both IMFs. In addition, different frequency modes can also be seen in the same IMF. With two extra channels of white noise, the decomposition of the same bivariate signal by the proposed NA-MEMD method is shown in Figure 3b. This demonstrates that mode mixing is significantly reduced, and each IMF only contains a single frequency mode.

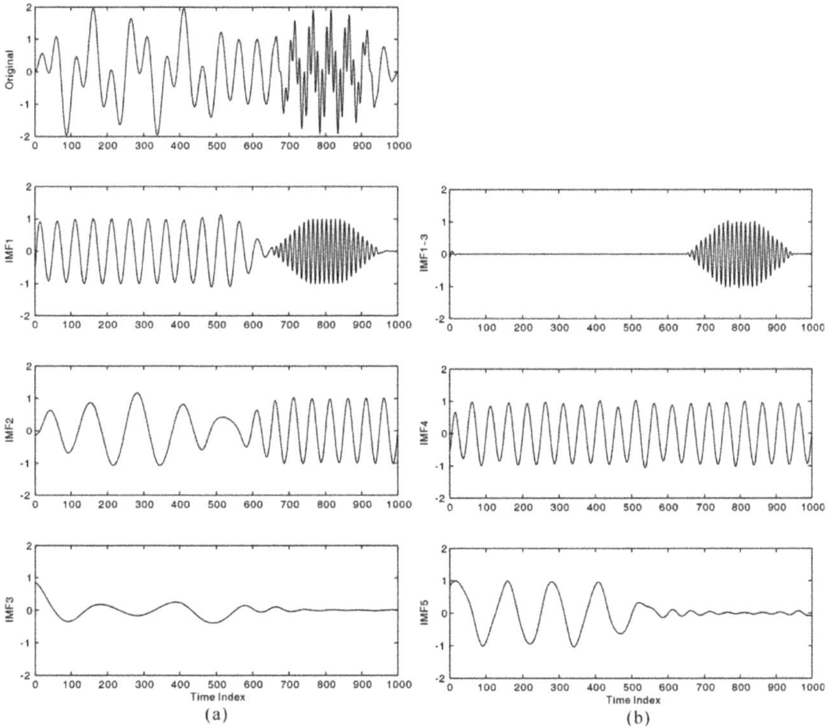

Figure 3. The EMD decompositions of a 20Hz sinusoid contaminated with two 8 Hz and 80 Hz intermittent signals: (**a**) IMFs of EMD method; (**b**) IMFs of the NA-MEMD method.

4. Ground Roll Attenuation via NA-MEMD

This three-component geophones record two horizontal components and one vertical component of the incident seismic data, which are an oscillation stack of different frequency sub-bands. The ground roll, which is located in the low-frequency range, is different from the reflection wave. The alignment of frequency sub-bands can be obtained from different channels of the multivariate signal by inserting them into noisy channels via the NA-MEMD method. As a result, the IMFs were aligned according to their frequency content. Then, we can utilize the low oscillation property to separate the ground roll from the reflection wave.

Attenuating the ground roll based on the NA-MEMD method includes two steps: The first step is to use NA-MEMD to decompose multicomponent seismic data into a set of IMFs. The IMFs of ground roll are present in high-index IMFs, and their amplitude spectra are computed. A complete estimate of IMFs is reduced to identify these high-index IMFs in that the ground roll is dominant. In the second step, we remove these high-index IMFs and reconstruct the remaining low-index IMFs to obtain the seismic data without ground roll. In this workflow, the key parameter is the number of IMFs of the high-index ground roll. The number choice is a trade-off between less damage to the reflection waves and more ground roll suppression.

4.1. *Synthetic Data*

In this section, we first test the proposed method on synthetic seismic data. The dataset contains a sweep signal with a 2–20 Hz frequency band to represent the ground roll and seismic wave simulated by finite difference for a three-layer model as shown in Table 1 [3,18]. According to the polarization property of ground roll, the phase of the horizontal component has a 90 degree lag compared to the vertical component. A 50 Hz Ricker wavelet was used for the body wave with a time sample interval of 1ms. The number of samples for each trace was 1001, and the number of traces for one shoot was 200. The model parameters are presented in Table 1. Figure 4 shows the ground roll attenuation results of the synthetic data obtained using the NA-MEMD method.

Table 1. Model parameters.

Layer	Thickness (m)	Poisson's Ratio	V_s (m/s)	V_P (m/s)	Density (kg/m^3)
1	3000	0.253	1150	2000	2010
2	1000	0.251	1730	3000	2200
3	4000	0.250	2310	4000	2350

Figure 4a,d represents the horizontal and vertical components of the synthetic data, respectively. Two noise-assisted signals are embedded into the synthetic data to help overcome mode mixing when using the MEMD method. We decomposed the model data into eight IMFs. According to the NA-MEMD principle, the first two IMFs are related to noises; therefore, they are discarded. Owing to the low-frequency characteristic of ground roll, the seventh and eighth IMFs dominated by the ground roll are directly subtracted from the decomposition. We reconstructed the low-index IMFs and obtained the final data with ground roll attenuation. Figure 4b,e displays the data after removing the ground roll of the horizontal component in Figure 4a and the vertical component in Figure 4d. It can be seen that the ground roll is well suppressed. From the ground roll separated in Figure 4c,f, we can see that the residual energy of the reflection wave is weak. During the removal of ground roll using the NA-MEMD method, the number of removed IMFs is a key parameter for the suppression effect. Figure 5 shows the sensitivity for the seismic wave separation with different numbers of subtracted IMFs. We find that the more high-index IMFs are removed, the better ground roll suppression is achieved. Meanwhile, more damage occurs to the reflected wave. The parameter choice of removed IMFs is a trade-off between less damage to the reflection energy and more suppression for the ground roll.

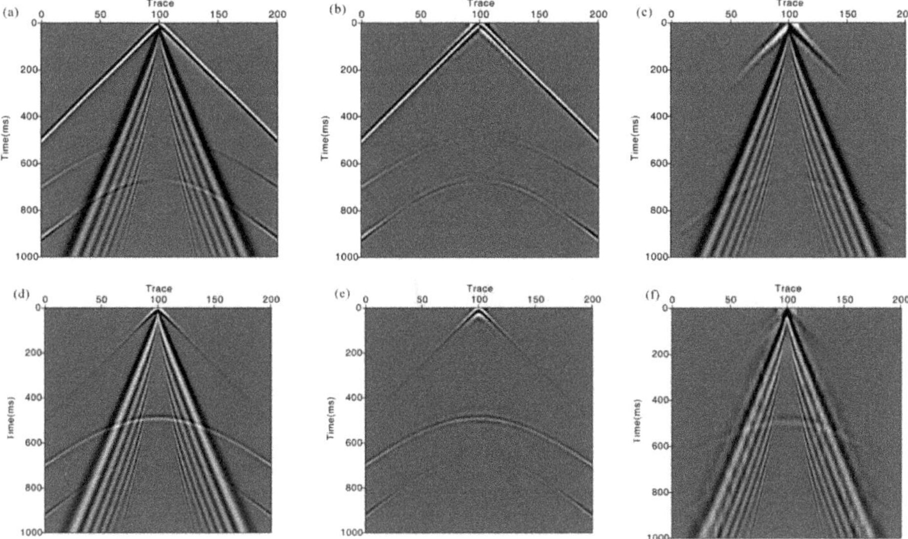

Figure 4. The ground roll attenuation results via NA-MEMD: (**a**) the horizontal component; (**b**) attenuated result of panel (**a**); (**c**) ground roll separated from panel (**a**); (**d**) the vertical component; (**e**) attenuated result of panel (**d**); (**f**) ground roll separated from panel (**d**).

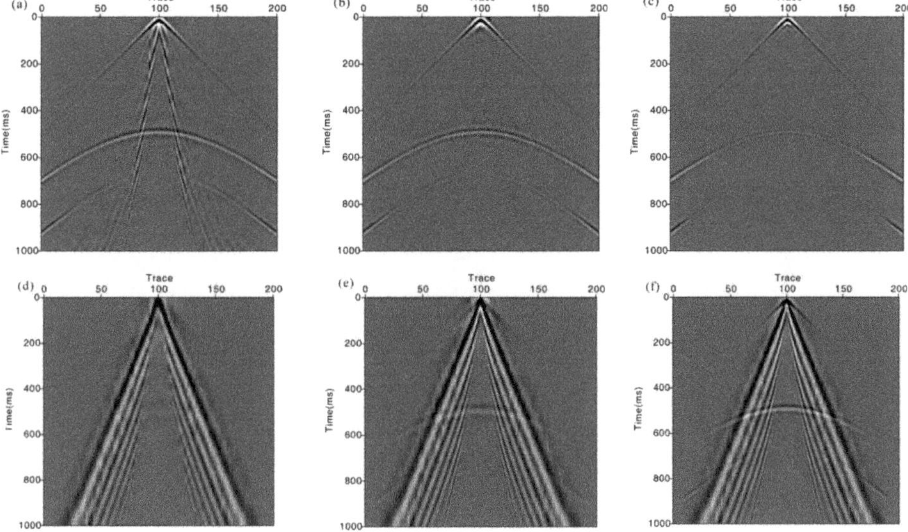

Figure 5. Comparison of the number of IMFs for the vertical component: reconstructed result using (**a**) the 3rd–7th IMFs; (**b**) The 3rd–6th IMFs; (**c**) the 3rd–5th IMFs; ground roll separated using (**d**) the 8th IMF; (**e**) the 7th and 8th IMFs; and (**f**) the 6th–8th IMFs.

4.2. Field Data

To further examine the ground roll attenuation performance of the NA-MEMD method, we processed multicomponent seismic data from land seismic acquisition. The data included two horizontal components and one vertical component. Figure 6a–c shows the raw seismic data that are recorded X, Y, and Z components from left to right, respectively.

It can be seen that seismic data in the near offset are contaminated by strong ground roll. We added three noise-assisted signals and decomposed data into eight IMFs. The ground roll with low velocity exists from the 112th to the 170th trace. Linear events with low frequency also ruin reflection waves. For each trace, we removed different numbers of IMFs. For the traces with ground rolls, we removed the first three highest index IMFs. For the rest, we removed only the first two highest index IMFs. Figure 6d–f displays the ground roll attenuation results via the NA-MEMD method for the raw three-component data, respectively. Figure 6g–i shows the ground roll and linear events separated with a low-frequency scale. From Figure 6d–f, we can see that the ground roll is successfully suppressed from the original data in Figure 6a–c.

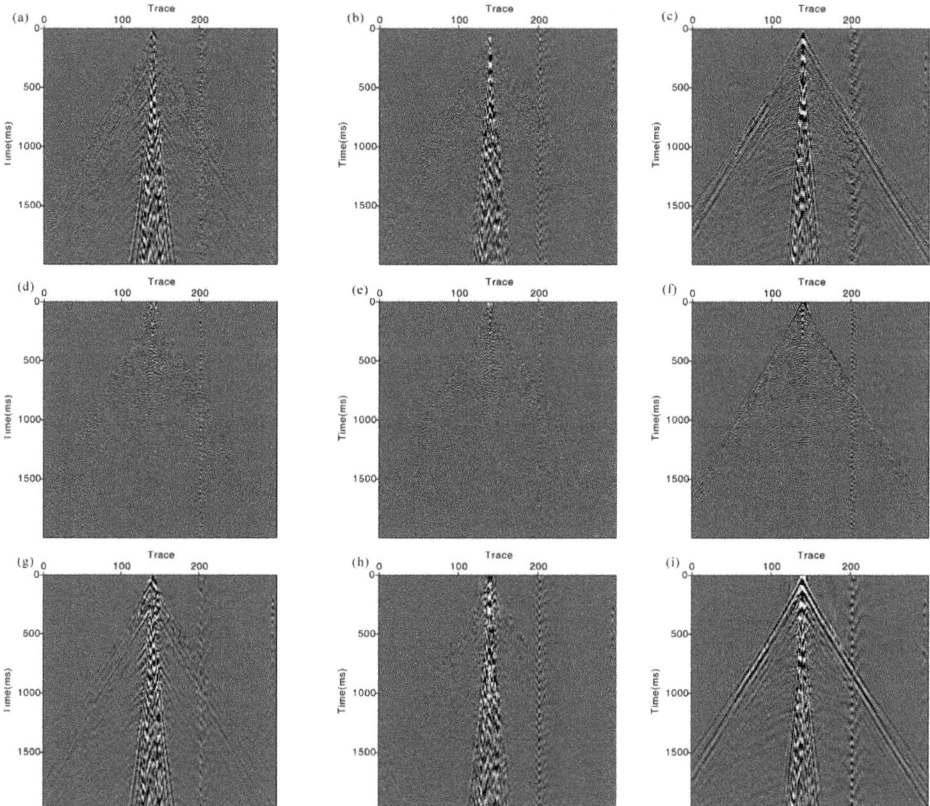

Figure 6. The ground roll attenuation results for field 3C data via NA-MEMD: (**a–c**) the raw data, (**d–f**) the attenuated results, and (**g–i**) the separated ground roll of three-component seismic data, respectively.

To further analyze the ground roll attenuation performance of the NA-MEMD method on field data, we compared it with the F-K filter method. Figure 7a,b displays the attenuation results and differences using F-K attenuation method for the vertical component, respectively. In Figure 6f,i, we can see that the ground roll is well attenuated by the NA-MEMD method. However, the F-K method suppresses the ground roll by cutting operation in the frequency-wavenumber domain, resulting in deletion of the low-frequency content of the reflection wave. Therefore, some weak reflection event energy remains in the difference section, as shown in Figure 7b. Figure 8a,d displays the waveform and amplitude spectra from the 90th and 135th traces, respectively. Although both methods successfully suppress the ground roll and linear interference, the NA-MEMD method preserves the reflection wave better in the low-frequency range (<10 Hz), which is important for later high-resolution imaging.

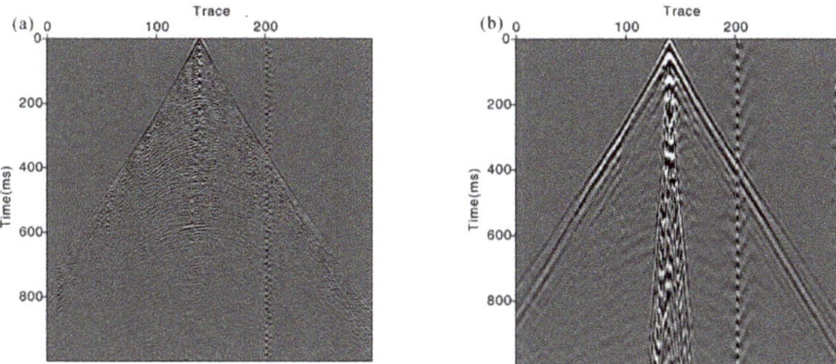

Figure 7. Attenuation results for vertical component in Figure 6c using F-K filter: (**a**) attenuated result; (**b**) separated ground roll.

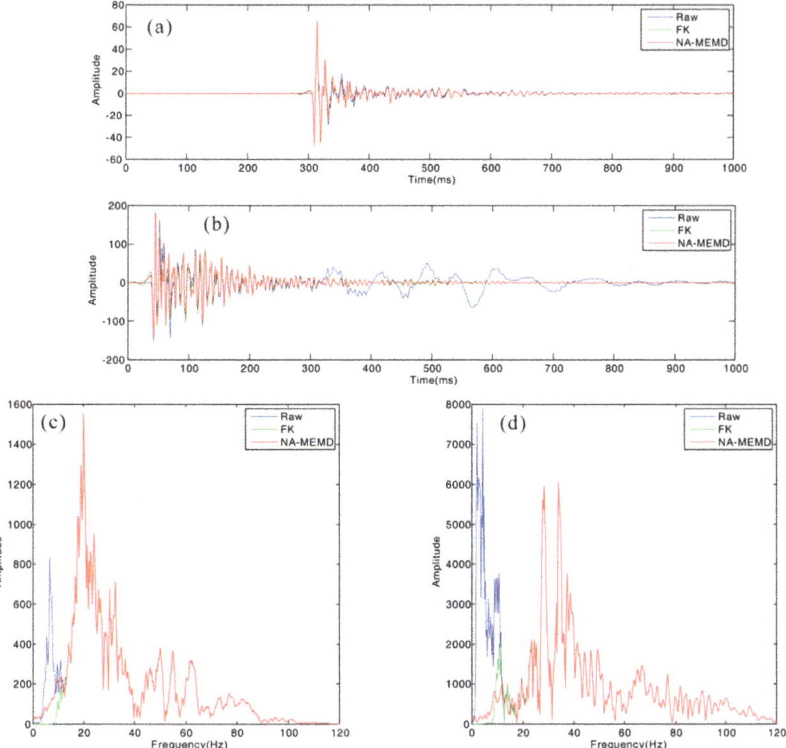

Figure 8. Waveform and amplitude spectra of raw data, FK and NA-MEMD attenuation: (**a**) 90th trace; (**b**) 135th trace; (**c**) spectra of (**a**); (**d**) spectra of (**b**).

5. Conclusions

We applied the NA-MEMD method to attenuate ground roll for multicomponent seismic data. The tests of the simulated multicomponent signal show that the NA-MEMD aligns similar modes present across multiple components and reduces the effect of mode mixing by introducing extra channels of multivariate noise. The performance of the NA-MEMD was analyzed for ground roll attenuation of multicomponent seismic data. We demonstrate that the NA-MEMD method can suppress ground roll and interference if low-frequency waves are present in different components of seismic data in the time domain. Compared to the F-K method, the NA-MEMD method can preserve the reflection wave better in low-frequency range. Synthetic and field seismic data examples show the effectiveness of the proposed NA-MEMD method for ground roll attenuation.

Author Contributions: Conceptualization, L.X. and Z.Z.; methodology, Z.Z.; software, L.X.; validation, L.X., Z.Z. and J.G.; formal analysis, L.X. and Z.Z.; investigation, Z.Z. and J.G.; writing—original draft preparation, L.X.; writing—review and editing, Z.Z. and J.G.; visualization, L.X.; supervision, Z.Z. and J.G.; project administration, Z.Z.; funding acquisition, Z.Z. and J.G. All authors have read and agreed to the published version of the manuscript.

Funding: This research was funded by the National Natural Science Foundation of China (grant numbers 41874165 and 41530321), and partly funded by the National Key R&D Program of China (grant number 2019YFC0312004).

Institutional Review Board Statement: Not applicable.

Informed Consent Statement: Not applicable.

Data Availability Statement: Not applicable.

Acknowledgments: The authors thank D. P. Mandic for the Matlab code of MEMD. The software package MEMD is available online at http://www.commsp.ee.ic.ac.uk/~mandic/research/emd.htm, (accessed on 15 February 2022).

Conflicts of Interest: The authors declare no conflict of interest.

References

1. Farfour, M.; Yoon, W.J. A review on multicomponent seismology: A potential seismic application for reservoir characterization. *J. Adv. Res.* **2016**, *7*, 515–524. [CrossRef] [PubMed]
2. Halliday, D.F.; Curtis, A.; Vermeer, P.; Strobbia, C.; Glushchenko, A.; van Manen, D.J.; Robertsson, J.O.A. Interferometric ground-roll removal: Attenuation of scattered surface waves in single-sensor data. *Geophysics* **2010**, *75*, Sa15–Sa25. [CrossRef]
3. Gazdova, R.; Vilhelm, J. DISECA—A Matlab code for dispersive waveform calculations. *Comput. Geotech.* **2011**, *38*, 526–531. [CrossRef]
4. Wang, Y.H. Antialiasing conditions in the delay-time Radon transform. *Geophys. Prospect.* **2002**, *50*, 665–672. [CrossRef]
5. Liu, Z.; Chen, Y.K.; Ma, J.W. Ground roll attenuation by synchrosqueezed curvelet transform. *J. Appl. Geophys.* **2018**, *151*, 246–262. [CrossRef]
6. Askari, R.; Siahkoohi, H.R. Ground roll attenuation using the S and x-f-k transforms. *Geophys. Prospect.* **2008**, *56*, 105–114. [CrossRef]
7. Naghizadeh, M.; Sacchi, M. Multicomponent f-x seismic random noise attenuation via vector autoregressive operators. *Geophysics* **2012**, *77*, V91–V99. [CrossRef]
8. Pinnegar, C.R. Polarization analysis and polarization filtering of three-component signals with the time-frequency S transform. *Geophys. J. Int.* **2006**, *165*, 596–606. [CrossRef]
9. Huang, N.E.; Shen, Z.; Long, S.R.; Wu, M.L.C.; Shih, H.H.; Zheng, Q.N.; Yen, N.C.; Tung, C.C.; Liu, H.H. The empirical mode decomposition and the Hilbert spectrum for nonlinear and non-stationary time series analysis. *Proc. R. Soc. A Math. Phys.* **1998**, *454*, 903–995. [CrossRef]
10. Gomez, J.L.; Velis, D.R. A simple method inspired by empirical mode decomposition for denoising seismic data. *Geophysics* **2016**, *81*, V403–V413. [CrossRef]
11. Bekara, M.; van der Baan, M. Random and coherent noise attenuation by empirical mode decomposition. *Geophysics* **2009**, *74*, V89–V98. [CrossRef]
12. Chen, Y.K.; Ma, J.T. Random noise attenuation by f-x empirical-mode decomposition predictive filtering. *Geophysics* **2014**, *79*, V81–V91. [CrossRef]
13. Chen, W.; Chen, Y.K.; Liu, W. Ground Roll Attenuation Using Improved Complete Ensemble Empirical Mode Decomposition. *J. Seism. Explor.* **2016**, *25*, 485–495.
14. Looney, D.; Mandic, D.P. Multiscale Image Fusion Using Complex Extensions of EMD. *IEEE Trans. Signal Process.* **2009**, *57*, 1626–1630. [CrossRef]
15. Rehman, N.; Mandic, D.P. Multivariate empirical mode decomposition. *Proc. R. Soc. A Math. Phys.* **2010**, *466*, 1291–1302. [CrossRef]

16. Mandic, D.P.; Rehman, N.U.; Wu, Z.H.; Huang, N.E. Empirical Mode Decomposition-Based Time-Frequency Analysis of Multivariate Signals. *IEEE Signal Process. Mag.* **2013**, *30*, 74–86. [CrossRef]
17. Rehman, N.U.; Mandic, D.P. Filter Bank Property of Multivariate Empirical Mode Decomposition. *IEEE Trans. Signal Process.* **2011**, *59*, 2421–2426. [CrossRef]
18. Graves, R.W. Simulating seismic wave propagation in 3D elastic media using staggered-grid finite differences. *Bull. Seismol. Soc. Am.* **1996**, *86*, 1091–1106.

applied sciences

MDPI

Article

Estimation of Relative Acoustic Impedance Perturbation from Reverse Time Migration Using a Modified Inverse Scattering Imaging Condition

Hong Liang [1,*], Houzhu Zhang [1] and Hongwei Liu [2]

[1] Aramco Americas: Aramco Research Center-Houston, Houston, TX 77084, USA
[2] Shandong Provincial Key Laboratory of Deep Oil and Gas, China University of Petroleum (East China), Qingdao 266580, China
* Correspondence: hong.liang@aramcoamericas.com

Abstract: Reverse Time Migration (RTM) is a preferred depth migration method for imaging complex structures. It solves the complete wave equation and can model all types of complex wave propagation with no dip limitation. Reverse time migration using the inverse scattering imaging condition produces structural images with an amplitude approximate to the reflectivity, which is a composite effect of the impedance and velocity changes in the acoustic media with variable velocity and density. In this study, we present a modified inverse scattering imaging condition to separate the effect of the impedance and velocity perturbations from the reflectivity. The proposed imaging condition is designed to predict the relative impedance perturbation by selecting near-angle reflections during common-shot RTM. We validate our approach on synthetic models and show that the proposed method can estimate reliable impedance perturbation.

Keywords: reverse time migration; modified inverse scattering imaging condition; acoustic impedance; inversion

Citation: Liang, H.; Zhang, H.; Liu, H. Estimation of Relative Acoustic Impedance Perturbation from Reverse Time Migration Using a Modified Inverse Scattering Imaging Condition. *Appl. Sci.* **2023**, *13*, 5291. https://doi.org/10.3390/app13095291

Academic Editor: Jianbo Gao

Received: 8 March 2023
Revised: 14 April 2023
Accepted: 20 April 2023
Published: 23 April 2023

1. Introduction

Conventional migration methods aim to create structural images of subsurface. Advances in the true-amplitude migration method further generate subsurface images with an amplitude approximate to the reflectivity of the subsurface reflectors. For acoustic cases with varying velocity and density, reflectivity is caused by the velocity and impedance changes across the interfaces. Acoustic impedance of the subsurface can be used for the direct interpretation of volume information, such as lithology and pore fill, allowing for target delineation. Those inferred rock properties can provide additional information for geologic interpretation and reservoir characterization, which may not be available from conventional seismic images [1,2]. Furthermore, relating acoustic impedance derived from seismic data to formation properties could have a significant impact on defining new potential drilling locations and optimizing well placement [3].

Earlier studies on ray+Born migration/inversion [4–7] solved the forward problem based on the Born approximation using Green's functions computed by ray theory, and implemented linearized inversion to recover the perturbed model parameters (velocity or acoustic impedance perturbation in acoustic cases; P-wave and S-wave impedance perturbations and density in elastic cases) from the observed data. However, ray-tracing based asymptotic theory is fundamentally flawed in simulating low frequency wave propagation, which is critical for an accurate estimation of media properties with blocky structures [8]. Bleistein et al. [9] extended the method by using more general Green's functions, other than the asymptotic forms. Zhang et al. [10] further developed the amplitude-preserving RTM to predict both impedance and velocity perturbations from angle-domain common-image gathers. However, RTM angle gathers for the purpose of impedance inversion can be computationally expensive. Here, we propose a modified inverse scattering imaging condition for RTM, in order to output the relative impedance perturbation from the stacked images without explicitly computing angle gathers.

In this paper, we first give an overview of model parameters estimation (relative impedance and velocity perturbations) from the observed data using common-shot RTM, in accordance with Zhang et al. [10]. Then we derive the modified inverse scattering imaging condition for the relative impedance perturbation estimation for the acoustic case with variable velocity and density. The conventional inverse scattering imaging condition was designed to reduce RTM artifacts caused by the correlation of source and receiver wavefields propagating in the same direction [11], such as backscattered and turning wave energy. The proposed modified imaging condition employs an exponential weighting function to the conventional inverse scattering imaging condition to select near-angle reflections, from which the relative impedance perturbation can then be estimated. Finally, we validate the proposed method on synthetic examples.

2. Relative Impedance Perturbation Estimation from RTM Using a Modified Inverse Scattering Imaging Condition

We first describe the algorithm for velocity and impedance inversion using RTM, and then propose a modified inverse scattering imaging condition for the relative impedance estimation.

2.1. Theory and Algorithm

In an isotropic acoustic medium, the wave equation is as follows [8,10]:

$$\begin{cases} \left(\frac{1}{v_0^2}\frac{\partial^2}{\partial t^2} - \rho_0 \nabla \frac{1}{\rho_0} \cdot \nabla\right) p_0(\mathbf{x};t;\mathbf{x}_s) = \delta(\mathbf{x}-\mathbf{x}_s)\delta(t) \\ d(\mathbf{x}_r;t;\mathbf{x}_s) = p_0(\mathbf{x}=\mathbf{x}_r;t;\mathbf{x}_s) \end{cases} \quad (1)$$

where v_0 and ρ_0 are velocity and density, respectively; $p_0(\mathbf{x};t;\mathbf{x}_s)$ is the pressure wavefield at any location of $\mathbf{x}$ due to a source located at $\mathbf{x}_s$; and $d(\mathbf{x}_r;\mathbf{x}_s;t)$ is the recorded data, i.e., the measured value of the pressure wavefield at a receiver position $\mathbf{x} = \mathbf{x}_r$.

For a second medium with small perturbations in velocity (δv) and density ($\delta\rho$) compared to the previous medium, the velocity and density are $v_0 + \delta v$ and $\rho_0 + \delta\rho$, respectively. The pressure wavefield in the perturbed medium $p_0 + \delta p$ satisfied the same acoustic wave equation as follows:

$$\begin{cases} \left(\frac{1}{(v_0+\delta v)^2}\frac{\partial^2}{\partial t^2} - (\rho_0+\delta\rho) \nabla \frac{1}{\rho_0+\delta\rho} \cdot \nabla\right) (p_0+\delta p) = \delta(\mathbf{x}-\mathbf{x}_s)\delta(t) \\ \delta d(\mathbf{x}_r;t;\mathbf{x}_s) = \delta p(\mathbf{x}=\mathbf{x}_r;t;\mathbf{x}_s) \end{cases} \quad (2)$$

where δp is the wavefield perturbation, and the observed data perturbation is represented by δd.

Based on the Born approximation, the following equation for δp can be derived by subtracting Equation (1) from Equation (2):

$$\left(\frac{1}{v_0^2}\frac{\partial^2}{\partial t^2} - \rho_0 \nabla \frac{1}{\rho_0} \cdot \nabla\right) \delta p(\mathbf{x};t;\mathbf{x}_s) \approx \left(\frac{2\delta v}{v_0^3}\frac{\partial^2}{\partial t^2} - \left(\nabla \frac{\delta\rho}{\rho_0}\right) \cdot \nabla\right) p_0(\mathbf{x};t;\mathbf{x}_s) \quad (3)$$

By following Zhang et al. [10] and using the asymptotic approximation, we obtained the ray-based relationship between δd, δv and $\delta\rho$:

$$\delta d(\mathbf{x}_r;\omega;\mathbf{x}_s) = -\int \frac{2\omega^2}{\mathrm{v}_0(\mathbf{x})^2}\left(\frac{\delta v(\mathbf{x})}{v_0(\mathbf{x})} + \cos^2\theta \frac{\delta\rho(\mathbf{x})}{\rho_0(\mathbf{x})}\right) \times A(\mathbf{x}_s;\mathbf{x}_r;\mathbf{x}) e^{i\omega T(\mathbf{x}_s;\mathbf{x}_r;\mathbf{x})} d\mathbf{x} \quad (4)$$

where $T(\mathbf{x}_s;\mathbf{x}_r;\mathbf{x}) = \tau(\mathbf{x};\mathbf{x}_s) + \tau(\mathbf{x}_r;\mathbf{x})$ and $A(\mathbf{x}_s;\mathbf{x}_r;\mathbf{x}) = A(\mathbf{x};\mathbf{x}_s)A(\mathbf{x}_r;\mathbf{x})$ represent the travertine summation and the amplitude product of the Green's function from the source location $\mathbf{x}_s$ to the image point $\mathbf{x}$ and reflected back to the receiver location $\mathbf{x}_r$, respectively, and θ is the subsurface reflection angle. Equation (4) represents the forward modeling formulation under a high-frequency assumption. The detailed derivations of Equations (3) and (4) can be found in [10].

Following a similar method to [5] and [8], we can invert Equation (4) for the composite model parameter perturbation, $\frac{\delta v(\mathbf{x})}{v_0(\mathbf{x})} + \cos^2\theta \frac{\delta\rho(\mathbf{x})}{\rho_0(\mathbf{x})}$. In 2D, the composite model parameter perturbation can be described in terms of the perturbed wavefield as follows (detailed derivation is given in Appendix A):

$$\sin^2\theta \frac{\delta v}{v_0} + \cos^2\theta \frac{\delta(\rho v)}{\rho_0 v_0} = -\iint\iint \frac{32\cos^2\theta'}{|\omega|} \frac{\cos\beta_r}{v(\mathbf{x}_r)} \frac{\cos\beta_s}{v(\mathbf{x}_s)} A(\mathbf{x};\mathbf{x}_s) A(\mathbf{x}_r;\mathbf{x}) \times(\theta' - \theta)\delta d(\mathbf{x}_r;\omega;\mathbf{x}_s) e^{-i\omega T(\mathbf{x}_s;\mathbf{x}_r;\mathbf{x})} d\mathbf{x}_r d\mathbf{x}_s d\omega d\theta' \tag{5}$$

where β_s and β_r represent the takeoff angles at the source and receivers, respectively.

Within the framework of amplitude-preserving RTM, the asymptotic forms of $p_F(\mathbf{x};\omega;\mathbf{x}_s)$ and $p_B(\mathbf{x};\mathbf{x}_s;\omega)$ for 2D acoustic case are given as follows [10,12]:

$$\overset{*}{p}_F(\mathbf{x};\omega;\mathbf{x}_s) = -2\frac{\cos\beta_s}{v(\mathbf{x}_s)}\sqrt{|\omega|} A(\mathbf{x};\mathbf{x}_s) e^{-i\omega\tau(\mathbf{x};\mathbf{x}_s) - i(\frac{\pi}{4})sgn(\omega)} \tag{6}$$

and

$$p_B(\mathbf{x};\omega;\mathbf{x}_s) = 2\int \frac{\cos\beta_r}{v(\mathbf{x}_r)}\sqrt{|\omega|} A(\mathbf{x}_r;\mathbf{x}) e^{-i\omega\tau(\mathbf{x}_r;\mathbf{x}) - i(\frac{\pi}{4})sgn(\omega)} \delta d(\mathbf{x}_r;\omega;\mathbf{x}_s) d\mathbf{x}_r \tag{7}$$

Substituting Equations (6) and (7) for the terms on the right-hand side of Equation (5), we can obtain:

$$\sin^2\theta \frac{\delta v}{v_0} + \cos^2\theta \frac{\delta(\rho v)}{\rho_0 v_0} = \iiint \frac{8\cos^2\theta'}{\omega^2} \delta(\theta' - \theta) \overset{*}{p}_F(\mathbf{x};\omega;\mathbf{x}_s) p_B(\mathbf{x};\omega;\mathbf{x}_s) d\omega d\theta' dx_s \tag{8}$$

The left-hand side of Equation (8) is the composite form of relative velocity perturbation and impedance perturbation. The right-hand side has the same form as RTM. This shows that the composite parameter on the left-hand side can be estimated using the RTM framework. In order to invert each individual parameter, Zhang et al. [10] pointed out that the near-angle stacked image can output impedance perturbation, $\frac{\delta(\rho v)}{\rho_0 v_0}$, while the far-angle stacked image can be used to estimate the velocity perturbation, $\frac{\delta v}{v_0}$. They separated the effects of impedance and velocity by first generating RTM angle-domain common-image gathers, and then using stacked images within different angle sections for velocity and impedance estimations. However, it could be computationally expensive to compute RTM angle gathers, especially in 3D.

Note that the right-hand side of Equation (8) shares a similar form as the true amplitude imaging principle proposed by Kiyashchenko et al. [13], where the $\cos^2\theta$ term is approximated by the ray theoretical slowness vectors in the Fourier domain and computed by applying the time derivatives and spatial gradients of wavefields. The time derivatives and spatial gradients of wavefields were also utilized in the inverse scattering imaging condition proposed by Whitmore and Crawley [11]. Next, we investigated the inverse scattering imaging condition, and further proposed a modified inverse scattering imaging condition to output the near-angle stacked images without computing the angles. Our method can reduce the computational costs compared with the method using angle gathers, and still utilize the imaging capabilities of RTM for the relative impedance estimation.

The inverse scattering imaging condition proposed by Whitmore and Crawley [11]:

$$\mathrm{I}(\mathbf{x};\mathbf{x}_s) = I_\nabla(\mathbf{x};\mathbf{x}_s) + B(\mathbf{x};\mathbf{x}_s) I_{dt}(\mathbf{x};\mathbf{x}_s) \tag{9a}$$

where

$$I_\nabla(\mathbf{x};\mathbf{x}_s) = \int \nabla p_F(\mathbf{x};t;\mathbf{x}_s) \cdot \nabla p_B(\mathbf{x};t;\mathbf{x}_s) dt \tag{9b}$$

and

$$I_{dt}(\mathbf{x};\mathbf{x}_s) = -\int \frac{1}{v^2(\mathbf{x})}\frac{\partial p_F(\mathbf{x};t;\mathbf{x}_s)}{\partial t}\frac{\partial p_B(\mathbf{x};t;\mathbf{x}_s)}{\partial t}dt \tag{9c}$$

and where $B(\mathbf{x};\mathbf{x}_s)$ is a weighting function to attenuate backscattered energy. Note that here we used the backpropagated wavefields $p_B(\mathbf{x};t;\mathbf{x}_s)$ in terms of wavefield propagation time t (instead of $T-t$, where T is the maximum extrapolation time), and used the relationship $\frac{\partial p_B(\mathbf{x};T-t;\mathbf{x}_s)}{\partial t} = -\frac{\partial p_B(\mathbf{x};t;\mathbf{x}_s)}{\partial t}$.

The relationship between the product of the time derivatives of p_F and p_B, and the products of their spatial gradients is given by the following equation [11,14,15]:

$$-\frac{\partial p_F}{\partial t}\frac{\partial p_B}{\partial t}\frac{\cos(2\theta)A(\mathbf{x})}{v^2(\mathbf{x})} = \nabla p_F{\cdot}\nabla p_B(\mathbf{x};t;\mathbf{x}_s) \tag{10}$$

where $A(\mathbf{x})$ is the parameter to compensate for far field approximation. We can ignore it by assuming far field approximation and obtain the following equations:

$$\nabla p_F(\mathbf{x};t;\mathbf{x}_s){\cdot}\nabla p_B(\mathbf{x};t;\mathbf{x}_s) \approx -\frac{\cos(2\theta)}{v^2(\mathbf{x})}\frac{\partial p_F(\mathbf{x};t;\mathbf{x}_s)}{\partial t}\frac{\partial p_B(\mathbf{x};t;\mathbf{x}_s)}{\partial t} \tag{11a}$$

$$-\frac{1}{v^2(\mathbf{x})}\frac{\partial p_F(\mathbf{x};t;\mathbf{x}_s)}{\partial t}\frac{\partial p_B(\mathbf{x};t;\mathbf{x}_s)}{\partial t} + \nabla p_F(\mathbf{x};t;\mathbf{x}_s){\cdot}\nabla p_B(\mathbf{x};t;\mathbf{x}_s) \approx -\frac{2\cos^2\theta}{v^2(\mathbf{x})}\frac{\partial p_F(\mathbf{x};t;\mathbf{x}_s)}{\partial t}\frac{\partial p_B(\mathbf{x};t;\mathbf{x}_s)}{\partial t} \tag{11b}$$

$$-\frac{1}{v^2(\mathbf{x})}\frac{\partial p_F(\mathbf{x};t;\mathbf{x}_s)}{\partial t}\frac{\partial p_B(\mathbf{x};t;\mathbf{x}_s)}{\partial t} - \nabla p_F(\mathbf{x};t;\mathbf{x}_s){\cdot}\nabla p_B(\mathbf{x};t;\mathbf{x}_s) \approx -\frac{2\sin^2\theta}{v^2(\mathbf{x})}\frac{\partial p_F(\mathbf{x};t;\mathbf{x}_s)}{\partial t}\frac{\partial p_B(\mathbf{x};t;\mathbf{x}_s)}{\partial t} \tag{11c}$$

Choosing $\mathrm{B}(\mathbf{x}) = 1$, Equation (9a) can be expressed in the frequency domain as:

$$\begin{aligned} &I_\nabla(\mathbf{x};\mathbf{x}_s) + I_{dt}(\mathbf{x};\mathbf{x}_s) \\ &= \int \left(-\frac{1}{v^2(\mathbf{x})}\frac{\partial p_F(\mathbf{x};t;\mathbf{x}_s)}{\partial t}\frac{\partial p_B(\mathbf{x};t;\mathbf{x}_s)}{\partial t} + \nabla p_F(\mathbf{x};t;\mathbf{x}_s){\cdot}\nabla p_B(\mathbf{x};t;\mathbf{x}_s)\right)dt \\ &\approx -\int \frac{2\cos^2\theta}{v^2(\mathbf{x})}\frac{\partial p_F(\mathbf{x};t;\mathbf{x}_s)}{\partial t}\frac{\partial p_B(\mathbf{x};t;\mathbf{x}_s)}{\partial t}dt \\ &= -\int \frac{2\omega^2\cos^2\theta}{v^2(\mathbf{x})}p_F^*(\mathbf{x};\omega;\mathbf{x}_s)p_B(\mathbf{x};\omega;\mathbf{x}_s)d\omega \end{aligned} \tag{12}$$

Equation (12) is similar to the energy norm imaging condition proposed by Rocha et al. [16] which was used to attenuate reflections with opening angles close to 180°. Comparing Equations (8) and (12), we can see that the composite parameter of the relative impedance and velocity perturbation ($\sin^2\theta\frac{\delta v}{v_0} + \cos^2\theta\frac{\delta(\rho v)}{\rho_0 v_0}$) can be computed using the inverse scattering imaging condition (Equation (12)) after the proper preprocessing of source and receiver wavefields. To estimate the impedance perturbation, we must further separate the effects of impedance and velocity. Inspired by Rocha et al. [16], who proposed an exponential weighting function to select far-angle reflections for the purpose of tomographic inversion, we applied the following weighting function for Equation (12) to select near-angle reflections [15]:

$$I_{small\ angle} = \int w\left(-\frac{1}{v^2}\frac{\partial p_F}{\partial t}\frac{\partial p_B}{\partial t} + \nabla p_F{\cdot}\nabla p_B\right)dt, \text{with } w = e^{-\frac{\alpha\left(-\frac{1}{v^2}\frac{\partial p_F}{\partial t}\frac{\partial p_B}{\partial t} - \nabla p_F\cdot\nabla p_B\right)}{-\frac{2}{v^2}\frac{\partial p_F}{\partial t}\frac{\partial p_B}{\partial t}}},\ \alpha > 1 \tag{13}$$

where the weighting term w is approximate to $e^{-\alpha sin^2\theta}$ following Equation (11c). This weighting function equals to 1 when $\theta = 0°$, and rapidly approaches 0 when θ increases, which is designed to select small reflection angles. More details about this weighting function and the choice of α are presented in the later section. Equation (8) shows that small-angle reflection produces an estimate of the impedance perturbation. Therefore, from Equations (8) and (13), we derived our proposed imaging condition for the relative impedance estimation from RTM:

$$\frac{\delta(\rho v)}{\rho_0 v_0} = -\iint 4v^2 w\left(-\frac{1}{v^2}\frac{\partial p'_F}{\partial t}\frac{\partial p'_B}{\partial t} + \nabla p'_F \cdot \nabla p'_B\right) dt d\mathrm{x_s}, \text{with } w = e^{-\frac{\alpha\left(-\frac{1}{v^2}\frac{\partial p'_F}{\partial t}\frac{\partial p'_B}{\partial t} - \nabla p'_F \cdot \nabla p'_B\right)}{-\frac{2}{v^2}\frac{\partial p'_F}{\partial t}\frac{\partial p'_B}{\partial t}}} \tag{14}$$

where p'_F and p'_B are forward and backward wavefields, respectively, scaled by $1/\omega^2$ in the frequency domain.

Figure 1 shows the workflow of our proposed method of estimating the relative impedance perturbation from RTM using the modified inverse scattering imaging condition. Compared with conventional RTM, the only difference is that we replaced the conventional cross-correlation imaging condition with our proposed imaging condition in Equation (14).

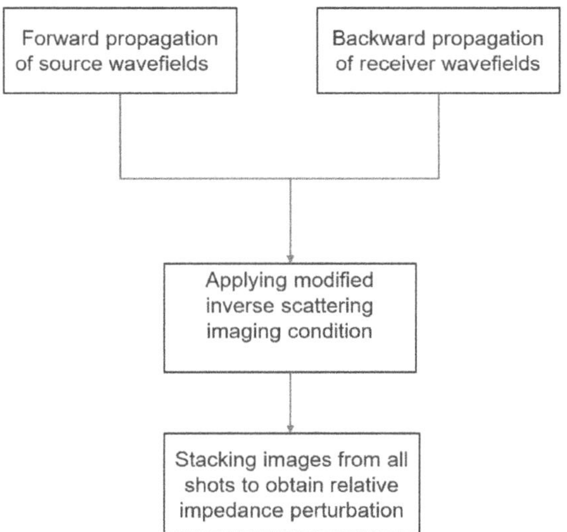

Figure 1. Workflow of the proposed method.

2.2. Comparison of Imaging Conditions

To illustrate how the proposed imaging condition in Equation (13) attenuates large angle reflections, we performed simple experiments on the prestack impulse response. The synthetic data were generated by a two-reflector layered model. Figure 2a shows the impulse response using I_{dt} for a single trace at an offset of 4000 m, from which we can see the image of two reflection events, as well as the backscattered events. The imaging condition $I_{\mathrm{dt}} + I_{\nabla}$ attenuated backscattered events, as shown in Figure 2b. By using the proposed imaging condition in Equation (13), Figure 2c only preserved the small-angle reflections. Figure 2d–f show the corresponding imaging comparison for a single trace at an offset of 0 m (one source and one receiver, both at 0 m on the surface). Using our proposed small-angle imaging condition, Figure 2f preserved all the zero-angle reflections available in Figure 2d, while attenuating the backscattered energy, which is mostly notable at location 0 m.

Figure 3 compares the imaging results of the single trace at zero offset using Equations (13) and (14). From the comparison, we can see that the factor $1/\omega^2$ in Equation (14) boosts the low-frequency components in Figure 3b.

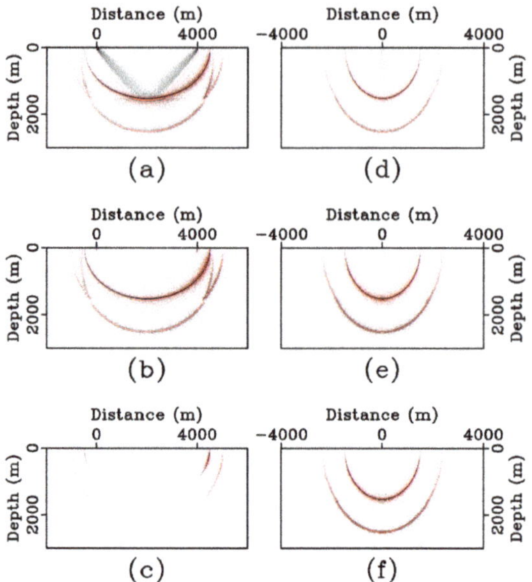

Figure 2. Imaging results of a single trace (offset = 4000 m) showing (**a**) I_{dt}; (**b**) $I_{dt} + I_{\nabla}$; and (**c**) $I_{small\ angle}$; Imaging results of a single trace (offset = 0 m) showing (**d**) I_{dt}; (**e**) $I_{dt} + I_{\nabla}$; and (**f**) $I_{small\ angle}$.

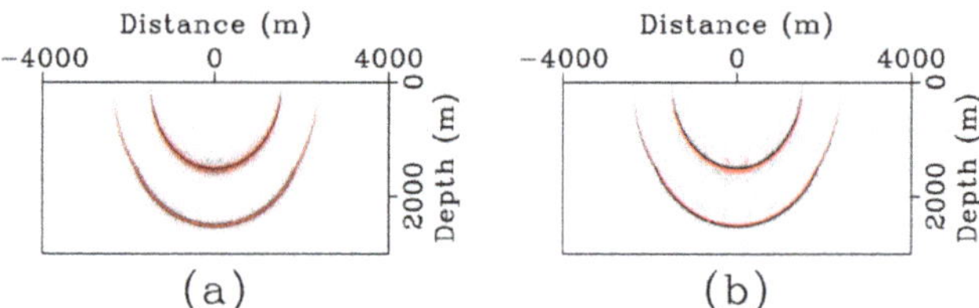

Figure 3. Comparison of imaging results of a single trace (offset = 0 m) (**a**) using Equation (13) and (**b**) using Equation (14).

3. Numerical Examples

In this section, we demonstrate the effectiveness of the proposed imaging condition for the impedance perturbation estimation (Equation (14)) on synthetic data. The first example was designed to demonstrate the effectiveness of the proposed method on a laterally invariant model with uncorrelated velocity and density. Figure 4a,b show the velocity and density model, respectively. Figure 4c shows the true relative impedance perturbation. We used the velocity and density in the first layer as the background velocity and density, respectively. We first subtracted the background impedance from the exact impedance and then analytically calculated the relative impedance perturbation. Figure 5a,b show the filtered true velocity and impedance perturbation, respectively. Figure 5c illustrates the impedance perturbation estimated by our proposed method, which matches well with the true results. Note that the depth errors in the estimated acoustic impedance are due to the fact that we used a constant migration velocity.

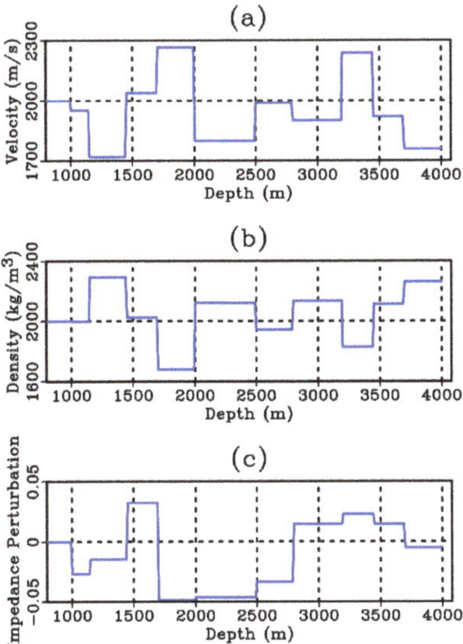

Figure 4. (**a**) Velocity model; (**b**) density model; and (**c**) true relative impedance perturbation.

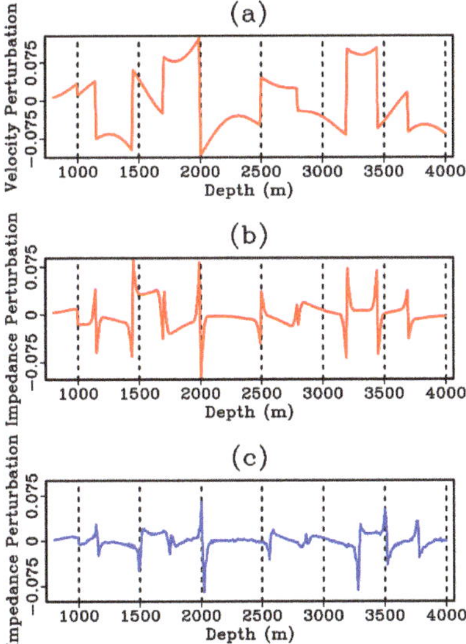

Figure 5. (**a**) Filtered true relative velocity perturbation; (**b**) filtered true relative impedance perturbation; and (**c**) estimated relative impedance perturbation.

Next, we demonstrated the acoustic impedance estimation on the Sigsbee2b model [17], and focused on the sediment areas with fine impedance structure. We performed acoustic finite difference modeling for synthetic data generation using a broadband source wavelet ($[f_1, f_2, f_3, f_4]$= [0, 2, 56, 60] Hz). The exact velocity model is shown in Figure 6a, from which the density model was generated using a predefined relationship of the two parameters. A $V(z)$ velocity model and a constant density model were used for migration. We analytically calculated the exact impedance perturbation and applied the band-pass filtering to the exact result according to the bandwidth of input data. The filtered true impedance perturbation is shown in Figure 6b. In comparison, Figure 6c illustrates the impedance perturbation results estimated by our proposed method. Figure 6d displays the overlay of the two images (Figure 6b,c). The detailed comparison between Figure 6b,c at the three different locations are shown in Figure 7. An amplitude calibration was performed using reflections from the water bottom. In Figure 7, the filtered true impedance perturbation is shown in red and the estimated impedance perturbation is shown in blue. The numerical results demonstrate the overall good match between the estimated and the true impedance perturbation.

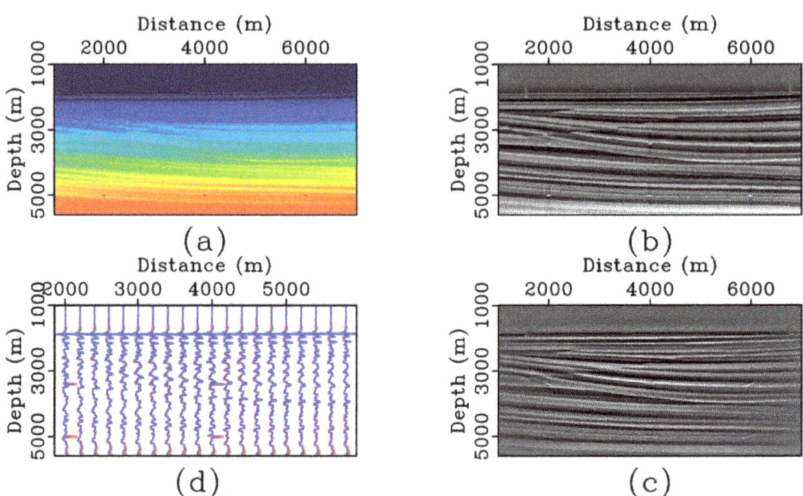

Figure 6. (**a**) Velocity model for synthetic data generation; (**b**) filtered true impedance perturbation; (**c**) impedance perturbation estimated by the proposed method; and (**d**) overlay plot of the two images (red: true, blue: estimated).

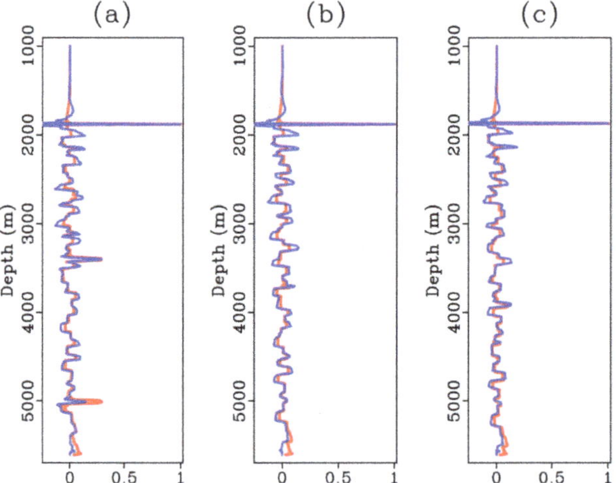

Figure 7. Comparison of filtered true impedance perturbation (red) and estimated results (blue) at three horizontal locations: 2000 m (**a**); 3500 m (**b**); and 5000 m (**c**).

In the third example, we demonstrated the proposed method on a 2D transition zone model. The exact velocity model (labels show grid numbers) is illustrated in Figure 8a, and the density model had similar structural characteristics (not shown here). As in the previous example, synthetic data were generated using a broadband source wavelet. A smoothed version of the true velocity was used for migration, with the assumption of a constant density. Figure 8b shows the filtered true impedance perturbation, while Figure 8c illustrates the estimated results using the proposed method. Detailed trace comparisons between the two are shown in Figure 9 at the three different horizontal locations ($X = 1900$, 2200, 3000). Figure 9b (at $X = 2200$) and Figure 9c (at $X = 3000$) show a slightly better match than Figure 9a ($X = 1900$, inside the island) due to better illumination. The comparison shows the true and the estimated results match quite well, demonstrating the effectiveness of our method on complex models.

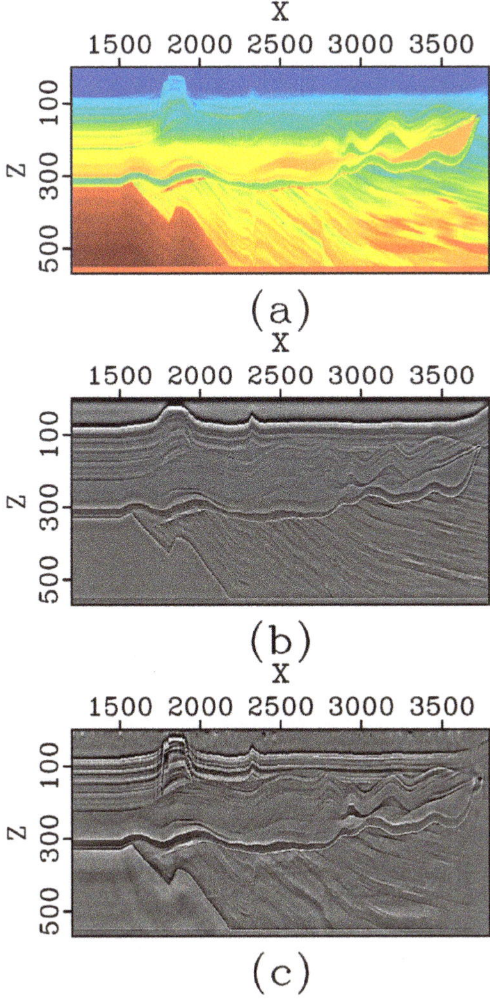

Figure 8. (**a**) A 2D transition zone velocity model; (**b**) true relative impedance perturbation after band-pass filtering; and (**c**) estimated relative impedance perturbation computed using the proposed method.

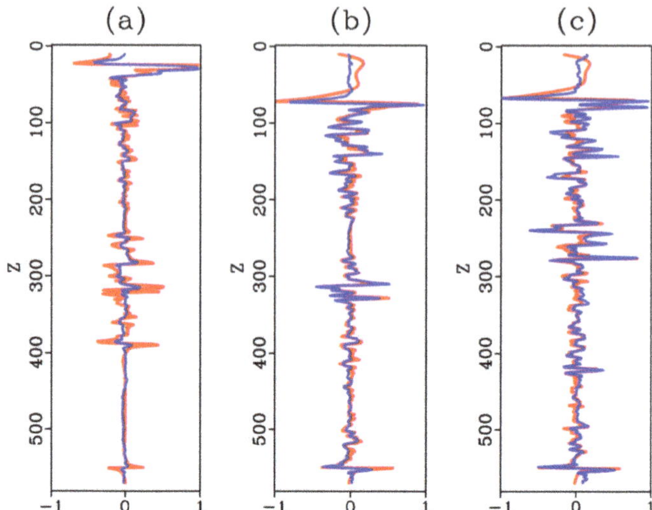

Figure 9. Comparison of filtered true impedance perturbation (red) and estimated results (blue) at three locations: (**a**) X = 1900; (**b**) X = 2200; and (**c**) X = 3000.

4. Discussion

In Equation (11b), we can see that the summation of the time derivative and the spatial gradient images had an extra term of $\cos^2\theta$ compared to the time derivative image. When we designed the small-angle imaging condition in Equation (13), we mentioned that the exponential term approximated to $e^{-\alpha \sin^2\theta}$. Since this exponential weighting was applied at each time step, we can consider the proposed imaging condition as the time derivative imaging condition weighted by the term $\cos^2\theta e^{-\alpha \sin^2\theta}$. Figure 10 shows the analysis of this term by choosing different parameter α, compared to the term $\cos^2\theta$. In the figure, we can see that a larger α attenuated the large reflections more rapidly than a small α; on the other hand, the attenuation curve was smoother for a small α, which generated less artifacts than a large α. In all our examples in the previous section, we empirically chose $\alpha = 5$ based on a compromise between the preservation of accuracy, and artifacts reduction. Figure 11 shows the sensitivity analysis of the inversion results using different α ($\alpha = 0, 5, 10, 20$) for the layered model (Figure 4). In the figure, we can see that for $\alpha = 0$, the velocity and impedance perturbation cannot be distinguished, while for a larger α ($\alpha = 10$ and $\alpha = 20$), the shallow part is problematic. In future studies, we may consider the functions of different forms to attenuate large reflections while still preserving the imaging quality.

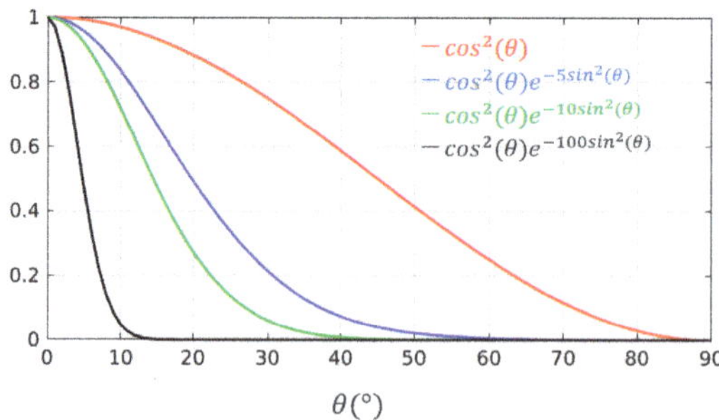

Figure 10. Analysis of different angle-dependent functions.

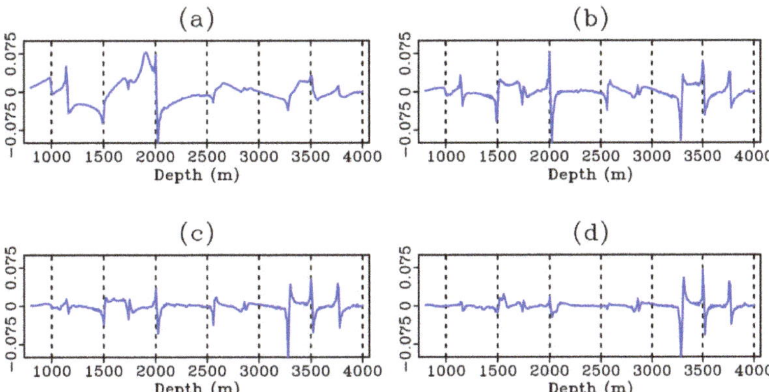

Figure 11. Estimated relative impedance perturbation for the layered model using different α in the weighting function in Equation (14): (**a**) $\alpha = 0$; (**b**) $\alpha = 5$; (**c**) $\alpha = 10$; and (**d**) $\alpha = 20$.

In all our numerical examples, broadband sources were used for synthetic data generation. Broad bandwidth played an important role in seismic inversion. Increasing the bandwidth at the high-frequency end improved seismic resolution, and adding more low frequencies in the data reduced the side lobes of the seismic wavelet [18]. Figure 12 demonstrates the impact of low frequencies on the inversion result. For the Sigsbee2b model, we applied a band-pass filter (f_1 = 3 Hz, f_2 = 5 Hz, f_3 = 56 Hz, f_4 = 60 Hz) to remove low frequencies in the input data, and then estimated the impedance perturbation using our proposed method. The estimated impedance perturbations with and without low frequencies are shown in Figures 12a and 12b, respectively. Figure 12a is the same as Figure 6c, and is repeated here for comparison. Figure 12c shows the trace comparison of Figure 12a,b at the horizontal location of 3500 m. The results demonstrate that low frequencies in the data affect the long-wavelength structures in the image. Low frequencies are crucial for both velocity and impedance inversions [8], and retaining low frequencies in the recorded data depends on broadband seismic acquisition and data processing.

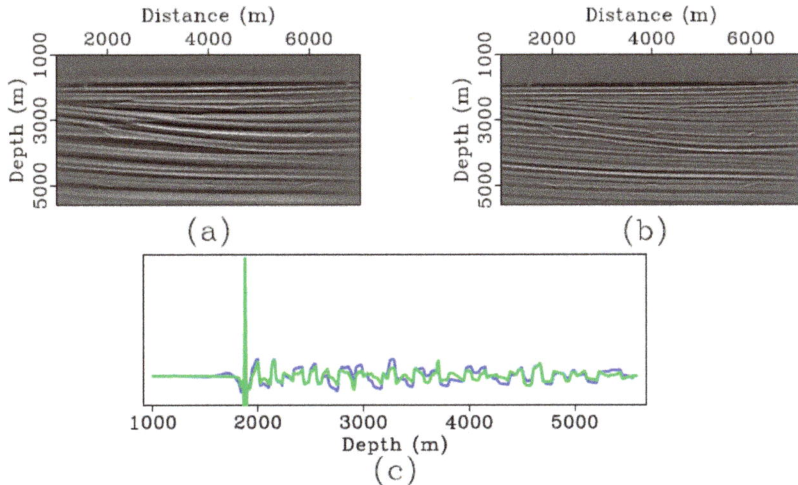

Figure 12. Comparison of estimated impedance perturbation with (**a**) and without (**b**) low frequencies in input data. (**c**) Trace comparison of estimated impedance perturbation with (blue) and without (green) low frequencies at the horizontal location of 3500 m.

As in conventional RTM, the proposed method assumes primaries only in the data, so de-multiple is a necessary preprocessing step in the case of strong multiples in the data. Even though our method was derived from the framework of amplitude-preserving RTM, the inversion was based on the Born approximation (Equation (4)). Therefore, transmission losses due to the overburden effects were also a challenge for our proposed method. The proposed method could be extended to finite-frequency based inversion [19] or FWI approaches [20] in the future work.

5. Conclusions

We proposed a weighted inverse scattering imaging condition for common-shot RTM, which outputs an estimation of the relative impedance perturbation. The construction of the weighting function was based on the two separate images from the conventional inverse scattering imaging condition. The proposed imaging condition was designed to select near-angle reflections during imaging, and therefore can separate the effect of impedance perturbation from velocity perturbation, for acoustic cases with variable velocities and densities. We demonstrated the effectiveness of the modified inverse scattering imaging condition using synthetic examples and showed that our method can produce reliable estimations of impedance perturbations.

Author Contributions: Conceptualization, H.L. (Hong Liang) and H.Z.; methodology, H.L. (Hong Liang); software, H.L. (Hong Liang), H.Z. and H.L. (Hongwei Liu); validation, H.L. (Hong Liang); formal analysis, H.L. (Hong Liang); investigation, H.L. (Hong Liang); resources, H.L. (Hongwei Liu); data curation, H.L. (Hong Liang); writing—original draft preparation, H.L. (Hong Liang); writing—review and editing, H.Z. and H.L. (Hongwei Liu); visualization, H.L. (Hong Liang); supervision, H.Z. and H.L. (Hongwei Liu); project administration, H.L. (Hongwei Liu); funding acquisition, H.Z. and H.L. (Hongwei Liu). All authors have read and agreed to the published version of the manuscript.

Funding: This research received no external funding.

Institutional Review Board Statement: Not applicable.

Data Availability Statement: Synthetic data used in this paper can be obtained from the corresponding author.

Acknowledgments: We use the Madagascar [21] for figure plotting in this paper.

Conflicts of Interest: The authors declare no conflict of interest.

Appendix A Derivation of Equation (5)

Equation (4) describes the relationship between the perturbation in the data and the perturbation in the model. Following [22], the inversion of the model perturbations from Equation (4) can be written as follows:

$$\frac{\delta v}{v_0} + \cos^2\theta \frac{\delta \rho}{\rho_0} = \iiint \mathbf{B}(\mathbf{x}_r;\omega;\mathbf{x}_s;\mathbf{x};\theta)\delta d(\mathbf{x}_r;\mathbf{x}_s;\omega)e^{-i\omega T(\mathbf{x}_s;\mathbf{x}_r;\mathbf{x})}d\mathbf{x}_r d\mathbf{x}_s d\omega \tag{A1}$$

where $\mathbf{B}$ is an inverse operator.

Substituting Equation (4) into Equation (A1), we have:

$$\begin{aligned}\frac{\delta v}{v_0} + \cos^2\theta\frac{\delta\rho}{\rho_0} = -\iiint \mathbf{B}(\mathbf{x}_r;\omega;\mathbf{x}_s;\mathbf{x};\theta)\times \\ \int \frac{2\omega^2}{v(\mathbf{x}')^2}\left(\frac{\delta v(\mathbf{x}')}{v(\mathbf{x}')} + \cos^2\theta'\frac{\delta\rho(\mathbf{x}')}{\rho(\mathbf{x}')}\right) \\ \times A(\mathbf{x}_s;\mathbf{x}_r;\mathbf{x}')e^{i\omega(T(\mathbf{x}_s;\mathbf{x}_r;\mathbf{x}')-T(\mathbf{x}_s;\mathbf{x}_r;\mathbf{x}))}d\mathbf{x}'d\mathbf{x}_r d\mathbf{x}_s d\omega\end{aligned} \tag{A2}$$

In 2D, we change the variables from $(\mathbf{x}_r, \mathbf{x}_s, \omega)$ to $(\mathbf{k}, \theta)$ and rewrite Equation (A2) as follows:

$$\begin{gathered}\frac{\delta v}{v_0} + \cos^2\theta\frac{\delta\rho}{\rho_0} = -\iint \mathbf{B}(\mathbf{x}_r;\omega;\mathbf{x}_s;\mathbf{x};\theta)\delta(\theta' - \theta)\times \\ \int \frac{2\omega^2}{v(\mathbf{x}')^2}\left(\frac{\delta v(\mathbf{x}')}{v(\mathbf{x}')} + \cos^2\theta\frac{\delta\rho(\mathbf{x}')}{\rho(\mathbf{x}')}\right)\times A(\mathbf{x}_s;\mathbf{x}_r;\mathbf{x}')e^{i\mathbf{k}(\mathbf{x}'-\mathbf{x})}\left|\frac{\partial(\mathbf{x}_s,\mathbf{x}_r,\omega)}{\partial(\mathbf{k},\theta)}\right|d\mathbf{x}'d\mathbf{k}d\theta'\end{gathered} \tag{A3}$$

The inverse operator $\mathbf{B}(\mathbf{x}_r;\omega;\mathbf{x}_s;\mathbf{x};\theta)$ must satisfy the following equation according to Equation (A3):

$$\frac{2\omega^2}{v^2}A(\mathbf{x}_s;\mathbf{x}_r;\mathbf{x})\mathbf{B}(\mathbf{x}_r;\omega;\mathbf{x}_s;\mathbf{x};\theta)\left|\frac{\partial(\mathbf{x}_s,\mathbf{x}_r,\omega)}{\partial(\mathbf{k},\theta)}\right| = -\left(\frac{1}{2\pi}\right)^2 \tag{A4}$$

The Jacobian is derived in Appendix B. Substituting the Jacobian in Equation (A16) into Equation (A4), we can obtain the following:

$$\begin{aligned}\mathbf{B}(\mathbf{x}_r;\omega;\mathbf{x}_s;\mathbf{x};\theta) &= -\left(\frac{1}{2\pi}\right)^2\frac{v^2}{2\omega^2}\times 64\pi^2|\omega||\mathbf{q}|^2A(\mathbf{x}_s;\mathbf{x}_r;\mathbf{x})\frac{\cos\beta_s}{v(\mathbf{x}_s)}\frac{\cos\beta_r}{v(\mathbf{x}_r)} \\ &= -\frac{32\cos^2\theta}{|\omega|}A(\mathbf{x};\mathbf{x}_s)A(\mathbf{x}_r;\mathbf{x})\frac{\cos\beta_s}{v(\mathbf{x}_s)}\frac{\cos\beta_r}{v(\mathbf{x}_r)}\end{aligned} \tag{A5}$$

By substituting Equation (A5) into Equation (A1), and by using the following relation:

$$\frac{\delta(\rho v)}{\rho_0 v_0} = \frac{\delta v}{v_0} + \frac{\delta\rho}{\rho_0} \tag{A6}$$

We can obtain the desired relationship in Equation (5):

$$\sin^2\theta\frac{\delta v}{v_0} + \cos^2\theta\frac{\delta(\rho v)}{\rho_0 v_0} = -\iint\iint\frac{32\cos^2\theta'}{|\omega|}\frac{\cos\beta_r}{v(\mathbf{x}_r)}\frac{\cos\beta_s}{v(\mathbf{x}_s)}A(\mathbf{x};\mathbf{x}_s)A(\mathbf{x}_r;\mathbf{x}) \tag{A7}$$

Appendix B Derivation of the Inverse of the Jacobian in Equation (A4)

In this appendix, we derive the inverse of the Jacobian $\left|\frac{\partial(\mathbf{x}_s,\mathbf{x}_r,\omega)}{\partial(\mathbf{k},\theta)}\right|$ following Appendix B from [9].

From the definition of the vector $\mathbf{q}$ in Figure A1, we have $\mathbf{q} = \mathbf{p}_s + \mathbf{p}_r$. The wavenumber vector $\mathbf{k}$ is defined as $\mathbf{k} = \omega\mathbf{q} = \omega(\mathbf{p}_s + \mathbf{p}_r)$. In the isotropic case, we have $|\mathbf{k}| = \frac{2|\omega|\cos\theta}{v}$, and $|\mathbf{q}| = \frac{2\cos\theta}{v}$, where θ is the reflection angle and v is the wave propagation velocity at the image point. By using the polar representation $(|\mathbf{k}|,\phi)$ for the vector $\mathbf{k}$, we obtain:

$$\begin{aligned}\left|\frac{\partial(\mathbf{k},\theta)}{\partial(\mathbf{x}_s,\mathbf{x}_r,\omega)}\right| &= \left|\frac{\partial(\mathbf{k},\theta)}{\partial(|\mathbf{k}|,\phi,\theta)}\right|\left|\frac{\partial(|\mathbf{k}|,\phi,\theta)}{\partial(\mathbf{x}_s,\mathbf{x}_r,\omega)}\right| \\ = |\mathbf{k}|\left|\frac{\partial(|\mathbf{k}|\phi,,\theta)}{\partial(\mathbf{x}_s,\mathbf{x}_r,\omega)}\right| &= |\omega||\mathbf{q}|\left|\frac{\partial(\mathbf{k})}{\partial(\omega)}\right|\left|\frac{\partial(\phi,\theta)}{\partial(\mathbf{x}_s,\mathbf{x}_r)}\right| = |\omega||\mathbf{q}|^2\left|\frac{\partial(\phi,\theta)}{\partial(\mathbf{x}_s,\mathbf{x}_r)}\right|\end{aligned} \tag{A8}$$

Using the relation:

$$\begin{cases}\phi = \frac{\alpha_s+\alpha_r}{2} \\ \theta = \frac{\alpha_s-\alpha_r}{2}\end{cases} \tag{A9}$$

We can have:

$$\left|\frac{\partial(\phi,\theta)}{\partial(\mathbf{x}_s,\mathbf{x}_r)}\right| = \left|\frac{\partial(\phi,\theta)}{\partial(\alpha_s,\alpha_r)}\right|\left|\frac{\partial(\alpha_s,\alpha_r)}{\partial(\mathbf{x}_s,\mathbf{x}_r)}\right| = \left|\frac{\partial(\alpha_s)}{\partial(\mathbf{x}_s)}\right|\left|\frac{\partial(\alpha_r)}{\partial(\mathbf{x}_r)}\right| \tag{A10}$$

Plugging Equation (A10) into Equation (A11), we can obtain:

$$\left|\frac{\partial(\mathbf{k},\theta)}{\partial(\mathbf{x}_s,\mathbf{x}_r,\omega)}\right| = |\omega||\mathbf{q}|^2\left|\frac{\partial(\alpha_s)}{\partial(\mathbf{x}_s)}\right|\left|\frac{\partial(\alpha_r)}{\partial(\mathbf{x}_r)}\right| \tag{A11}$$

From Figure A1, we can see that the angle ϕ is fixed for a given image point $\mathbf{x}$. Thus, the angles θ and α_s differ only by a constant angle: $\theta = \alpha_s - \phi$. So:

$$\left|\frac{\partial \theta}{\partial \alpha_s}\right| = 1 \tag{A12}$$

$$\left|\frac{\partial \theta}{\partial \mathbf{x}_s}\right| = \left|\frac{\partial \theta}{\partial \alpha_s}\right|\left|\frac{\partial \alpha_s}{\partial \mathbf{x}_s}\right| = \left|\frac{\partial \alpha_s}{\partial \mathbf{x}_s}\right| \tag{A13}$$

The expression for the left-hand side of the above equation can be found in [9] as Equation (7):

$$\left|\frac{\partial \theta}{\partial \mathbf{x}_s}\right| = 8\pi A^2(\mathbf{x}_s, \mathbf{x})\frac{\cos\beta_s}{v(\mathbf{x}_s)} \tag{A14}$$

Similarly, we can have:

$$\left|\frac{\partial \theta}{\partial \mathbf{x}_r}\right| = 8\pi A^2(\mathbf{x}_r, \mathbf{x})\frac{\cos\beta_r}{v(\mathbf{x}_r)} \tag{A15}$$

From Equations (A11) and (A14)–(A16), we have:

$$\begin{aligned} \left|\frac{\partial(\mathbf{k},\theta)}{\partial(x_s,\mathbf{x}_r,\omega)}\right| &= |\omega||\mathbf{q}|^2\left|\frac{\partial(\alpha_s)}{\partial(\mathbf{x}_s)}\right|\left|\frac{\partial(\alpha_r)}{\partial(\mathbf{x}_r)}\right| \\ &= |\omega||\mathbf{q}|^2 \times 8\pi A^2(\mathbf{x}_s,\mathbf{x})\frac{\cos\beta_s}{v(\mathbf{x}_s)} \times 8\pi A^2(\mathbf{x}_r,\mathbf{x})\frac{\cos\beta_r}{v(\mathbf{x}_r)} \\ &= 64\pi^2|\omega||\mathbf{q}|^2 A^2(\mathbf{x}_s,\mathbf{x})A^2(\mathbf{x}_r,\mathbf{x})\frac{\cos\beta_s}{v(\mathbf{x}_s)}\frac{\cos\beta_r}{v(\mathbf{x}_r)} \end{aligned} \tag{A16}$$

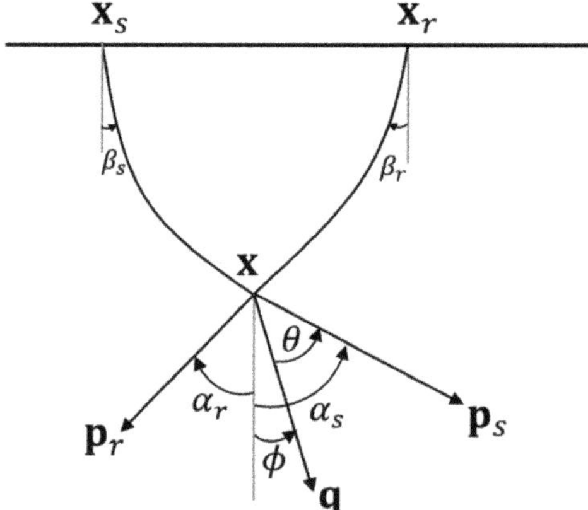

Figure A1. Coordinates of the 2D ray approximation. $\mathbf{x}$ is the image point, and $\mathbf{x}_s$ and $\mathbf{x}_r$ are the source and receiver position, respectively; vectors $\mathbf{p}_s$ and $\mathbf{p}_r$ define the specular ray parameters at the image point from the source and receiver, respectively; vector $\mathbf{q}$ is defined as the sum of the source ray parameter $\mathbf{p}_s$ and receiver ray parameter $\mathbf{p}_r$; in the specular cases, $\mathbf{q}$ coincides with the reflector normal vector;α_s, α_r, and ∅ are the angles with respect to the vertical of the vectors $\mathbf{p}_s$, $\mathbf{p}_r$ and $\mathbf{q}$, respectively; θ is the reflection angle with respect to the normal, and 2θ is the angle between $\mathbf{p}_s$ and $\mathbf{p}_r$; β_s and β_r are the takeoff angles at the source and receivers, respectively (figure adapted from [9]).

References

1. Latimer, R.; Davidson, R.; van Riel, P. An interpreter's guide to understanding and working with seismic-derived acoustic impedance data. *Lead. Edge* **2000**, *19*, 242–256. [CrossRef]
2. Routh, P.; Neelamani, R.; Lu, R.; Lazaratos, S.; Braaksma, H.; Hughes, S.; Saltzer, R.; Stewart, J.; Naidu, K.; Averill, H.; et al. Impact of high-resolution FWI in the Western Black Sea: Revealing overburden and reservoir complexity. *Lead. Edge* **2017**, *36*, 60–66. [CrossRef]
3. Barclay, F.; Bruun, A.; Rasmussen, K.B.; Alfaro, J.C.; Cooke, A.; Cooke, D.; Salter, D.; Godfrey, R.; Lowden, D.; McHugo, S.; et al. Seismic inversion: Reading between the lines. *Oilfield Rev.* **2008**, *1*, 42–63.
4. Beylkin, G. Imaging of discontinutes in the inverse scattering problem by inversion of a causal generalized Radon transform. *J. Math. Phys.* **1985**, *26*, 99–108. [CrossRef]
5. Jin, S.; Madariaga, R.; Virieux, J.; Lambaré, G. Two-dimensional asymptotic iterative elastic inversion. *Geophys. J. Int.* **1992**, *108*, 575–588. [CrossRef]
6. Forgues, E.; Lambaré, G. Parameterization study for acoustic and elastic ray+Born inversion. *J. Seism. Explor.* **1997**, *6*, 253–277.
7. Thierry, P.; Operto, S.; Lambare, B. Fast 2-d ray+born migration/inversion in complex media. *Geophysics* **1999**, *64*, 162–181. [CrossRef]
8. Bleistein, N.; Cohen, J.K.; Stockwell, J.W. *Mathematics of Multidimensional Seismic Imaging, Migration, and Inversion*; Springer: New York, NY, USA, 2001.
9. Bleistein, N.; Zhang, Y.; Zhang, G.; Gray, S.H. Migration inversion: Think image point coordinates, process in acquisition surface coordinates. *Inverse Probl.* **2005**, *21*, 1715–1744. [CrossRef]
10. Zhang, Y.; Ratcliffe, A.; Roberts, G.; Duan, L. Amplitude-preserving reverse time migration: From reflectivity to velocity and impedance inversion. *Geophysics* **2014**, *79*, S271–S283. [CrossRef]
11. Whitmore, N.D.; Crawley, S. Applications of RTM inverse scattering imaging conditions. In Proceedings of the 82nd Annual International Meeting, SEG, Expanded Abstracts, Las Vegas, NV, USA, 4–9 November 2012; pp. 1–6. [CrossRef]
12. Zhang, Y.; Xu, S.; Bleistein, N.; Zhang, G. True-amplitude, angle-domain, common-image gathers from one-way wave-equation migrations. *Geophysics* **2007**, *72*, S49–S58. [CrossRef]
13. Kiyashchenko, D.; Plessix, R.; Kashtan, B.; Troyan, V. A modified imaging principle for true-amplitude wave-equation migration. *Geophys. J. Int.* **2007**, *168*, 1093–1104. [CrossRef]
14. Liang, H.; Zhang, H. Wavelet-domain reverse time migration image enhancement using inversion-based imaging condition. *Geophysics* **2019**, *84*, S401–S409. [CrossRef]
15. Liang, H.; Zhang, H.; Liu, H. Reverse time migration for acoustic impedance inversion using inversion-based imaging condition. In Proceedings of the 89th Annual International Meeting, SEG, Expanded Abstracts, San Antonio, TX, USA, 15–20 September 2019; pp. 4365–4369. [CrossRef]
16. Rocha, D.; Tanushev, N.; Sava, P. Acoustic wavefield imaging using the energy norm. *Geophysics* **2016**, *81*, S151–S163. [CrossRef]
17. Paffenholz, J.; McLain, B.; Zaske, J.; Keliher, P. Subsalt multiple attenuation and imaging: Observations from the Sigsbee2B synthetic dataset. In Proceedings of the 72nd Annual International Meeting, SEG, Expanded Abstracts, Salt Lake City, UT, USA, 6–11 October 2012; pp. 2122–2125. [CrossRef]
18. Fons ten Kroode, F.; Bergler, S.; Corsten, C.; de Maag, J.W.; Strijbos, F.; Tijhof, H. Broadband seismic data—The importance of low frequencies. *Geophysics* **2013**, *78*, WA3–WA14. [CrossRef]
19. Zhang, H.; Liang, H.; Baek, H.; Zhao, Y. Computational aspects of finite-frequency traveltime inversion kernels. *Geophysics* **2021**, *86*, R109–R128. [CrossRef]
20. Qin, B.; Lambare, G. Joint inversion of velocity and density in preserved-amplitude full-waveform inversion. In Proceedings of the 86th Annual International Meeting, SEG, Expanded Abstracts, Dallas, TX, USA, 16–21 October 2016; pp. 1325–1330. [CrossRef]
21. Fomel, S.; Sava, P.; Vlad, I.; Liu, Y.; Bashkardin, V. Madagascar: Open-source software project for multidimensional data analysis and reproducible computational experiments. *J. Open Res. Softw.* **2013**, *1*, e8.
22. Bleistein, N. On the imaging of reflectors in the earth. *Geophysics* **1987**, *52*, 1426–1436. [CrossRef]

Article

Reverse Time Migration Imaging Using SH Shear Wave Data

Chengyao Zhou [1,2], Wenjie Yin [2], Jun Yang [2], Hongmei Nie [2] and Xiangyang Li [1,*]

1 State Key Laboratory of Petroleum Resources and Prospecting, China University of Petroleum (Beijing), Beijing 102249, China
2 BGP Inc., China National Petroleum Corporation, Beijing 100083, China
* Correspondence: xy1962li@163.com; Tel.: +86-10-8973-2259

Featured Application: Imaging through gas clouds using SH shear wave.

Abstract: In this paper, we discussed the reverse time migration imaging of compressional wave (P-wave) and horizontally polarized shear wave (SH shear wave) seismic data, together with P- and SH shear wave constrained velocity model building. In the Sanhu area in Qaidam Basin, there are large areas of gas clouds, which leads to poor P-wave seismic imaging. The P and SH shear wave seismic data of a co-located survey line with the same acquisition geometry were used to access their imaging capability using reverse time migration. We first estimated the change in P-wave and SH shear wave velocity ratio using pre-stack time migration (PSTM) for constraining the overall depth domain velocity model. Additionally, we then used an eighth-order finite difference scheme for P-wave reverse time migration on a variable grid and used the sixth-order combined compact difference (CCD) wave field simulation method for SH shear wave reverse time migration on a regular grid. The results show that the constrained velocity model produces a good match in the overall geological structure shown in the P-wave and SH shear wave reverse time migration results. However, in the gas cloud areas, SH shear wave reverse time migration has obvious imaging advantages, which can clearly image the structure inside the gas clouds.

Keywords: shear wave; reverse time migration; numerical simulation; Sanhu depression; Qaidam Basin

Citation: Zhou, C.; Yin, W.; Yang, J.; Nie, H.; Li, X. Reverse Time Migration Imaging Using SH Shear Wave Data. *Appl. Sci.* **2022**, *12*, 9944. https://doi.org/10.3390/app12199944

Academic Editors: Guofeng Liu, Zhifu Zhang and Xiaohong Meng

Received: 9 August 2022
Accepted: 30 September 2022
Published: 3 October 2022

1. Introduction

Currently, the main means of oil-gas exploration are mainly based on the P-wave (compressional wave). A series of mature technologies developed for P-wave exploration, including data acquisition, data pre-processing, pre-stack time migration, and pre-stack depth migration, have made great progress and are still making progress. However, with the extension of exploration to low permeability, deep and unconventional reservoirs and the increasingly complex exploration targets, relying solely on P-wave technology is facing more and more challenges. Through an analysis of the limitations of P-wave exploration and the potential of shear wave exploration, Gou et al. [1] proposed the effectiveness and applicability of SH shear wave (horizontally polarized shear wave) exploration technology and conducted SH shear wave data acquisition and processing tests in an exploration area in Western China [2].

Generally speaking, the resolution of seismic exploration is a quarter of the wavelength, which depends on the velocity and frequency. As the wave velocity of subsurface media is objective and certain, the resolution can only be improved by increasing the frequency. In terms of tight reservoirs or deep carbonate reservoirs, the P-wave velocity can be higher than 5000 m/s, and the effective frequency will be reduced to about 30 Hz with the attenuation of deep high-frequency components, which is very unfavorable for the identification of deep thin layers [3,4]. The further improvement of the effective frequency band in the deep layer will face great challenges in source excitation. If the frequency band is consistent, the wave speed of the shear wave is usually about half that of the P-wave, and the resolution can reach twice that of the P-wave. Shear-wave exploration is one of the effective ways to improve the resolution [5–7].

In recent years, with the rapid development of computer calculation, reverse time migration has gradually become an industrial applied imaging technology used to solve the imaging of various complex structures. Compared with that of the P-wave, the shear wave velocity is low and the frequency band is wide, such as carbonate rocks (Takougang et al., 2020) [8], which can improve the resolution of the formation; the shear wave propagation is only related to the rock frame, which can be used to accurately construct the subsurface structure. In terms of the near offset, the shear wave has a higher signal-to-noise ratio than the P-wave, so it has higher inversion precision [9–16]. To improve the accuracy and efficiency of shear-wave pre-stack depth migration, it is necessary to apply a high-precision finite-difference scheme to shear wave seismic imaging on the basis of the low-speed characteristics of the shear wave, and this method needs to maintain good accuracy of spatial difference and time difference under the condition of a large spatial grid and time step. Thus, migration algorithm selection is one of the subjects discussed in this paper.

Since the mid-1990s, reverse time migration has been applied to multi-component wave seismic data excited from an elastic-wave source [17–20], and it has overcome calculation problems and interference artifacts in P- and shear wave simulation. In recent years, most studies have focused on the P- and shear wave decoupled method to separate P- and shear waves [21–28]. In these studies, there is no case of the direct use of shear wave data for reverse time migration, but the elastic wave field was used to separate P- and shear waves or decoupled P- and shear wave fields obtained from the elastic-wave equation. This paper will focus on the direct use of SH shear wave source data for reverse time migration imaging and together with corresponding procedures for velocity analysis and velocity model building.

The study area, Sanhu, referred to in this paper, is located in the Qaidam Basin. The reservoir in the area is mainly consisted of thin sand-mudstone interbed and unconsolidated sandstone reservoirs, and this area is the largest Quaternary biogas production area discovered in China. Affected by the reservoir gas, the reservoir imaging is seriously blurred and distorted. To determine the characteristics of the study area, in this paper, the SH shear wave velocity model in the depth domain is established by using the SH shear wave data in the survey, and the subsurface reflecting interface in this area is imaged by using the SH shear wave reverse time migration imaging technology [29,30]. During depth domain velocity model building, we obtain the P-to-SH shear wave velocity ratio in the area from the time domain P-wave and SH shear wave imaging results, which is used to constrain the overall P- and SH shear wave depth domain velocity model building.

2. Application Background and Method Principles

2.1. Survey Background

The survey studied in this paper is from the Sanhu area of the Qaidam Basin, and its surface is mostly composed of aeolian monadnock, sand dunes, sandbag, saline-alkali land, hard-alkali land, salt marsh, etc. The Sanhu area is a biogas enrichment area. Affected by subsurface gas, as shown in Figure 1 the energy and frequency absorption of P-wave seismic data in the seismic anomaly area is serious, the data signal-to-noise ratio is low, and the structure is distorted [31]. It is difficult to obtain reliable imaging of low-relief structures.

At the same time, there are unique geological characteristics of the Quaternary system in the Sanhu area, such as weak diagenesis, a thin sand-mudstone interbed, an indiscriminate hydrocarbon source and reservoir, as well as the succession and relay of gas generation. Under this unique geological background, a high resolution is required for imaging the structure in this area, and SH shear wave imaging has a high stratigraphic resolution due to its low velocity and wide frequency band.

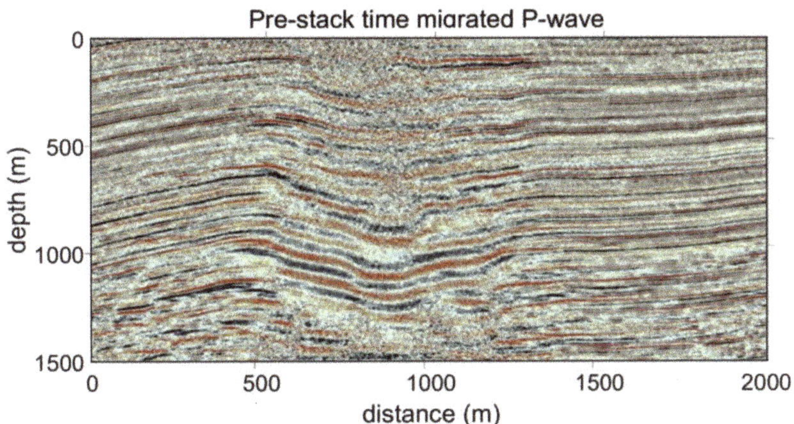

Figure 1. Pre-stack time migrated P-wave seismic section in a gas-cloud region of Sanhu.

2.2. Principle of Reverse Time Migration

The principle of SH shear wave reverse time migration (RTM) used in this paper is briefly introduced below:

Reverse time migration based on the solution of two-way wave equation is the most advanced pre-stack depth migration imaging. After an accurate velocity model is obtained, SH shear wave reverse time migration imaging (SH-RTM) generally includes three steps [32]:

(1) The source wave field is obtained by using the source constructed manually or extracted from actual data, and the corresponding model is numerically simulated to obtain the source wave field $S(x,z,t)$, where x,z is the space vector.

(2) Using the seismic data obtained at the receiver, the reverse continuation propagation passes through the same velocity model, and the corresponding receiver wave field $R(x,z,t)$ is obtained, where the position of the receiver is x_r, z_r.

(3) We can then apply appropriate imaging conditions, such as cross-correlation, we obtain the $I(x,z)$ (reverse-time migration image results):

$$I(x,z) = \int_0^T S(x,z,t)R(x,z,t)dt \tag{1}$$

where $S(x,z,t)$ is the source wavefield obtained by forward modeling, $R(x,z,t)$ is the receiver wavefield obtained by reverse continuation simulation at the same time under the same velocity model, and t is the total propagation time.

It can be seen from Equation (1) that the final result of reverse time migration is affected by the accuracy of the source wave field $S(x,z,t)$ and the receiver wave field $R(x,z,t)$. Therefore, in order to obtain accurate imaging of the target area, it is necessary to make good use of high-quality SH shear wave data and apply them to reverse time migration.

2.3. Depth Domain Velocity Model Building for SH Shear Wave Data

The depth domain velocity model building is very important for the effect of depth domain imaging, and P-wave and SH shear wave data are simultaneously acquired in this study area [33]. In depth-domain imaging, the reflector of the subsurface structure, the P-wave and SH shear wave, should be matched [34]. Generally, in the case of the combination of shear wave or converted wave and P-wave imaging, a priority will be given to determine the P-wave velocity (V_P) during the model establishment, and then a P-to-S wave velocity ratio will be estimated and used together with the V_P (P-wave velocity) to establish the model of SH shear wave velocity (V_S). In this study area, due to the gas clouds and the difficulty of obtaining effective signals of P-waves, it is difficult to establish a velocity model of V_P. While the pure SH shear wave can propagate through the rock frame without being affected by gas and fluid, the pure SH shear wave data can be directly used to establish a model of the SH shear wave velocity vs. (SH shear wave velocity).

First of all, we separately used the SH shear wave and P-wave velocity models. The establishment of P-wave velocity model refers to the establishment of SH shear wave velocity model since SH shear wave data in the region has a better signal-to-noise ratio in the survey and is more credible in the absorption area. Then, the velocity ratios of the P- and SH shear waves in this region are estimated, and the obtained ratios are taken as constraints. In order to better match the imaging of the P-wave and SH shear wave in the depth domain, we also need to estimate the change in the velocity ratio of the P-wave and SH shear wave by using the pre-stack time migration image. There are many methods of depth domain velocity model building, and we mainly use layer- and grid-based tomography. After the establishment of the two models, the spatial P-wave-to-SH shear wave ratio mentioned above is used to obtain the interlayer travel time in the depth domain model, calculate the proportion of model correction, and apply it to the correction of the P-wave model. In the actual constraint process, the velocity correction is about 1–3%, which can be attributed to weak anisotropic media.

2.4. Principle of Combined Compact Difference Scheme

The combined compact difference (CCD) to suppress the numerical dispersion caused by the spatial step size are enclosed in this paper to solve the problem of low SH shear wave velocity; the formula is as follows [35,36]:

$$\begin{cases} a_1\left(f'_{i+1}+f'_{i-1}\right)+f'_i+hb_1\left(f''_{i+1}-f''_{i-1}\right)=\frac{1}{h}\sum\limits_{m=1}^{n}c_m\left(f_{i+m}-f_{i-m}\right) \\ \frac{1}{h}a_2\left(f'_{i+1}-f'_{i-1}\right)+f''_i+b_2\left(f''_{i+1}+f''_{i-1}\right)=\frac{1}{h^2}\sum\limits_{m=1}^{n}d_m\left(f_{i+m}-2f_i+f_{i-m}\right) \end{cases} \quad (2)$$

In Equation (2), h is the grid spacing, a, b, c, d are the difference coefficient matrices; f is the function value of node i; f'_i and f''_i represent the first- and second-order derivatives of node i, respectively; f_{i+m}, f_{i-m} represent the function values of node i successively m nodes forward and m nodes backward; ${f'}_{i+1}, {f'}_{i-1}$ represent the first-order derivatives of node i successively one node forward and one node backward, respectively; ${f''}_{i+1}, {f''}_{i-1}$ represent the second derivative of i node one node forward and one node backward. The wave field of the SH shear wave can be simulated with the above method applied to the numerical simulation of SH shear wave propagation under the condition of a two-dimensional medium.

In this paper, we use the three-point sixth order format of Equation (2) for the SH shear wave reverse time migration, that is, Formula (3).

$$\begin{cases} \frac{7}{16}\left(f'_{i+1}+f'_{i-1}\right)+f'_i-\frac{h}{16}\left(f''_{i+1}-f''_{i-1}\right)=\frac{15}{16h}\left(f_{i+1}-f_{i-1}\right) \\ \frac{9}{8h}\left(f'_{i+1}-f'_{i-1}\right)+f''_i-\frac{1}{8}\left(f''_{i+1}+f''_{i-1}\right)=\frac{3}{h^2}\left(f_{i+1}-2f_i+f_{i-1}\right) \end{cases} \quad (3)$$

3. Analysis of Combined Compact Difference Scheme

In numerical simulations, if the size of the spatial grid is too large, it will cause large solution errors and produce numerical dispersion [37,38]. The simple harmonic wave solution is introduced to have the velocity ratio curve at different θ values in the isotropic case; θ is the angle between the wave's propagation direction and the x-axis. It is used to compare the spatial dispersion suppression effects of CCD with the traditional finite difference scheme. It is shown as follows:

In Figure 2, the ratio of numerical wave velocity to true velocity is defined as:

$$\lambda=\frac{v_{num}}{v} \quad (4)$$

where v_{num} is the numerical wave velocity, and v is the true wave velocity. It is shown in Figure 2 that with the velocity ratio curves of the CCD, the other two different formats at different θ values. The Courant numbers ($\alpha = v\Delta t/h$) are 0.25, the horizontal axis $\varphi \in [0, \pi]$ is the product of the wavenumber and the spatial step, and the y-coordinate is the velocity ratio λ, with 1 meaning that the numerical wave speed is consistent with the theoretical wave speed and indicating that the numerical dispersion is low. It also indicates that CCD has the best suppression effects.

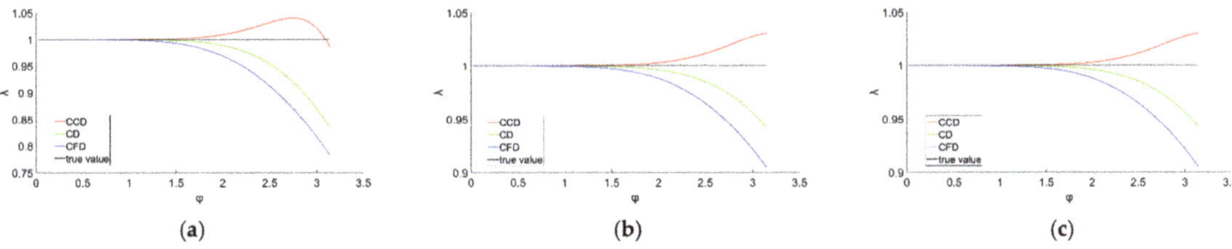

Figure 2. Velocity ratio curves for different numerical simulation methods. The blue curve is the traditional central difference scheme, the green curve is the traditional implicit difference scheme, the red curve is the CCD difference scheme and the black line is the velocity ratio constant of 1: (**a**) $\theta = 0, \alpha = 0.25$; (**b**) $\theta = \pi/6, \alpha = 0.25$; (**c**) $\theta = \pi/3, \alpha = 0.25$.

Additionally, the simulation error is calculated by simulating the two-dimensional plane harmonic initial value problem to analyze and compare the numerical simulation accuracy of the CCD and CFD (centered finite difference scheme).

In Figure 3, the simulation results show that the accuracy is relatively high for the SH shear wave simulation results with the adoption of the CCD format, and the numerical simulation of the seismic wave field with a large sampling time can be performed [39].

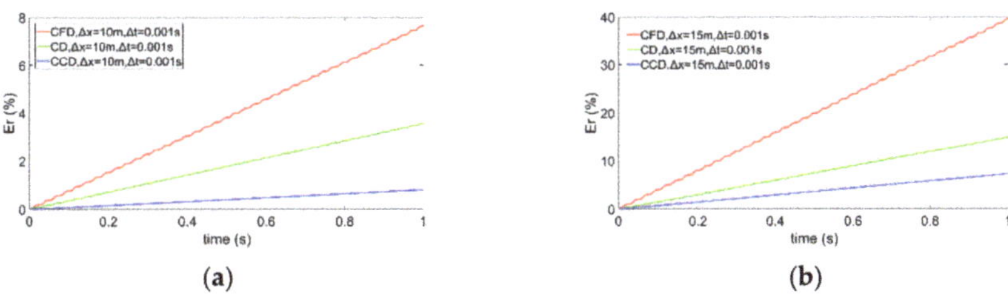

Figure 3. Relative errors of numerical simulation for different schemes and gird size: (**a**) Δx = 10 m, Δt = 0.001 s; (**b**) Δx = 15 m, Δt = 0.001 s.

4. Real Data Application

4.1. Data Characteristics

Figure 4 shows a single-shot record of the P-wave source in the study area; theoretically, the seismic records generated by P wave sources do not have shear wave fields. As shown in Figure 4a, with the X-component wave field seismic record of the P-wave source, the wave field is relatively complex, the converted wave signal is relatively weak and submerged in a large amount of scattering noise, and there is obvious leaked pure shear wave information in the near trace. The Y-component wave field seismic record is shown in Figure 4b; there is basically no converted wave information in this component, and the near trace is dominated by noise with weak pure shear wave information. The Z-component wave field seismic record is shown in Figure 4c; the P-wave reflection hyperbola characteristics of the shallow part are relatively obvious in this wave field, the near-trace is dominated by noise without pure shear wave information, and the deep information is submerged in a large amount of scattered noise. The reason why the converted wave signal of the P-wave field is relatively weak and the effective information of the wave field is submerged in a large amount of

scattered noise is likely due to the fact that the area is a biogas enrichment area in relatively deep strata based on the geological characteristics of the area, which is also the reason why the use of P-wave data for reverse time migration imaging in this area is not ideal.

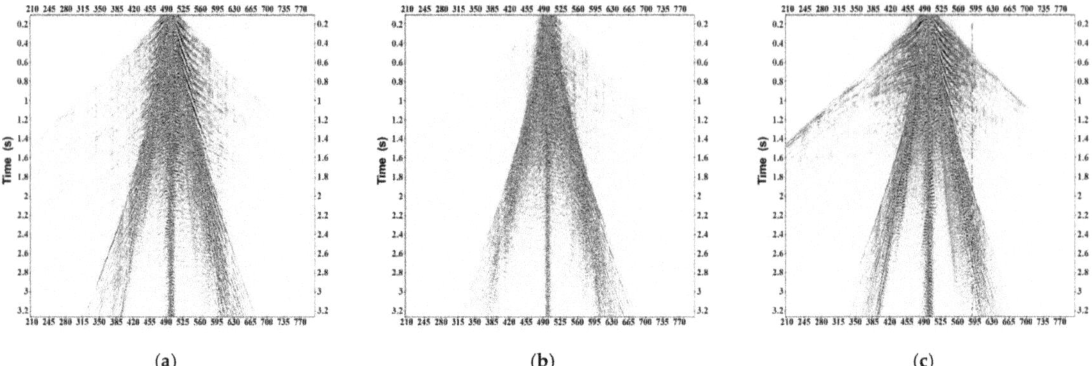

Figure 4. (**a**) The X-component wave field seismic record of the P-wave source. (**b**) The Y-component wave field seismic record of the P-wave source. (**c**) The Z-component wave field seismic record of the P-wave source.

Figure 5 shows the corresponding single-shot record obtained from the SH shear wave source (i.e., the y-direction concentrated force source) in this area. As shown in Figure 5a, the X-component data, there should be no SH shear wave information in isotropic media, but there is relatively weak SH shear wave signal leakage in actual data. The most likely cause of SH shear wave signal leakage is due to shear-wave splitting, which is not the focus of this article. The seismic record of the Z-component wave field in Figure 5c is dominated by noise without SH shear wave information, which is consistent with the theory of the SH shear wave propagation. In theory, the wave field obtained by concentrating force sources in the y-direction only has wave field information in the Y-component in isotropic media. Figure 5b is the seismic record of the y-component wave field, with a high signal-to-noise ratio and obvious SH shear wave information. Overall, the resolution of SH shear wave seismic records is higher than that of P-wave seismic record, comparing Figure 5b with Figure 4c. SH shear wave source data can also have a high signal-to-noise ratio in the deep gas cloud region due to its insensitivity to gas, and it has great advantages in geological structure characterization and reverse time migration imaging research in this area.

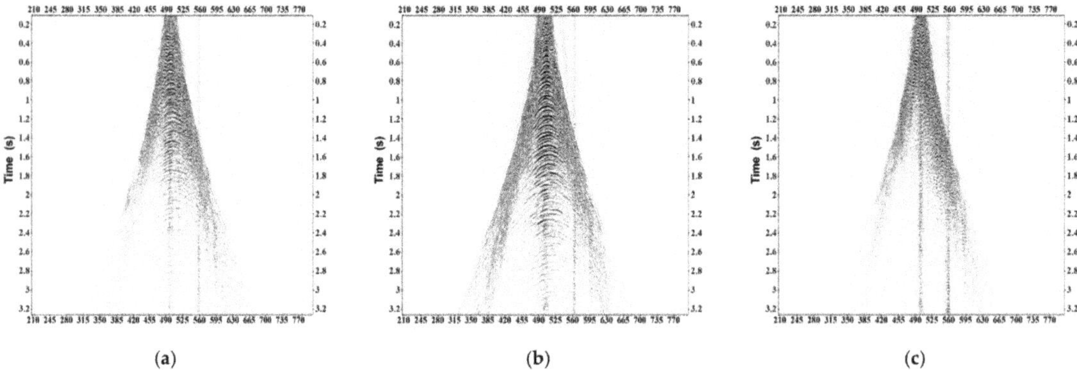

Figure 5. (**a**) The X-component wave field seismic record of the SH shear wave source. (**b**) The Y-component wave field seismic record of the SH shear wave source. (**c**) The Z-component wave field seismic record of the SH shear wave source.

It can be seen through the analysis of shot records that the data quality of the SH-Y component is relatively good. This research focuses on imaging research on SH-Y component data. This component is perpendicular to the survey line direction.

4.2. Depth Domain Imaging Matching and Velocity Model Building

The depth domain velocity model of the P-wave and SH shear wave have been established, respectively, with a velocity model building workflow including layer- and grid-based tomography. It is shown in Figure 6 with the depth domain velocity models of the P-wave and SH shear wave, respectively, that they have a certain similarity in the spatial variation of the velocity field; the velocity of the P-wave is relatively low at the position of CMP (common middle point) 1200–2500, and it is relevant to the gas enrichment in this area, which can also be seen in the pre-stack depth migration results shown later.

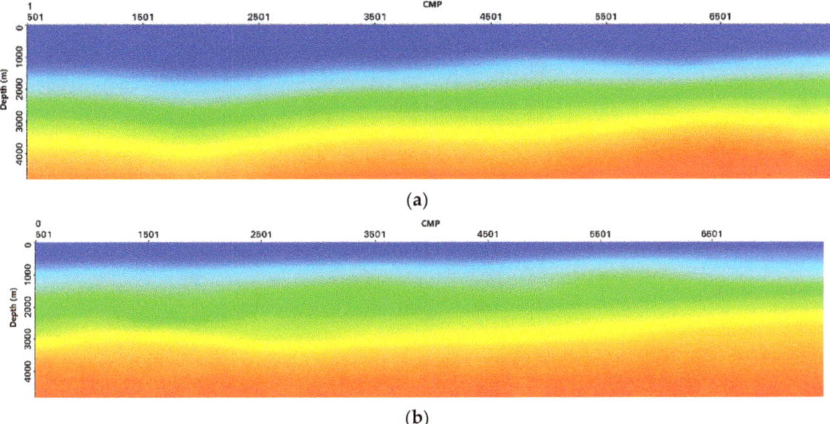

Figure 6. (**a**) The depth domain velocity model of the P-wave with a velocity model building workflow including layer- and grid-based tomography. (**b**) The corresponding depth domain velocity model of the SH shear wave.

The imaging of the reflector of the P-wave and the strong reflector of the SH shear wave should be located at the same depth because the pre-stack depth migration images show the wave impedance interface of subsurface media. It is very difficult to match the depth domain imaging positions of the P-wave and SH shear wave in practical applications due to the low signal-to-noise ratio of P-wave data, the possibility of anisotropy, and various uncertainties in the area. A relatively easy-to-implement process has been designed to attempt to match the imaging depth of the P-wave and SH shear wave. Our idea is to constrain V_P and vs. because the most significant factor affecting the imaging depth is the velocity field. The specific method is that a set of strong reflectors in the survey are selected; they are picked up in the migration images of the P-wave and SH shear wave, respectively; and their average velocities are calculated from the surface to the reflectors. The specific formula is $\overline{V} = \frac{\Delta Z}{\sum t_i}$, in which, for $\sum t_i = \sum\limits_i \frac{2\Delta z_i}{V_i}$, $\Delta Z = \sum \Delta z_i$, similar calculations for both the P-wave and SH shear wave have been made, the imaging position of the strong reflector in the depth domain is determined by this average velocity, and the average velocity ratio obtained should also be consistent with the P-wave and SH shear wave velocity in the survey [40]. The existing depth domain velocity can be corrected if the velocity ratio of SH shear wave and SH shear wave in the space of the survey is obtained. The pre-stack time migration images have been used and the corresponding reflectors have been picked up to solve this problem, the pre-stack time images of the P-wave and SH shear wave approximately represent the vertical two-way travel time of the P-wave and SH shear wave with zero offset, and the travel time ratio is obtained to further constrain the V_S/V_P ratio.

Back to the first step, a corresponding proportional constraint has been made on the P-wave velocity field with a correction of about 1–3%, and the weak anisotropy parameter is utilized to solve the residual moveout of imaging caused by this correction. In fact, the numerical model verifies that isotropic and anisotropic imaging profiles are close to each other under weak anisotropy (<5%); therefore, anisotropy is ignored in this study [41]. In Figure 7, the horizon selected for the P-wave and SH shear wave in the time domain is used to calculate the P-wave and SH shear wave velocity ratio. Figure 8 shows the P-wave and SH shear wave velocity ratio.

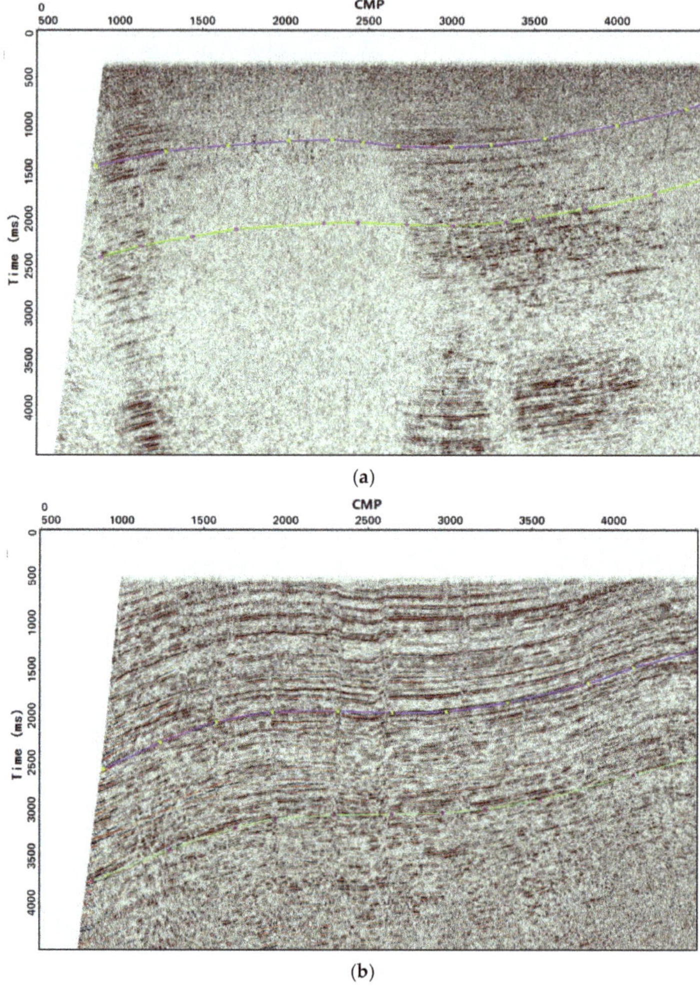

Figure 7. Pre-stack time migration (PSTM) sections: (**a**) P-wave and (**b**) SH shear wave.

For ease of understanding, we built a flow chart of velocity modeling, as shown in Figure 9. Figure 9 is about the P wave and SH shear wave velocity modeling process in this paper. In this figure, for the convenience of display, the S wave means SH shear wave. We can see that, after the preliminary modeling, we need to go through layer-based tomography and grid-based tomography, and the velocity ratio is obtained simultaneously with the velocity modeling. After the velocity ratio is obtained, it is used to constrain the velocity model of P wave to get the accurate velocity model.

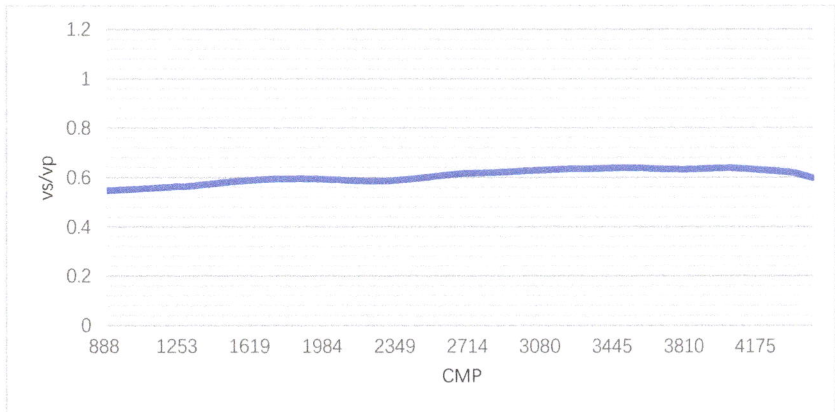

Figure 8. V_S/V_P ratio in the survey.

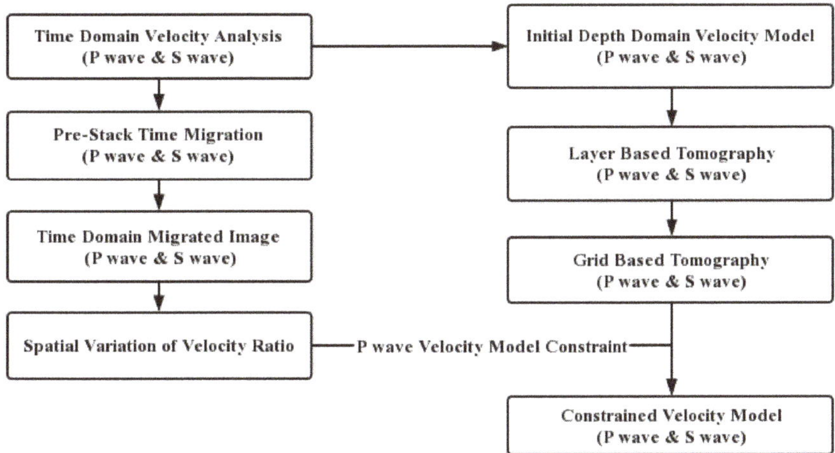

Figure 9. Velocity modeling flowchart.

4.3. RTM Results of P and SH Shear Wave

The reverse time migration of P-wave and SH shear wave data were carried out, respectively; there are 1519 shots and 8472 detection points in this survey, and the same acquisition geometry was adopted for both the P- and SH shear wave. The eighth-order finite difference scheme with a variable grid was adopted for P-wave reverse time migration for seismic wave simulation, the imaging condition is the cross-correlation imaging condition, and we use Laplace filtering to remove the low-frequency noise of the imaging. The SH shear wave reverse time migration was realized on a regular grid with the sixth-order CCD (Formula (3)) wave field simulation method adopted, an interval of 2 m for the grid size in the z-direction and an interval of 5 m for the grid size in the x-direction were adopted, and Laplace filtering was adopted to remove the low-frequency noise of the imaging to ensure stability and suppress alias frequencies in the wave field continuation process. Figure 10 shows the pre-stack depth migration images of the P-wave and SH shear wave. It can be seen in the image that there is no clear reflector for the P-wave in the area of CMP (common middle point) 1200–2500, which is due to the gas cloud in the area; the P-wave is significantly absorbed and has difficulty in passing through the area and reflecting to the surface when the compression wave goes through the area, but the SH shear wave can propagate normally in the area because it propagates through the rock frame and is not affected by the gas. Therefore, SH shear wave reverse time migration can clearly image through this area and has obvious imaging advantages in this area over P-wave.

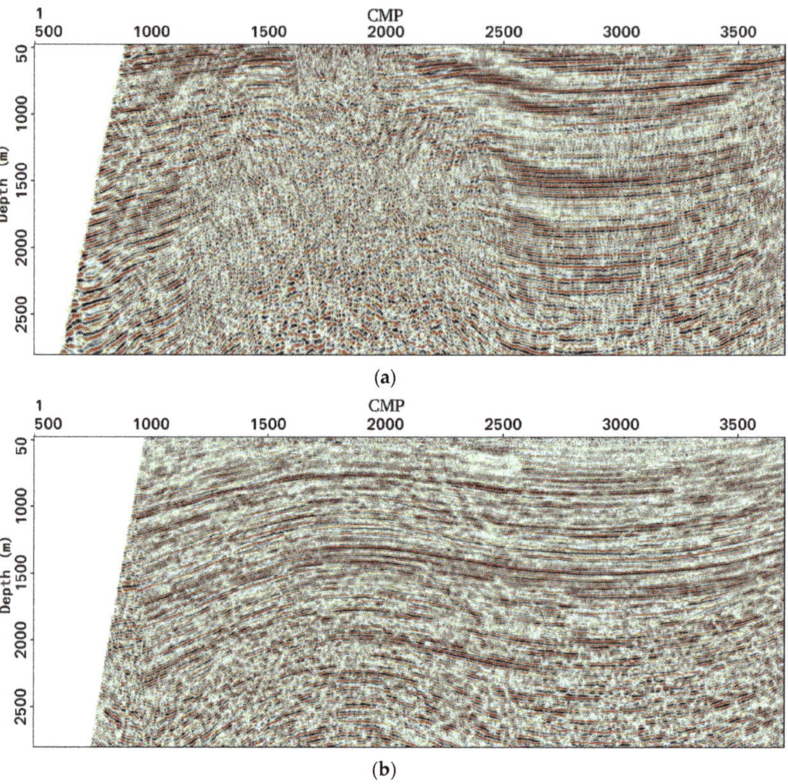

Figure 10. Pre-stack depth migration sections using RTM: (**a**) the P-wave, $\Delta x = 5$ m, $\Delta z = 2$ m; (**b**) the SH shear wave, $\Delta x = 5$ m, $\Delta z = 2$ m.

5. Conclusions

The depth domain pre-stack migration practice was performed for P-wave and SH shear wave seismic data in the Sanhu area of the Qaidam Basin in Qinghai Province, including depth domain velocity model building, P-wave to SH shear wave velocity ratio estimation, and reverse time migration (RTM). The layer- and grid-based tomography methods were used to build the depth velocity model. Additionally, the pre-stack time migration results of the P-wave and SH shear wave were utilized to approximately calculate the distribution of the P-wave and SH shear wave velocity ratio in the study area, and the ratio was used to constrain the P-wave and SH shear wave velocity model building so that the reflectors of the P-wave and SH shear wave pre-stack depth migration results were matched in the depth domain. An eighth-order finite difference scheme was used for P-wave RTM on a variable grid, and a sixth-order CCD was used for SH shear wave RTM on a regular grid. The results showed that the SH shear wave results have obvious imaging advantages compared with the P-wave results in the gas-cloud region, which verified the accuracy of the SH shear wave RTM algorithm and demonstrated great potential in seismic imaging application.

Author Contributions: Conceptualization, C.Z.; Formal analysis, J.Y.; Funding acquisition, X.L.; Investigation, C.Z., W.Y. and H.N.; Methodology, C.Z. and W.Y.; Project administration, J.Y. and X.L.; Resources, J.Y. and H.N.; Software, C.Z.; Supervision, X.L.; Validation, H.N.; Writing—original draft, C.Z. and W.Y.; Writing—review & editing, X.L. All authors have read and agreed to the published version of the manuscript.

Funding: The study is supported by the Science and Technology Research and Development Project of CNPC (2021DJ3506) and (2021ZG03) and R&D Department of China National Petroleum Corporation (Investigations on fundamental experiments and advanced theoretical methods in geophysical prospecting applications, 2022DQ0604-02).

Institutional Review Board Statement: Not applicable.

Informed Consent Statement: Not applicable.

Data Availability Statement: Not applicable.

Conflicts of Interest: The authors declare no conflict of interest.

References

1. Gou, L.; Zhang, S.; Li, X. Improving the Effectiveness of Shear-wave Seismic Exploration through Geophysical Technology Innovation. *Pet. Sci. Technol. Forum* **2021**, *40*, 12–19.
2. Shao, Z.; He, S.; Hou, L.; Wang, Y.; Tian, C.; Liu, X.; Zhou, Y.; Hao, M.; Lin, C. Dynamic Accumulation of the Quaternary Shale Biogas in Sanhu Area of the Qaidam Basin, China. *Energies* **2022**, *15*, 4593. [CrossRef]
3. Bouchaala, F.; Guennou, C. Estimation of viscoelastic attenuation of real seismic data by use of ray tracing software: Application to the detection of gas hydrates and free gas. *Comptes Rendus Geosci.* **2012**, *344*, 57–66. [CrossRef]
4. Matsushima, J.; Ali, M.Y.; Bouchaala, F. A novel method for separating intrinsic and scattering attenuation for zero-offset vertical seismic profiling data. *Geophys. J. Int.* **2017**, *211*, 1655–1668. [CrossRef]
5. Li, X.; Zhang, S. Forty years of shear-wave splitting in seismic exploration: An overview. *Geophys. Prospect. Pet.* **2021**, *60*, 190–209.
6. Gaiser, J.E. 3-D converted shear wave rotation with layer stripping. U.S. Patent US5610875, 11 March 1997.
7. Wu, Y.; He, Z.; He, J.; Deng, Z.; Wang, Y.; Yin, W. P-wave and S-wave joint acquisition technology and its application in Sanhu Area. In *SEG Technical Program Expanded Abstracts 2018*; Society of Exploration Geophysicists: Tulsa, OK, USA, 2018; pp. 1–5.
8. Takougang, E.M.T.; Ali, M.Y.; Bouzidi, Y.; Bouchaala, F.; Sultan, A.A.; Mohamed, A.I. Characterization of a carbonate reservoir using elastic full-waveform inversion of vertical seismic profile data. *Geophys. Prospect.* **2020**, *68*, 1944–1957. [CrossRef]
9. Dai, F.; Zhang, F.; Li, X.-Y. SH-SH wave inversion for S-wave velocity and density. *Geophysics* **2022**, *87*, A25–A32. [CrossRef]
10. Zhang, F. Simultaneous inversion for S-wave velocity and density from the SV-SV wave. *Geophysics* **2021**, *86*, R187–R195. [CrossRef]
11. LI, X.Y. Processing P-P and P-S waves in multicomponent sea-floor data for azimuthal anisotropy: Theoryandoverview. *Oil Gas Sci. Technol.* **1998**, *53*, 607–620.
12. Zhang, F.; Li, X. Inversion of the reflected SV-wave for density and S-wave velocity structures. *Geophys. J. Int.* **2020**, *221*, 1635–1639. [CrossRef]
13. Cheng, J.W.; Zhang, F.; Li, X.Y. Nonlinear amplitude inversion using a hybrid quantum genetic algorithm and the exact zoeppritz equation. *Pet. Sci.* **2022**, *v*, 1048–1064. [CrossRef]
14. Cheng, J.W.; Zhang, F.; Li, X.Y. Seismic Amplitude Inversion for Orthorhombic Media Based on a Modified Reflection Coefficient Approximation. *Surv. Geophys.* **2022**, *v*, 1–39. [CrossRef]
15. Deng, Z.; Li, C.; Chen, G.; Yang, J.; Wang, R.; Hu, Y.; An, S.; Wang, H.; Du, Z. The application of pure shear wave seismic data for gas reservoir delineation. In *SEG Technical Program Expanded Abstracts 2019*; Society of Exploration Geophysicists: Tulsa, OK, USA, 2019; pp. 2690–2694.
16. Brian, R.; Larry, L. A Gassmann consistent rock physics template. *CSEG Rec.* **2013**, *38*, 22–30.
17. Li, X.Y. Fractured reservoir delineation using multicomponent seismic data. *Geophys. Prospect.* **1997**, *45*, 39–64. [CrossRef]
18. Chang, W.; McMechan, G.A. 3-D elastic prestack, reverse-time depth migration. *Geophysics* **1994**, *59*, 597–609. [CrossRef]
19. Du, Q.Z.; Zhu, Y.T.; Ba, J. Polarity reversal correction for elastic reverse time migration. *Geophysics* **2012**, *77*, S31–S41. [CrossRef]
20. Dai, N.; Wu, W.; Zhang, W.; Wu, X. TTI RTM using variable grid in depth. *Int. Pet. Technol. Conf.* **2011**, *v*, 1–7.
21. Zhang, J.; Tian, Z.; Wang, C. P- and S-wave separated elastic wave equation numerical modeling using 2D staggered-grid. In Proceedings of the 77th Annual International Meeting, SEG, San Antonio, TX, USA, 14 September 2007; pp. 2104–2109.
22. Xiao, X.; Leaney, W.S. Local vertical seismic profiling (VSP) elastic reverse-time migration and migration resolution: Salt-flank imaging with transmitted P-to-S waves. *Geophysics* **2010**, *75*, S35–S49. [CrossRef]
23. Gu, B.; Li, Z.; Ma, X.; Liang, G. Multi-component elastic reverse time migration based on the P and S separating elastic velocity-stress equation. *J. Appl. Geophys.* **2015**, *112C*, 62–78. [CrossRef]
24. Wang, W.; McMechan, G.A.; Zhang, Q. Comparison of two algorithms for isotropic elastic P and S decomposition in the vector domain. *Geophysics* **2015**, *80*, T147–T160. [CrossRef]

25. Du, Q.; Gong, X.; Zhang, M.; Zhu, Y.; Fang, G. 3D PS-wave imaging with elastic reverse-time migration. *Geophysics* **2014**, *79*, S173–S184. [CrossRef]
26. Zhang, Q.; McMechan, G.A. 2D and 3D elastic wavefield vector decomposition in the wavenumber domain for VTI media. *Geophysics* **2010**, *75*, D13–D26. [CrossRef]
27. Du, Q.Z.; Guo, C.F.; Zhao, Q.; Gong, X.; Wang, C.; Li, X.Y. Vector-based elastic reverse time migration based on scalar imaging condition. *Geophysics* **2017**, *82*, S111–S127. [CrossRef]
28. Nguyen, B.D.; McMechan, G.A. Five ways to avoid storing source wavefield snapshots in 2D elastic prestack reverse time migration. *Geophysics* **2015**, *80*, S1–S18. [CrossRef]
29. Tian, J.; Zeng, X.; Wang, W.; Shaosheng, Z.; Zeqing, G.; Hua, K. The detection of biogas in unconsolidated sandstone formation of the Qua-ternary in Qaidam Basin. *Geophys. Prospect. Pet.* **2016**, *55*, 408–413.
30. Sun, P.; Guo, Z.-Q.; Zhang, L.; Tian, J.-X.; Zhang, S.-S.; Zeng, X.; Kong, H.; Yang, J. Biologic gas accumulation mechanism and exploration strategy in Sanhu area, Qaidam Basin. *Nat. Gas Geosci.* **2013**, *24*, 494–504.
31. Chen, G.; Deng, Z.; Jiang, T.; Junyong, Z.; Xuejiao, Y.; Chengye, Q.; Xiaoping, X. Application of PP-wave and SS-wave joint interpretation technology in gas cloud area. *Lithol. Reserv.* **2019**, *31*, 79–87.
32. Zhu, T.; Harris, J.M.; Biondi, B. Q-compensated reverse-time migration. *Geophysics* **2014**, *79*, S77–S87. [CrossRef]
33. Helene, H.V.; Martin, L. Simultaneous inversion of PP and PS seismic data. *Geophysics* **2006**, *71*, R1–R10.
34. Ruixue, S.; Ayse, K.; Christopher, J. High resolution seismic reflection PP and PS imaging of the bedrock surface below glacial deposits in Marsta, Sweden. *J. Appl. Geophys.* **2022**, *198*, 1–13.
35. Chu, P.C.; Fan, C.W. A three-point combined compact difference scheme. *J. Comput. Phys.* **1998**, *140*, 370–399. [CrossRef]
36. Dong, L.; Zhan, J. Combined super compact finite difference scheme and application to simulation of shallow water equations. *Chin. J. Comput. Mech.* **2008**, *25*, 791–796.
37. Sengupta, T.K.; Ganeriwal, G.; De, S. Analysis of central and upwind compact schemes. *J. Comput. Phys.* **2003**, *192*, 677–694. [CrossRef]
38. Wu, G.C.; Wang, H.Z. Analysis of numerical dispersion in wave 2 field simulation. *Prog. Geophys.* **2005**, *20*, 58–65.
39. Zhou, C.; Wu, W.; Sun, P.; Yin, W.; Li, X. The Combined Compact Difference Scheme Applied to Shear-Wave Reverse-Time Migration. *Appl. Sci.* **2022**, *12*, 7047. [CrossRef]
40. Brian, R.; Hedlinz, K.; Hilterman, F.J.; Lines, R. Tutorial Fluid-property discrimination with AVO: A Biot-Gassmann perspective. *Geophysics* **2003**, *68*, 29–39.
41. Jin, S. Tying PS to PP depth section: Two examples of anisotropic prestack depth imaging of 4C data. In *SEG Technical Program Expanded Abstracts 2002*; Society of Exploration Geophysicists: Tulsa, OK, USA, 2002.

Article

The Combined Compact Difference Scheme Applied to Shear-Wave Reverse-Time Migration

Chengyao Zhou [1,2], Wei Wu [2], Pengyuan Sun [2], Wenjie Yin [2] and Xiangyang Li [1,*]

[1] State Key Laboratory of Petroleum Resources and Prospecting, China University of Petroleum (Beijing), Beijing 102249, China; fgccc01@163.com

[2] BGP Inc., China National Petroleum Corporation, Beijing 100088, China; wu.wei@cnpc.com.cn (W.W.); sunpengyuan@cnpc.com.cn (P.S.); yinwenjie@cnpc.com.cn (W.Y.)

[*] Correspondence: xy1962li@163.com; Tel.: +86-10-8973-2259

Abstract: In this paper, the combined compact difference scheme (CCD) and the combined super-compact difference scheme (CSCD) are used in the numerical simulation of the shear-wave equation. According to the Taylor series expansion and shear-wave equation, the fourth-order discrete scheme of the displacement field is established; then, the CCD and CSCD schemes are used to calculate the spatial derivative of the displacement field. Additionally, the accuracy, dispersion, and stability of the CCD and CSCD are analyzed, and numerical simulation analyses are carried out using 1D uniform models. Lastly, based on the processing of artificial boundary reflection using PML boundary conditions, shear-wave reverse-time migrations are carried out using synthetic data. The results show that (1) CCD and CSCD have smaller truncation errors, higher simulation precision, and lower numerical dispersion than other normal difference schemes; (2) CCD and CSCD can use the coarse grid and larger time step to calculate, with less memory and high computational efficiency; (3) finally, the result of the shear-wave reverse-time migration of the 2D synthetic data model show that the reverse-time migration imaging is clear, and the proposed method for shear-wave reverse-time migration is practical and effective.

Keywords: shear wave; reverse-time migration; combined supercompact difference scheme; combined compact difference scheme; numerical simulation

Citation: Zhou, C.; Wu, W.; Sun, P.; Yin, W.; Li, X. The Combined Compact Difference Scheme Applied to Shear-Wave Reverse-Time Migration. *Appl. Sci.* **2022**, *12*, 7047. https://doi.org/10.3390/app12147047

Academic Editor: Filippos Vallianatos

Received: 18 May 2022
Accepted: 6 July 2022
Published: 12 July 2022

1. Introduction

At present, reverse-time migration imaging simulation is an important means to explore the morphology of underground media. As the reverse-time migration imaging method is based on the two-way wave equation, which can accurately describe the dynamic and kinematic characteristics of a seismic wavefield propagating underground, reverse-time migration has no inclination angle limit and can adapt to the imaging of complex structural areas, especially for structures with clear lateral velocity changes [1–4]. The finite difference scheme is widely used in the numerical simulation of the elastic-wave equation because of its simplicity and flexibility, high calculation efficiency, and small memory requirement [5–13]. On the one hand, with the development of multicomponent seismic exploration in recent years, particularly shear-wave seismic exploration, in order to minimize computational costs, there is a need to increase the time and spatial steps used in finite difference modeling while maintaining sufficient accuracy during numerical simulation [14]. On the other hand, if the traditional finite difference scheme is used for numerical simulation, small time and spatial steps are required to achieve sufficient accuracy. Compared with the traditional difference scheme, the compact difference scheme has the same accuracy and high stability. The development of compact difference schemes can be traced back to 1989 when Dennis and Hudson (1989) first proposed spatial fourth-order compact schemes for Navier–Stokes equations [15]. Lele (1992) studied the Pade scheme and proposed a symmetric compact difference scheme for solving the first and second derivatives, and the accuracy of the scheme can reach that of the spectral method [16]. Chu et al. (2000) used the combined compact difference scheme (CCD) for the convection–diffusion equation [17].

Since the mid-1990s, reverse-time migration has been applied to multicomponent wave seismic data excited from P-wave sources [18–20], having overcome calculation problems and interference artifacts in P- and S-wave simulations. During reverse-time migration, the P- and S-wave vectors of the source wavefield and the P- and S-wave vectors of the receiver wavefield are obtained and then imaged. In general, P-and S-wave wavefields are obtained either via the Helmholtz decomposition or by using the pled elastic-wave equation. These approaches are all designed for P-wave and elastic-wave sources, focusing mainly on how to retain the phase, amplitude, wavefield and vector characteristics of the wavefield efficiently during wavefield separation [21–28]. In recent years, because of a breakthrough in S-wave vibrator technology, pure S-wave seismic data can be obtained via artificial excitation [29,30]. As the wavelet length of an S-wave is shorter than that of a P-wave for the same frequency bandwidth, the resolution of its wavefield is higher, and its advantage for imaging beneath gas clouds area is unmatched by a P-wave [31]. Using the S-wave wavefield generated by an S-wave source for reverse-time migration imaging (RTM) is a key step for processing shear-wave data. Due to the characteristics of shear-wave wavelength, higher accuracy is required in numerical simulation. To ensure the accuracy of shear-wave RTM, a smaller time and spatial step than P-wave is needed, which reduces the calculation efficiency.

To improve the accuracy and efficiency of shear-wave RTM, based on the characteristics of the shear-wave velocity model, we used the combined compact difference (CCD) and combined supercompact difference scheme (CSCD) for shear-wave (SH) RTM, with larger spatial grid conditions. Wang Shuqiang (2002) applied the compact difference scheme to the numerical simulation of seismic wavefields [32]. Most recent studies of the finite difference scheme focus on the compressional wavefield from an explosion source, but few have studied the finite difference scheme based on the shear wavefield of the shear-wave source (SH-wave).

This paper aims at the problem of low shear-wave velocity by introducing the supercompact difference scheme to suppress the numerical dispersion caused by the large spatial step. We introduce the basic concept of shear-wave reverse-time migration, followed by the implementation methods of the combined compact difference scheme (CCD) and combined supercompact difference scheme (CSCD); the numerical simulation accuracy of CCD and CSCD is also discussed, and the accuracy of CCD and CSCD is compared with the traditional finite difference scheme. Finally, the method is applied to synthetic data to verify the accuracy and efficiency of the algorithm. Furthermore, extending the present research to take into the viscoelastic behavior of media will have a great potential for imaging oil and gas reservoirs [33].

2. Principle of RTM and Combined Compact Difference Scheme

2.1. Principle of RTM

The technique of reverse-time migration imaging (RTM) is composed of the following three steps [34]:

(1) The source wavefield is obtained by using the source constructed manually or extracted from actual data, and the corresponding model is numerically simulated to obtain the source wavefield $S(x,z,t)$, where x,z is the space vector.

(2) Using the seismic data obtained at the receiver, the reverse continuation propagation passes through the same velocity model, and the corresponding receiver wavefield $R(x,z,t)$ is obtained, where the position of the receiver is x_r, z_r.

(3) Applying appropriate imaging conditions, such as cross-correlation, we obtain

$$I(x,z) = \int_0^T S(x,z,t)R(x,z,t)dt \tag{1}$$

where $S(x,z,t)$ is the source wavefield obtained via forward modeling, $R(x,z,t)$ is the receiver wavefield obtained at the same time via reverse continuation simulation under the same velocity model, and t is the total propagation time.

It can be seen from Equation (1) that the final result of reverse-time migration is affected by the accuracy of the source wavefield $S(x,z,t)$ and the receiver wavefield $R(x,z,t)$. Notably, the method to improve the accuracy of the source wavefield $S(x,z,t)$ can also be applied to the receiver wavefield $R(x,z,t)$. Thus, a high-precision finite difference scheme is applied to generate the source and receiver wavefields $S(x,z,t)$ and $R(x,z,t)$, which will give an ideal result of reverse-time migration. In this paper, we adopted cross-correlation imaging conditions; in addition, the method in this paper can also be applied to some imaging conditions developed by researchers in recent years [35–37].

2.2. Principle of Combined Compact Difference Scheme

For the construction of the compact difference scheme, Lele extended Hermite's equation in 1992 and created the compact difference scheme as follows:

$$\begin{aligned} cf'_{i-1} + f'_i + cf'_{i+1} &= a_2 \tfrac{f_{i+2}-f_{i-2}}{4h} + a_1 \tfrac{f_{i+1}-f_{i-1}}{2h} \\ cf''_{i-1} + f''_i + cf''_{i+1} &= a_2 \tfrac{f_{i+2}-2f_i+f_{i-2}}{4h^2} + a_1 \tfrac{f_{i+1}-2f_i+f_{i-1}}{h^2} \end{aligned} \tag{2}$$

In Equation (2), h is the grid spacing, a, c are the difference coefficient matrices. f is the function value of node i. f'_i and f''_i represent the first- and second-order derivatives of node i, respectively; $f_{i+1}, f_{i+2}, f_{i-1}, f_{i-2}$ represent the function values of node i successively two nodes forward and two nodes backward; f'_{i+1}, f'_{i-1} represent the first-order derivatives of node i successively one node forward and one node backward, respectively; f''_{i+1}, f''_{i-1} represent the second derivative of i node one node forward and one node backward, respectively.

Following [38], the wave equation of a shear wave in a two-dimensional medium is

$$\frac{\partial^2 V_y}{\partial t^2} = (1+2*\gamma)vs_0{}^2\frac{\partial^2 V_y}{\partial x^2} + vs_0{}^2\frac{\partial^2 V_y}{\partial z^2} + \rho F_y \tag{3}$$

where $V_y(x,z)$ is the displacement component of the shear wave. $vs_0(x,z)$ is the shear-wave velocity in the vertical direction, and ρF_y is the shear-wave source, that is, the concentrated force source excited in the y direction on the surface. If the input medium is isotropic, the value of the anisotropic parameter $\gamma = 0$.

Expanding the above shear-wave equation to represent the time of $n+1$ and $n-1$, we obtain

$$V_{y(i,j)}^{n+1} = V_{y(i,j)}^{n} + \Delta t\left(\frac{\partial V_y}{\partial t}\right)_{i,j}^{n} + \frac{(\Delta t)^2}{2}\left(\frac{\partial^2 V_y}{\partial t^2}\right)_{i,j}^{n} + \frac{(\Delta t)^3}{6}\left(\frac{\partial^3 V_y}{\partial t^3}\right)_{i,j}^{n} + \frac{(\Delta t)^4}{24}\left(\frac{\partial^4 V_y}{\partial t^4}\right)_{i,j}^{n} + O\left(\Delta t^5\right) \tag{4}$$

$$V_{y(i,j)}^{n-1} = V_{y(i,j)}^{n} - \Delta t\left(\frac{\partial V_y}{\partial t}\right)_{i,j}^{n} + \frac{(\Delta t)^2}{2}\left(\frac{\partial^2 V_y}{\partial t^2}\right)_{i,j}^{n} - \frac{(\Delta t)^3}{6}\left(\frac{\partial^3 V_y}{\partial t^3}\right)_{i,j}^{n} + \frac{(\Delta t)^4}{24}\left(\frac{\partial^4 V_y}{\partial t^4}\right)_{i,j}^{n} + O\left(\Delta t^5\right) \tag{5}$$

By adding these two equations, omitting the higher-order term, and substituting it into the shear-wave equation, the fourth-order accuracy difference scheme of displacement field time can be obtained as follows:

$$\begin{aligned} V_{y(i,j)}^{n+1} &= 2V_{y(i,j)}^{n} - V_{y(i,j)}^{n-1} + (1+2\gamma)(\Delta t v)^2\left(\tfrac{\partial^2 V_y}{\partial x^2}\right)_{i,j}^{n} + \left(\tfrac{\partial^2 V_y}{\partial z^2}\right)_{i,j}^{n} \\ &+ (1+2\gamma)^2\tfrac{(\Delta t v)^4}{12}\left(\tfrac{\partial^4 V_y}{\partial x^4}\right)_{i,j}^{n} + (1+2\gamma)\tfrac{(\Delta t v)^4}{6}\left(\tfrac{\partial^4 V_y}{\partial x^2\partial z^2}\right)_{i,j}^{n} + \tfrac{(\Delta t v)^4}{12}\left(\tfrac{\partial^4 V_y}{\partial z^4}\right)_{i,j}^{n} \end{aligned} \tag{6}$$

Equation (6) can be used to advance the time step of the shear wavefield in a two-dimensional medium. The equation contains the second and fourth derivatives of the sum of displacement pairs in the *horizontal and vertical* directions, and the sum of displacement pairs in the *horizontal and vertical* directions can be differentiated by the finite difference scheme.

Chu (1998) and others constructed a combined compact difference (CCD) scheme with higher accuracy as follows:

$$\begin{cases} a_1\left(f'_{i+1}+f'_{i-1}\right)+f'_i+hb_1\left(f''_{i+1}-f''_{i-1}\right)=\frac{1}{h}\sum\limits_{m=1}^{n}c_m\left(f_{i+m}-f_{i-m}\right) \\ \frac{1}{h}a_2\left(f'_{i+1}-f'_{i-1}\right)+f''_i+b_2\left(f''_{i+1}+f''_{i-1}\right)=\frac{1}{h^2}\sum\limits_{m=1}^{n}d_m\left(f_{i+m}-2f_i+f_{i-m}\right) \end{cases} \tag{7}$$

Compared with CCD, the supercompact difference scheme needs fewer adjacent nodes to obtain the high-order accuracy approximations of the first and second derivatives at the same time step. The first and second derivatives in the above equation are coupled, which can be obtained at the same time, increasing the fidelity of the waveform.

For the CCD scheme, it is assumed that the number of longitudinal and transverse nodes of the model is m and n, and h is the size of the spatial grid. We use Equation (7) to find the spatial partial derivatives $\partial^2 V_y/\partial z^2$ and $\partial^2 V_y/\partial x^2$ in Equation (3) and then express the results as

$$AE = BV_y,\ FC = V_y D \tag{8}$$

where A and C are the difference coefficient matrices at the left end of the CCD Equation (7), and the sizes are $2m \times 2m$ and $2n \times 2n$, respectively. E and F are the first- and second-order derivative matrices of the displacement field space to be solved, and the sizes are $2m \times n$ and $m \times 2n$, respectively. B and D are the difference coefficient matrices at the right end of Equation (7), and the sizes are $2m \times m$ and $n \times 2n$, respectively. V_y is the displacement field matrix, and the size is $m \times n$. These matrices are expressed as

$$A=\begin{bmatrix} 1 & 0 & a_1 & b_1h & & & & & & \\ 0 & 1 & a_2/h & b_2 & & & & & & \\ a_1 & -b_1h & 1 & 0 & a_1 & b_1h & & & & \\ -a_2/h & b_2 & 0 & 1 & a_2/h & b_2 & & & & \\ & & & & & & \ddots & \ddots & \ddots & \ddots \\ & & & & & & a_1 & -b_1h & 1 & 0 \\ & & & & & & -a_2/h & b_2 & 0 & 1 \end{bmatrix} \tag{9}$$

$$B=\begin{bmatrix} 0 & \sum\limits_{m=1}^{n}c_m/h & & & \\ (-2\sum\limits_{m=1}^{n}d_m)/h^2 & (\sum\limits_{m=1}^{n}d_m)/h^2 & & & \\ -\sum\limits_{m=1}^{n}c_m/h & 0 & \sum\limits_{m=1}^{n}c_m/h & & \\ (\sum\limits_{m=1}^{n}d_m)/h^2 & (-2\sum\limits_{m=1}^{n}d_m)/h^2 & (\sum\limits_{m=1}^{n}d_m)/h^2 & & \\ & \ddots & \ddots & \ddots & \\ & & & -\sum\limits_{m=1}^{n}c_m/h & 0 \\ & & & (\sum\limits_{m=1}^{n}d_m)/h^2 & (-2\sum\limits_{m=1}^{n}d_m)/h^2 \end{bmatrix} \tag{10}$$

$$C = \begin{bmatrix} 1 & 0 & a_1 & -a_2/h & & & \\ 0 & 1 & -b_1 h & b_2 & & & \\ a_1 & a_2/h & 1 & 0 & & & \\ b_1 h & b_2 & 0 & 1 & & & \\ & & a_1 & a_2/h & & & \\ & & b_1 h & b_2 & & & \\ & & & & \ddots & a_1 & -a_2/h \\ & & & & \ddots & -b_1 h & b_2 \\ & & & & \ddots & 1 & 0 \\ & & & & \ddots & 0 & 1 \end{bmatrix} \tag{11}$$

$$D = \begin{bmatrix} 0 & (-2\sum\limits_{m=1}^{n} d_m)/h^2 & -\sum\limits_{m=1}^{n} c_m/h & (\sum\limits_{m=1}^{n} d_m)/h^2 & & & \\ \sum\limits_{m=1}^{n} c_m/h & (\sum\limits_{m=1}^{n} d_m)/h^2 & 0 & (-2\sum\limits_{m=1}^{n} d_m)/h^2 & & & \\ & & \sum\limits_{m=1}^{n} c_m/h & (\sum\limits_{m=1}^{n} d_m)/h^2 & & & \\ & & & & \ddots & -\sum\limits_{m=1}^{n} c_m/h & (\sum\limits_{m=1}^{n} d_m)/h^2 \\ & & & & \ddots & 0 & (-2\sum\limits_{m=1}^{n} d_m)/h^2 \end{bmatrix} \tag{12}$$

$$E = \begin{bmatrix} \left(\frac{\partial V_y}{\partial z}\right)_{1,1} & \left(\frac{\partial V_y}{\partial z}\right)_{1,2} & \cdots & \left(\frac{\partial V_y}{\partial z}\right)_{1,n-1} & \left(\frac{\partial V_y}{\partial z}\right)_{1,n} \\ \left(\frac{\partial^2 V_y}{\partial z^2}\right)_{1,1} & \left(\frac{\partial^2 V_y}{\partial z^2}\right)_{1,2} & \cdots & \left(\frac{\partial^2 V_y}{\partial z^2}\right)_{1,n-1} & \left(\frac{\partial^2 V_y}{\partial z^2}\right)_{1,n} \\ \vdots & \vdots & \vdots & \vdots & \vdots \\ \left(\frac{\partial V_y}{\partial z}\right)_{m,1} & \left(\frac{\partial V_y}{\partial z}\right)_{m,2} & \cdots & \left(\frac{\partial V_y}{\partial z}\right)_{m,n-1} & \left(\frac{\partial V_y}{\partial z}\right)_{m,n} \\ \left(\frac{\partial^2 V_y}{\partial z^2}\right)_{m,1} & \left(\frac{\partial^2 V_y}{\partial z^2}\right)_{m,2} & \cdots & \left(\frac{\partial^2 V_y}{\partial z^2}\right)_{m,n-1} & \left(\frac{\partial^2 V_y}{\partial z^2}\right)_{m,n} \end{bmatrix} \tag{13}$$

$$F = \begin{bmatrix} \left(\frac{\partial V_y}{\partial x}\right)_{1,1} & \left(\frac{\partial^2 V_y}{\partial x^2}\right)_{1,1} & \cdots & \left(\frac{\partial V_y}{\partial x}\right)_{1,n} & \left(\frac{\partial^2 V_y}{\partial x^2}\right)_{1,n} \\ \left(\frac{\partial V_y}{\partial x}\right)_{2,1} & \left(\frac{\partial^2 V_y}{\partial x^2}\right)_{2,1} & \cdots & \left(\frac{\partial V_y}{\partial x}\right)_{2,n} & \left(\frac{\partial^2 V_y}{\partial x^2}\right)_{2,n} \\ \vdots & \vdots & \vdots & \vdots & \vdots \\ \left(\frac{\partial V_y}{\partial x}\right)_{m-1,1} & \left(\frac{\partial^2 V_y}{\partial x^2}\right)_{m-1,1} & \cdots & \left(\frac{\partial V_y}{\partial x}\right)_{m-1,n} & \left(\frac{\partial^2 V_y}{\partial x^2}\right)_{m-1,n} \\ \left(\frac{\partial V_y}{\partial x}\right)_{m,1} & \left(\frac{\partial^2 V_y}{\partial x^2}\right)_{m,1} & \cdots & \left(\frac{\partial V_y}{\partial x}\right)_{m,n} & \left(\frac{\partial^2 V_y}{\partial x^2}\right)_{m,n} \end{bmatrix} \tag{14}$$

$$V_y = \begin{bmatrix} V_{y(1,1)} & V_{y(1,2)} & \cdots & V_{y(1,n-1)} & V_{y(1,n)} \\ V_{y(2,1)} & V_{y(2,2)} & \cdots & V_{y(2,n-1)} & V_{y(2,n)} \\ \vdots & \vdots & \vdots & \vdots & \vdots \\ V_{y(m-1,1)} & V_{y(m-1,2)} & \cdots & V_{y(m-1,n-1)} & V_{y(m-1,n)} \\ V_{y(m,1)} & V_{y(m,2)} & \cdots & V_{y(m,n-1)} & V_{y(m,n)} \end{bmatrix} \tag{15}$$

From Equation (8), $E = A^{-1}BV_y$, the odd number behavior is $\partial V_y/\partial z$, and the even number behavior is $\partial^2 V_y/\partial z^2$. Similarly, the sum can also be obtained from $\partial V_y/\partial x$ and $\partial^2 V_y/\partial x^2$ by transposing the displacement field.

At present, this method has been applied to acoustic forward modeling [39]. However, for shear-wave reverse-time migration, it is necessary to adapt to the situation of low-transverse wave velocity and take into account the accuracy and efficiency of calculations.

Based on CCD, we introduce a combined supercompact difference scheme (CSCD), and its equation configuration is as follows [40]:

$$\begin{cases} f_i' + a_{11}\left(f_{i+1}' + f_{i-1}'\right) + a_{12}h\left(f_{i+1}'' - f_{i-1}''\right) + a_{13}h^2\left(f_{i+1}''' + f_{i-1}'''\right) + a_{14}h^3\left(f_{i+1}^{(4)} - f_{i-1}^{(4)}\right) \\ = \frac{a_1}{h}\left(f_{i+1} - f_{i-1}\right) + \frac{a_2}{h}\left(f_{i+2} - f_{i-2}\right) \ldots + \frac{a_n}{h}\left(f_{i+n} - f_{i-n}\right) \\ f_i'' + a_{21}h^{-1}\left(f_{i+1}' - f_{i-1}'\right) + a_{22}\left(f_{i+1}'' + f_{i-1}''\right) + a_{23}h\left(f_{i+1}''' - f_{i-1}'''\right) + a_{24}h^2\left(f_{i+1}^{(4)} + f_{i-1}^{(4)}\right) \\ = \frac{b_1}{h^2}\left(f_{i+1} + f_{i-1} - 2f_i\right) + \frac{b_2}{h^2}\left(f_{i+2} + f_{i-2} - 2f_i\right) \ldots + \frac{b_n}{h^2}\left(f_{i+n} + f_{i-n} - 2f_i\right) \\ f_i''' + a_{31}h^{-2}\left(f_{i+1}' + f_{i-1}'\right) + a_{32}h^{-1}\left(f_{i+1}'' - f_{i-1}''\right) + a_{33}\left(f_{i+1}''' + f_{i-1}'''\right) + a_{34}h\left(f_{i+1}^{(4)} - f_{i-1}^{(4)}\right) \\ = \frac{c_1}{h^3}\left(f_{i+1} - f_{i-1}\right) + \frac{c_2}{h^3}\left(f_{i+2} - f_{i-2}\right) \ldots + \frac{c_n}{h^3}\left(f_{i+n} - f_{i-n}\right) \\ f_{i+1}^{(4)} + a_{41}h^{-3}\left(f_{i+1}' - f_{i-1}'\right) + a_{42}h^{-2}\left(f_{i+1}'' + f_{i-1}''\right) + a_{43}h^{-1}\left(f_{i+1}''' - f_{i-1}'''\right) + a_{44}\left(f_{i+1}^{(4)} + f_{i-1}^{(4)}\right) \\ = \frac{d_1}{h^4}\left(f_{i+1} + f_{i-1} - 2f_i\right) + \frac{d_2}{h^4}\left(f_{i+2} + f_{i-2} - 2f_i\right) \ldots + \frac{d_n}{h^4}\left(f_{i+n} + f_{i-n} - 2f_i\right) \end{cases} \tag{16}$$

In this paper, we focused on the three-point format of Equation (16).

Using the CSCD scheme to calculate the spatial partial derivative of the equation is similar to that of CCD, except for the difference coefficient matrix. The difference coefficient matrix at the left end needs to be expanded, and its sizes are $4m \times 4m$ and $4n \times 4n$, respectively. E and F are changed to the first-, second-, third-, and fourth-order derivative matrices of the displacement field space to be solved, with the sizes of $4m \times n$ and $m \times 4n$, respectively. B and D are the difference coefficient matrices at the right end of the equation, and the size is changed to $4m \times m$ and $n \times 4n$.

3. Analysis of CCD and CSCD

3.1. Analysis of Truncation Error

In this section, the truncation errors of the second derivative calculated by these schemes are compared. The results are shown in Table 1. α, β, a, b, c are the difference coefficient matrices. It can be seen from Table 1 that, although the three methods can achieve a certain order of spatial accuracy, the truncation error is relatively different. The truncation error of the traditional difference scheme (central difference scheme) in calculating the second-order partial derivative is greater than that of the CCD and CSCD.

Table 1. Truncation errors in various difference schemes for the second-order derivative calculations.

		α	β	a	b	c	**Truncation Error**
$\frac{\partial^2 u}{\partial x^2}$	CFD	0	0	3/2	−3/5	1/10	$\approx 1.78 \times 10^{-3}(O(h^6))$
	CD	2/11	0	12/11	3/11	0	$\approx -4.2163 \times 10^{-4}(O(h^6))$
	CCD			/			$-4.9603 \times 10^{-5}(O(h^6))$
	CSCD			/			$\approx 4 \times 10^{-8}(O(h^8))$

3.2. Dispersion Analysis

In numerical simulations, if the size of the spatial grid is too large, it will cause large solution errors and produce numerical dispersion [41,42]. Therefore, the analysis of the dispersion relationship is not simply an important method to evaluate the advantages and disadvantages of numerical simulation methods, but it is also an important way to determine the size of the spatial grid [43].

If we use a simple harmonic solution $f_i^n = Aexp[I(ihk - kv'n\Delta t)]$, we can obtain the following equations for CCD:

$$\begin{cases} I[(a_1 + a_1)\cos\beta + 1]\beta' + [(b_1 + b_1)\cos\beta](-1)(\beta'')^2 \\ = \sum\limits_{m=1}^{n} [(c_m + c_m)\cos\beta] \\ 2a_2(I\sin\beta)\beta' + (1 - 2b_2\cos\beta)(-1)(\beta'')^2 = \sum\limits_{m=1}^{n} ((d_m + d_m)\cos\beta - 1) \end{cases} \tag{17}$$

and the following equations for CSCD:

$$\begin{cases} (1 + \frac{187}{128}\cos(\beta))\beta' + \frac{47}{128}\sin(\beta)\beta'' - \frac{6}{128}\cos(\beta)\beta''' - \frac{2}{768}\sin(\beta)\beta^{(4)} = \frac{315}{128}\sin(\beta) \\ -\frac{325}{64}\sin(\beta)\beta' + (\frac{69}{64}\cos(\beta) + 1)\beta'' + \frac{11}{96}\sin(\beta)\beta''' - \frac{1}{192}\cos(\beta)\beta^{(4)} = 10(\cos(\beta) - 1) \\ \frac{315}{16}\cos(\beta)\beta' - \frac{105}{16}\sin(\beta)\beta'' + (1 + \cos(\beta))\beta''' + \frac{1}{16}\sin(\beta)\beta^{(4)} = -\frac{315}{16}\sin(\beta) \\ \frac{645}{4}\sin(\beta)\beta' - \frac{165}{4}\cos(\beta)\beta'' - 5\sin(\beta)\beta''' + \frac{1}{4}(\cos(\beta) + 1)\beta^{(4)} = -240(\cos(\beta) - 1) \end{cases} \tag{18}$$

where $\beta = hk, \beta' = hk', \beta'' = hk'', \beta''' = hk^{(3)}, \beta^{(4)} = hk^{(4)}$ Solving the equations of CCD gives,

$$\beta''_{CCD} = \sqrt{\frac{81 - 48\cos\beta - 33\cos 2\beta}{48 + 40\cos\beta + 2\cos 2\beta}} \tag{19}$$

In this equation, $\beta = kh$, k is the wavenumber, h is the spatial gird size, and β''_{CCD} is the second-order modified wavenumber of the CCD scheme.

If this is substituted into the two-dimensional shear-wave equation, Equation (20) becomes

$$\beta''_{CCD(x)} = \sqrt{\frac{81 - 48\cos\beta_x - 33\cos 2\beta_x}{48 + 40\cos\beta_x + 2\cos 2\beta_x}}\ \beta''_{CCD(z)} = \sqrt{\frac{81 - 48\cos\beta_z - 33\cos 2\beta_z}{48 + 40\cos\beta_z + 2\cos 2\beta_z}} \tag{20}$$

where $\beta_x = \beta cos\theta, \beta_z = \beta sin\theta$, and θ is the angle between the wave's propagation direction and the x-axis. The modified wavenumber of the corresponding derivatives of CSCD can be solved by substituting the numerical solution into the Equation Group (18) obtained by the simple harmonic solution without writing the corresponding analytical expression.

The modified wavenumber solution of the finite difference scheme can then be substituted into the finite difference scheme of the shear-wave equation, yielding.

$$\begin{aligned} 2\cos(kv_{num}\Delta t) &= 2 - (v\Delta t/h)^2\left[(1+2\gamma)(\beta''_x)^2 + (\beta''_z)^2\right] \\ &+(v\Delta t/h)^4\left[(1+2\gamma)^2(\beta''_x)^4 + 2(1+2\gamma)(\beta''_x\beta''_z)^2 + (\beta''_z)^4\right]/12 \end{aligned} \tag{21}$$

where v_{num} is the numerical wave velocity, v is the true wave velocity, Δt is the time step, $v\Delta t/h$ is the Courant number, and $I = \sqrt{-1}$. Equation (21) shows that the dispersion relationship for the shear-wave equation is related to the value of spatial step, propagation speed, and time step. Based on the above dispersion relation, the corresponding $kv_{num}\Delta t$ can be solved after β is determined. The ratio of numerical wave velocity to true velocity is defined as

$$\lambda = \frac{v_{num}}{v} = \frac{kv_{num}\Delta t}{kv\Delta t} = \frac{kv_{num}\Delta t}{k\frac{v\Delta t}{h}h} = \frac{kv_{num}\Delta t}{\frac{v\Delta t}{h}\beta} \tag{22}$$

Ideally, if there is no numerical dispersion, then the velocity ratio λ is equal to one. The larger the velocity ratio, the more serious the numerical dispersion. For comparison, we calculate the dispersion relation between velocity ratio and kh for the central difference scheme (CFD), compact difference scheme (CD), CCD, and CSCD difference scheme. Figure 1 shows the velocity ratio curve for different θ for the above four methods in the isotropic ($\gamma = 0$) condition.

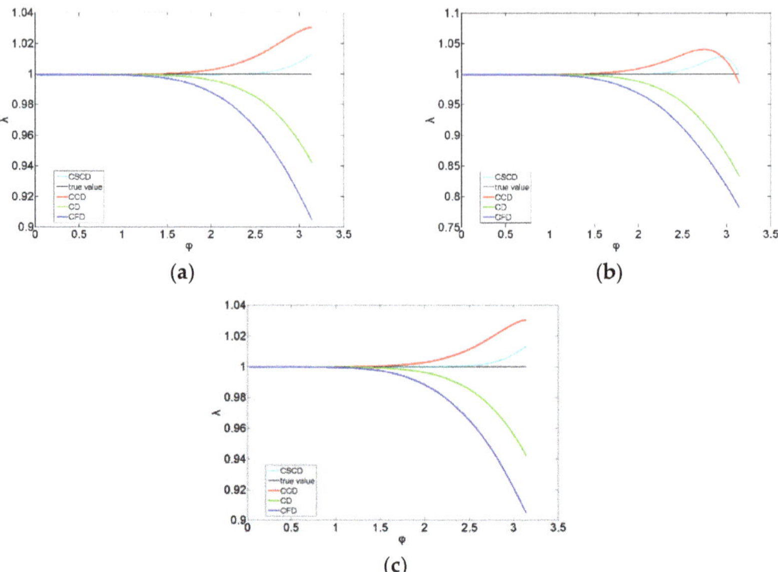

Figure 1. Velocity ratio curves for different numerical simulation methods. The blue curve is the traditional central difference scheme, the green curve is the traditional implicit difference scheme, the red curve is the CCD difference scheme, the blue curve is the CSCD difference scheme, and the black line is the velocity ratio constant of 1: (**a**) $\theta = 0, \alpha = 0.25$; (**b**) $\theta = \pi/6, \alpha = 0.25$; (**c**) $\theta = \pi/3, \alpha = 0.25$.

Figure 1a–c show the velocity ratio curves for the CCD and CSCD schemes and the other two difference schemes using different θ. The Courant numbers ($\alpha = v\Delta t/h$) are 0.25, the horizontal axis $\varphi \in [0, \pi]$ is the product of the wavenumber and the spatial step, and the vertical axis is the velocity ratio. A velocity ratio of one indicates that the numerical wave velocity is the same as the theoretical wave velocity, which shows that the method can suppress numerical dispersion better; otherwise, it indicates that the numerical dispersion of the method is worse. It can be seen that the dispersion phenomenon of the four methods is gradually intensified with the decrease in the number of spatial sampling points. The numerical dispersion of CSCD, CCD, and CD schemes is smaller than that of the CFD scheme, as their dispersion curves are closer to one. CSCD shows the best dispersion suppression, followed by CCD.

3.3. The Numerical Simulation Accuracy Analysis

To compare the numerical simulation accuracy of the models, we calculate the simulation error by simulating the two-dimensional plane harmonic initial value problem and then compare the numerical simulation accuracy of the CCD, CSCD, and CFD schemes. The initial value problem of the two-dimensional plane wave can be expressed as

$$\begin{cases} \frac{\partial^2 p_{sh}(t,x,z)}{\partial t^2} = v^2 \frac{\partial^2 p_{sh}(t,x,z)}{\partial x^2} + v^2 \frac{\partial^2 p_{sh}(t,x,z)}{\partial z^2} \\ p_{sh}(0,x,z) = \cos\left(-\frac{2\pi f_0}{v} \cdot \cos\theta \cdot x - \frac{2\pi f_0}{v} \cdot \sin\theta \cdot z\right) \\ \frac{\partial p_{sh}(0,x,z)}{\partial t} = -2\pi f_0 \sin\left(-\frac{2\pi f_0}{v} \cdot \cos\theta \cdot x - \frac{2\pi f_0}{v} \cdot \sin\theta \cdot z\right) \end{cases} \tag{23}$$

where v is the plane wave velocity, θ is the angle between the propagation direction and the x-axis, and f_0 is the frequency of the simple harmonic plane wave.

The analytical solution for the above initial value problem is

$$p_{sh}(t,x,z) = \cos\left[2\pi f_0\left(t - \frac{x}{v/\cos\theta} - \frac{z}{v/\sin\theta}\right)\right]. \tag{24}$$

To simulate the two-dimensional shear wavefield, we specify the following: $f_0 = 20$ Hz and $\theta = \pi/4$, the wave velocity used is 1000 m/s, the length and depth of the model are 2000 m, the length of the vertical and horizontal grids are the same, and the sampling time is 1 s. The relative error of numerical simulation is calculated under different spatial mesh lengths and time steps, and it is defined as

$$E_r(\%) = \left\{ \frac{1}{\sum\limits_{i=1}^{N}\sum\limits_{j=1}^{N}\left[p_{sh}(t_n, x_i, z_j)\right]^2} \sum_{i=1}^{N}\sum_{j=1}^{N}\left[p_{sh}{}_{(i,j)}^{n} - p_{sh}(t_n, x_i, z_j)\right]^2 \right\}^{\frac{1}{2}} \times 100 \tag{25}$$

where $p_{sh}{}_{(i,j)}^{n}$ is the numerical solution, and $p_{sh}(t_n, x_i, z_j)$ is an analytical solution. We then compare the relative error curves of the CCD, CSCD, CD, and CFD schemes under different spatial steps (10 m, 15 m) and fixed time steps of 1 ms (Δt = 0.001 s), as shown in Figure 2. It can be seen from Figure 2 that the relative error gradually increases with increasing spatial grid length, time step, and simulation time. When the spatial step is 15 m and the time step is 1 ms, the maximum relative error of the CCD scheme is 8%, which is much smaller when compared with the maximum relative error of the CFD scheme (39%). When the smaller spatial gird size (10 m) is adopted, the accuracy is significantly improved, and the relative error is only 0.8%. The shear-wave simulation result using the CCD scheme has high accuracy and can handle the numerical simulation of the seismic wavefield with long sampling times. The simulation accuracy of the CSCD scheme is even higher than the CCD scheme. For CSCD, when the spatial gird size is 15 m, and the time step size is 1 ms, the maximum relative error is only 0.05%, further reducing to 0.0036% when a small spatial step (10 m) is used.

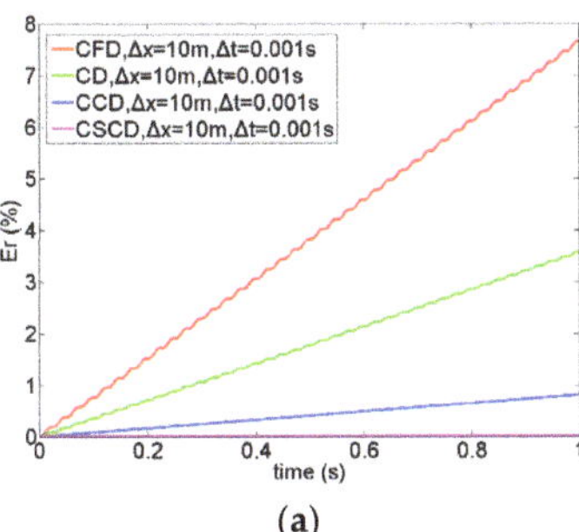

(a)

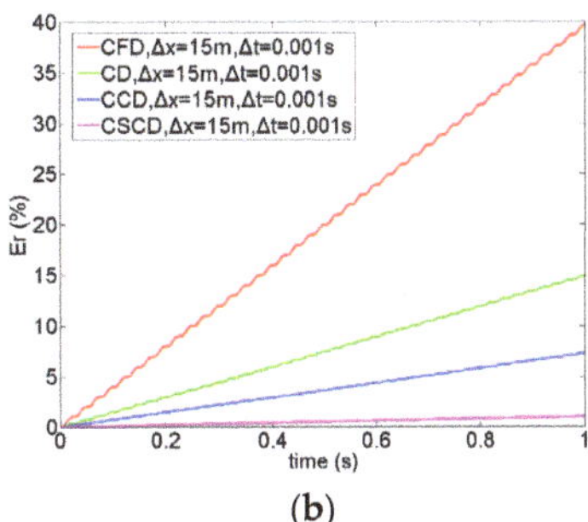

(b)

Figure 2. Relative errors of numerical simulation for different schemes and gird size: (**a**) Δx = 10 m, Δt = 0.001 s; (**b**) Δx = 15 m, Δt = 0.001 s.

3.4. Comparison of Spatial Dispersion Suppression Effect

As the shear wave has the characteristics of low propagation speed, a one-dimensional low-speed homogeneous medium is constructed below. The velocity of the homogeneous medium model is 1000 m/s, the excitation position is located in the center of the model, and the source is a 10 Hz Rayleigh wavelet. Different space steps are set with a time step of 1 ms, the seismic records of 6 s are recorded, and the results are shown in Figures 3–5. As shown in Figure 3, the traditional CFD scheme has numerical dispersion when the gird size is 9 m, and the dispersion increased substantially when the step increases to 10 m (Figure 5). In contrast, as shown in Figure 4, the corresponding seismic record calculated with the

CCD scheme shows no numerical dispersion, and there is still no numerical dispersion when the gird size increases by 10 m (Figure 5).

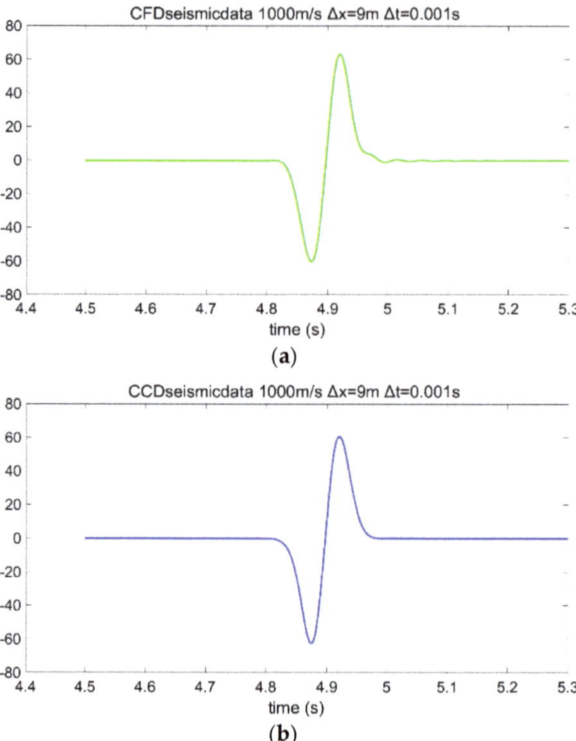

Figure 3. (**a**) CFD is used to simulate the numerical simulation of seismic records in the 1D homogeneous medium model, $\Delta x = 9$ m; (**b**) CCD is used to simulate the numerical simulation of seismic records in the 1D homogeneous medium model, $\Delta x = 9$ m.

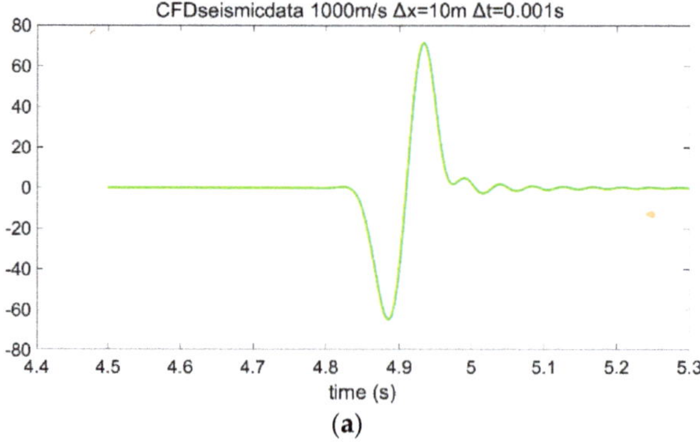

Figure 4. *Cont.*

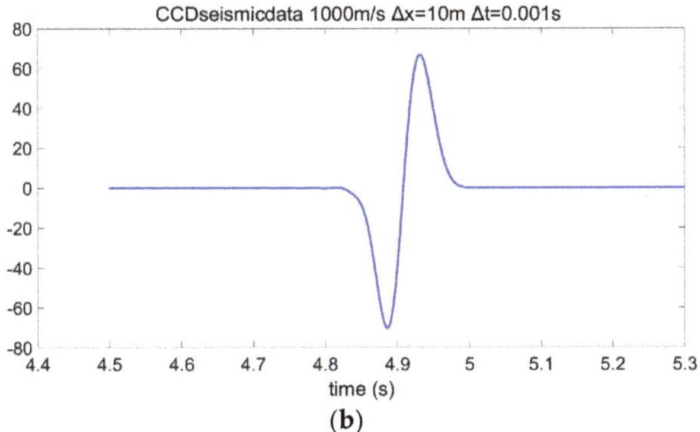

Figure 4. (**a**) CFD is used to simulate the numerical simulation of seismic records in the 1D homogeneous medium model, Δx = 10 m; (**b**) CCD is used to simulate the numerical simulation of seismic records in the 1D homogeneous medium model, Δx = 10 m.

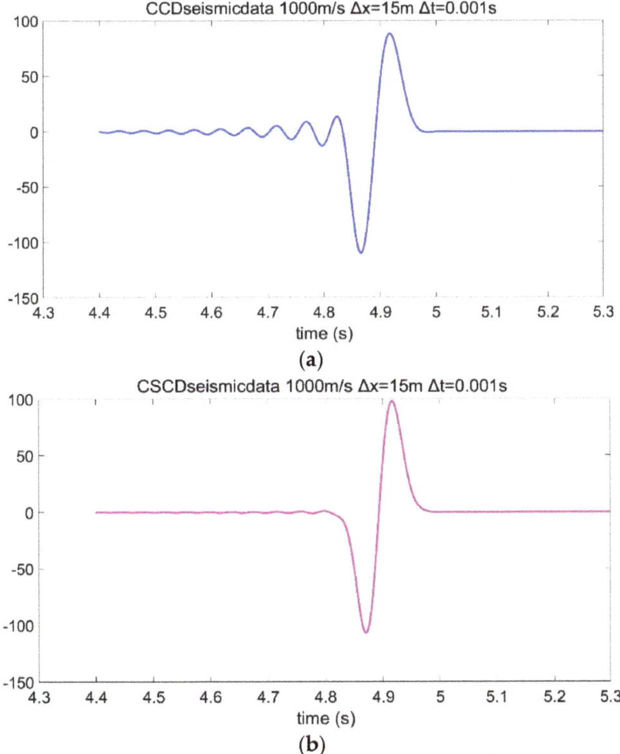

Figure 5. (**a**) CCD is used to simulate the numerical simulation of seismic records in the 1D homogeneous medium model, Δx = 15 m; (**b**) CSCD is used to simulate the numerical simulation of seismic records in the 1D homogeneous medium model, Δx = 15 m.

Figure 5a,b compare the numerical dispersion between the CCD and CSCD schemes when the step increases to 15 m. As shown in Figure 5a, at the speed of 1000 m/s, when the gird size continues to increase to 15 m, the CCD scheme shows dispersion, while the CSCD scheme shows no numerical dispersion at the same step. This confirms that the

CSCD scheme improves the compactness of the CCD scheme and can better suppress the numerical dispersion so that a coarser grid may be used in the numerical simulation.

3.5. Time Dispersion Suppression Comparison

Seismic waves propagate in time and space. The numerical dispersion caused by grid discretization includes both spatial dispersion and time dispersion. The accuracy of time extrapolation of the traditional high-order difference (2 m, 2) scheme is only second order. When a large time step is adopted, there will be obvious time dispersion.

We use the same model to evaluate time dispersion as used for spatial dispersion. To evaluate time dispersion, it is necessary to eliminate the influence of numerical dispersion caused by the spatial gird size. Therefore, the spatial grid size is set to 5 m, and we increase the time step to 4 ms. The calculated seismic records of CCD and CFD are shown in Figure 6a,b.

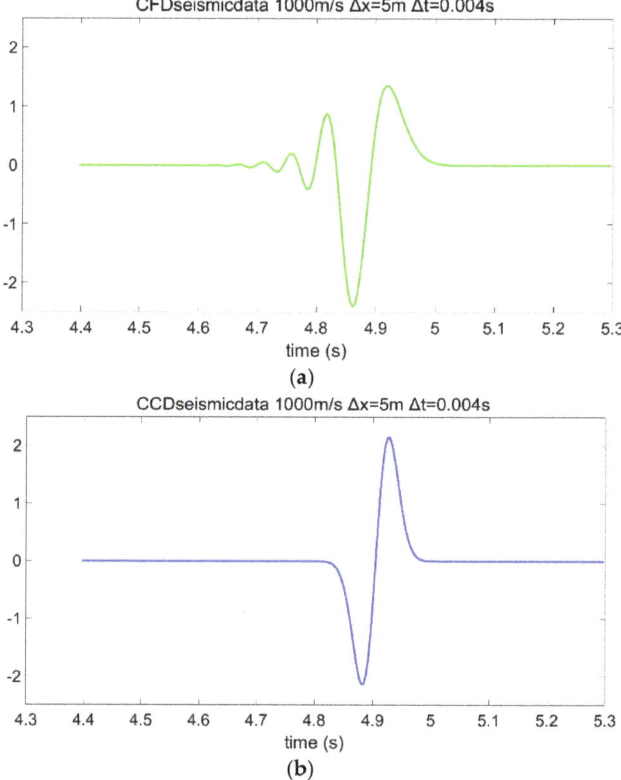

Figure 6. (**a**) CFD is used to simulate the numerical simulation of seismic records in the 1D homogeneous medium model, $\Delta t = 0.004$ s; (**b**) CCD is used to simulate the numerical simulation of seismic records in the 1D homogeneous medium model, $\Delta t = 0.004$ s.

The seismic record simulated using the CFD scheme shows serious and obvious dispersion, while the result of the CCD scheme shows almost no such dispersion. As the CCD scheme proposed in this paper uses the fourth-order difference operator to approximate the time partial derivative, it has better stability than the CFD scheme, which can only use the second-order difference operator to approximate the time partial derivative. The CSCD scheme, by contrast, uses the same fourth-order difference operator to approximate the time partial derivative, which shows even higher accuracy than the CCD scheme.

Figure 7a,b compare the seismic records using the fourth- and second-order difference operators to approximate the time partial derivative from the CCD scheme. It can be seen that when using the same finite difference scheme to suppress the numerical dispersion caused by the spatial gird size, the fourth-order operators show smaller numerical dispersion caused by increasing time step size than the second-order operators, which provides a theoretical basis for the use of large spatial gird size and large time step size for shear-wave simulation.

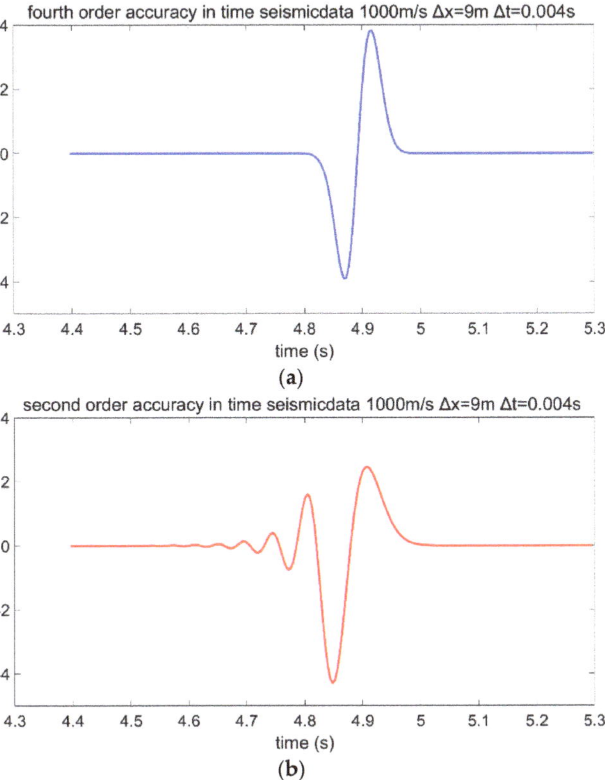

Figure 7. (**a**) CCD is used to simulate the numerical simulation of seismic records in the 1D homogeneous medium model with fourth-order accuracy in time, dt = 0.004 s; (**b**) CCD is used to simulate the numerical simulation of seismic records in the 1D homogeneous medium model with second-order accuracy in time, dt = 0.004 s.

4. Shear-Wave (SH) RTM

4.1. Implementation of SH-RTM

The steps of shear-wave reverse-time migration, similar to those of traditional reverse-time migration, are as follows:

1. Forward extrapolating of the source wavefield: starting from a given or estimated source wavefield, we solve the equation to forward propagate the source wavefield. Thus, for a source emitting at source positions x_s, z_s,

$$V_y(x_s, z_s, t) = S(x_s, z_s, t) \tag{26}$$

2. Shear-wave reverse continuation: For receiver wavefield propagation, we reverse the R of the seismic receiver recorded in time and then set the initial receiver position as the initial boundary condition. We then use the selected finite difference scheme

to solve the shear-wave equation (Equation (5)) iteratively to obtain the receiver wavefield. As shown in the following equation, where x_r, z_r is the position of the source transmitter and receiver, and T is the total duration of forward propagation.

$$V_y(x_r, z_r, t) = R(x_r, z_r, T - t). \tag{27}$$

3. Imaging conditions of shear-wave application: the last step is to use the cross-correlation of source wavefield and receiver wavefield obtained in the previous two steps to obtain the image of the underground structure.

$$I(x, z) = \int_0^T S(x, z, t) R(x, z, t) dt. \tag{28}$$

4.2. Shear-Wave Reverse-Time Migration in Marmousi Models

A set of two-dimensional Marmousi models are set up, as shown in Figures 8 and 9, where the S-wave velocity (Figure 9) is modified by the ratio of horizontal to vertical P-wave velocity (Figure 8). The model size is 121 × 401 grid points, the spatial grid size Δx = 10 m, the time step is 1 ms, the Ricker wavelet of 20 Hz is excited, and the sampling time is 4 s. With these initial conditions, the reverse-time migration imaging is then carried out.

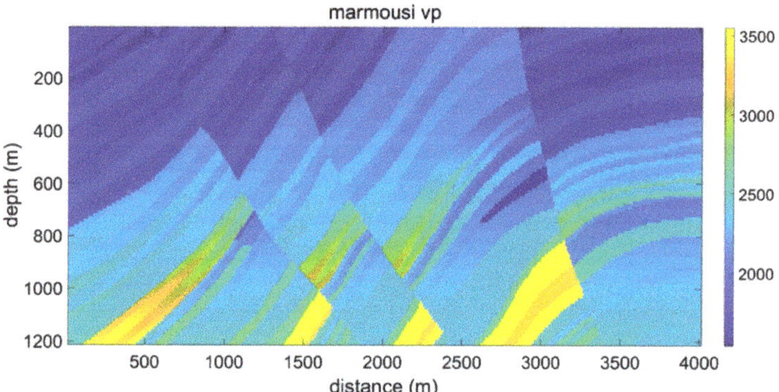

Figure 8. Marmousi model: acoustic velocity model.

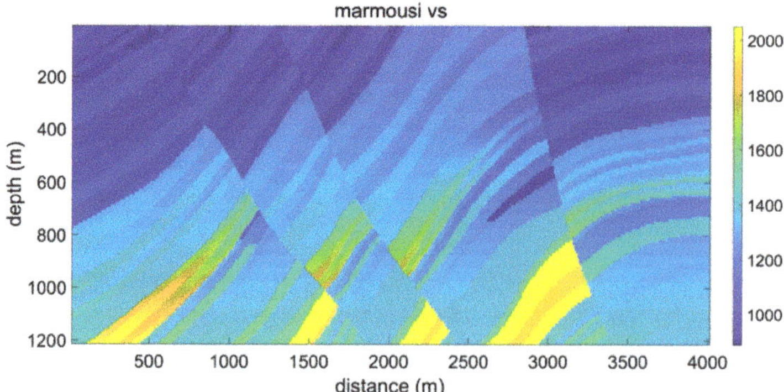

Figure 9. Marmousi model: shear-wave velocity model.

We use the acoustic-wave equation and P-wave source loading method for the reverse-time migration of the P-wave velocity (Figure 10), and the shear-wave equation and shear-wave source loading method for the reverse-time migration of the S-wave velocity (Figure 11). The result of using shear-wave velocity for reverse-time migration (Figure 11) has clearer structural definitions and higher resolution compared with the result of acoustic reverse-time migration using P-wave velocity in Figure 10. We then continue with the SH reverse-time migration imaging experiment, by increasing the spatial gird size, setting $\Delta x = 15$ m, get new S-wave velocity Marmousi model (Figure 12) and the sampling time to 4000 ms. The images for different differential methods are then compared. The following figures (Figures 13–15) show the reverse-time migration results of the Marmousi model in the shear-wave equation using different finite difference schemes.

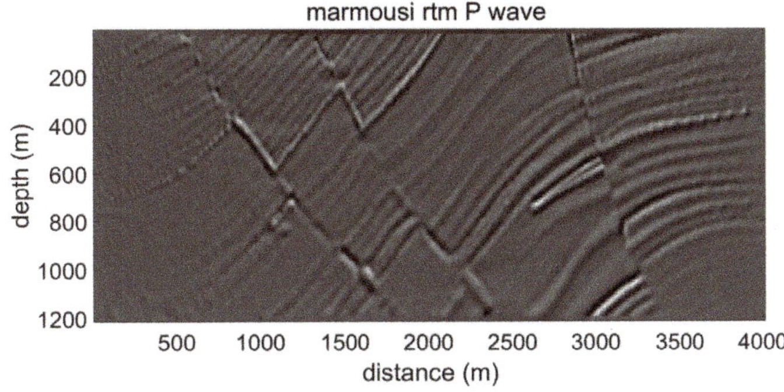

Figure 10. The reverse-time migration result with Marmousi acoustic velocity model.

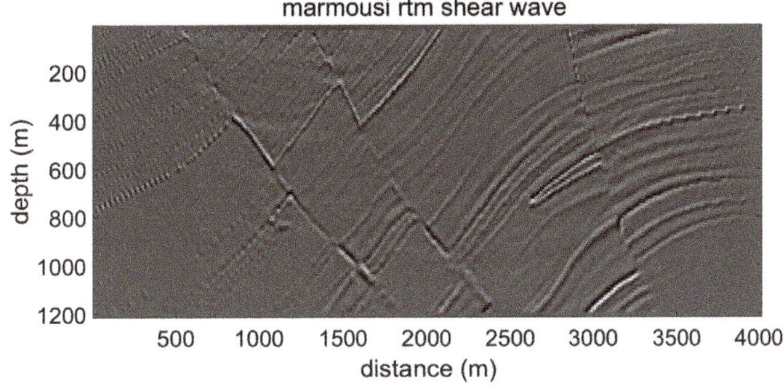

Figure 11. The reverse-time migration result with Marmousi shear-wave velocity model.

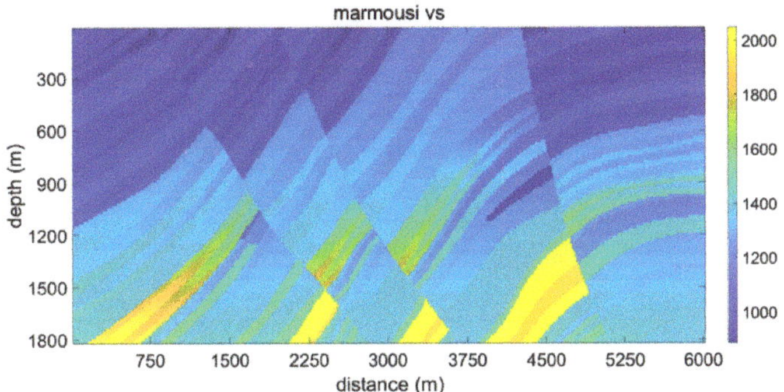

Figure 12. Marmousi model: shear-wave velocity model.

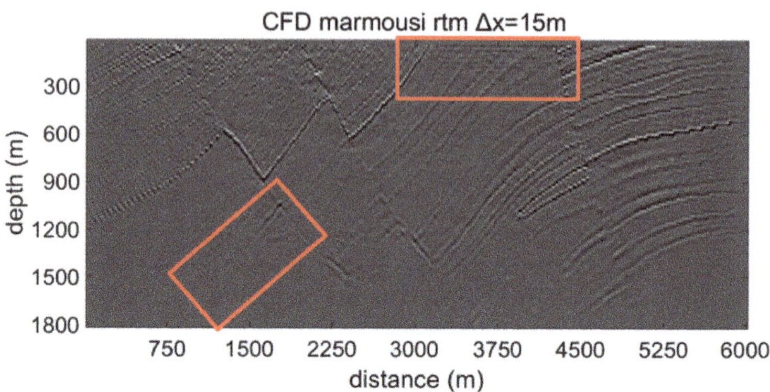

Figure 13. CFD is used for the reverse-time migration result with Marmousi shear-wave velocity model, Δx = 15 m.

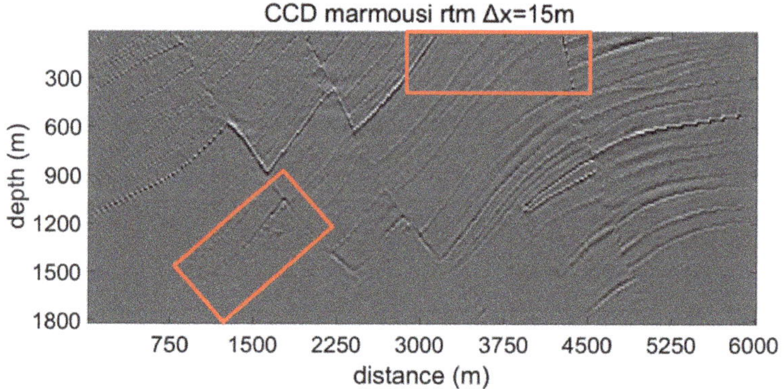

Figure 14. CCD is used to the reverse-time migration result with Marmousi shear-wave velocity model, Δx = 15 m.

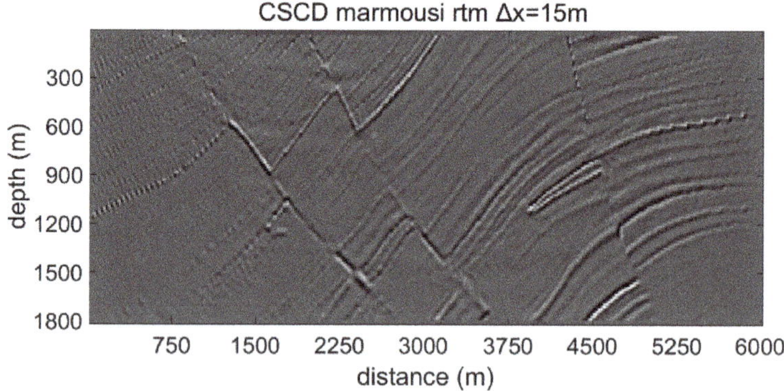

Figure 15. CSCD is used for the reverse-time migration result with Marmousi shear-wave velocity model, Δx = 15 m.

As shown in Figures 13 and 14, the CCD scheme yields better imaging than the CFD scheme, where the red boxes mark the area of improvement, due to higher accuracy in accounting for numerical dispersion caused by the spatial gird size. Compared with

Figures 13 and 14, the CSCD scheme produces the best imaging results of the Marmousi model (Figure 15).

5. Conclusions

In this paper, we proposed a fourth-order time difference solution for the shear-wave equation using the supercompact difference scheme and the combined supercompact difference scheme. We carried out a detailed comparison of their accuracy in numerical simulations and reverse-time migrations. Thus, we reached the following conclusions:

The combined compact difference scheme (CCD) has the characteristics of low numerical dispersion, high stability, and simulation accuracy. It is suitable for the numerical simulation of the seismic wavefield with large space and time steps. It provides an effective method for simulating shear-wave propagation and implementing shear-wave reverse-time migration.

The combined supercompact difference scheme (CSCD) is extended and optimized from the combined compact difference scheme (CCD). Compared with its prototype, the supercompact difference scheme, the combined supercompact difference scheme further suppresses the numerical dispersion caused by the increase in spatial step length and is suitable for numerical simulation with even larger spatial step length.

Finally, we carried out the reverse-time migration imaging of the Marmousi model of shear-wave velocity under isotropic conditions. The combined supercompact difference scheme (CSCD) yields the best shear-wave imaging results of the Marmousi model when compared with the other methods. These results reveal the potential for further extending the supercompact difference scheme and the combined supercompact difference scheme forward simulation of shear waves in complex media such as two-dimensional or three-dimensional anisotropic media and viscoelastic media. It is worth highlighting that the method presented in this paper is restricted to isotropic media at the moment. However, these results reveal the potential for further extending the supercompact difference scheme and the combined supercompact difference scheme forward simulation of shear-wave in complex media such as two-dimensional or three-dimensional anisotropic or viscoelastic media. Of course, the method in this paper also has its limitations, such as low computational efficiency, making it very challenging to apply to three-dimensional media, which is also the focus of our subsequent research.

Author Contributions: Conceptualization, C.Z.; Funding acquisition, X.L.; Formal analysis, W.Y.; Investigation, C.Z.; Methodology, C.Z., X.L. and P.S.; Project administration, W.W., X.L. and P.S.; Resources, W.Y.; Software, C.Z. and W.Y.; Supervision, P.S.; Validation, W.W.; Visualization, C.Z.; Writing—original draft, C.Z.; Writing—review & editing, W.W. and X.L. All authors have read and agreed to the published version of the manuscript.

Funding: The study is supported by the Science and Technology Research and Development Project of CNPC (2021DJ3506) and (2021ZG03) and R&D Department of China National Petroleum Corporation (Investigations on fundamental experiments and advanced theoretical methods in geophysical prospecting applications, 2022DQ0604-02).

Institutional Review Board Statement: Not applicable.

Informed Consent Statement: Not applicable.

Data Availability Statement: Not applicable.

Conflicts of Interest: The authors declare no conflict of interest.

References

1. Gray, S.H.; Etgen, J.; Dellinger, J.; Whitmore, D. Seismic migration problems and solutions. *Geophysics* **2001**, *66*, 1622–1640. [CrossRef]
2. Yilmaz, O. Seismic data analysis. *Soc. Expl. Geophys.* **2001**, 1230–1280.
3. Leveille, J.P.; Jones, I.F.; Zhou, Z.Z.; Wang, B.; Liu, F. Subsalt imaging for exploration, production, and development: A review. *Geophysics* **2017**, *76*, WB3–WB20. [CrossRef]

4. Liu, H.; Dai, N.; Niu, F.; Wu, W. An explicit time evolution method for acoustic wave propagation. *Geophysics* **2014**, *79*, T117–T124. [CrossRef]
5. Huang, J.Q.; Li, Z.C.; Huang, J.P.; Zhang, J.; Sun, W. Lebedev grid high-order finite-difference modeling and elastic wave-mode separation for TTI media. *Oil Geophys. Prospect.* **2017**, *52*, 915–927.
6. Amundsen, L.; Ørjan, P. Time step n-tupling for wave equations. *Geophysics* **2017**, *82*, T249–T254. [CrossRef]
7. Chu, C.L.; Stoffa, P.L. An implicit finite-difference operator for the Helmholtz equation. *Geophysics* **2012**, *77*, T97–T107. [CrossRef]
8. Jiang, Z.D.; Fan, C.W.; Li, X.Z.; Deng, G.J.; Chen, K.; Li, Y.S. Numerical simulation of marine ghost wave based on high order finite difference method with variable grids. *Prog. Geophys.* **2021**, *36*, 365–373. [CrossRef]
9. Li, B.; Liu, Y.; Sen, M.K.; Ren, Z. Time-space-domain mesh-free finite difference based on least squares for 2D acoustic-wave modeling. *Geophysics* **2017**, *82*, T143–T157. [CrossRef]
10. Liu, L.B.; Duan, P.R.; Zhang, Y.Y.; Tian, K.; Tan, M.Y.; Li, Z.C.; Dou, J.Y.; Li, Q.Y. Overview of mesh-free method of seismic forward numerical simulation. *Prog. Geophys.* **2020**, *35*, 1815–1825. [CrossRef]
11. Wang, E.J.; Liu, Y.; Sen, M.K. Effective finite-difference modelling methods with 2-D acoustic wave equation using a combination of cross and rhombus stencils. *Geophys. J. Int.* **2016**, *206*, 1933–1958. [CrossRef]
12. Zhang, B.Q.; Zhou, H.; Chen, H.M.; Shen, S.B. Time-space domain high-order finite difference method for seismic wave numerical simulation based on new stencils. *Chin. J. Geophys.* **2016**, *59*, 1804–1814. [CrossRef]
13. Duan, Y.T.; Sava, P. Scalar imaging condition for elastic reverse time migration. *Geophysics* **2015**, *80*, S127–S136. [CrossRef]
14. Adams, N.A.; Shariff, K.A. High-resolution hybrid compact-ENO scheme for Shock-Turbulence interaction problems. *J. Comput. Phys.* **1996**, *127*, 27–51. [CrossRef]
15. Dennis, S.C.R.; Hundson, J.D. Compact h4 finite-difference approximations to operators of Navier-Stokes type. *J. Comput. Phys.* **1989**, *85*, 390–416. [CrossRef]
16. Lele, S.K. Compact finite difference scheme with spectral-like resolution. *J. Comput. Phys.* **1992**, *103*, 16–42. [CrossRef]
17. Chu, P.C.; Fan, C.W. A three-point combined compact difference scheme. *J. Comput. Phys.* **1998**, *140*, 370–399. [CrossRef]
18. Chang, W.; McMechan, G.A. 3-D elastic prestack, reverse-time depth migration. *Geophysics* **1994**, *59*, 597–609. [CrossRef]
19. Du, Q.Z.; Zhu, Y.T.; Ba, J. Polarity reversal correction for elastic reverse time migration. *Geophysics* **2012**, *77*, S31–S41. [CrossRef]
20. Dai, N.; Wu, W.; Zhang, W.; Wu, X. TTI RTM using variable grid in depth. In Proceedings of the International Petroleum Technology Conference, Bangkok, Thailand, 7–9 February 2012; pp. 1–7.
21. Zhang, J.; Tian, Z.; Wang, C. P- and S-wave separated elastic wave equation numerical modeling using 2D staggered-grid. In Proceedings of the 77th Annual International Meeting, SEG, San Antonio, TX, USA, 14 September 2007; pp. 2104–2109.
22. Xiao, X.; Leaney, W.S. Local vertical seismic profiling (VSP)elastic reverse-time migration and migration resolution: Salt-flank imaging with transmitted P-to-S waves. *Geophysics* **2010**, *75*, S35–S49. [CrossRef]
23. Gu, B.; Li, Z.; Ma, X.; Liang, G. Multi-component elastic reverse time migration based on the P and S separating elastic velocity-stress equation. *J. Appl. Geophys.* **2015**, *112C*, 62–78. [CrossRef]
24. Wang, W.; McMechan, G.A.; Zhang, Q. Comparison of two algorithms for isotropic elastic P and S decomposition in the vector domain. *Geophysics* **2015**, *80*, T147–T160. [CrossRef]
25. Du, Q.; Gong, X.; Zhang, M.; Zhu, Y.; Fang, G. 3D PS-wave imaging with elastic reverse-time migration. *Geophysics* **2014**, *79*, S173–S184. [CrossRef]
26. Zhang, Q.; McMechan, G.A. 2D and 3D elastic wavefield vector decomposition in the wavenumber domain for VTI media. *Geophysics* **2010**, *75*, D13–D26. [CrossRef]
27. Du, Q.Z.; Guo, C.F.; Zhao, Q.; Gong, X.; Wang, C.; Li, X.Y. Vector-based elastic reverse time migration based on scalar imaging condition. *Geophysics* **2017**, *82*, S111–S127. [CrossRef]
28. Nguyen, B.D.; McMechan, G.A. Five ways to avoid storing source wavefield snapshots in 2D elastic prestack reverse time migration. *Geophysics* **2015**, *80*, S1–S18. [CrossRef]
29. Li, X.; Zhang, S. Forty years of shear-wave splitting in seismic exploration: An overview. *Geophys. Prospect. Pet.* **2021**, *60*, 190–209.
30. Bouchaala, F.; Ali, M.Y.; Matsushima, J. Compressional and shear wave attenuations from walkway VSP and sonic data in an offshore Abu Dhabi oilfield. *Comptes Rendus Géosci.* **2021**, *353*, 337–354. [CrossRef]
31. Wu, Y.; He, Z.; Hu, J.; Deng, Z.; Wang, Y.; Yin, W. P-wave and S-wave Joint Acquisition Technology and Its Application in Sanhu Area. In Proceedings of the 88th Annual International Meeting, Anaheim, CA, USA, 14 October 2018; pp. 1–5.
32. Wang, S.Q.; Yang, D.H.; Yang, K.D. Compact finite difference scheme for elastic equations. *J. Tsinghua Univ.* **2002**, *42*, 1128–1131.
33. Bouchaala, F.; Guennou, C. Estimation of viscoelastic attenuation of real seismic data by use of ray tracing software: Application to the detection of gas hydrates and free gas. *Comptes Rendus Géosci.* **2012**, *344*, 57–66. [CrossRef]
34. Zhu, T.; Harris, J.M.; Biondi, B. Q-compensated reverse-time migration. *Geophysics* **2014**, *79*, S77–S87. [CrossRef]
35. Liu, F.; Zhang, G.; Morton, S.A.; Leveille, J.P. An effective imaging condition for reverse-time migration using wavefield decomposition. *Geophysics* **2011**, *76*, S29–S39. [CrossRef]
36. Moradpouri, F.; Moradzadeh, A.; Pestana, R.N.C.; Ghaedrahmati, R.; Soleimani, M. An improvement in wave-field extrapolation and imaging condition to suppress RTM artifacts. *Geophysics* **2017**, *82*, S403–S409. [CrossRef]
37. Moradpouri, F.; Morad-Zadeh, A.; Pestana, R.C.; Soleimani, M. Seismic reverse time migration using a new wave field extrapolator and a new imaging condition. *Acta Geophys.* **2016**, *64*, 1673–1690. [CrossRef]

38. Virieux, J. SH-wave propagation in heterogeneous media: Velocity-stress finite-difference method. *Geophysics* **1984**, *49*, 1933–1942. [CrossRef]
39. Wang, Y.; Shi, H.G.; Zhou, C.Y.; Gui, Z.X. Numerical simulation of 2D seismic wave-field used combined compact difference scheme. *Chin. J. Geophys.* **2018**, *61*, 4568–4583.
40. Dong, L.; Zhan, J. Combined super compact finite difference scheme and application to simulation of shallow water equations. *Chin. J. Comput. Mech.* **2008**, *25*, 791–796.
41. Blayo, E. Compact finite difference schemes for ocean models. *J. Comput. Phys.* **2000**, *164*, 241–257. [CrossRef]
42. Sengupta, T.K.; Ganeriwal, G.; De, S. Analysis of central and upwind compact schemes. *J. Comput. Phys.* **2003**, *192*, 677–694. [CrossRef]
43. Wu, G.C.; Wang, H.Z. Analysis of numerical dispersion in wave2field simulation. *Prog. Geophys.* **2005**, *20*, 58–65.

 applied sciences

Article

Elastic Reverse Time Migration for Weakly Illuminated Structure

Liwei Song [1,2,3,4], Ying Shi [1,2,3,5,*], Wei Liu [1,2,3,5] and Qiang Zhao [1,2,3]

1 State Key Laboratory of Shale Oil and Gas Enrichment Mechanisms and Effective Development, Beijing 100083, China; zhidao90@163.com (L.S.); m18345970967@163.com (W.L.); zhaoqiang.syky@sinopec.com (Q.Z.)
2 Sinopec Key Laboratory of Seismic Elastic Wave Technology, Beijing 100083, China
3 Sinopec Petroleum Exploration and Production Research Institute, Beijing 100083, China
4 School of Physics and Electronic Engineering, Northeast Petroleum University, Daqing 163318, China
5 Key Laboratory of Continental Shale Hydrocarbon Accumulation and Efficient Development, Northeast Petroleum University, Daqing 163318, China
* Correspondence: shiying@nepu.edu.cn

Abstract: One of the most effective techniques to obtain PP and PS images is elastic reverse time migration which employs multi-component seismic data. The two types of complementary images play an important role in reducing blind spots in seismic exploration. However, the migration image of deep structures is always blurred due to the shielding effect of overburden rock on seismic waves. To overcome this issue, we develop an elastic reverse time migration approach for insufficient illumination. This approach contains two crucial elements. The first is that we derive an elastic wave equation to extract the wavefields associated with the exploration target using the staining algorithm. Secondly, we develop an inner product imaging condition with a filter to mute migrated artifacts. The filter, consisting of two vectors, determines which part of the wavefield is contributed to imaging. Synthetic examples exhibit that the proposed elastic reverse time migration method can improve the signal-to-noise ratio of PP and PS images of weakly illuminated structures.

Keywords: reverse time migration; stained elastic wave equation; weak illumination

Citation: Song, L.; Shi, Y.; Liu, W.; Zhao, Q. Elastic Reverse Time Migration for Weakly Illuminated Structure. *Appl. Sci.* **2022**, *12*, 5264. https://doi.org/10.3390/app12105264

Academic Editors: Guofeng Liu, Xiaohong Meng and Zhifu Zhang

Received: 26 March 2022
Accepted: 20 May 2022
Published: 23 May 2022

1. Introduction

Elastic reverse time migration (ERTM) possesses the ability to reposition the multi-component seismic data into underground structure information, thus providing a scientific basis for further seismic interpretation and hydrocarbon development [1,2]. Particularly, subsalt has become a significant hydrocarbon exploration target due to the appearance of salt beds as regional overburden rock [3]. However, owing to the shielding effect generated by the complex overburden rock on seismic signals, the imaging results of the subsurface geological structure are often unsatisfactorily [4,5]. Therefore, it is essential to apply the elastic migration technique for insufficient illumination areas. The wide-azimuth measurement, illumination compensation and acquisition aperture correction approaches have been developed to enhance the imaging quality of shadow regions [6–8]. In addition, the staining algorithm can highlight the wavefields associated with weakly illuminated structures among the full wavefields [9–11], which is a powerful means to realize target-oriented imaging within the ERTM framework.

ERTM images produced by multi-component seismic data always involve crosstalks due to the reason that P- and S-waves are concurrently excited when the waves propagate to a reflecting interface [12]. To suppress the crosstalk artifacts, the elastic wavefield should be adequately decoupled before imaging. A common approach is to apply the curl and divergence operators to the coupled wavefield [13]. However, the decoupled S-wave presenting the polarity reversal phenomenon ultimately destroys the continuity of the reflection lineups of multi-shot stacked profiles. Du et al. [14] argued that the polarity of the decoupled S-wave is reversed on both sides of the vertical incident direction of the P-wave. With the assistance of the Poynting vector, a sign factor can be estimated to correct the polarity of PS imaging [15]. However, the divergence and curl operators will change the amplitude and phase of wavefields. As a consequence, it is difficult for ERTM to obtain a definite physical meaning from the converted wave imaging profile. To address this problem, the pure P- and S-wave can be effortlessly decoupled by solving the vector-decoupled wave equation [16].

When applying the correlation imaging condition, the noises inevitably accompanying valuable information will debase the quality of migration profiles, especially on the interface where the velocity varies dramatically. Although the high-pass filter can suppress migration noises, it alters the waveform of the original image [17]. Another image denoising approach whereby the high angle information of the common imaging gathers in the angle domain is eliminated has advantages in amplitude-preserving [18,19]. Liu et al. [20] first suggested that the migration noises are related to the wave propagation path through the scattering point, and finally proposed an imaging condition with wavefield decomposition to remove the noises. Unfortunately, the initial wavefield decomposition method is realized in the wavenumber-frequency domain, which may result in a huge burden in computation, storage and input/output. Fei et al. [21] imported a cost-effective scheme to decompose the wavefield based on the Hilbert transform. Further, Wang et al. [22] decomposed the wavefield into several components according to the propagating directions. Particularly, the Poynting vector can indicate the direction of energy transfer, thereby flexibly realizing wavefield decomposition [23].

The paper is arranged as follows: after the introduction section, the vector-decoupled wave equation is reviewed. On this basis, a novel equation is derived to extract the wavefield associated with insufficient illumination. Next, an inner product imaging condition with filters is described. Finally, some synthetic examples are employed to demonstrate the effectiveness of our approach.

2. Methodology

2.1. Vector-Decoupled Wave Equation

The velocity-stress formula in 2D isotropic medium is expressed as,

$$\begin{cases} \rho\frac{\partial \boldsymbol{V}}{\partial t} = \mathcal{L}\boldsymbol{\sigma} \\ \frac{\partial \boldsymbol{\sigma}}{\partial t} = \rho \boldsymbol{A}\mathcal{L}^T\boldsymbol{V} \end{cases} \tag{1}$$

where ρ denotes the density of media, $\boldsymbol{V} = [v_x, v_z]^T$ is the particle-velocity vector, t signifies the time variable, $\boldsymbol{\sigma} = [\tau_{xx}, \tau_{zz}, \tau_{xz}]^T$ denotes the stress vector and $\mathcal{L}$ is a spatial partial derivative operator expressed as:

$$\mathcal{L} = \begin{bmatrix} \partial_x & 0 & \partial_z \\ 0 & \partial_z & \partial_x \end{bmatrix} \tag{2}$$

x and z, respectively represent the horizontal and vertical directions, and $\boldsymbol{A}$ is a parameter matrix as shown below:

$$\boldsymbol{A} = \begin{bmatrix} c_P^2 & c_P^2 - 2c_S^2 & 0 \\ c_P^2 - 2c_S^2 & c_P^2 & 0 \\ 0 & 0 & c_S^2 \end{bmatrix} \tag{3}$$

where c_P and c_S are P-wave and S-wave velocity in the propagation medium. Equation (1) can be named as vector-coupled wave equation (VCWE).

When the wavefield is simulated according to Equation (1), the P- and S-waves are coupled in the wavefield of particle velocity. If the coupled wavefield is directly applied to the ERTM, the imaging profile will suffer from crosstalk. To tackle this problem,

Wang et al. [24] introduced a scalar P-wave stress τ^P to decouple the wavefield, and the formulation is written as,

$$\begin{cases} \frac{\partial \tau^P}{\partial t} = \rho c_P^2 \left(\frac{\partial v_x}{\partial x} + \frac{\partial v_z}{\partial z} \right) \\ \rho \frac{\partial v_x^P}{\partial t} = \frac{\partial \tau^P}{\partial x} \\ \rho \frac{\partial v_z^P}{\partial t} = \frac{\partial \tau^P}{\partial z} \end{cases} \tag{4}$$

where v_x^P and v_z^P are the decoupled P-waves in x- and z-directions. The residual S-waves are defined as $v_x^S = v_x - v_x^P$ and $v_z^S = v_z - v_z^P$. The decoupled wavefield can preserve the same physical properties as the coupled wavefield. The combination of Equations (1) and (4) is the vector-decoupled wave equation (VDWE).

2.2. Stained VDWE

To extract the wavefield associated with the weakly illuminated structure, we establish a governing equation. According to the principle of staining algorithm [9], the variables $\boldsymbol{V}$, $\boldsymbol{\sigma}$ and $\boldsymbol{A}$ in VDWE are extended to complex domain, and possess the following forms:

$$\boldsymbol{V} = \overline{\boldsymbol{V}} + i\widetilde{\boldsymbol{V}} = [\overline{v_x}, \overline{v_z}]^T + i[\widetilde{v_x}, \widetilde{v_z}]^T, \tag{5}$$

$$\boldsymbol{\sigma} = \overline{\boldsymbol{\sigma}} + i\widetilde{\boldsymbol{\sigma}} = [\overline{\tau_{xx}}, \overline{\tau_{zz}}, \overline{\tau_{xz}}]^T + i[\widetilde{\tau_{xx}}, \widetilde{\tau_{zz}}, \widetilde{\tau_{xz}}]^T, \tag{6}$$

$$\boldsymbol{A} = \overline{\boldsymbol{A}} + i\widetilde{\boldsymbol{A}} = \begin{bmatrix} (\overline{c_P} + i\widetilde{c_P})^2 & (\overline{c_P} + i\widetilde{c_P})^2 - 2(\overline{c_S} + i\widetilde{c_S})^2 & 0 \\ (\overline{c_P} + i\widetilde{c_P})^2 - 2(\overline{c_S} + i\widetilde{c_S})^2 & (\overline{c_P} + i\widetilde{c_P})^2 & 0 \\ 0 & 0 & (\overline{c_S} + i\widetilde{c_S})^2 \end{bmatrix} \tag{7}$$

where $i = \sqrt{-1}$, variables with $-$ and $\sim$ symbols, respectively, represent the real and imaginary parts. Note that setting c_P and c_S is the core step in extracting the wavefield. $\overline{c_P}$ and $\overline{c_S}$ are consistent with the P- and S-wave velocity. $\widetilde{c_P}$ and $\widetilde{c_S}$ are obtained by a dimensionless multiplied $\overline{c_P}$ and $\overline{c_S}$ near the exploration target, while $\widetilde{c_P}$ and $\widetilde{c_S}$ are equal to 0 in the unconcerned area. The location where $\widetilde{c_P}$ and $\widetilde{c_S}$ are non-zero is referred to as the stained target.

By substituting Equations (5)–(7) into the VDWE, we obtain,

$$\begin{cases} \rho \frac{\partial (\overline{\boldsymbol{V}} + i\widetilde{\boldsymbol{V}})}{\partial t} = \mathcal{L}(\overline{\boldsymbol{\sigma}} + i\widetilde{\boldsymbol{\sigma}}) \\ \frac{\partial (\overline{\boldsymbol{\sigma}} + i\widetilde{\boldsymbol{\sigma}})}{\partial t} = \rho \left(\overline{\boldsymbol{A}} + i\widetilde{\boldsymbol{A}} \right) \mathcal{L}^T \left(\overline{\boldsymbol{V}} + i\widetilde{\boldsymbol{V}} \right) \end{cases} \tag{8}$$

$$\begin{cases} \frac{\partial \left(\overline{\tau^P} + i\widetilde{\tau^P} \right)}{\partial t} = \rho (\overline{c_P} + i\widetilde{c_p})^2 \left(\frac{\partial (\overline{v_x} + i\widetilde{v_x})}{\partial x} + \frac{\partial (\overline{v_z} + i\widetilde{v_z})}{\partial z} \right) \\ \rho \frac{\partial \left(\overline{v_x^P} + i\widetilde{v_x^P} \right)}{\partial t} = \frac{\partial \left(\overline{\tau^P} + i\widetilde{\tau^P} \right)}{\partial x} \\ \rho \frac{\partial \left(\overline{v_z^P} + i\widetilde{v_z^P} \right)}{\partial t} = \frac{\partial \left(\overline{\tau^P} + i\widetilde{\tau^P} \right)}{\partial z} \end{cases} \tag{9}$$

Taking the real parts of Equations (8) and (9), we acquire an equation expressed as,

$$\begin{cases} \rho \frac{\partial \overline{\boldsymbol{V}}}{\partial t} = \mathcal{L} \overline{\boldsymbol{\sigma}} \\ \frac{\partial \overline{\boldsymbol{\sigma}}}{\partial t} = \rho \left(\overline{\boldsymbol{A}} \mathcal{L}^T \overline{\boldsymbol{V}} - \widetilde{\boldsymbol{A}} \mathcal{L}^T \widetilde{\boldsymbol{V}} \right) \\ \frac{\partial \overline{\tau^P}}{\partial t} = \rho \alpha \left(\frac{\partial \overline{v_x}}{\partial x} + \frac{\partial \overline{v_z}}{\partial z} \right) - 2\rho\beta \left(\frac{\partial \widetilde{v_x}}{\partial x} + \frac{\partial \widetilde{v_z}}{\partial z} \right) \\ \rho \frac{\partial \overline{v_x^P}}{\partial t} = \frac{\partial \overline{\tau^P}}{\partial x} \\ \rho \frac{\partial \overline{v_z^P}}{\partial t} = \frac{\partial \overline{\tau^P}}{\partial z} \end{cases} \tag{10}$$

where $\alpha = \overline{c_P}^2 - \widetilde{c_P}^2$ and $\beta = \overline{c_P}\widetilde{c_P}$. The real parts of S-waves are defined as $\overline{v_x^S} = \overline{v_x} - \overline{v_x^P}$ and $\overline{v_z^S} = \overline{v_z} - \overline{v_z^P}$. The imaginary parts of Equations (8) and (9) can be written as,

$$\begin{cases} \rho \frac{\partial \widetilde{\boldsymbol{V}}}{\partial t} = \mathcal{L} \widetilde{\boldsymbol{\sigma}} \\ \frac{\partial \widetilde{\boldsymbol{\sigma}}}{\partial t} = \rho \left(\overline{\boldsymbol{A}} \mathcal{L}^T \widetilde{\boldsymbol{V}} - \widetilde{\boldsymbol{A}} \mathcal{L}^T \overline{\boldsymbol{V}} \right) \\ \frac{\partial \widetilde{\tau^P}}{\partial t} = \rho \alpha \left(\frac{\partial \widetilde{v_x}}{\partial x} + \frac{\partial \widetilde{v_z}}{\partial z} \right) + 2\rho\beta \left(\frac{\partial \overline{v_x}}{\partial x} + \frac{\partial \overline{v_z}}{\partial z} \right) \\ \rho \frac{\partial \widetilde{v_x^P}}{\partial t} = \frac{\partial \widetilde{\tau^P}}{\partial x} \\ \rho \frac{\partial \widetilde{v_z^P}}{\partial t} = \frac{\partial \widetilde{\tau^P}}{\partial z} \end{cases} \tag{11}$$

The imaginary parts of S-waves are defined as $\widetilde{v_x^S} = \widetilde{v_x} - \widetilde{v_x^P}$ and $\widetilde{v_z^S} = \widetilde{v_z} - \widetilde{v_z^P}$. Equations (10) and (11) jointly constitute the stained VDWE (SVDWE). According to Equation (11), we find that $\widetilde{\boldsymbol{V}}$ is only excited when $\overline{\boldsymbol{V}}$ arrives at the stained target. The imaginary part $\widetilde{\boldsymbol{V}}$ is named as stained wavefield. The term $\widetilde{\boldsymbol{A}}\mathcal{L}^T\overline{\boldsymbol{V}}$ can be considered as the source signature of $\widetilde{\boldsymbol{V}}$. $\widetilde{\boldsymbol{A}}$ and β are dimensionless due to the principle of $\widetilde{c_P}$ and $\widetilde{c_S}$ setting. It is evident that Equation (10) is approximately equivalent to the VDWE.

2.3. Inner Product Imaging Condition with Filters

Imaging condition can convert seismic data into structure images through the focus of the source and receiver wavefield. For the ERTM, the inner product imaging condition has an advantage in terms of definite physical meanings. However, applying such imaging condition inevitably results in migration artifacts because of backscatter. To suppress the unwanted seismic events in the migration profile, we add filters to the imaging conditions. The final images can be expressed as:

$$\widetilde{I^{PP}} = \sum_{t=1}^{T} \left(F_{So}^P \widetilde{\boldsymbol{V}_{So}^P} \right) \cdot \left(F_{Re}^P \widetilde{\boldsymbol{V}_{Re}^P} \right) \tag{12}$$

$$\widetilde{I^{PS}} = \sum_{t=1}^{T} \left(F_{So}^P \widetilde{\boldsymbol{V}_{So}^P} \right) \cdot \left(F_{Re}^S \widetilde{\boldsymbol{V}_{Re}^S} \right) \tag{13}$$

where $\widetilde{I^{PP}}$ and $\widetilde{I^{PS}}$ are PP and PS images, the symbol $\cdot$ denotes the inner product operation. $\widetilde{\boldsymbol{V}}$ with the subscripts *So* and *Re* denotes the stained wavefields of seismic sources and receivers. The superscripts *P* and *S* represent P- and S-wave, respectively. *T* is the maximum time of seismic records. F_{So}^P is a filter with the mathematical form shown below:

$$F_{So}^P = \begin{cases} 1, & \widetilde{\boldsymbol{D}_{So}^P} \cdot \boldsymbol{n}_{So}^P > 0 \\ 0, & \widetilde{\boldsymbol{D}_{So}^P} \cdot \boldsymbol{n}_{So}^P < 0 \end{cases} \tag{14}$$

where $\widetilde{\boldsymbol{D}_{So}^P} = \left(\widetilde{d_x^P}, \widetilde{d_z^P} \right)$ is the Poynting vector, $\boldsymbol{n}_{So}^P$ is the direction vector. If the result of the inner product between $\widetilde{\boldsymbol{D}_{So}^P}$ and $\boldsymbol{n}_{So}^P$ is positive (negative), then the wavefield is retained (muted). Other filters have a similar expression. The Poynting vector can be calculated by:

$$\begin{cases} \widetilde{d_a^P} = -\widetilde{\tau^P}\widetilde{v_a^P} \\ \widetilde{d_a^S} = -\left[\widetilde{\tau_{ab}} - \widetilde{\tau^P}\delta(a-b) \right] \widetilde{v_b^S} \end{cases} \tag{15}$$

where a and b are the x- or z-component of the Poynting vectors, and δ is the Dirac function. From formula (15), the Poynting vector is 0 if either particle-velocity and stress is 0. There is a position between the crest and the trough where the wavefield is 0, but the Poynting vector at that position should not be 0. To improve the stability of equation (15), the local least squares strategy [25] is applied.

3. Numerical Examples

In the first example, we employ a layered velocity model to probe the performance of the above equations. The P-wave migration velocity model is sketched in Figure 1a. The S-wave migration velocity is obtained using an empirical formula $c_S = c_P/\sqrt{3}$. The density is set as 2000 kg/m^3. The model is dispersed by 650×400 grid numbers, with a grid size of 10×10 m^2. A Ricker wavelet with the dominant frequency of 25 Hz is set as the source signature at the location of (1 km, 0). The wavefields are simulated by the finite difference method with the second-order in time and tenth-order in space. In addition, we add a perfectly matched layer around the simulated area to absorb spurious reflections from the truncated boundary.

Based on the VCWE, the x-component of the particle-velocity is shown in Figure 2a, in which the P- and S-waves are coupled together. When the coupled wavefield is extrapolated by solving Equation (4), it can be completely decoupled into pure P- and S-waves (Figure 2b,c). Notably, the seismic energy of the deep structure information is relatively weak in the decoupled wavefield.

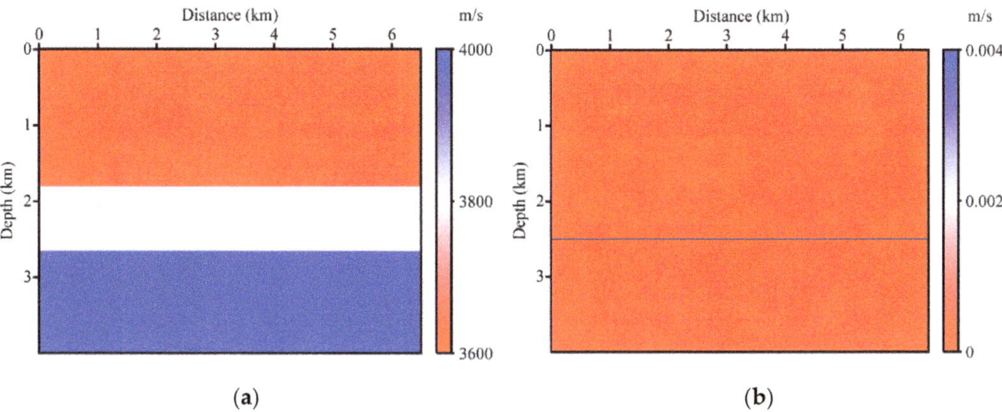

Figure 1. Migration model: (**a**) P-wave velocity; (**b**) Stained target.

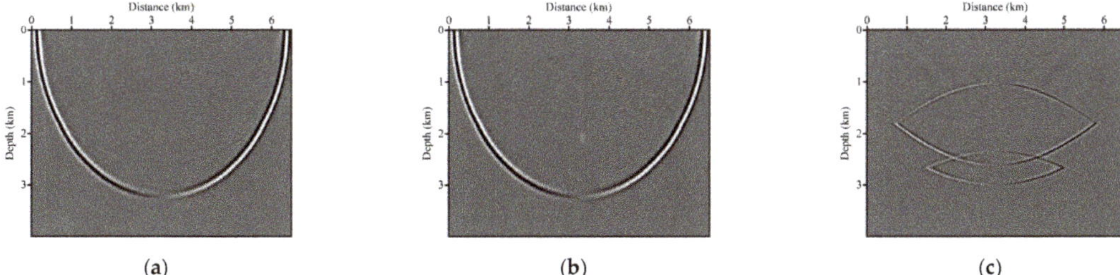

Figure 2. x-component of particle-velocity wavefield: (**a**) Coupled wave; (**b**) Pure P-wave; (**c**) Pure S-wave.

To reduce the shielding effect of the overburden rock on seismic waves, we simulate the wavefield based on SVDWE. Here, the second interface is set as the stained target. $\widetilde{c_P}$ is obtained by $\overline{c_P}$ multiplied 10^{-6} in the target area while is zero in other locations. The setting of $\widetilde{c_S}$ is similar to that of $\widetilde{c_P}$. Considering the similarity between $\widetilde{c_P}$ and $\widetilde{c_S}$ models, only one of them is shown in Figure 1b. By solving SVDWE, the stained wavefield $\widetilde{V}$ is generated after the $\overline{V}$ hits the stained target. The stained wavefield $\widetilde{V}$ only contains the reflection and transmission about the second interface as exhibited in Figure 3. Consequently, we infer that imaging with the stained wave can reduce interference from non-target structures.

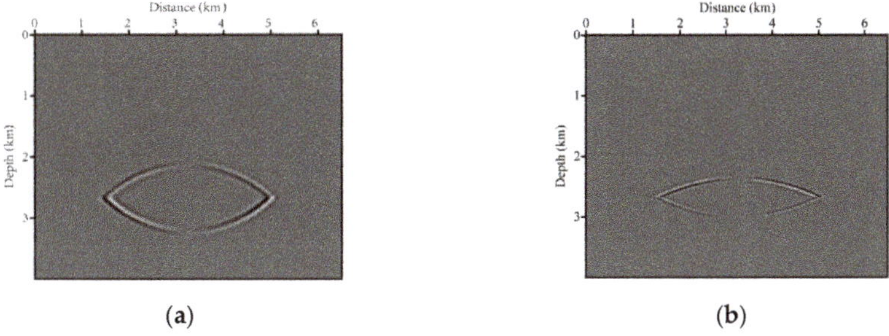

Figure 3. x-component of the stained wavefield: (**a**) Pure P-wave; (**b**) Pure S-wave.

Furthermore, we utilize a contrastive approach to illustrate the adaptability of the ERTM for different equations. We evenly deploy 10 shots and 650 receivers on the surface of the layered velocity model. The total length of seismic records is 3 s, with a step time of 1ms. Figure 4 shows the multi-shot stacked images based on VDWE. Moreover, we implement the ERTM using the SVDWE, the PP and PS images are sketched in Figure 5. The settings of the stained target are the same as those of previous numerical experiments. Thanks to the stained wavefields, the primary information in the images is the second interface. However, some unwanted migration artifacts may interfere with subsequent interpretations.

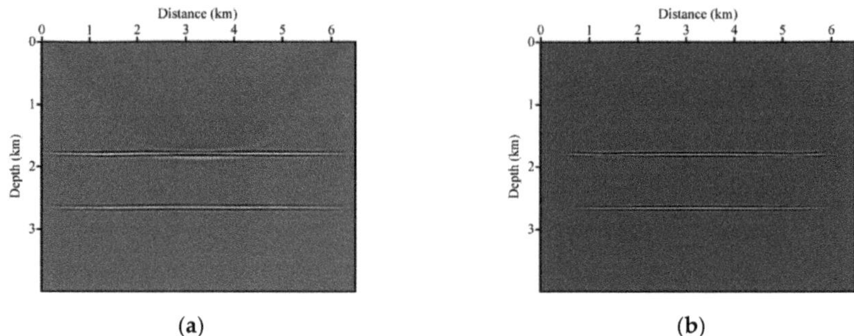

Figure 4. ERTM image obtained based on the VDWE: (**a**) PP imaging; (**b**) PS imaging.

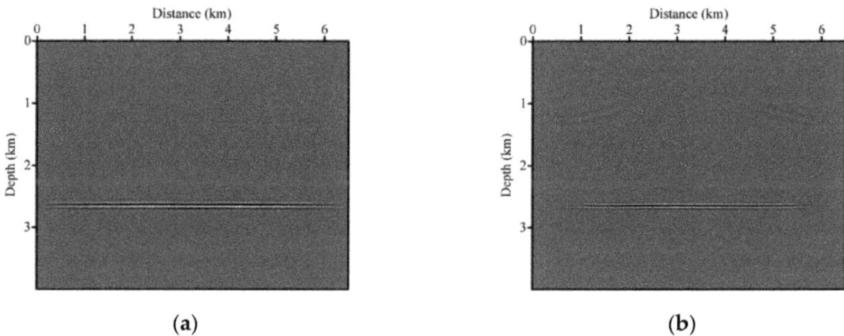

Figure 5. ERTM image based on the SVDWE: (**a**) PP imaging; (**b**) PS imaging.

Next, we demonstrate that the proposed imaging condition plays an important role in eliminating migration artifacts. The filters in the imaging condition need to be prepared in advance, and their properties are determined by the direction vector and Poynting vector. For the snapshot shown in Figure 3a, the Poynting vector (Figure 6) can be obtained using Equation (15). Here, we exhibit four filtered wavefields (Figure 7) that, respectively, correspond to different direction vectors: $\boldsymbol{n}_{So}^{P} = (1,\ 0)$, $\boldsymbol{n}_{So}^{P} = (-1,\ 0)$, $\boldsymbol{n}_{So}^{P} = (0, -1)$, and $\boldsymbol{n}_{So}^{P} = (0, 1)$.

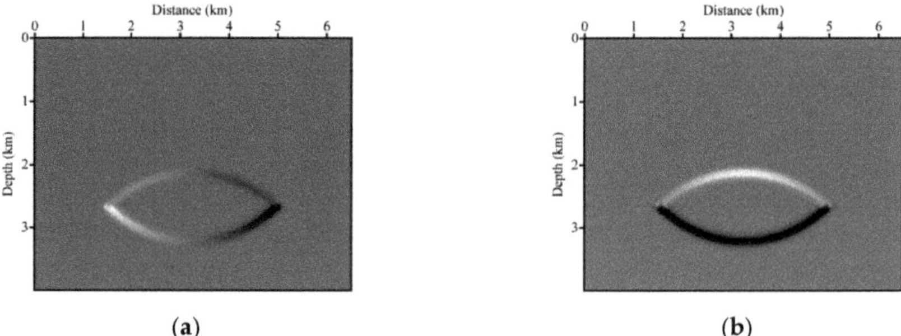

Figure 6. Poynting vector: (**a**) x-component; (**b**) z-component.

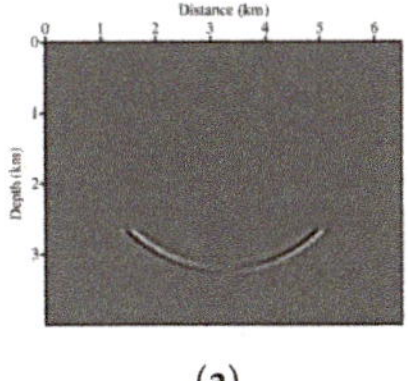

(**a**)

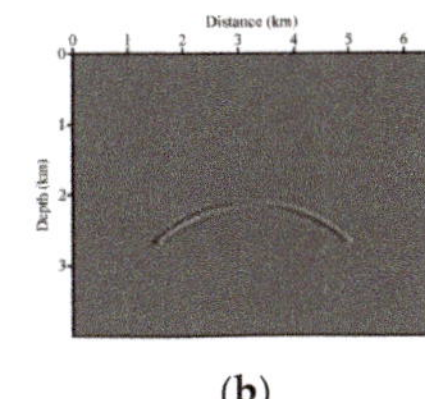

(**b**)

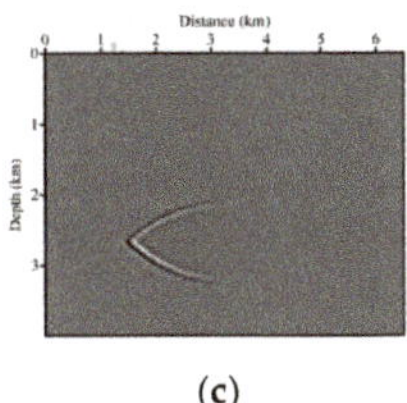

(**c**)

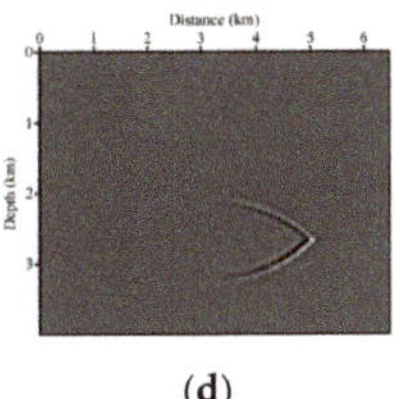

(**d**)

Figure 7. Filtered wavefield: (**a**) downgoing; (**b**) upgoing; (**c**) Leftgoing; (**d**) rightgoing.

For flat strata, images with fewer artifacts can be obtained in the inner product of the upgoing stained source wavefield and the down-going stained receiver wavefield (Figure 8). Figure 9 quantitatively compares PP and PS images, and the two sets of single-channel data are extracted from Figure 8 (from 2 km to 3 km in depth), and then mapped to the wavenumber domain through the Fourier transform. It is prominent that PS images are characterized by abundant high wavenumber information, robustly confirming that PS images have a higher spatial resolution than PP images. The basic reason for such a phenomenon is that S-wave travels slower than P-wave in the subsurface. Namely, the converted S-waves possess a shorter wavelength.

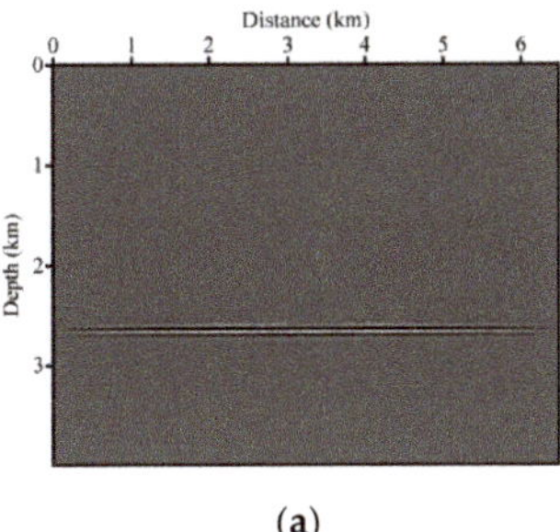

(**a**)

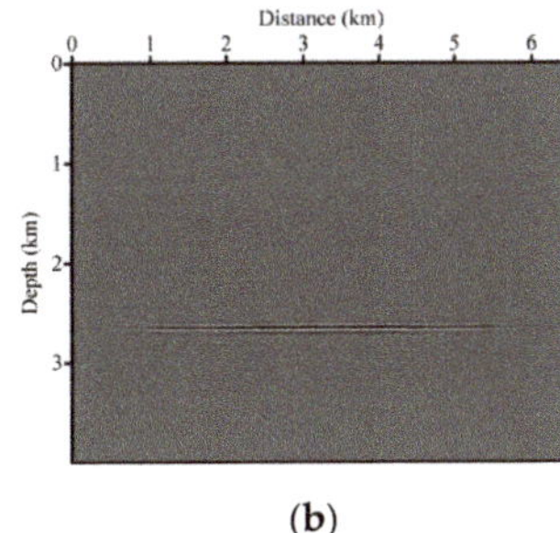

(**b**)

Figure 8. ERTM profile with fewer artifacts compared with Figure 5: (**a**) PP imaging; (**b**) PS imaging.

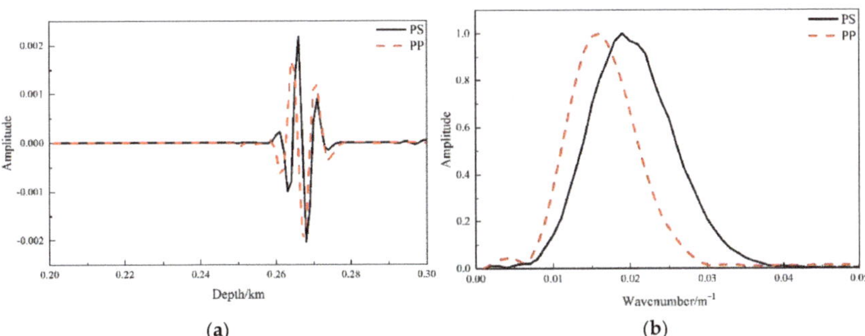

(**a**) (**b**)

Figure 9. Comparison between PP and PS images: (**a**) Single-channel data; (**b**) Wavenumber spectrum.

In this example, a salt model is applied to evaluate the efficiency of the proposed approach in complex geologic regions. The true P-wave velocity model is exhibited in Figure 10a, and S-wave velocity is also calculated using the previous empirical formula. The model size is 400 × 400 grids, with a grid size of 10 × 10 m^2 in both horizontal and vertical directions. A total of 40 sources are evenly excited on the surface of the model. Each source has 400 receivers with a spacing of 10 m. The maximum recording time is 6 s. Figure 11 shows the PP and PS images obtained by the conventional ERTM, and it is obvious that the shallow structure and the salt boundary can be effectively identified. However, the high-velocity overburden rock leads to insufficient illumination for the deep subsurface, which ultimately blurs the subsalt image. We set the stained target in Figure 10b. Figure 12 displays the PP and PS images with the proposed imaging condition. We can see that the two complementary images are favorable for the identification of subsalt faults.

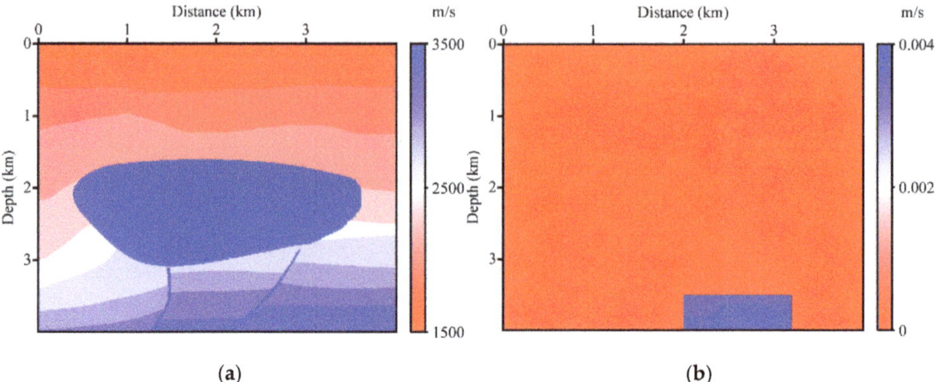

Figure 10. Migration model: (**a**) True P-wave velocity; (**b**) Stained target.

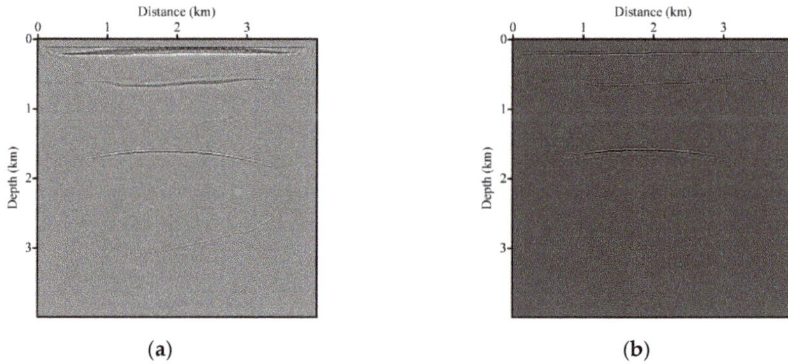

Figure 11. ERTM images based on the VDWE: (**a**) PP imaging; (**b**) PS imaging.

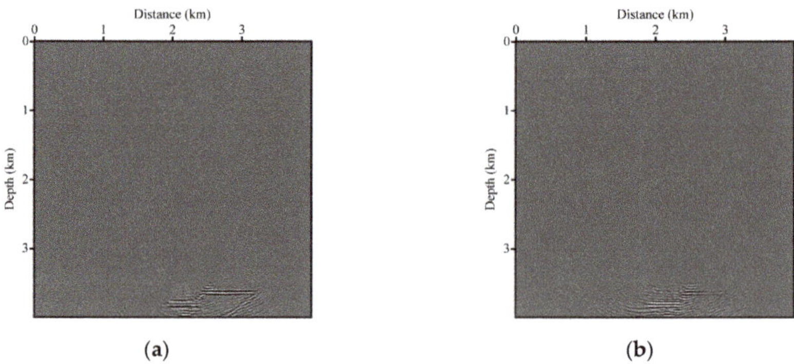

Figure 12. ERTM images based on the SVDWE: (**a**) PP imaging; (**b**) PS imaging.

In the last experiment, a field data is utilized to confirm the effectiveness of our method. Figure 13 displays the migration velocity model that has 940 × 300 grid points with a 20 × 20 m^2 element. There are 240 sources unevenly distributed at the free surface. Figure 14 show some shot gathers whose maximum recording time is 5 s. Since we only have the vertical component of the seismic record where P-wave information dominates, the valuable PS images are not obtained. The PP images based on VDWE are sketched in Figure 15, from which we can see the main structures. To enhance the quality of imaging in weakly illuminated areas, we set the box in Figure 13 as the stained targets. By using the proposed method, the migration events (Figure 16) relevant to the stained targets are highlighted, and the reflections of nontarget structures are muted.

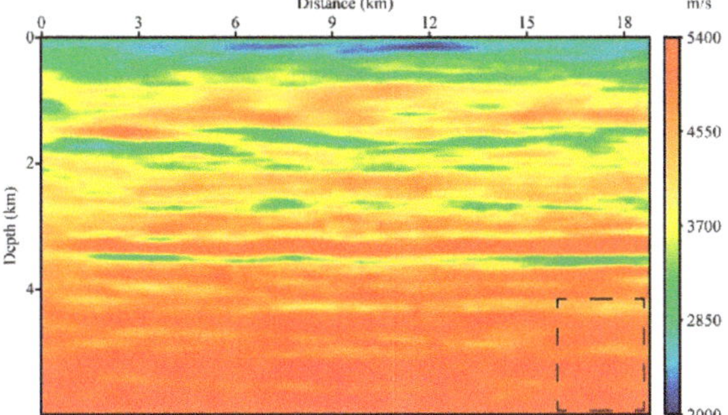

Figure 13. P-wave velocity.

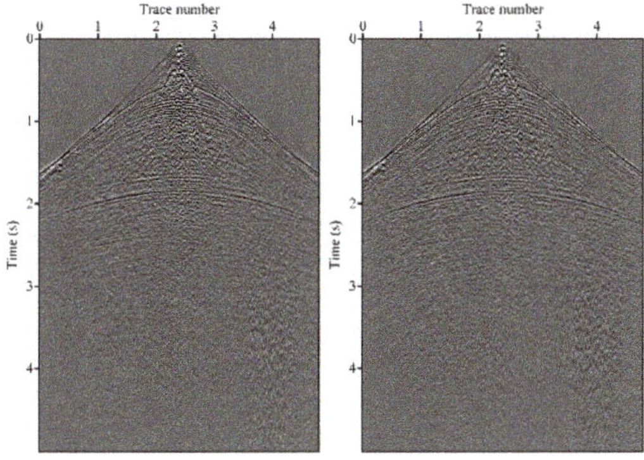

Figure 14. Two shot gathers.

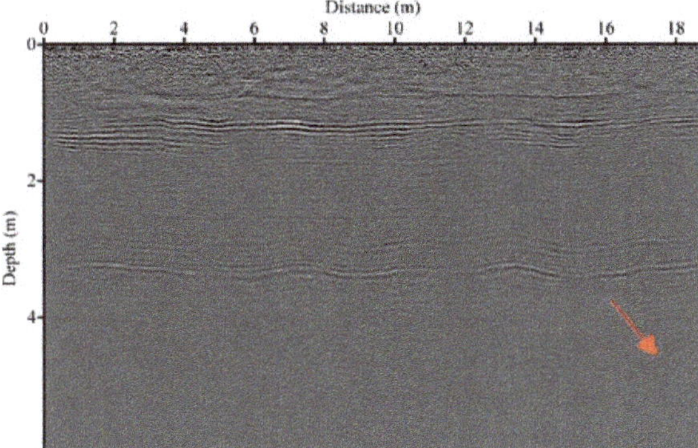

Figure 15. PP imaging based on VDWE.

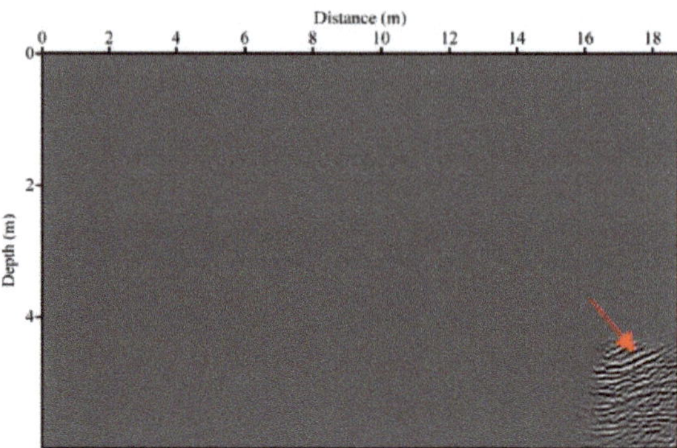

Figure 16. PP imaging based on SVDWE.

4. Discussion

Our approach shows the satisfying performance for imaging weakly illuminated structures. When an overall migrated profile is acquired, it is proper to achieve locally enhanced imaging by artificially setting exploration targets in advance. Since the SVDWE contains two systems of equations, the algorithm of such an approach is twice as computationally intensive as the conventional one. The staining strategy is intrinsically related to the imaging results and is worthy of further investigation. The filters in the inner product imaging condition screen different wavefields according to the direction of wave propagation. Theoretically, it is possible to achieve wavefield decomposition in an arbitrary direction. However, in this paper, we decompose the wavefield into up- and down-going categories under the assumption that the target structure is relatively flat. To enhance the imaging adaptability for complex geological situations, the wavefield screening method according to stratigraphic dip angle needs to be persistently explored.

5. Conclusions

We have presented a new ERTM technique to produce the weakly illuminated structure. To extract the elastic wavefield related to the target, a stained vector-decoupled wave equation is adopted. The filter in the inner product imaging condition can suppress the migration artifact, improving the resolution of ERTM results. Numerical results show that our method can generate better amplitude for the weakly illuminated structure, providing possible technical support for reducing the risk of hydrocarbon development.

Author Contributions: Conceptualization, Y.S. and L.S.; methodology, Y.S.; software, W.L.; validation, Y.S. and L.S.; formal analysis, W.L.; investigation, L.S.; resources, Y.S.; data curation, Y.S.; writing—original draft preparation, L.S.; writing—review and editing, L.S.; visualization, W.L.; supervision, Y.S. and Q.Z.; project administration, Y.S. and Q.Z.; funding acquisition, Y.S. and L.S. All authors have read and agreed to the published version of the manuscript.

Funding: This research was funded by the National Natural Science Foundation of China (U19B6003-004), the National Key R&D Program of China (2018YFA0702505), the National Natural Science Foundation of China (41930431) and the University Nursing Program for Young Scholars with Creative Talents in Heilongjiang Province (Grant No. UNPYSCT-2020149).

Institutional Review Board Statement: Not applicable.

Informed Consent Statement: Not applicable.

Data Availability Statement: Not applicable.

Conflicts of Interest: The authors declare no conflict of interest.

References

1. Baysal, E.; Dan, D.K.; Sherwood, J. Reverse-Time Migration. *Geophysics* **1983**, *48*, 1514–1524. [CrossRef]
2. McMechan, G. Migration by extrapolation of time-dependent boundary values. *Geophys. Prospect.* **1983**, *31*, 413–420. [CrossRef]

3. Leveille, J.P.; Jones, I.F.; Zhou, Z.Z.; Wang, B.; Liu, F.Q. Subsalt imaging for exploration, production, and development: A review. *Geophysics* **2011**, *76*, WB3–WB20. [CrossRef]
4. Liu, Y.; Chang, X.; Jin, D.; He, R.; Sun, H.; Zheng, Y. Reverse time migration of multiples for subsalt imaging. *Geophysics* **2011**, *76*, WB209–WB216. [CrossRef]
5. Vigh, D.; Kapoor, J.; Moldoveanu, N.; Li, H. Breakthrough acquisition and technologies for subsalt imaging. *Geophysics* **2011**, *76*, WB41–WB51. [CrossRef]
6. Regone, C. Using 3D finite-difference modeling to design wide-azimuth surveys for improved subsalt imaging. In Proceedings of the 2006 SEG International Exposition and Annual Meeting, New Orleans, LA, USA, 1–6 October 2006.
7. Gherasim, M.; Albertin, U.; Nolte, B.; Askim, O.; Trout, M.; Hartman, K. Wave-equation angle-based illumination weighting for optimized subsalt imaging. In Proceedings of the 2010 SEG International Exposition and Annual Meeting, Denver, CO, USA, 17–22 October 2010.
8. Cao, J.; Wu, R. Fast acquisition aperture correction in prestack depth migration using beamlet decomposition. *Geophysics* **2009**, *74*, S67–S74. [CrossRef]
9. Chen, B.; Jia, X. Staining algorithm for seismic modeling and migration. *Geophysics* **2014**, *79*, S121–S129. [CrossRef]
10. Li, Q.; Jia, X. Generalized staining algorithm for seismic modeling and migration. *Geophysics* **2017**, *82*, T17–T26. [CrossRef]
11. Song, L.; Shi, Y.; Chen, S. Target-oriented reverse time migration in transverse isotropy media. *Acta Geophys.* **2021**, *69*, 125–134. [CrossRef]
12. Sun, R.; McMechan, G.; Lee, C.; Chow, J.; Chen, C. Prestack scalar reverse-time depth migration of 3D elastic seismic data. *Geophysics* **2006**, *71*, S199–S207. [CrossRef]
13. Yan, J.; Sava, P. Isotropic angle-domain elastic reverse-time migration. *Geophysics* **2008**, *73*, 229–239. [CrossRef]
14. Du, Q.; Zhu, Y.; Ba, J. Polarity reversal correction for elastic reverse time migration. *Geophysics* **2012**, *77*, S31–S41. [CrossRef]
15. Li, Z.; Ma, X.; Fu, C.; Liang, G. Wavefield separation and polarity reversal correction in elastic reverse time migration. *J. Appl. Geophys.* **2016**, *127*, 56–67. [CrossRef]
16. Zhang, W.; Shi, Y. Imaging conditions for elastic reverse time migration. *Geophysics* **2019**, *84*, S95–S111. [CrossRef]
17. Zhang, Y.; Sun, J. Practical issues in reverse time migration: True amplitude gathers, noise removal and harmonic source encoding. *First Break* **2009**, *27*, 53–59. [CrossRef]
18. Sava, P.C.; Fomel, S. Angle-domain common-image gathers by wavefield continuation methods. *Geophysics* **2003**, *68*, 1065–1074. [CrossRef]
19. Liu, Q.; Zhang, J. Efficient dip-angle adcig estimation using poynting vector in acoustic RTM and its application in noise suppression. *Geophys. Prospect.* **2018**, *66*, 1714–1725. [CrossRef]
20. Liu, F.; Zhang, G.; Morton, S.A.; Leveille, J.P. An effective imaging condition for reverse-time migration using wavefield decomposition. *Geophysics* **2011**, *76*, S29–S39. [CrossRef]
21. Fei, T.; Luo, Y.; Yang, J.; Liu, H.; Qin, F. Removing false images in reverse time migration: The concept of de-primary. *Geophysics* **2015**, *80*, S237–S244. [CrossRef]
22. Wang, Y.; Zheng, Y.; Xue, Q.; Chang, X.; Tong, F.; Luo, Y. Reverse time migration with Hilbert transform based full wavefield decomposition. *Chin. J. Geophys.* **2016**, *59*, 4200–4211.
23. Guo, X.; Shi, Y.; Wang, W.; Jing, H.; Zhang, Z. Wavefield decomposition in arbitrary direction and an imaging condition based on stratigraphic dip. *Geophysics* **2020**, *85*, S299–S312. [CrossRef]
24. Wang, W.; McMechan, G. Vector-based elastic reverse time migration. *Geophysics* **2015**, *80*, 245–258. [CrossRef]
25. Lu, Y.; Liu, Q.; Zhang, J.; Yang, K.; Sun, H. Poynting and polarization vectors based wavefield decomposition and their application on elastic reverse time migration in 2D transversely isotropic media. *Geophys. Prospect.* **2019**, *67*, 1296–1311. [CrossRef]

Article

Least-Squares Reverse Time Migration in Imaging Domain Based on Global Space-Varying Deconvolution

Bo Li, Minao Sun *, Chen Xiang and Yingzhe Bai

SINOPEC Geophysical Research Institute, Nanjing 210003, China; libo.swty@sinopec.com (B.L.); xiangch.swty@sinopec.com (C.X.); baiyz.swty@sinopec.com (Y.B.)
* Correspondence: sunma.swty@sinopec.com; Tel.: +86-151-2102-9645

Abstract: The classical least-squares migration (LSM) translates seismic imaging into a data-fitting optimization problem to obtain high-resolution images. However, the classical LSM is highly dependent on the precision of seismic wavelet and velocity models, and thus it suffers from an unstable convergence and excessive computational costs. In this paper, we propose a new LSM method in the imaging domain. It selects a spatial-varying point spread function to approximate the accurate Hessian operator and uses a high-dimensional spatial deconvolution algorithm to replace the common-used iterative inversion. To keep a balance between the inversion precision and the computational efficiency, this method is implemented based on the strategy of regional division, and the point spread function is computed using only one-time demigration/migration and inverted individually in each region. Numerical experiments reveal the differences in the spatial variation of point spread functions and highlight the importance to use a space-varying deconvolution algorithm. A 3D field case in Northwest China can demonstrate the effectiveness of this method on improving spatial resolution and providing better characterizations for small-scale fracture and cave units of carbonate reservoirs.

Keywords: least-squares migration; imaging domain; point spread function; global space-varying deconvolution

Citation: Li, B.; Sun, M.; Xiang, C.; Bai, Y. Least-Squares Reverse Time Migration in Imaging Domain Based on Global Space-Varying Deconvolution. *Appl. Sci.* **2022**, *12*, 2361. https://doi.org/10.3390/app12052361

Academic Editor: Amerigo Capria

Received: 29 November 2021
Accepted: 19 February 2022
Published: 24 February 2022

1. Introduction

High-resolution seismic imaging is a critical tool to acquire information on underground structures from observed seismic data [1,2]. Reverse time migration (RTM) has no assumption for the high-frequency approximation in traditional ray methods and shows good performance in lateral velocity varying media [3,4]. However, as the real data may have limited borehole diameter for observation, irregular observation systems, and limited wavelet frequency band, and thus images obtained from RTM have limited resolution and poor amplitude fidelity, providing inaccurate estimates of underground reflectivity [5].

By using the reverse Hessian operator on the conventional migration, the least-squares migration (LSM) allows for higher resolution, fewer migration artifacts, and better fidelity [6]. Since the Hessian matrix is too large in scale and the direct inversion process is not realistic, the LSM method is always recast as a data-driven linear optimization problem. Due to the good performance on improving resolution, it has been investigated in seismic imaging for acoustic-wave single component data [7–18] and elastic-wave multiple component data [19–22]. Besides, the elastic least-squares migration is extended to obtain multiparameter images, such as P- and S-wave velocity and density, to cope with the trade-off effects [23,24]. However, since this data-domain LSM has a high dependency on the accuracy of wavelet and velocity, most achievements for LSM are mainly obtained in theoretical experiments. Furthermore, in 3D data processing in practice, this data-domain LSM always suffers from unstable convergence and excessive computational cost, which make it quite difficult in practical applications.

In contrast, the imaging-domain LSM aims to find a reasonable approximation for the Hessian instead of performing the expensive iterative data-fitting process. The diagonal Hessian operator is commonly used to enhance the amplitude balance by considering the illumination for observation, providing similar results as the classical amplitude-preserving migration [25]. In horizontally layered media, the Green function for migration can be also used as a deconvolution operator, thus processing the results from migration [26,27]. An unsteady-phase filter can be estimated using the images from the front and rear iterations in data-domain LSM thus replacing the inversion in Hessian matrix to calibrate and filter the migration results [28–31]. Point spread function (PSF) from one scattering point is consistent with a row of elements in the Hessian matrix, which physically describes the impact of single-point-scattered energy on underground media, including local illumination characteristics of the observation geometry, the space variation in velocity model, and the band-limiting effect in seismic wavelet and received data [32,33]. These approximations for Hessian in imaging domain LSM can be also used as preconditioners for the data domain LSM, which, after multiple iterations, can more quickly approximate the real reflection coefficient [34–37].

However, as PSFs are of significant space-varying characters, it is rough to apply a space-invariant PSF from a single scattering point on the whole area. The novelty of our method is that it introduces a regional division strategy in the image-domain least-squares migration based on a global space-varying PSF. As a result, it can divide the global PSF into sub-blocks, use a high-dimensional space-varying deconvolution in each sub-block, and eventually perform the data reduction for all the sub-blocks. Since the continuity of velocity model and illumination is much better in local sub-blocks, the proposed method can provide high-resolution images and have computation efficiency.

2. Methodology and Principle

2.1. Basic Theory on LSM

The forward problem in the classical migration imaging can be expressed as:

$$d = Lm, \tag{1}$$

where, L is the linear operator in the linear Born simulation, which expresses the relation between underground reflectivity model m and seismic scattering data d. The migration images can be obtained by projecting the adjoint Born operator L^T into the seismic scattering data d:

$$I = L^T d. \tag{2}$$

The adjoint Borning operators L^T can correctly reverse the propagation effects on the travel time and phase. However, due to problems such as limited observation apertures and limited data frequency band, the adjoint operator is not a good approximation for the inverse Born operators, which leads to unpreserved amplitude and low spatial resolution.

The LSM enables to provide high-fidelity images by minimizing the difference between synthetic and observed data, and the misfit function of the least-squares migration can be defined as,

$$C(m) = \frac{1}{2}||Lm - d_{obs}||_2^2. \tag{3}$$

When the objective function reaches the minimum, the reflectivity model m satisfies the following equation:

$$I = \left(L^T L\right) = Hm. \tag{4}$$

In this equation, H is the Hessian operator, which is the second derivative of the misfit function (3). This equation is always named as the Newton normal equation, which establishes the basic principle for the LSM in the imaging domain and physically reveals that the migration image obtained by the conventional migration method is the Hessianblurring version of the subsurface reflectivity model m. Thus, the seeking reflectivity model can be realized using the generalized inverse under the sense of the norm L_2.

$$m = \left(L^T L\right)^{-1} I = H^{-1} I. \tag{5}$$

2.2. *LSM in Imaging Domain*

2.2.1. The Global Space-Varying Point Spread Function

According to the acoustic-wave equation, Hessian can be expressed in the frequency domain as:

$$H(x_1, x_2, \omega) = L(x_1, \omega; x_s, x_r)^T L(x_2, \omega; x_s, x_r), \tag{6}$$

where x_s, x_r are the positions of shots and receivers, and L denotes the sensitivity kernel, satisfying

$$L(x, \omega; x_s, x_r) = f(\omega) G(x, x_s, \omega) G(x, x_r, \omega), \tag{7}$$

where, $f(\omega)$ is a seismic wave, $G(x_1, x_s, \omega)$ is the Green function excited from the shot position x_s.

The PSF is commonly used as the approximation for Hessian, including a row of elements in the Hessian matrix x_0.

$$K(x_2, x_0, \omega) = L(x_2, \omega; x_s, x_r)^T [L(x_0, \omega; x_s, x_r) \delta(x - x_0)]. \tag{8}$$

Here, $K(x_2, x_0, \omega)$ reflects the PSF of the perturbation at the point x_0.

Substitute Equation (8) into Equation (6), mark the point spread function at x_i as $\boldsymbol{K}_i$, and Hessian can be expressed as,

$$\boldsymbol{H} = \begin{bmatrix} \boldsymbol{K}_1 & \cdots & \boldsymbol{K}_i & \cdots & \boldsymbol{K}_m \end{bmatrix}^T. \tag{9}$$

Substitute the above equation into Equation (4) to obtain:

$$\boldsymbol{I}_i = \boldsymbol{K}_i \boldsymbol{m}. \tag{10}$$

The reflectivity model can be obtained by

$$\boldsymbol{m} = \frac{\boldsymbol{I}_i}{(\boldsymbol{K}_i + \alpha)}, \tag{11}$$

where α is the regularization parameter. The Equations (10) and (11) can be also expressed in the space domain as:

$$\boldsymbol{I}_i = \boldsymbol{K}_i * \boldsymbol{m}, \tag{12}$$

and:

$$\boldsymbol{m} = \boldsymbol{I}_i \otimes \boldsymbol{K}_i, \tag{13}$$

where, $*$ and $\otimes$ represent the operation of convolution and deconvolution, respectively. It reveals that, the migration image at x_i is the superposition of the convolution of the global reflectivity model and the point spread function at x_i; on the contrary, the reflectivity model at x_i can be computed within a high-dimensional deconvolution of the corresponding point spread functions.

2.2.2. Global Space-Varying Deconvolution

Equation (8) shows that the point spread function is exactly the image of the scattered wave radiated from the scattering point, which has a limited distribution in the space. Thus, the image-domain LSM can be based on the framework of the region division strategy, where the global point spread function can be divided into n blocks and the real impact of each point spread function is constrained within local neighborhoods s_n to the scattering point x_i, $i = 1, \ldots, s_n$,

$$K(x_n, x_i, \omega) = L(x_n, \omega; x_s, x_r)^T [L(x_i, \omega; x_s, x_r)\delta(x - x_i)]. \tag{14}$$

Furthermore, within the neighborhood, the media velocity and illumination conditions are slowly varied, thus the PSFs for each position in s_n can be assumed to be the same, and the reflectivity model for the neighborhood can be expressed as

$$\boldsymbol{m}_n = \sum_{i \in s_n} \boldsymbol{I}_i \otimes \boldsymbol{K}_i = \boldsymbol{I}_n \otimes \boldsymbol{K}_i. \tag{15}$$

Equation (15) means the regional reflectivity model for the neighborhood s_n can be obtained by performing local deconvolution of the regional RTM image and the point spread function. The entire model can be obtained by stacking all the regional parts, and the global space-varying PSF can be computed using only one-time demigration and migration.

3. Analysis of Numerical Experiments

This section focuses on analyzing the physical characteristics of the point spread function. Figure 1 shows the global point spread function for the 2D Marmousi model. We can find the variations of observation illumination, wave propagation, and limited wavelet frequency-band at different positions. The propagation effect is mainly reflected in amplitude performance. Compared to the PSFs in the shallow model, those in the deep show much worse focusing with increasing attenuation radius. Since the RTM and PSF results are computed under the same observation system, the deconvolution can remove the influence of limited illumination angles in RTM results. Moreover, the seismic wavelets should match the one in RTM as they may have a great influence on point spread functions with respect to phase, amplitude, and attenuation radius.

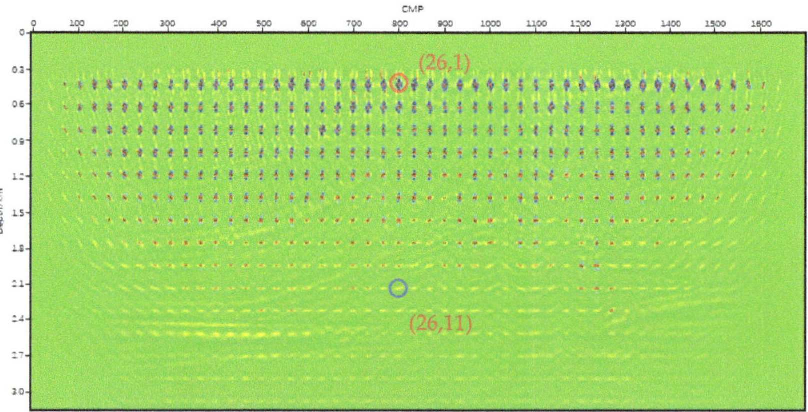

Figure 1. The global point spread function in the 2D Marmousi model.

Figure 2 displays the inverted images using different point spread functions (red and blue circles in Figure 1). Even with the same wavelet, the images using the PSFs show significant variations in the space. The PSF at grid (26, 1) has wider illumination angles and better focusing than the one at grid (26, 11), and thus it can more accurately reflect the illumination characteristics of shallow structures in the model. As a consequence, the above PSF allows for better descriptions of the shallow structures but provides inaccuracy in the deep model. In contrast, due to limited illumination angle and wavefield propagation effect, the image obtained by the PSF at grid (26, 11) is much closer to an RTM result.

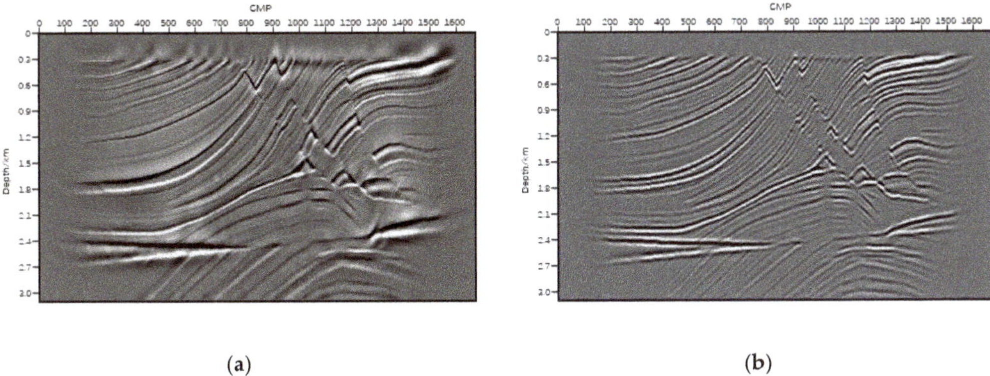

(a) (b)

Figure 2. Results from LSM in imaging domain by using single point spread function. (**a**) the point spread function at grid (26, 1), the shallow details are similar as reflectivity model; (**b**) the point spread function at grid (26, 11), the image is much closer to RTM result.

4. Numerical Experiment

The method is tested on the 3D Overthrust model. The model is 16 km, 5 km, and 3 km in size, and the migration velocity is obtained by Gaussian smoothing with a window of 40 m. The Ricker wavelet with the main frequency of 30 Hz is used in this experiment. As shown in Figure 3, the RTM result has an insufficient spatial resolution, and the amplitude of the reflectors attenuates rapidly as the depth increases.

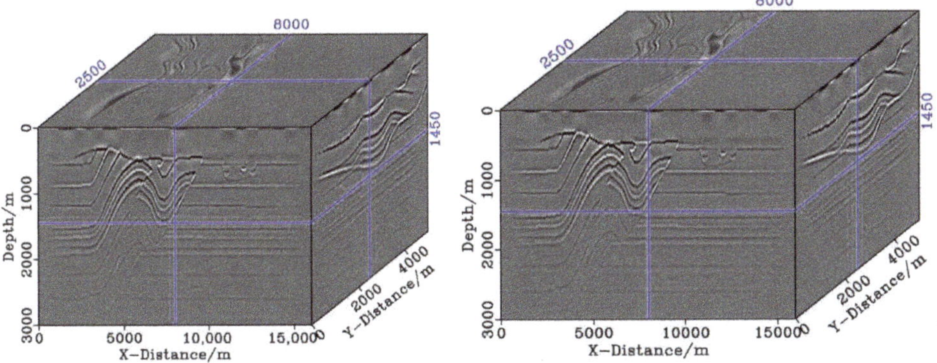

Figure 3. The 3D data volume of RTM in the 3D Overthrust model.

Then we compute the PSF using the same wavelet. Based on the migration velocity, the sampling interval (400 m, 400 m, 400 m) of the global PSF is set for the scattering model. To give intuitive insight into PSF, we present the X-Z tangent plane for point spread functions in Figure 4. As the geological structure and the illumination condition change, we can easily find that the difference of global PSFs in morphology is not negligible.

Figure 5 shows the results from the proposed imaging-domain LSM. Compared with the RTM results, the LSM results show higher resolution, fewer migration noise, more balanced amplitude in deep and better focusing on diffracted structures (as shown in Figure 6). To compare the spatial resolution between these results, we compute the vertical wavenumber spectrum of Figures 3 and 5 for the whole model and present them in Figure 7. From the vertical wavenumber spectrum, we can find that the result from the LSM method has a broader frequency-band, especially for mid-high wave numbers. It means that the proposed LSM method can estimate the Hessian operator and remove those Hessian effects on the RTM results, which thus provides a more accurate reflectivity model.

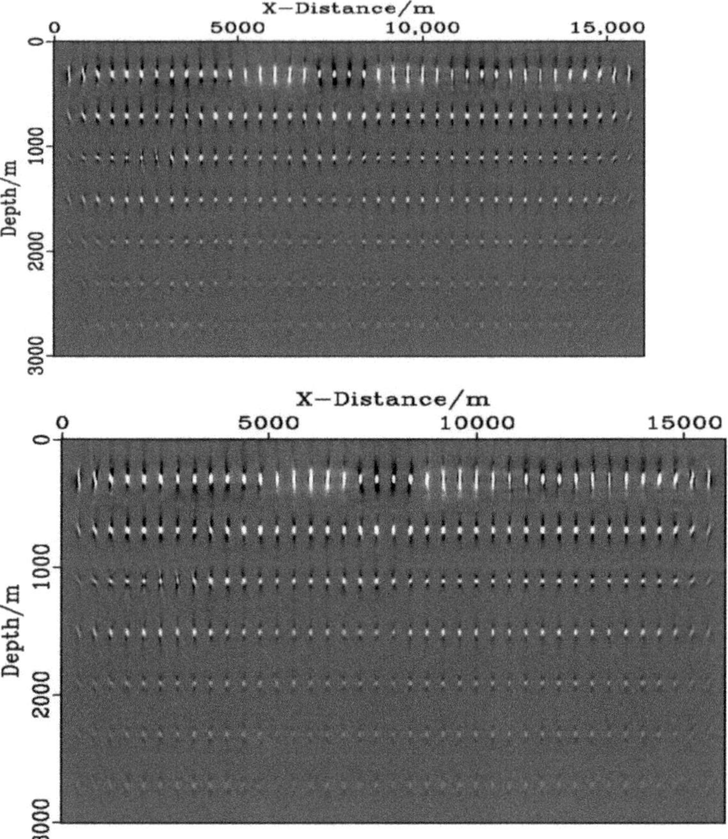

Figure 4. The 2D Display of the X-Z tangent plane of the point spread function.

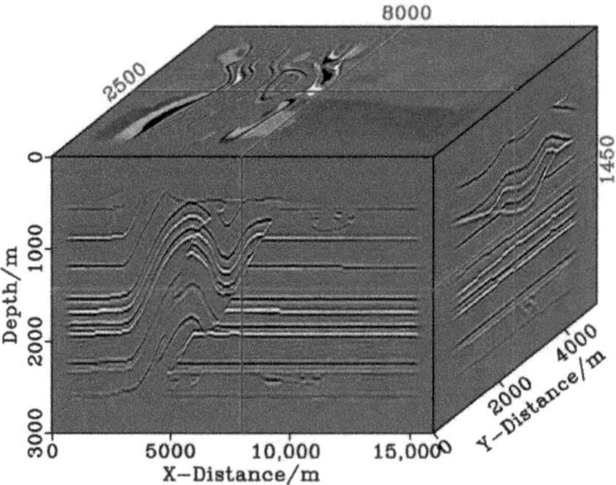

Figure 5. *Cont.*

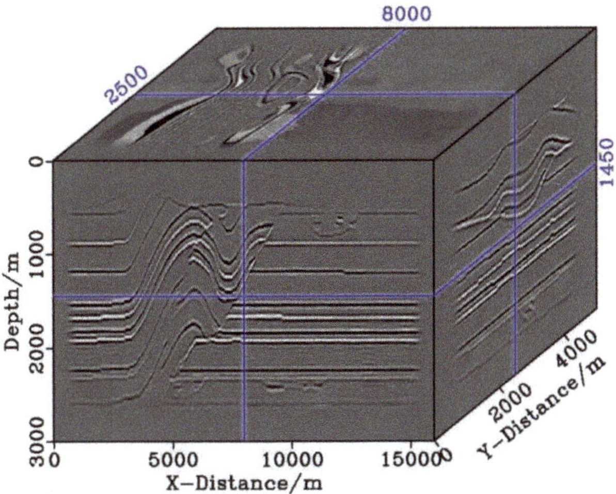

Figure 5. The 3D data volume of LSM in the 3D Overthrust model.

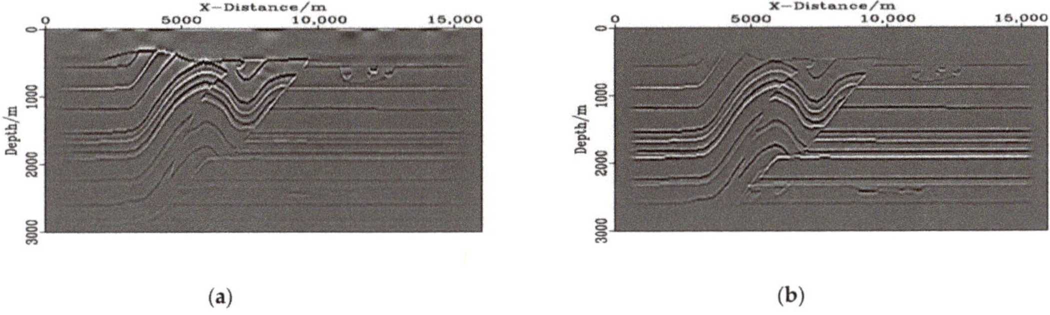

Figure 6. The 2D display of the X-Z tangent planes results from (a) RTM; (b) LSM.

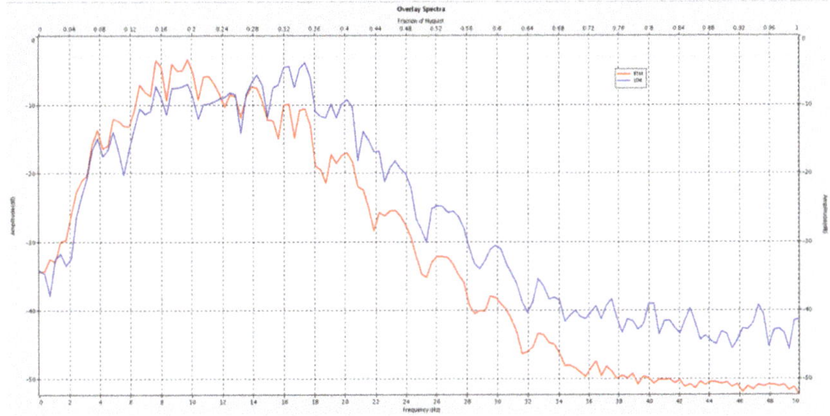

Figure 7. The comparison of vertical-wave-number spectrum between RTM (red) and LSM (blue) from Figures 3 and 5. The unit of y-axis is decibel, and x-axis indicates the wavenumber component.

5. Field Case

The target area is a part of the Tarim oil field in Northwest China, which is located in the middle of the North Tarim Uplift Belt to the east of the Halahatang depression. It has a large span, the target layer is deeply buried, the physical properties of the matrix are poor, and the type and scale of fracture caves are seriously affected by karst weathering. From the current development stage, the positioning accuracy of small and medium-sized karst cave cannot meet the requirements of resolution analysis of venting leakage around the well. In this work area, the geophone array recorded 33,000 shots at 5 m depth covering an area of 81 km^2. The main conundrum to be tackled in this work area is high-resolution imaging of Ordovician carbonate fracture-cavity reservoirs, with the depth ranging from 5.5 to 8 km. The migration velocity is built by a 3D travel time tomography as shown in Figure 8. The maximum cut-off frequency of the wavelet is 70 Hz.

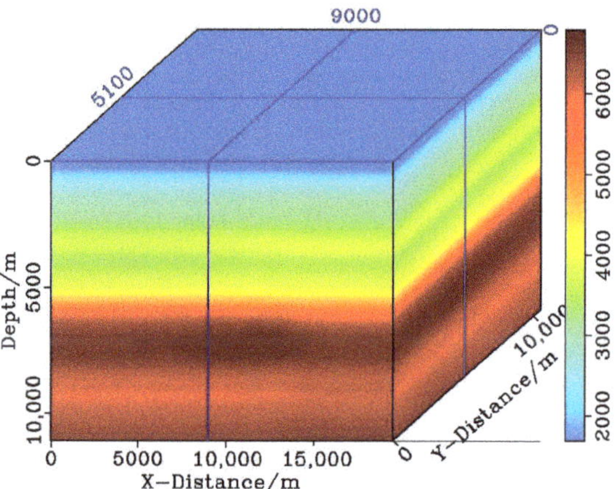

Figure 8. Migration velocity for Tarim oil field.

Figure 9 shows the results from RTM and LSM at the inline index of 200. Compared with the RTM results (Figure 9a), the LSM results (Figure 9b) provide higher spatial resolution, finer reflection events, more balanced amplitude, better focus ability, and more accurate fractured-cavity imaging. Figure 10 shows the zoomed region in Figure 9. We can find that the LSM result provides finer descriptions of fractured-cavity imaging and the boundaries between adjacent fractured-cavities, and the fractured-cavity bodies with small sizes in Ordovician carbonatite can be presented with higher resolution. Moreover, the deep fracture in the LSM result (Figure 10b) is better retrieved, which demonstrates that the proposed method is effective in fracture description. Figure 11 shows the vertical wave-number spectrum corresponding to the imaging results. The wave-number distribution of the LSM result is much broader, which lifts up at the low and mid-high sides of wave numbers. Besides, LSM can clearly provide information on cavities in the fracture zones, provide a high-quality scientific basis to analyze the development of fracture-controlled fractured-cavities and give guidance to the interpretation of reservoirs. To give an intuitive sight of the caves in Figure 10 denoted by red arrows, we compute the amplitude attributes from the RTM and LSM results and present a regional transverse section at the depth of these caves in Figure 12. We can find that the LSM results can distinguish the adjacent caves and provide better focusing, highlighting the proposed method can provide a high resolution in the horizontal direction and give a great help for reservoir interpretation and development. However, due to the limited observation aperture and shallow desert area, conventional RTM cannot provide correct images. The logging result located at the arrows in Figure 12 can also support the migration results, highlighting the effectiveness of the proposed method.

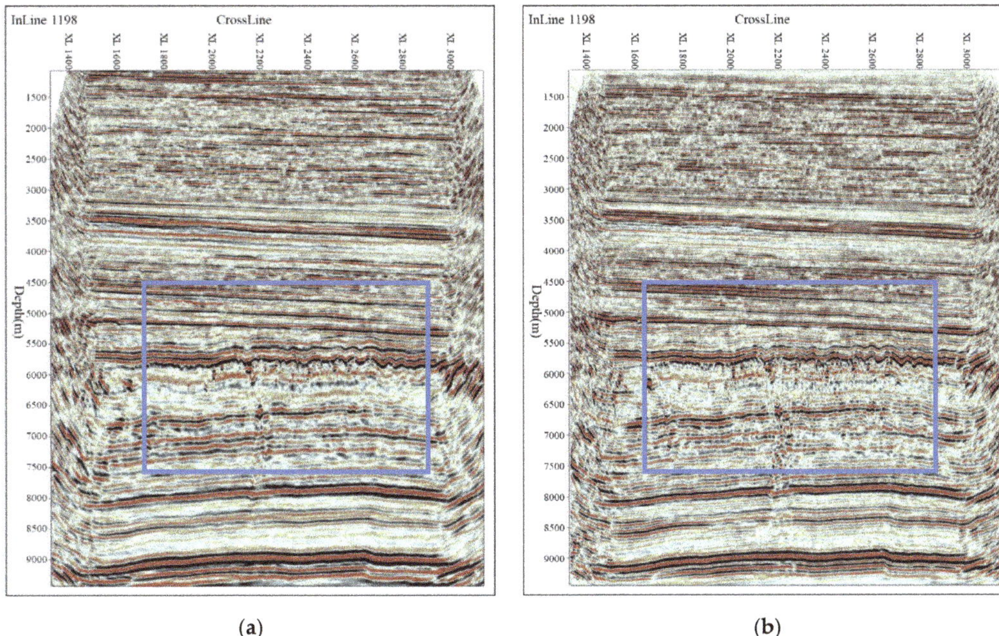

Figure 9. Comparison between migration results for CDP index of 1198. (**a**) results from RTM; (**b**) results from LSM.

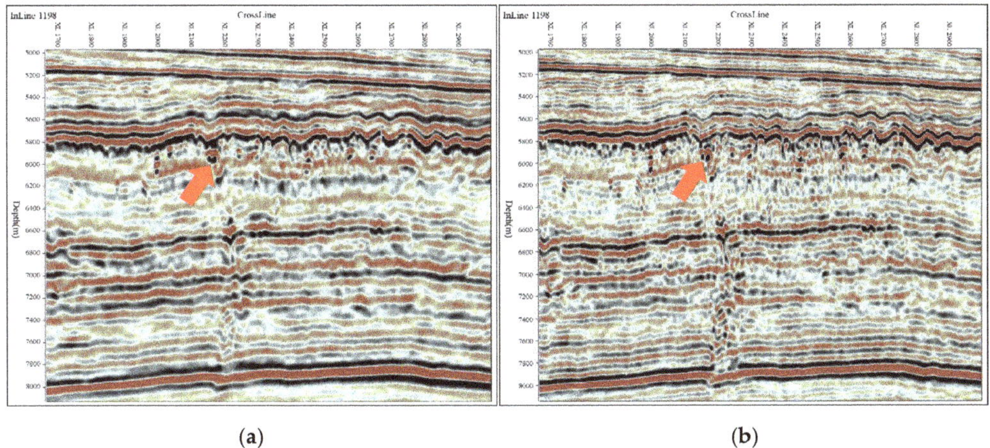

Figure 10. Locally zoomed-in reservoir sections marked in Figure 9. (**a**) results from RTM; (**b**) results from LSM.

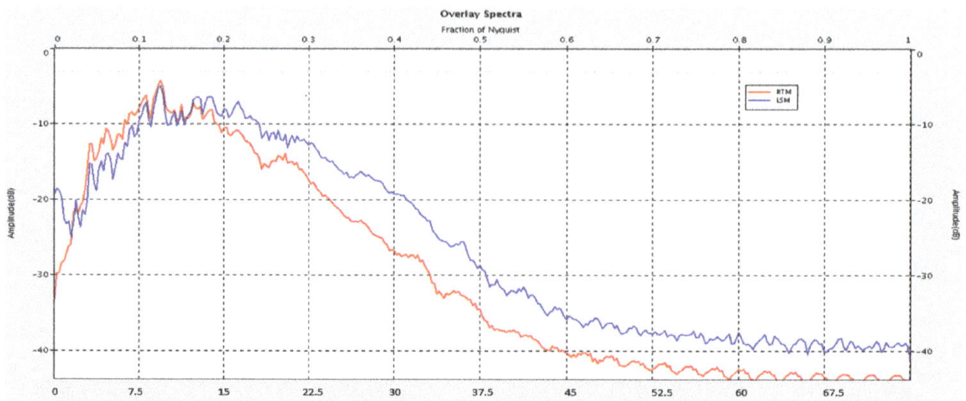

Figure 11. Comparison in vertical-wave-number spectrum (RTM: red line; LSM: blue line). The unit of y-axis is decibel, and x-axis indicates the wavenumber component.

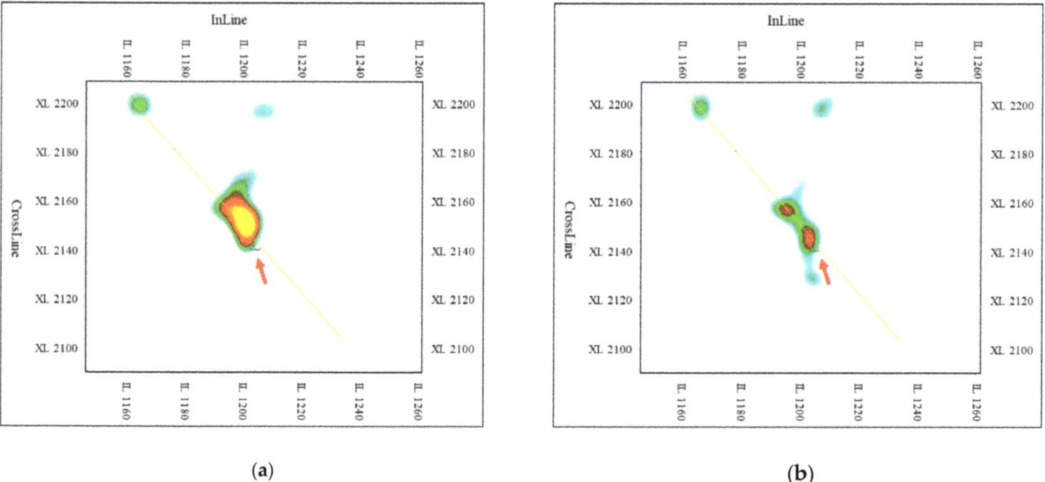

Figure 12. Comparison between results of amplitude attributes. (**a**) zoomed results from RTM; (**b**) zoomed results from LSM. The target caves are corresponding to the ones marked by red arrows in Figure 10.

6. Conclusions

Utilization of the point spread function and the global space-varying high-dimension deconvolution algorithm effectively solves the demand of excessive computation in the iterative data-fitting process of classical least-squares migration and alleviates the high dependency on the accuracy of seismic wavelet and velocity model. Both the theoretical analysis and the model experiment demonstrate that the global space-varying point spread function can effectively approximate the Hessian and the high-dimensional spatial deconvolution algorithm can replace the deburring effects from the Hessian operator. The application on a field 3D data suggests that the method is effective in improving the imaging resolution, especially for caves and fractures, and provide higher-quality seismic images for ultra-deep fractured-cavity reservoirs.

Author Contributions: Conceptualization, B.L.; methodology, M.S., Y.B.; software, M.S., C.X. and B.L.; validation, M.S., C.X.; formal analysis, B.L., Y.B.; investigation, M.S., B.L.; resources, M.S., C.X. and B.L.; writing—original draft preparation, M.S.; writing—review and editing, M.S., C.X. and B.L.; supervision, B.L. All authors have read and agreed to the published version of the manuscript.

Funding: This work was supported by the National Key Research and Development Program of China (No. 2017YFB0202900), and China Postdoctoral Science Foundation (Grant No. 2020M671539).

Institutional Review Board Statement: Not applicable.

Informed Consent Statement: Not applicable.

Data Availability Statement: Not applicable.

Conflicts of Interest: The authors declare no conflict of interest.

References

1. Moreno, W.; Galeano, I. Solution of high velocity anomalies imperceptible to seismic resolution, by means of synthetic models, Penobscot Field, Canada. *Rud. Geološko-Naft. Zb.* **2019**, *34*, 71–82. [CrossRef]
2. Moreno, W.; Galeano, I. Identification of high velocity anomalies, imperceptible to seismic resolution, by integration of seismic attributes, in the Penobscot Field, Canada. *Rud. Geološko-Naft. Zb.* **2019**, *34*, 13–26. [CrossRef]
3. Baysal, E.; Dan, D.K.; Sherwood, J.W.C. Reverse time migration. *Geophysics* **1983**, *48*, 1514–1524. [CrossRef]
4. McMechan, G.A. Migration by extrapolation of time-dependent boundary values. *Geophys. Prospect.* **1983**, *31*, 413–420. [CrossRef]
5. Nemeth, T.; Wu, C.; Schuster, G.T. Least-squares migration of incomplete reflection data. *Geophysics* **1999**, *64*, 208–221. [CrossRef]
6. Schuster, G.T. Least-Squares Cross-Well Migration. *SEG Tech. Program Expand. Abstr.* **1993**, *BG 4.6*, 110–113.
7. Tang, T.B. Target-oriented wave-equation least-squares migration/inversion with phase-encoded Hessian. *Geophysics* **2009**, *74*, WCA95–WCA107. [CrossRef]
8. Mandy, W.; Shuki, R.; Biondo, B. Least-squares reverse time migration/inversion for ocean bottom data: A case study. *SEG Tech. Program Expand. Abstr.* **2011**, 2369–2373.
9. Dai, W.; Fowler, P.; Schuster, G.T. Multi-source least-squares reverse time migration. *Geophys. Prospect.* **2012**, *60*, 681–695. [CrossRef]
10. Dai, W.; Schuster, G.T. Plane-wave least-squares reverse-time migration. *Geophysics* **2013**, *78*, S165–S177. [CrossRef]
11. Dai, W.; Wang, X.; Schuster, G.T. Least-squares migration of multisource data with a deblurring filter. *Geophysics* **2011**, *76*, R135–R146. [CrossRef]
12. Dutta, G.; Schuster, G.T. Wave-equation Q tomography. *Geophysics* **2016**, *81*, R471–R484. [CrossRef]
13. Dutta, G.; Schuster, G.T. Attenuation compensation for least-squares reverse time migration using the visco acoustic-wave equation. *Geophysics* **2014**, *79*, S251–S262. [CrossRef]
14. Zhang, S.; Luo, Y.; Schuster, G.T. Shot- and angle-domain wave-equation traveltime inversion of reflection data: Synthetic and field data examples. *Geophysics* **2015**, *80*, U47–U59. [CrossRef]
15. Zhang, S.; Luo, Y.; Schuster, G.T. Shot- and angle-domain wave-equation traveltime inversion of reflection data: Theory. Geophysics. *Geophysics* **2015**, *80*, S79–S92. [CrossRef]
16. Zhang, Y.; Biondi, B. Moveout-based wave-equation migration velocity analysis. *Geophysics* **2013**, *78*, U31–U39. [CrossRef]
17. Zhang, Y.; Duan, L.; Xie, Y. A stable and practical implementation of least-squares reverse time migration. *Geophysics* **2015**, *80*, V23–V31. [CrossRef]
18. Zhang, Y.; Ratcliffe, A.; Roberts, G.; Duan, L. Amplitude-preserving reverse time migration: From reflectivity to velocity and impedance inversion. *Geophysics* **2014**, *79*, S271–S283. [CrossRef]
19. Duan, Y.; Guitton, A.; Sava, P. Elastic least-squares reverse time migration. *Geophysics* **2017**, *82*, 4152–4157. [CrossRef]
20. Feng, Z.; Schuster, G.T. Elastic least-squares reverse time migration. *Geophysics* **2017**, *82*, S143–S157. [CrossRef]
21. Ren, Z.; Liu, Y.; Sen, M.K. Least-squares reverse time migration in elastic media. *Geophys. J. Int.* **2017**, *208*, 1103–1125. [CrossRef]
22. Chen, K.; Sacchi, M.D. Elastic least-squares reverse time migration via linearized elastic full waveform inversion with pseudo-Hessian preconditioning. *Geophysics* **2017**, *82*, 1–89. [CrossRef]
23. Sun, M.; Dong, L.; Yang, J.; Huang, C.; Liu, Y. Elastic least-squares reverse time migration with density variations. *Geophysics* **2018**, *83*, 1–62. [CrossRef]
24. Sun, M.; Jin, S.; Yu, P. Elastic least-squares reverse-time migration based on a modified acoustic-elastic coupled equation for OBS four-component data. *IEEE Trans. Geosci. Remote Sens.* **2021**, *59*, 9772–9782. [CrossRef]
25. Chavent, G.; Clément, F.; Gómez, S. Automatic determination of velocities via migration-based traveltime waveform inversion: A synthetic data example. *SEG Tech. Program Expand. Abstr.* **1994**, *13*, 1179–1182.
26. Hu, J.X.; Schuster, G.T. Poststack migration deconvolution. *Geophysics* **2001**, *66*, 939–952. [CrossRef]
27. Yu, J.H.; Schuster, G.T. Migration deconvolution vs. least squares migration. *SEG Tech. Program Expand. Abstr.* **2003**, 1047–1050.
28. James, E.R. Illumination-based normalization for wave-equation depth migration. *Geophysics* **2003**, *68*, 1371–1379.

29. Antoine, G. Amplitude and kinematic corrections of migrated images for nonunitary imaging operators. *Geophysics* **2004**, *69*, 1017–1024.
30. Symes, W.W. Migration velocity analysis and waveform inversion. *Geophys. Prospect.* **2008**, *56*, 765–790. [CrossRef]
31. Fletcher, R.P.; Nichols, D.; Bloor, R.; Coates, R.T. Least-squares migration—Data domain versus image domain using point spread functions. *Lead. Edge* **2016**, *35*, 157–162. [CrossRef]
32. Chen, C.; Hansen, H.H.G.; Hendriks, G.A.G.M.; Menssen, J.; Lu, J.-Y.; De Korte, C.L. Point Spread Function Formation in Plane-Wave Imaging: A Theoretical Approximation in Fourier Migration. *IEEE Trans. Ultrason. Ferroelectr. Freq. Control.* **2020**, *67*, 296–307. [CrossRef] [PubMed]
33. Jensen, K.; Lecomte, I.; Gelius, L.J.; Kaschwich, T. Point-spread function convolution to simulate prestack depth migrated images: A validation study. *Geophys. Prospect.* **2021**, *69*, 1571–1590. [CrossRef]
34. Alejandro, A.V.; Biondo, B.; Antoine, G. Target-oriented wave-equation inversion. *Geophysics* **2006**, *71*, A35–A38.
35. Ren, H.R.; Wu, R.S.; Wang, H.Z. Least square migration with Hessian in the local angle domain. *SEG Tech. Program Expand. Abstr.* **2009**, 3010–3014.
36. Ren, H.R.; Wu, R.S.; Wang, H.Z. Frequency domain wave equation based angular Hessian for amplitude correction. *SEG Tech. Program Expand. Abstr.* **2010**, 3145–3314.
37. Bai, J.; Yilmaz, O. Image-domain least-squares reverse-time migration through point spread functions. *SEG Tech. Program Expand. Abstr.* **2020**, 3063–3067.

applied sciences

MDPI

Article

Research on Initial Model Construction of Seismic Inversion Based on Velocity Spectrum and Siamese Network

Luping Sun *, Ling Ding and Xiangchun Wang

School of Geophysics and Information Technology, China University of Geosciences, Beijing 100083, China
* Correspondence: sunluping@cugb.edu.cn; Tel.: +86-158-1037-1315

Abstract: The initial model plays an important role in seismic inversion. Generally, the initial model is constructed by lateral extrapolation of parameters under horizons constraints. However, without horizon data, initial modeling becomes a challenging task. Velocity spectrum is a 2D image that can reflect the characteristics of the formations. We regard the problem of establishing the initial model as the problem of similarity analysis of seismic lateral characteristics and propose a method of establishing the initial inversion model based on velocity spectrum and Siamese network. Firstly, the lateral variation of formation characteristics is tracked on velocity spectra generated by common depth point (CDP) gathers. Then, the target tracking results at different CDP positions are obtained with the triple Siamese network. Finally, the discrete inversion parameters are extrapolated along the tracking paths to obtain the initial inversion model. The Siamese network can quickly obtain the similarity of 2D images and does not need manual labels. The theoretical and practical results show that our method can efficiently generate the initial model that conforms to the seismic structure and stratigraphic characteristics without the constraint of interpreted horizon data.

Keywords: velocity spectrum; triple structure Siamese network; lateral automatic tracking; horizon constraint framework; initial model of inversion

Citation: Sun, L.; Ding, L.; Wang, X. Research on Initial Model Construction of Seismic Inversion Based on Velocity Spectrum and Siamese Network. *Appl. Sci.* **2022**, *12*, 10593. https://doi.org/10.3390/app122010593

Academic Editor: José A. Peláez

Received: 31 August 2022
Accepted: 19 October 2022
Published: 20 October 2022

1. Introduction

Establishing the initial model is one of the necessary and key steps of the model-based inversion method, which has a great impact on the inversion effect. Traditionally, the method of generating the initial model is to extrapolate the inversion parameters observed from the wells or vertical seismic profile (VSP) along the interfaces picked from seismic profiles [1,2] or to extrapolate the seismic attributes [3–7]. In these methods, seismic horizons are necessary and manual interpretations are required. Horizon picking is one of the most time-consuming and labor-intensive steps [8]. Automatic horizon interpretation technology has been developing. Many geophysicists have proposed a variety of algorithms to improve the efficiency and accuracy of automatic horizons tracking [9–14], but these methods still need some prior information, such as manually setting up seed points or carrying out manual interpretation of large sets of strata. It is difficult to establish the initial model directly without manual interpretation. In addition, seismic inversion usually requires the initial model to cover the entire 3D region, but in practice, the automatic interpretation of the entire data volume is difficult to achieve. Therefore, there is a need for an efficient and accurate seismic extrapolation method to establish an initial model with the condition of lack of horizons.

Neidall et al. [15] presented the concept of velocity spectrum. The seismic gathers are automatically scanned with equal velocity intervals, and the velocity spectra are generated by stacking energy or similarity coefficient. Compared with artificial horizon interpretation, the velocity spectra are easier to generate. In conventional seismic data processing, stacking or migration velocity is obtained by velocity analysis on velocity spectra. The hyperbolas in the prestack seismic gathers reflect the characteristics of the interfaces, and the velocity spectrum contains the position and velocity information of each hyperbola in prestack gathers. Compared with the gathers, the features on velocity spectrum are more focused. The velocity spectrum is a 2D image, which can be used to identify the changes of the lateral characteristics of the formation.

Deep learning technology has shown much better performance than traditional neural network methods in speech and visual tasks, such as image classification [16], semantic segmentation [17], and image segmentation [18]. This is mainly due to the strong feature expression ability of deep networks such as convolutional neural network (CNN). In the field of exploration geophysics, deep learning has been successfully applied to seismic processing and interpretation [19,20]. However, most of the current applications need to manually make and label samples first, and then extract the corresponding type of feature information from the seismic images. For the establishment of the initial inversion model, it is still unable to get rid of the dependence on horizons. Deep learning has the potential to provide more global information. As an important part of computer vision, visual target tracking technology has been developed rapidly. The Siamese network uses the idea of similarity learning, describes the tracking problem as an object matching problem, and judges the position of objects by comparing the similarity of objects [21,22].

In this paper, we transformed the parameter extrapolation problem in constructing the initial inversion model into a 2D image tracking problem and proposed a method for constructing the initial model based on velocity spectrum and Siamese network. Firstly, the velocity spectra generated by CDP gathers were used to track the lateral variation of formation characteristics. Then, the target tracking results at different CDP positions were obtained by combining the similarity of the triple Siamese network analog velocity spectrum. Finally, the discrete inversion parameters were extrapolated along the tracking path to obtain the initial model. We proposed an improved triple Siamese structure, which adds an update branch to solve the lateral variation of velocity spectrum characteristics during target tracking. This improvement takes dynamic update into tracking. In order to verify the applicability of the method, we carried out tests on theoretical and practical data. The results show that it can automatically obtain the extrapolation paths and can be used to establish the initial model of seismic inversion.

2. Materials and Methods

2.1. Velocity Spectrum

The velocity spectrum [15] is a 2D image generated by scanning and stacking (or correlating) energy along different velocities on each CDP gather using the similarity criterion. The x and y axes are the velocity and two-way travel time, respectively. The velocity spectrum is composed of a series of energy clusters, and the velocity information is determined according to the abscissa position of the energy clusters. Most of the current research on velocity spectrum focuses on improving the resolution of velocity spectrum and then improving the accuracy of velocity pickup [23–29]. The application of deep learning in velocity analysis almost focuses on the automatic pickup of energy clusters in velocity spectrum [30–32].

In addition to the position and velocity information that can be used for velocity analysis in conventional seismic processing, velocity spectrum is also a 2D image that can reflect formation characteristics. Velocity spectra also have advantages in identifying lateral characteristics of formation. The reasons are as follows: firstly, the interference of complex underground structures and noise always leads to incomplete, staggered, and amplitude varying hyperbolas, while the velocity spectrum scans and superimposes the events along the hyperbola, which can attenuate the non-hyperbolic noise to a certain extent. Secondly, in most cases, the energy clusters in velocity spectrum corresponds to the interfaces. Because the formation interfaces have relatively coherent structure in the underground space, the velocity spectra of adjacent CDP points have similarity, and the characteristics of the formation interface can be obtained from the velocity spectra by this similarity. The velocity spectrum Formula (1) is as follows:

$$S = \frac{\sum_{j=0}^{M}\left(\sum_{i=1}^{N} A_{i,j}\right)^2}{I\sum_{j=0}^{M}\sum_{i=1}^{N}\left(A_{i,j}\right)^2} \quad (1)$$

where M is the length of the time window, N is the offset length, I is the number of seismic traces in the CDP gather, and $A_{i,j}$ is the amplitude at offset i and time j. According to Formula (1), if all seismic traces are the same, S equals to 1. If each seismic trace is a random value, S approaches 0. Only when the scanning velocity is equal to the normal move-out (NMO) velocity, the waveforms of each trace are the most similar, in-phase stacking is realized within the time window, and S is close to 1. Here, we use real seismic data after conventional data processing and prestack migration to show the consistency between energy clusters in velocity spectrum and formation interfaces. Figure 1a is the common reflection point (CRP) gather at the location of CDP number 30 (CDP 30), which has undergone NMO removal processing with v = 1500 m/s. Figure 1b is the velocity spectrum generated from the gather shown in Figure 1a. The x and y axes are the velocity and two-way travel time, respectively. The center of the energy clusters in velocity spectrum corresponds to the maximum value of stacking energy. Figure 1c is the stack section. It can be seen that the centers of the energy clusters correspond to the events on the seismic section. Each energy cluster corresponds to an obvious formation interface. Figure 2a–c are velocity spectra of CDP 32, CDP 34, and CDP 36, respectively. The three velocity spectra have high similarity. With the similarity of the energy clusters in lateral, the positions of the formation interfaces can be obtained by tracking the energy clusters.

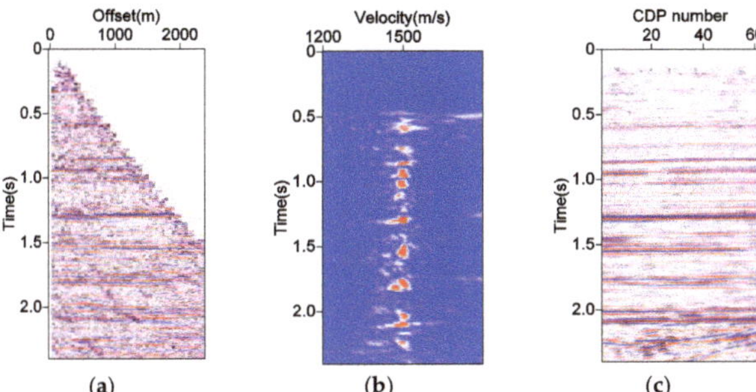

Figure 1. Consistency between energy clusters in velocity spectrum and formation interfaces. (**a**) CDP gather at CDP 30; (**b**) velocity spectrum generated from the gather shown in (**a**); (**c**) stack section.

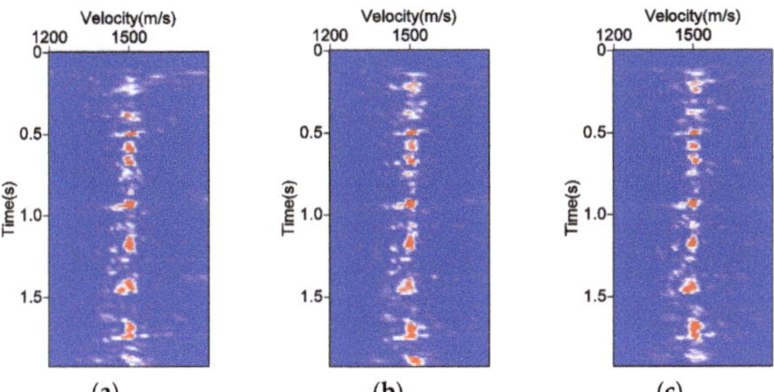

Figure 2. Similarity of adjacent velocity spectra. (**a**) Velocity spectrum of CD P32; (**b**) velocity spectrum of CDP 34; (**c**) velocity spectrum of CDP 36.

2.2. The Triple Structure Siamese Network for Velocity Spectra Lateral Target Tracking

For lateral target tracking in velocity spectra, the prior materials are the images of the energy clusters in the given windows. The tracking algorithm needs to overcome the challenges of target deformation, background interference, scale change, and angle rotation, and also needs to take into account the accuracy and efficiency. The Siamese network can find a set of parameters, so that the similarity of input images is large when they belong to the same category and small when they belong to different categories. Another advantage of the Siamese network is that the input is a pair of images rather than an image, so it can naturally increase the amount of training data and make full use of limited datasets to train, which is very important in the field of target tracking. In this paper, we improved the traditional Siamese network structure by adding an input update branch. A triple structure Siamese network for velocity spectra lateral target tracking is presented. We set the target area as a positive sample, and the background area and other morphological energy clusters as negative samples. Firstly, the tracking algorithm extracts the features of the target through a series of convolution operations and trains the classifier. Next, the well-trained classifier is used to find the most similar region in the velocity spectra of different CDP positions. Finally, the lateral extrapolation path of inversion parameters can be obtained through the change trend of the target positions. The added update branch can take the prediction result of the current position as an input and update the initial target according to the current tracking result, so as to adapt to the lateral change of the velocity spectrum.

2.2.1. Triple Siamese Network Structure

The traditional Siamese network always uses the initial image as the tracking target. However, the velocity spectrum varies in the lateral direction. With the deepening of the tracking process, the information contained in the initial sample is insufficient to track the subsequent targets. We solve this problem by adding an update of target features in the tracker. The velocity spectrum samples at the initial position provide the most basic characteristics of the target and play a leading role in the tracking process. With the lateral movement of the target position, tracking by fusing the previous position features is better than using only the initial samples. Based on this consideration, we proposed a triple structure Siamese network based on the traditional Siamese network. We added an update branch of the current target and used the current prediction result as the target to track the velocity spectrum of the next position. During the tracking process, the current target will be updated with the result of target tracking. The structure of the triple velocity spectrum tracking network is shown in Figure 3.

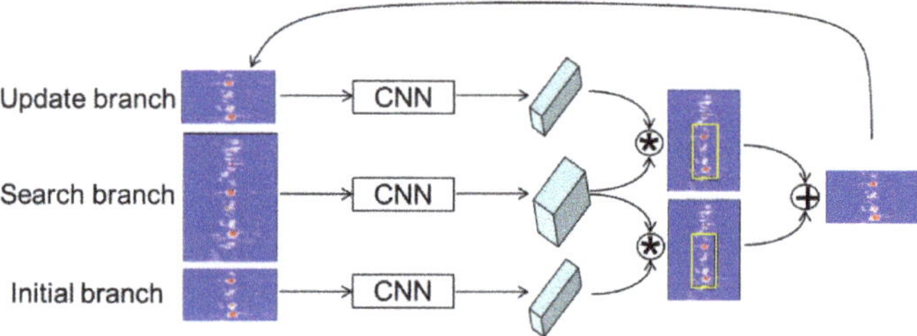

Figure 3. Structure of the triple velocity spectrum tracking network.

Different from the traditional Siamese network, the triple network consists of three branches: an initial branch I with an initial target as an input, a search branch S with a search area as an input, and an update branch U with a tracking result at the previous location as an input. The backbone models in the three branches share the same CNN architecture. Through the same network model, the responses of I, S, and U are $\varphi(I)$, $\varphi(S)$, and $\varphi(U)$, which are embedded into the feature space of subsequent tasks. I is the initial branch, and the specified initial velocity spectrum is used as the sample. I remains unchanged throughout the tracking process. U is the update branch, that is, the previous target tracking result. U will be updated with the tracking process. In order to embed the information of these branches, we use the feature map of the updated branch and the feature map of the search area to perform cross-correlation operations. Similar to the traditional Siamese network, the triple network uses a full convolution network for feature extraction, and each channel also generates a corresponding mapping response R. Since there are three branches, the mapping results of two channels are generated, which are:

$$R_1 = \varphi(S) * \varphi(I) \tag{2}$$

$$R_2 = \varphi(S) * \varphi(U) \tag{3}$$

where $*$ represents a cross-correlation operation. R_1 is the mapping result of the initial branch and the search branch, and R_2 is the mapping result of the update branch and the search branch. In order to use the two-mapping information, the mapping results are fused weighted R_f:

$$R_f = a_1 R_1 + a_2 R_2 \tag{4}$$

where a_1 and a_2 are the weight coefficients, and $a_1 + a_2 = 1$.

2.2.2. Weight Coefficients

The weight coefficient is determined by the similarity between the current target image and the previously saved image. Here, we use a perceptual Hash algorithm to judge the similarity [33]. We know that the high-frequency information in an image can provide details, and the low-frequency information can describe the frame. Hash algorithm is a method to detect the similarity between images with the low-frequency information of images. First, the image is down-sampled to 8×8 to remove high-frequency information and discard the difference caused by different sizes. Next, the image is converted into a gray-scale image, and its gray-scale average value is calculated. Further, the gray value of each pixel in the image is compared with the average gray value. If the gray value is greater than or equal to the average value, it is recorded as 1, and if the gray value is less than the average value, it is recorded as 0. Finally, these 64 values are combined to form the Hash fingerprint of this image. This method is fast and efficient and is not affected by image size or scale. The main process of determining the weight coefficient with Hash algorithm is as follows:

Step 1: calculate the Hash fingerprint of the initial target image and each predicted image.

Step 2: calculate the number *Num* of different points of the Hash fingerprint between the target image and the predicted image.

Step 3: determine the weight coefficients a_1 and a_2 according to the value of *Num*. There are several situations. If $Num < 5$, the current target is considered to be similar to the initial target image, replace the initial target with the current target, and set $a_1 = 0$, $a_2 = 1$. If $5 < Num < 10$, perform image fusion and set $a_1 = 0.5$, $a_2 = 0.5$. If $Num > 10$, there is a large difference between the current target and the initial target image, the current image is discarded and set $a_1 = 1$, $a_2 = 0$, that is, the initial target image will not be updated. The reasonable threshold (5 and 10 or other number) can be determined by manual test.

2.2.3. Loss Function and Network Parameters

In the mapping result, the points within the search area are positive samples, and the points outside the area are negative samples. The loss function for each point in the mapping result is:

$$L_p(x, s) = \lg(1 + e^{-xs}) \tag{5}$$

where s is the true value of the point, x is the label of the point, and $x \in \{+1, -1\}$. The overall loss of the mapping result is the average of the losses of all points, that is:

$$L_{ALL}(x,s) = \sum_{D} L_p(x(z), s(z)) \tag{6}$$

where z is the position, $x(z)$ is expressed as:

$$x(z) = \begin{cases} 1 & h|z-c| \leq R \\ -1 & others \end{cases} \tag{7}$$

where h is the step size of the network, c is the center point, and R is the radius of the search area.

The weight coefficients of the similarity discriminant function are solved by the gradient descent method to minimize the error between the sample labels x and the similarity discriminant function. The specific parameters of the network are shown in Table 1. The maximum pooling layer is used after the first two convolution layers, respectively. The Relu nonlinear activation function is used after each convolution layer except the last layer. The batch normalization (BN) layer is embedded after each linear layer. There is no padding in the network.

Table 1. Network parameters.

Network Layers	**Conv1**	**Pool1**	**Conv2**	**Pool2**	**Conv3**	**Conv4**	**Conv5**
Convolution kernel	11 × 11	3 × 3	7 × 7	3 × 3	3 × 3	3 × 3	3 × 3
Step size	2	2	2	2	1	1	1

2.3. Workflow

After the initial target image on the velocity spectrum is given, all subsequent images are compared with the initial target image for similarity. The Siamese network performs full convolution on the search image. In order to find the position of the target in the searched image, all possible positions can be tested exhaustively, and the image with the greatest similarity to the target can be determined. The triple Siamese structure actually calculates the cross-correlation between the two input branches and the search branch, determines the weight coefficient of the branch fusion according to the similarity of the image, and adapts to the lateral change of the velocity spectrum by updating the initial target. The main implementation steps of the method are as follows:

Step 1: generate velocity spectrum images of all positions.

Step 2: extract the target image feature H_i at the specified position and within the specified window by using the initial branch.

Step 3: extract the image feature H_c of the search area at the current position by using the search branch.

Step 4: calculate the cross-correlation between the features H_i and H_c to obtain the target response R_1.

Step 5: extract the target image feature H_r at the specified position and within the specified window by using the update branch.

Step 6: calculate the cross-correlation between the features H_r and H_c to obtain the target response R_2, and take R_2 as the current position tracking result.

Step 7: determine the fusion weight coefficients a_1 and a_2, and obtain $R_f = a_1 R_1 + a_2 R_2$ as the new input of the update branch.

Step 8: move the window, repeat steps 3–7 until the velocity spectrum of this position is traversed, and then end the current position task.

Step 9: move the position, repeat steps 2–8 until the velocity spectra of all positions are traversed, and then the whole task is completed.

3. Model Test

In order to verify the feasibility of the method, we carried out tests on theoretical model data and real seismic data. The lateral extrapolation path of inversion parameters is obtained by tracing the lateral position of velocity spectrum according to the method in this paper. In our tests, the interval velocity is used as the inversion parameter to check the extrapolation result. The tests were carried out on a PC equipped with Intel (R) i7 12,700 CPU, an RTX3060TI GPU, 16 GB RAM, and Ubuntu 19 operating system.

3.1. Theoretical Model

A theoretical model with four interfaces is designed to synthesize seismic data and test the method. The depth of the model is 5500 m and the sampling interval is 5 m. The width of the model is 1000 m, and 100 CDP points are set. The theoretical model of interval velocity is shown in Figure 4. CDP gathers are obtained through forward modeling, with the time length of 1 s and the maximum offset of 3100 m. Figure 5a–c are the CDP gathers at positions of CDP 30, CDP 50, and CDP 70, respectively. For each CDP gather, NMO removal processing is performed according to the root mean square velocity of 1500 m/s and a velocity spectrum is generated. Figure 6a–c are the velocity spectra of CDP 30, CDP 50, and CDP 70, respectively.

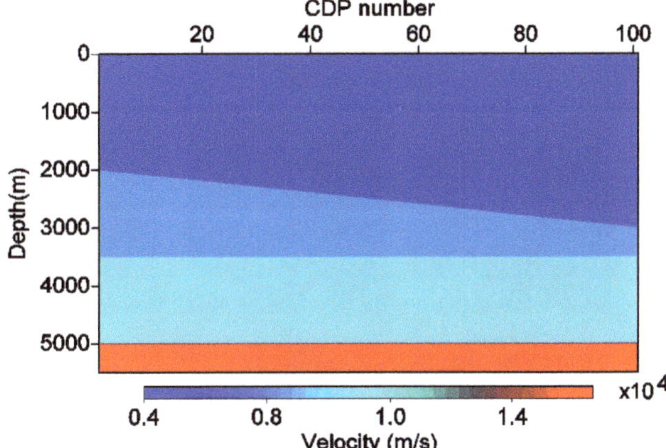

Figure 4. Theoretical model of interval velocity.

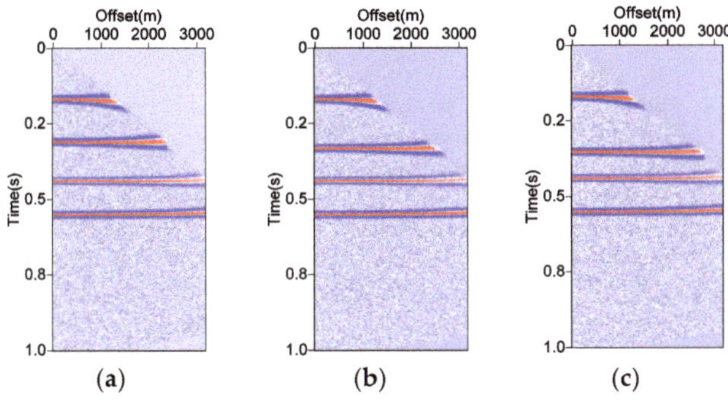

Figure 5. CDP gathers at different positions. (**a**) CDP 30; (**b**) CDP 50; (**c**) CDP 70.

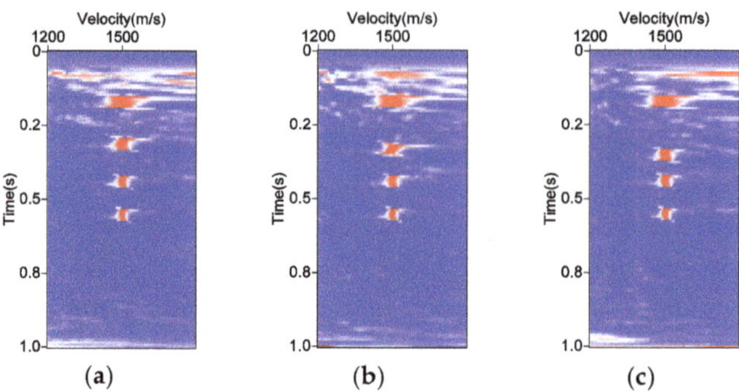

(a) (b) (c)

Figure 6. Velocity spectrum. (**a**) CDP 30; (**b**) CDP 50; (**c**) CDP 70.

3.2. Model Training

The velocity spectrum of CDP 50 is used as the tracking target of the training dataset. As shown in Figure 6b, there are four energy clusters. We take the energy cluster of 0.2–0.35 s (as shown in Figure 7a) as an example to explain the generation of a training dataset. Based on the energy cluster of 0.2–0.35 s, 100 training samples labeled as positive are generated by stretching, compression, and filtering. Figure 7b,c are examples of images after deformation and stretching, and Figure 7d–f are examples of images after filtering. The positive samples are set to 1.

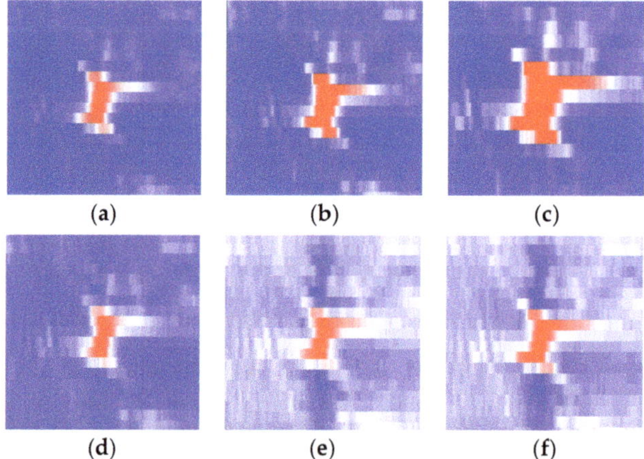

(a) (b) (c)

(d) (e) (f)

Figure 7. Examples of positive sample. (**a**) Tracking target image; (**b**) image after 1.5 times stretching; (**c**) image after 3 times stretching; (**d**) image after low-pass filtering; (**e**) image after bandpass filtering; (**f**) image after high pass filtering.

The energy clusters of 0.1–0.2 s, 0.35–0.5 s, and 0.5–0.65 s in Figure 6b are stretched, compressed, and filtered to generate 300 training samples marked as negative, as shown in Figure 8. The negative samples are set to 0. The training process has 500 iterations, and the learning rate is set to 0.0001. After this training process is completed, set the tracking target as the energy cluster of 0.35–0.5 s, and set the energy clusters of 0.1–0.2 s, 0.2–0.35 s, and 0.5–0.65 s as the negative samples, and then repeat the same training process.

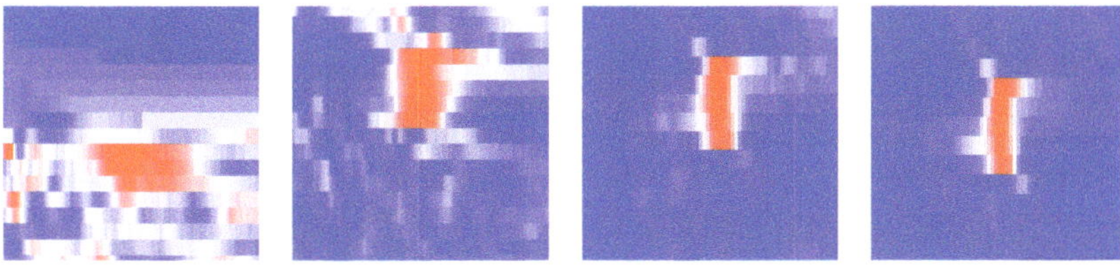

Figure 8. Examples of negative sample.

3.3. Tracking Results

The velocity spectrum images of all 100 CDP points are sequentially tracked. The tracking results of CDP 50 to CDP 55 are shown in Figure 9. The red rectangle is the tracking target, and the yellow rectangle is the velocity spectrum tracking results of different CDPs. The tracking path is formed by connecting the midpoint positions of the tracking results. As shown in Figure 10, the four color lines are the paths of the energy clusters to track laterally. It can be found by comparing Figure 4 that the four tracking paths are consistent with the four interfaces in the model. By extrapolating seismic parameters along the tracking path, the initial model conforming to the underground structure can be obtained.

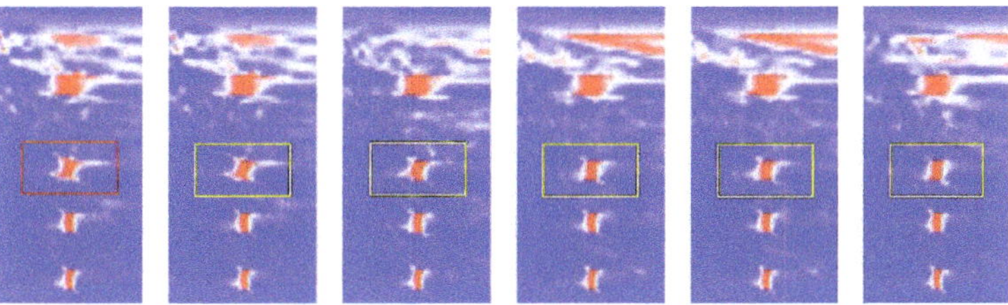

Figure 9. The tracking results of CDP 50 to CDP 55.

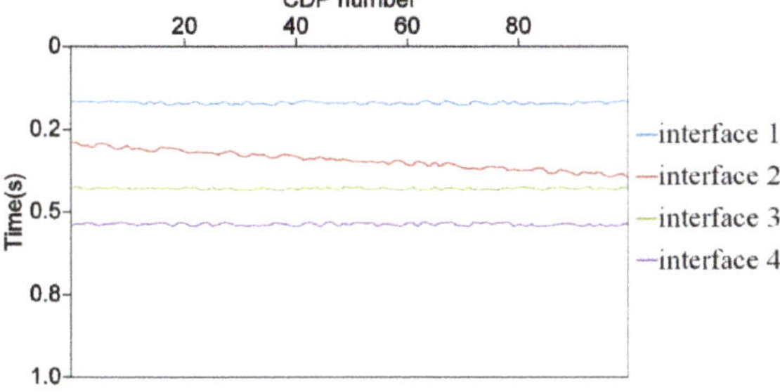

Figure 10. Result of tracking paths.

3.4. Initial Velocity Model

The interval velocity curve is extracted at the theoretical velocity model CDP 30 to simulate the data of a well. The layer velocity curve at CDP 30 is extrapolated under the paths constraint shown in Figure 10 to obtain a interval velocity model, as shown in Figure 11, which is consistent with the theoretical model shown in Figure 4. The vertical difference is because the theoretical velocity model is in depth domain and the extrapolation result is in time domain.

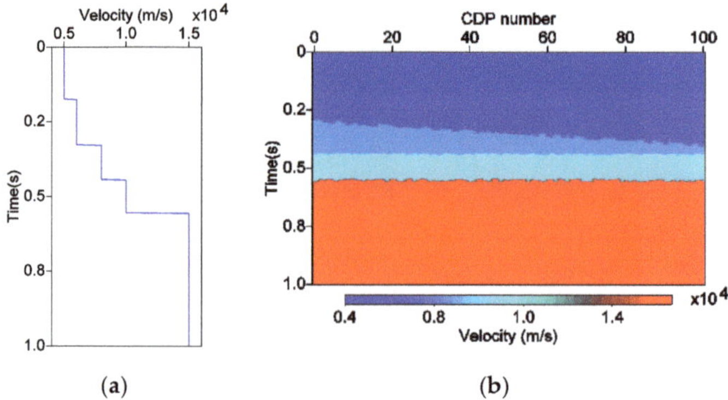

(a) (b)

Figure 11. (**a**) Interval velocity curve at CDP 30; (**b**) extrapolation velocity result under the paths constraint.

4. Real Data Applications

In the real data test, we used the land seismic data from eastern China. The CRP gathers after prestack time migration were used to generate velocity spectra. The number of CRP gathers was 500. The maximum offset was 2200 m. The longitudinal time was 2.5 s. The sampling interval was 2 ms. The stack section is shown in Figure 12a. Figure 12b is the CRP gather at CDP 203, and Figure 12c is the velocity spectrum at CDP 203.

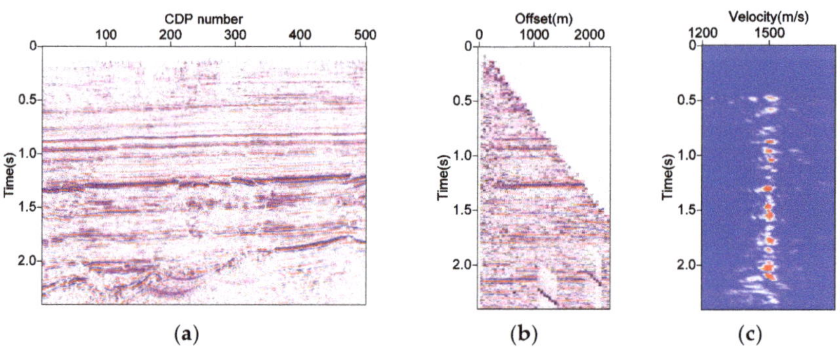

(a) (b) (c)

Figure 12. (**a**) Stack section; (**b**) CRP gather at CDP 203; (**c**) velocity spectrum at CDP 203.

The velocity spectrum at CDP 203 was used as the tracking target of the training data set. Since there was no obvious boundary between the energy clusters in the velocity spectrum in Figure 12, we used the sliding window to realize the target tracking. The time window length is given as 250 ms. The tracking windows are shown in Figure 13a. The real data samples were made in the same way as the theoretical data. The training process had 500 iterations, and the learning rate was set to 0.0001. The position of the window was changed and the same training process repeated until the tracking of the target in the whole time range was completed. The tracking results of all CDP points were connected to form the final tracking paths as shown in Figure 13b. Comparing Figure 13b with Figure 12a, the tracking paths are consistent with the interfaces. The faults between CDP 200–CDP 350 and 1.3–1.8 s can be correctly identified on the tracking results. Along the tracking path, the parameter to be inversed can be extrapolated from the well, and the initial inversion model consistent with the structure can be obtained.

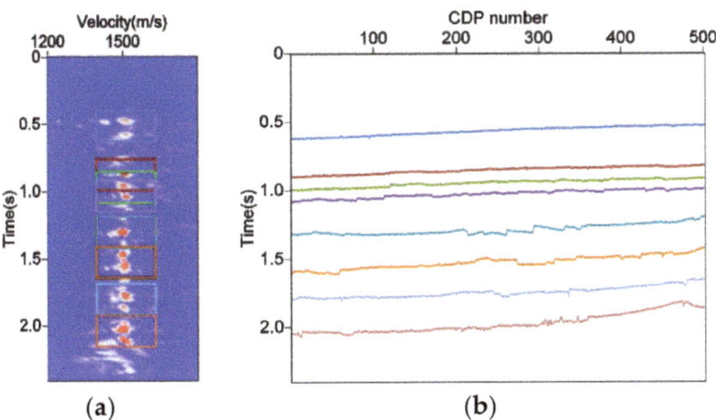

Figure 13. (**a**) Tracking windows; (**b**) the result of tracking paths.

5. Discussions

This method also has some application limitations. Because the characteristics of the energy clusters in the velocity spectrum are closely related to the reflection characteristics of the formation, the method can obtain ideal tracking results when the reflection characteristics from the same formation interface are relatively stable in lateral. In contrast, if the reflection characteristics of the formation change dramatically in lateral, the velocity spectrum tracking will be unstable and the results may be multi-solution. This requires manual intervention or adding constraints to obtain reliable tracking results, such as limiting the trend of the tracking path within a certain range or limit the maximum change of the tracking results of adjacent traces. In addition, if the signal-to-noise ratio of real seismic data is very low, the velocity spectrum features will not be significant, and it will be difficult to extract features and track targets.

In the velocity spectrum tracking of real seismic data, there may be difficulties such as unclear clusters features or background interference. In these cases, the accuracy of target tracking can be improved by changing the window length. The tracking task can be carried out with different window lengths, and the corresponding similarity results will be output for each window length. The results with the highest similarity can be extracted and combined as the final target tracking result.

The extrapolation path of geophysical parameters can be obtained by lateral tracking of velocity spectrum, which can be used as a constraint framework to construct the initial model. The essence of the method is the similarity of the characteristics of velocity spectrum in lateral, that is, the invariable part of the velocity spectrum. In fact, the characteristics of the velocity spectrum have certain changes in lateral, and this change also represents the lateral changes of geophysical parameters to a certain extent. In the absence of gooddata, how to make full use of the characteristics of velocity spectrum to provide more geophysical information will be the next possible research direction.

6. Conclusions

A lateral tracking method of velocity spectrum based on a triple Siamese network structure is proposed in this paper. With this method, the positions of the target image on the velocity spectrum of each CDP can be tracked. The results of the tracking paths can constrain the lateral extrapolation of seismic parameter to establish an initial inversion model, for example, an initial velocity model for prestack depth migration or P-wave, S-wave and density models for elastic parameters inversion. The method does not depend on the interpretation horizons and manual annotation samples. The theoretical and practical results show that the method can efficiently generate the initial model that conforms to the seismic structure and stratigraphic characteristics without the constraint of interpreted horizon data.

Author Contributions: Methodology, L.S.; validation, L.S. and L.D.; writing—original draft preparation, L.S.; writing—review and editing, L.D. and X.W. All authors have read and agreed to the published version of the manuscript.

Funding: This work was supported by the National Natural Science Foundation of China (grantno. 42204128).

Institutional Review Board Statement: Not applicable.

Informed Consent Statement: Not applicable.

Data Availability Statement: The data presented in this study are available on request from the corresponding author.

Conflicts of Interest: The authors declare no conflict of interest.

References

1. Mallet, J.L. *Geomodeling*; Oxford University Press: Oxford, UK, 2002.
2. Caumon, G.; Collon-Drouaillet, P.; de Veslud, C.L.C.; Viseur, S.; Sausse, J. Surface-based 3D modeling of geological structures. *Math. Geosci.* **2009**, *41*, 927–945. [CrossRef]
3. Hampson, D.; Todorov, T.; Russell, B. Using multi-attribute transforms to predict log properties from seismic data. *Explor. Geophys.* **2000**, *31*, 481–487. [CrossRef]
4. Hansen, T.M.; Mosegaard, K.; Pedersen-Tatalovic, R.; Uldall, A.; Jacobsen, N.L. Attribute-guided well-log interpolation applied to low-frequency impedance estimation. *Geophysics* **2008**, *73*, R83–R95. [CrossRef]
5. Dave, H. Image-guided 3D Interpolation of Borehole Data. In *SEG Technical Program Expanded Abstracts 2010*; Society of Exploration Geophysicists: Houston, TX, USA, 2010; pp. 1266–1270.
6. Hale, D.; Wu, X. Horizon volumes with interpreted constraints. *Geophysics* **2015**, *80*, IM21–IM33.
7. Karimi, P.; Fomel, S.; Zhang, R. Creating detailed subsurface models using predictive image-guided well-log interpolation. *Interpreatation* **2017**, *5*, T279–T285. [CrossRef]
8. Hoyes, J.; Cheret, T. A review of "global" interpretation methods for automated 3D horizon picking. *Lead. Edge* **2011**, *30*, 38–47. [CrossRef]
9. Harrigan, E.; Kroh, J.; Sandham, W.; Durrani, T. Seismic Horizon Picking Using an Artificial Neural Network. In Proceedings of the IEEE International Conference on Acoustics, Speech, and Signal Processing, San Francisco, CA, USA, 23–26 March 1992; pp. 105–108.
10. Leggett, M.; Sandham, W.A.; Durrani, T.S. 3D horizon tracking using artificial neural networks. *First Break* **1996**, *14*, 413–418. [CrossRef]
11. Huang, Y.; Di, H. Dip Interpolation for Improved Multitrace Seismic-Attribute Analysis. In Proceedings of the 2017 SEG International Exposition and Annual Meeting, Houston, TX, USA, 29 September 2017.
12. Fehmers, G.C.; Höcker, C.F. Fast structural interpretation with structure-oriented filtering. *Lead. Edge* **2002**, *21*, 238–243. [CrossRef]
13. Yu, Y. Automatic Horizon Picking In 3D Seismic Data Using Optical Filters And Minimum Spanning Tree (Patent Pending). In *SEG Technical Program Expanded Abstracts 2011*; Society of Exploration Geophysicists: Houston, TX, USA, 2011; pp. 965–969.
14. Wu, H.; Zhang, B.; Lin, T.; Cao, D.; Lou, Y. Semiautomated seismic horizon interpretation using the encoder-decoder convolutional neural network. *Geophysics* **2019**, *84*, B403–B417. [CrossRef]
15. Neidell, N.S.; Taner, M.T. Semblance and Other Coherency Measures for Multichannel Data. *Geophysics* **1971**, *36*, 482. [CrossRef]
16. Krizhevsky, A.; Sutskever, I.; Hinton, G.E. Imagenet classification with deep convolutional neural networks. *Adv. Neural Inf. Process. Syst.* **2012**, *25*, 1097–1105. [CrossRef]
17. He, K.; Gkioxari, G.; Dollár, P.; Girshick, R. Mask r-cnn. In Proceedings of the IEEE International Conference on Computer Vision, Venice, Italy, 22–29 October 2017; pp. 2961–2969.
18. Li, Y.; Qi, H.; Dai, J.; Ji, X.; Wei, Y. Fully Convolutional Instance-Aware Semantic Segmentation. In Proceedings of the IEEE Conference on Computer Vision and Pattern Recognition, Honolulu, HI, USA, 21–26 July 2017; pp. 2359–2367.
19. Wang, W.; George, A.M.; Ma, J.; Xie, F. Automatic velocity picking from semblances with a new deep-learning regression strategy: Comparison with a classification approach. *Geophysics* **2021**, *86*, 1942–2156. [CrossRef]
20. Araya-Polo, M.; Dahlke, T.; Frogner, C.; Zhang, C.; Poggio, T.; Hohl, D. Automated fault detection without seismic processing. *Lead. Edge* **2017**, *36*, 208–214. [CrossRef]
21. Bromley, J.; Guyon, I.; LeCun, Y.; Säckinger, E.; Shah, R. Signature verification using a "siamese" time delay neural network. *Int. J. Pattern Recognit. Artif. Intell.* **1993**, *7*, 669–688. [CrossRef]
22. Hadsell, R.; Chopra, S.; LeCun, Y. Dimensionality Reduction by Learning an Invariant Mapping. In Proceedings of the 2006 IEEE Computer Society Conference on Computer Vision and Pattern Recognition (CVPR'06), New York, NY, USA, 17–22 June 2006; pp. 1735–1742.
23. Vries, D.D.; Berkhout, A.J. Velocity analysis based on minimum entropy. *Geophysics* **1984**, *49*, 2132. [CrossRef]
24. Biondi, B.L.; Kostov, C. High-resolution velocity spectra using eigen structure methods. *Geophysics* **1989**, *54*, 832–842. [CrossRef]

25. Key, S.C.; Smithson, S.B. New approach to seismic-reflection event detection and velocity determination. *Geophysics* **1990**, *55*, 1057. [CrossRef]
26. Fomel, S. Velocity analysis using AB semblance. *Geophys. Prospect.* **2009**, *57*, 311–321. [CrossRef]
27. Luo, S.; Hale, D. Velocity analysis using weighted semblance. *Geophys. J. Soc. Explor. Geophys.* **2012**, *77*, U15–U22. [CrossRef]
28. Ursin, B.; da Silva, M.G.; Porsani, M.J. Generalized Semblance Coefficients Using Singular Value Decomposition. In Proceedings of the 13th International Congress of the Brazilian Geophysical Society & EXPOGEF, Rio de Janeiro, Brazil, 26–29 August 2013; pp. 1544–1549.
29. Ebrahimi, S.; Kahoo, A.R.; Chen, Y.; Porsani, M. A high-resolution weighted AB semblance for dealing with amplitude-variation-with-offset phenomenon. *Geophysics* **2017**, *82*, V85–V93. [CrossRef]
30. Araya-Polo, M.; Jennings, J.; Adler, A.; Dahlke, T. Deep-learning tomography. *Lead. Edge* **2018**, *37*, 58–66. [CrossRef]
31. Ma, Y.; Ji, X.; Fei, T.W.; Luo, Y. Automatic Velocity Picking with Convolutional Neural Networks. In Proceedings of the 2018 SEG International Exposition and Annual Meeting, Anaheim, CA, USA, 14–19 October 2018.
32. Zhang, H.; Zhu, P.; Gu, Y.; Li, X. Automatic Velocity Picking Based on Deep Learning. In *SEG Technical Program Expanded Abstracts*; Society of Exploration Geophysicists: Houston, TX, USA, 2019; pp. 2604–2608.
33. Schneider, M.; Chang, S.F. A Robust Content Based Digital Signature for Image Authentication. In Proceedings of the International Conference on Image Processing (ICIP), Lausanne, Switzerland, 19 September 1996; pp. 227–230.

Article

Seismic Structure Beneath the Molucca Sea Collision Zone from Travel Time Tomography Based on Local and Regional BMKG Networks

Gazali Rachman [1,2,*], Bagus Jaya Santosa [1], Andri Dian Nugraha [3,4], Supriyanto Rohadi [5], Shindy Rosalia [3,4], Zulfakriza Zulfakriza [3,4], Sungkono Sungkono [1], David P. Sahara [3,4], Faiz Muttaqy [6], Pepen Supendi [5], Mohamad Ramdhan [6], Ardianto Ardianto [7] and Haunan Afif [8]

1 Physics Department, Institut Teknologi Sepuluh Nopember (ITS), Surabaya 60111, Indonesia
2 Physics Education-FKIP, Pattimura University, Kota Ambon 97233, Indonesia
3 Global Geophysics Research Group, Faculty of Mining and Petroleum Engineering, Institut Teknologi Bandung, Bandung 40132, Indonesia
4 Center for Earthquake Science and Technology (CEST), Research Center for Disaster Mitigation, Institut Teknologi Bandung, Bandung 40132, Indonesia
5 Agency for Meteorology, Climatology and Geophysics (BMKG), Jakarta 83124, Indonesia
6 Research Center for Geological Disaster, National Research and Innovation Agency of Indonesia (BRIN), Bandung 40135, Indonesia
7 Geophysical Engineering Department, Faculty of Mining and Petroleum Engineering, Institut Teknologi Bandung, Bandung 40132, Indonesia
8 Center for Volcanology and Geological Hazards Mitigation, Geological Agency, Ministry of Energy and Mineral Resources, Bandung 40122, Indonesia
* Correspondence: gazali.rachman@gmail.com

Citation: Rachman, G.; Santosa, B.J.; Nugraha, A.D.; Rohadi, S.; Rosalia, S.; Zulfakriza, Z.; Sungkono, S.; Sahara, D.P.; Muttaqy, F.; Supendi, P.; et al. Seismic Structure Beneath the Molucca Sea Collision Zone from Travel Time Tomography Based on Local and Regional BMKG Networks. *Appl. Sci.* **2022**, *12*, 10520. https://doi.org/10.3390/app122010520

Academic Editors: Guofeng Liu, Zhifu Zhang and Xiaohong Meng

Received: 20 September 2022
Accepted: 14 October 2022
Published: 18 October 2022

Publisher's Note: MDPI stays neutral with regard to jurisdictional claims in published maps and institutional affiliations.

Abstract: The Molucca Sea Plate, and Sangihe and Halmahera plates have a complex tectonic setting and interact to create the Molucca Sea Collision Zone. We re-picked 1647 events recorded from 2010 to 2017 from 32 of The Agency for Meteorology, Climatology, and Geophysics (BMKG) stations and obtained P- and S-arrivals of ~17,628 phases. Hypocenter locations were determined using the software NonLinLoc. Then, we performed a travel time tomography in order to image the subsurface and approximate the Molucca Sea Plate subduction angle beneath Sulawesi's north arm, the relationship subduction zone and volcanic activity in Halmahera, and depth comparison of the Molucca Sea Plate. Our results show that (i) high Vp, high Vs, and low Vp/Vs are associated with the Molucca Sea Plate beneath Sulawesi's north arm, and the approximate subduction angle is ~40°. (ii) Low Vp, low Vs, and high Vp/Vs beneath the northern and southern Halmahera Volcanic Arc are associated with a possible magma source. Volcanoes in the north have experienced eruptions and are dormant in the south. This group of volcanoes is connected by partial melting above the Molucca Sea Plate subducts to the east. (iii) High Vp, high Vs, and low Vp/Vs are interpreted as double subduction of the Molucca Sea Plate. It is submerged deeper in the north compared with the south, which is nearer to the surface.

Keywords: travel time tomography; Molucca Sea Plate; magma source; partial melting

1. Introduction

The Molucca Sea region is a very complex area where three main plates, the Eurasian, Australian, and Pacific plates, interact and create a collision zone known as the Molucca Sea Collision Zone [1–5]. The activity of these plates and the subduction of the Molucca Sea Plate are the main factors causing the high level of seismicity in this area. Tectonic activity is also increased by the Molucca Sea Plate which completely submerges the Halmahera microplate [6] and the Sangihe microplate [7], as shown in Figure S1.

The Molucca Sea Collision Zone is located between the Halmahera microplate in the east and the Sangihe microplate in the west with the Molucca Sea in the center. One collision resulted in the formation of a high ridge in the middle of the two microplates, known as the Mayu-Talaud Ridge, which is characterized by intense, shallow earthquakes and an low gravity anomaly [3,5,7,8]. The collision is ~250 km apart [4] and convergently collides with the Molucca Sea Plate [4,9] at a movement of 1.5 cm/yr [9]. The Molucca Sea Plate has an inverted U- or V-shape [2,8,10–13] and has slab subducting to the west under the Sangihe Arc and subducts to the east under the Halmahera Arc. The subduction of the Molucca Sea Plate in both eastern and western directions makes for arc volcanisms in the Halmahera and Sangihe Arcs that run parallel to the subduction zone [8,13,14] with its two Wadati–Benioff Zone active of mantle earthquakes [4,7,10]. The collision of the Halmahera and Sangihe Arcs with the subduction of the Molucca Sea Plate remains active today [15–18].

The Molucca Sea is one of the most seismically active areas in the world [19], as shown in Figure 1. There are many seismic zones in this area, such as the seismic zone concentrated in the central Molucca Sea at a depth of <60 km beneath the Mayu-Talaud Ridge, the seismic zone beneath the Sangihe and Halmahera Arcs, the seismic zone around the Gorontalo Basin, along the Minahasa peninsula, around Morotai Island, the Sulawesi Sea, the Molucca-Sorong Fault, the Bacan-Sorong Fault, the Sula-Sorong Fault, the Sulawesi slab in the northern part of Sulawesi's north arm, and the area around the Philippine slab in the eastern part of Halmahera Island. The red dots in Figure 1 indicate shallow earthquakes; these are mostly concentrated in the central Molucca Sea area beneath the Mayu-Talaud Ridge and the Gorontalo Basin; the brown dots indicate intermediate earthquakes that are mostly concentrated beneath the Sangihe and the Halmahera Arcs; the yellowish blue dots indicate deep earthquakes that are concentrated in the Sulawesi Sea area.

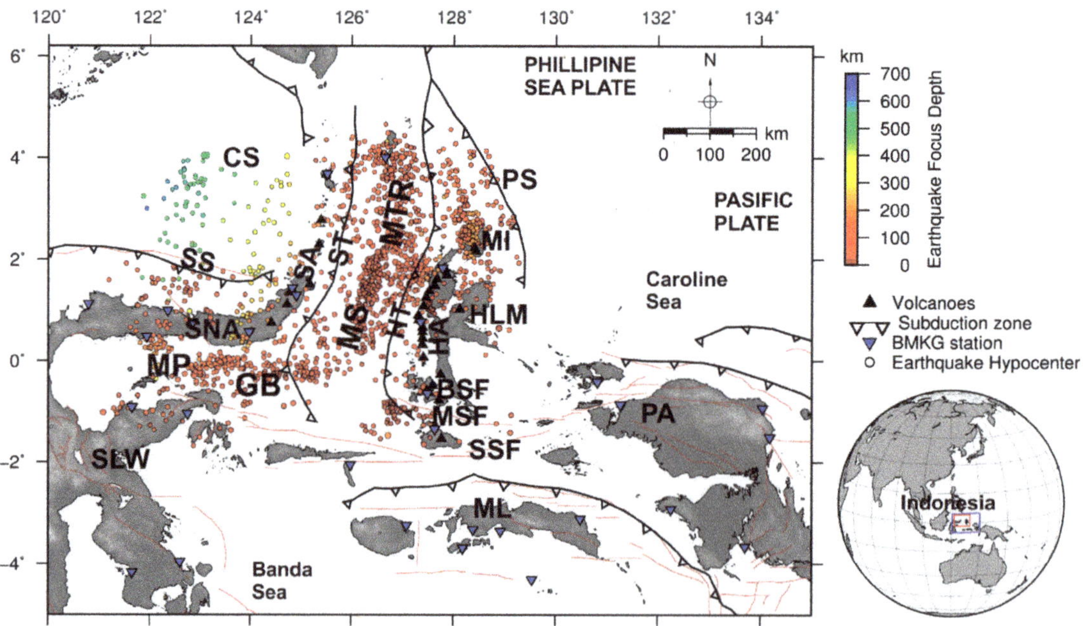

Figure 1. The horizontal distribution of seismic map of the research location around the Molucca Collision Zone. SLW: Sulawesi island; ML: Molucca island; PA: Papua island; HLM: Halmahera island; SNA: Sulawesi's north arm; SS: Sulawesi slab; PS: Philippine slab; HT and ST: Halmahera and Sangihe Trench; HA and SA: Halmahera and Sangihe volcanoes arc; CS: Celebes sea; MP: Minahassa peninsula; GB: Gorontalo Basin; MS: Molucca Sea; MTR: Mayu-Talaud Ridge; BSF, MSF, and SSF: Bacan, Molucca, and Sula Sorong Faults, respectively.

Numerous studies on seismicity that relate to the structure and geometry of the Molucca Sea Plate have been carried out by previous researchers. One such study covers the two Benioff zones that lie in opposite directions and form a double subduction pattern in the Molucca Sea Plate [5,7,8,10,13]. The subducted slab under the Sangihe Arc can be identified as far down as the mantle transition zone layer at a depth of ~650 km; the subducted slab under the Halmahera Arc reaches the solid upper asthenosphere layer at a depth of ~250 km [8]. According to Cardwell et al. [10], the subducting slab to the east under the Halmahera Arc subducts at an angle of ~45° and reaches a depth of ~230 km; Hall and Spakman [20] show that it reaches depths of ~400 km, which is in line with recent research [21] that used 3D earthquake data distribution plots from ISC-EHB to identify slab configurations that were not modeled by slab2. Hutchings and Mooney [12] showed that the subducted slab in the Sangihe Arc at depths of ~250 to 350 km has a seismic gap, while the tomography image of the slab appears to be continuous.

Tomographic studies have been carried out by previous researchers, including Puspito et al. [22] who used P-phase travel time data from ISC bulletins; their work shows a high-velocity zone that forms a double subduction of the Molucca Sea Plate. The slab is subducted to the west under the Sangihe Arc and reaches to the lower mantle. Meanwhile, the slab that subducts to the east under the Halmahera Arc reaches a depth of 400 km, which is conformable with the results of the study by Widiyantoro and Hilst [23]. This research was later updated by Widiyantoro [24] who shows that the subducted slab to the west reaches the lower mantle with a folded-looking slab. Huang et al. [25] also detected a high-velocity zone in the form of bipolar subduction. Meanwhile, Zhang et al. [26] showed that the Molucca Sea Plate has a positive velocity anomaly with unique asymmetry. Fan and Zhao [27,28] showed that the subduction of the Molucca Sea Plate ~5 Ma ago propagated from south to north; they were able to detect the Molucca Sea Plate using anisotropic tomography. Zenonos et al. [29], using data from ISC-EHB, show the high-velocity Vp and Vs of the double subduction of the Molucca Sea Plate in which the high Vp reached a depth of ~700 km beneath the Sangihe Arc and ~400 km beneath the Halmahera Arc. The high Vs reached a depth of ~500 km, which is in positive agreement with a study performed by Chen et al. [21] who used only P-wave data. Zenonos et al. [30] showed that negative Vp/Vs anomalies can reach ~160 km, and positive Vp/Vs anomalies can reach depth of ~200 km. Another study by Cao et al. [31] observed the mantle flow using seismic anisotropy. Their study shows that beneath the Halmahera Arc the sub-slab mantle flow is oriented oblique to the Halmahera trench, while the sub-slab mantle flow beneath the Sangihe Arc runs in a parallel direction to the Sangihe trench.

According to previous research, tomography is generally based on data from global teleseismic earthquakes within a broad region, producing a tomographic image with a broad resolution that displays only relatively large objects. Tomography around the Molucca Sea region usually only shows the main features of the Molucca Sea Plate [22,23,26,29]. Several local researchers in the Molucca Sea area have used different methods [2,17,18]. However, their results extend only up to the layer of the Earth's crust; the tomographic results have not shown anomalies of 3D velocity structures due to volcanic activity in the subduction zone. The research undertaken in this study made use of local and regional data from BMKG (Figure 1) to obtain more detailed features of the Molucca Sea region, especially the estimated subduction angle of the Molucca Sea Plate from south to north beneath Sulawesi's north arm; the relation between the subduction zone of the Molucca Sea Plate subducts to the east and its volcanic activity in the Halmahera Volcanics Arc, and the depth of submerged of the double subduction of the Molucca Sea Plate in the south and north.

2. Materials and Methods

2.1. Data

The earthquake data used in this research was obtained from the BMKG network of broadband stations in the form of waveform data from three-component seismograms (BHZ, BHN, and BHE) from 1 January 2010 to 31 December 2017. A total of 32 stations were used, covering the area 4.8° S–4.25° N and 120.5° E–134.5° E (Figures 1 and 2).

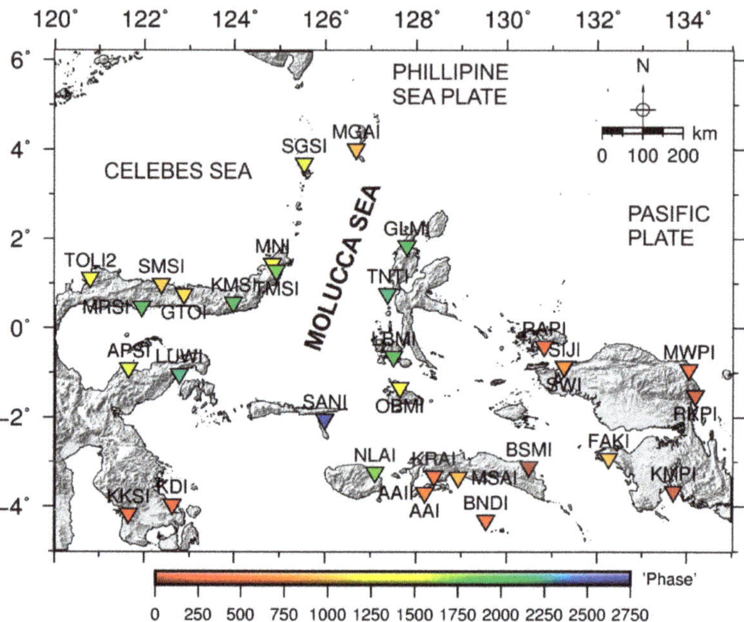

Figure 2. Arrival times distribution of P-and S-wave for all stations. Colored inverted triangles denote the number of phases per station.

Considering the distribution and locations of the regional stations, earthquakes with a magnitude ≥ 4 were chosen to facilitate the identification of P- and S-wave arrival times. A total of 1647 earthquake events have been recorded in the study area. Picking was performed manually to obtain P- and S-wave arrival times using SeisGram2K version 7.0 software [32]. P- and S-wave arrival times were determined in pairs at each station; the arrival times are clearly identified and recorded by at least four stations (Figure 3). The arrival times of a total of ~17,628 P-waves and ~17,628 S-waves were identified. The highest number of P- and S- phases were recorded in the southern part of the earthquake epicenter, while a smaller number of phases were recorded in the eastern part of Seram Island. The arrival times distribution for each station are shown in Figure 2, and its histogram in Figure S2.

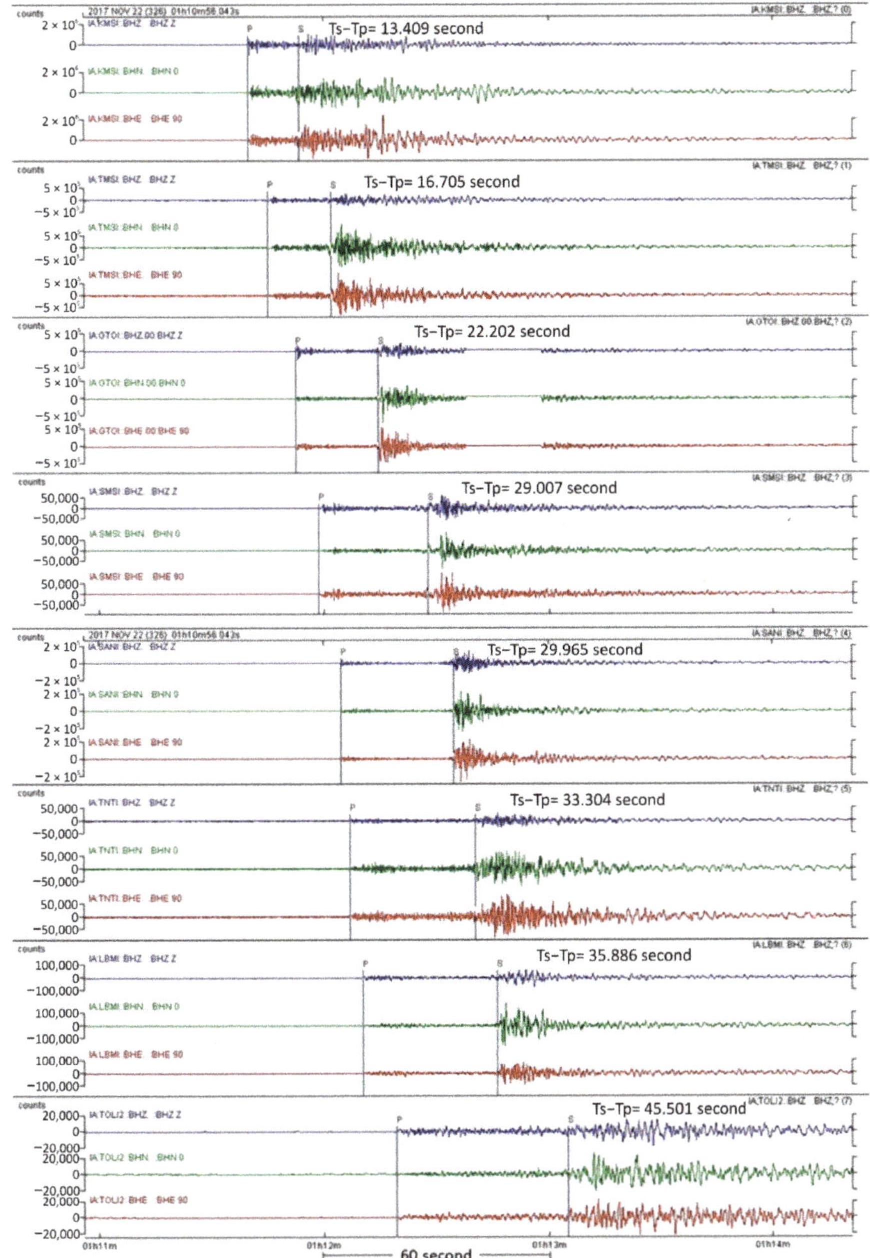

Figure 3. Example of the 22 November 2017 event waveform from a three-component seismogram (BHZ, BHN, and BHE) which was picked using SeisGram2K version 7.0 software [32]. The blue, green, and red colors from each station represent the vertical component (BHZ), the north–south component (BHN), and the east–west component (BHE). From top to bottom, the recorded seismograms are shown at the KMSI, TMSI, GTOI, SMSI, SANI, TNTI, LBMI, and TOLI2 stations. The vertical P and S lines show the picked arrival times for P- and S-waves, while Ts-Tp is the time difference between the arrival times of the S- and P-waves.

2.2. Methods

In NLLoc version 7.00 [33], the location algorithm adopts the inversion approach from Tarantola and Valette [34], and the event location procedure from Tarantola and Valette [34], Moser et al. [35], and Wittlinger et al. [36]. This algorithm generated the misfit function for each sample cell, estimated the probability density function (PDF), and determined the optimal hypocenter location. The location search process began by defining the search area in 3D by setting the boundaries of the x, y, z regions and then applying the 1-D AK135 velocity model [37,38]. The optimal hypocenter location for each event was determined by estimating the probability density function (PDF) using an oct-tree importance sampling.

The result of the hypocenter location from the NonLinLoc program with an equal amount of P- and S-waves used to process the travel time tomography inversion was performed using SIMULPS12 [39] to obtain the 3D structure of both Vp and Vp/Vs ratio and simultaneously hypocenters relocation by using pseudo-bending ray tracing [40], as shown in Figure S3. This program performs an inversion of Vp and Vp/Vs models from P- arrival times and S-P times directly, not Vp and Vs separately, due to differences in data quality between P- and S- arrival times [41,42]. The initial velocity model used in the inversion process is a 3D regional-global seismic velocity model (Vp) beneath Indonesia obtained from high-resolution tomographic imaging [23]. Meanwhile, the value of the Vp/Vs ratio was obtained from the Wadati Diagram (Figure S4), which is 1.76 and is consistently used for each layer.

We performed the inversion through three stages: Firstly, a damping analysis was performed to obtain the optimum damping parameter input by making variations in the data distribution, number, size, and grid nodes distance [41]. The damping value was obtained by comparing the variance model (km^2/s^2) and the variance data (s2) for both Vp and Vp/Vs values. The optimum value for Vp and Vp/Vs damping obtained was 150 and 130, respectively (Figure S5). This value was used to obtain a stable solution when performing the tomographic inversion and a checkerboard resolution test (CRT). Secondly, we ran a checkerboard resolution test using several grid spacings to define the best model parameterization used in this study. We decided to use a 60 $\times$ 60 km grid and a vertical grid size of 30 km after considering the results and dataset distribution. Thirdly, we performed the tomographic inversion using the damped least square where the norm of the perturbations model is weighted and combined with the misfit data squared. We set high residual data < 10 s at the early inversion. It is automatically discarded if a value greater than 10 s is found. We set the maximum number of iterations to 20 times and stopped at 15 iterations. The weighted Root Mean Square (RMS) residual value decreases each iteration until convergence. The stopping condition provided by an F-test indicates that variance reduction is insignificant after more iterations [39].

2.3. Resolution

Several methods were used to determine whether the inversion results are reliable for interpretation or not. Firstly, the Checkerboard Resolution Test (CRT) was used to test the resolution of the tomography model by making a synthetic velocity model. The CRT results showed a pattern of high and low-velocity anomalies relative to the 1D initial model, which is $\pm$10% relative to the 1D initial model. Secondly, the Derivative Weight Sum (DWS) calculated the density of ray paths that pass through the grid nodes by giving them a certain weight. A high DWS value represents a high-resolution zone (Figure S6). Thirdly, the Ray Hit Count (RHC) was used to locate the number of ray paths that pass through the grid nodes. An increased number of ray paths represents high resolution zones that can be well-interpreted (Figure S7). Fourthly, the Diagonal Resolution Element (DRE) was used, as shown in Figure S8, which relates to the resolution matrix R (damped least-squares problems); these have an R-value between 0 and 1 [39,43,44].

The focus of this research on the vertical cross-section for the resolution test and tomography inversion consisted of four areas in the north–south cross-section: profile A-A′ (longitude 123.80°) and profile B-B′ (longitude 127.58°), and the west–east cross-section, profile C-C′ (latitude 3.26°) and profile D-D′ (latitude 1.10°). The horizontal grid size used in this study was 60 km × 60 km; the vertical grid size ranged from 30 km to a depth of 330 km, as shown in Figure 4.

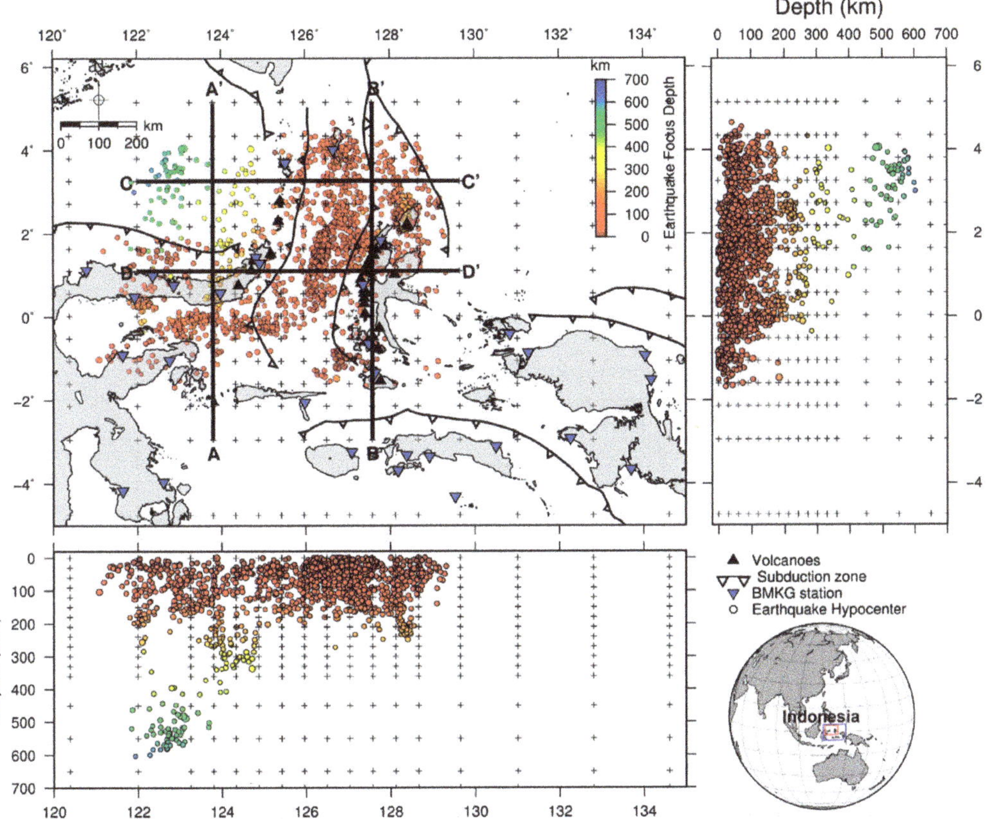

Figure 4. The horizontal and vertical seismic map of the research location in the Molucca Collision Zone. The bold black lines A-A′ (south–north), B-B′ (south–north), C-C′ (west–east), D-D′ (west–east) denote the areas of the vertical cross-section tomography. The black plus sign represents the distribution of the grid nodes for tomographic inversion; the colored dots indicate the distribution of earthquakes; the blue triangles are BMKG stations, and the black triangles signify volcanoes.

3. Results

Based on the re-picking phase and distribution of available BMKG stations, we determined the location of the hypocenter using the NonLinLoc Global mode. The estimated mean error in x, y, and z directions using the covariances values are 10,198 km, 14,081 km, and 19,035 km, respectively. The distribution frequency estimated error location in the x, y, and z directions is shown in Figure S9. The distribution between the number of events and the azimuthal gap is shown in Figure S10, and the epicentral distance distribution is shown in Figure S11. We limit the value of travel time residuals in the range of $-10 \leq$ initial travel time residuals (itr) ≤ 10 to control the quality of the input data for the tomographic inversion process. The bad picks can be traced in unit 20 of simulps output file, as shown in Figure S12. Before the inversion, the initial travel time residual values varied in the range of −9.993 s to 9.858 s, with a mean value of −1.111 s. After inversion, the final travel time residuals varied in the range of −7.148 to 6.197 s, with a mean value of 0.156 s, forming a Gaussian distribution centered around 0 s. Tomographic inversion achieved the convergence after the 15th iteration with the weighted Root Mean Square (RMS) residual reduced by 47%, initially from 0.82312 s down to 0.43871 s (Figure S13). The RMS residual before the inversion ranged from 0.06 to 1.26 s with an average value of 0.54 s. After inversion, the range of values was down to 0.04 to 0.83 s with a mean value of 0.36 s (Figure S14) or reduced by about 33%.

The Interpretation of the tomogram becomes valid and has a high confidence level by performing a resolution test of the seismic tomography model. Therefore, we present the CRT results shown in Figure 5 for Vp and Vp/Vs of profiles A-A′, B-B′, C-C′, and D-D′ and Figure S15 horizontal section for Vp and Vp/Vs with the recovered area is delineated by dashed purple lines. The CRT positive recovery for A-A′ is beneath the Gorontalo Basin, Sulawesi's north arm, and the Celebes Sea up to a depth of 210 km. The CRT positive recovery for the B-B′ profile is beneath the Halmahera Volcanic Arc up to a depth of 240 km; for profile C-C′ the CRT positive recovery is shown beneath the Celebes and Molucca Seas, and the Morotai Basin up to a depth of 210 km. For profile D-D′, the CRT positive recovery is beneath Sulawesi's north arm, the Molucca Sea, and Halmahera Island up to a depth of 300 km. These results are consistent with the density distribution of P- and S-rays as revealed by the DWS, RHC, and DRE results (Shown in Figures S6–S8); the highest Vp and Vp/Vs values are listed in Table S1. The primary resolution test we used to interpret the tomographic results was the CRT, supported by the DWS, RHC, and DRE results. Hence, this research focused on the travel time tomography results in four areas in the subduction zone beneath Sulawesi's north arm, the subduction zone beneath the Halmahera Volcano Arc, and the double subduction zone of the Molucca Sea.

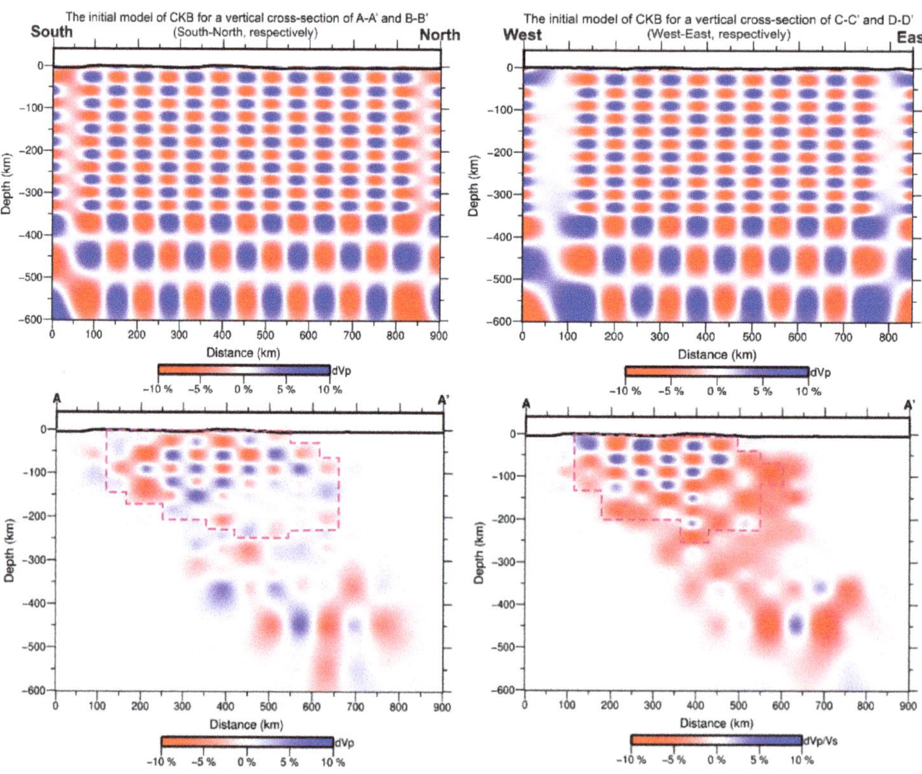

Figure 5. *Cont.*

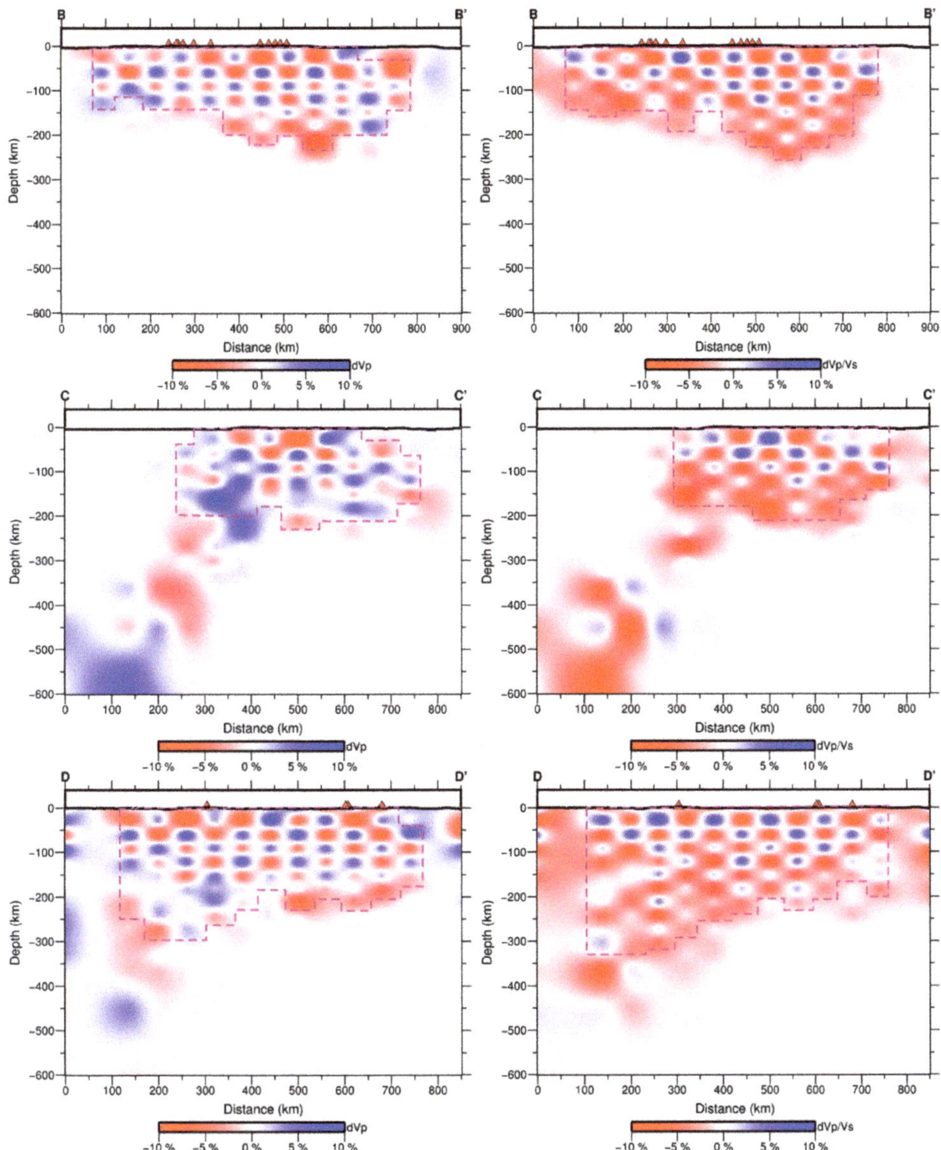

Figure 5. The vertical cross-section of CRT profiles. Profile A-A′ (south–north), B-B′ (south–north), C-C′ (west–east), and D-D′ (west–east) are vertical cross-sections of the CRT recovery model from the inversion process. The purple dashed line is the positive CRT recovery area for tomographic interpretation. The first column is Vp perturbation, the second column is Vp/Vs perturbation, and the red triangles signify volcanoes.

Tomographic inversion results of cross-section A-A′, as shown in Figure 6 as follows:

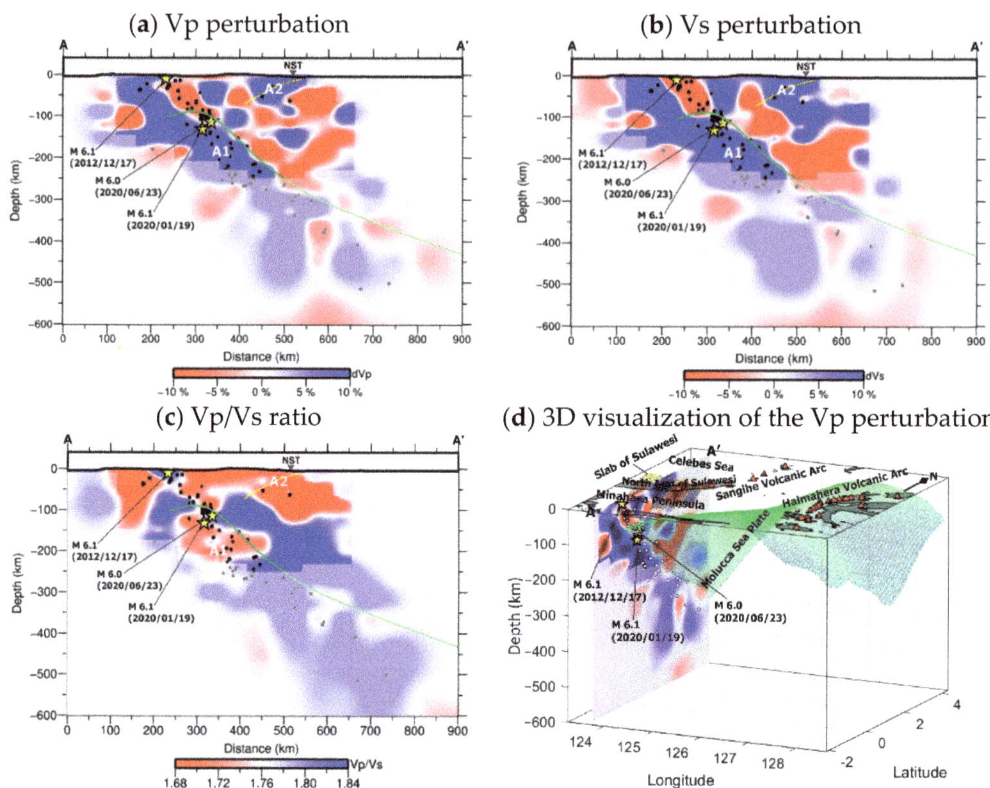

Figure 6. Vertical cross-section of A-A′ (south–north) of the Molucca Sea Plate, crossing the Minahasa Peninsula, Sulawesi's north arm, and the Celebes Sea. The red color indicates low velocity and low Vp/Vs ratio, and the blue color indicates high velocity and high Vp/Vs ratio. The A1 and A2 are areas for the interpretation of tomographic inversion results. The black dots signify earthquakes within a radius of 30 km ≤ A-A′ ≤ 30 km, and the yellow stars are significant earthquakes. The green line and the green three-dimensional surface plot are the slab2 models of the Molucca Sea Plate. The yellow line is the slab2 model of Sulawesi [45]. NST is the North Sulawesi Trench. The red triangles signify volcanoes. The blurred area is an area that is not well-resolved.

Tomographic inversion results of cross-section B-B′, as shown in Figure 7 as follows:

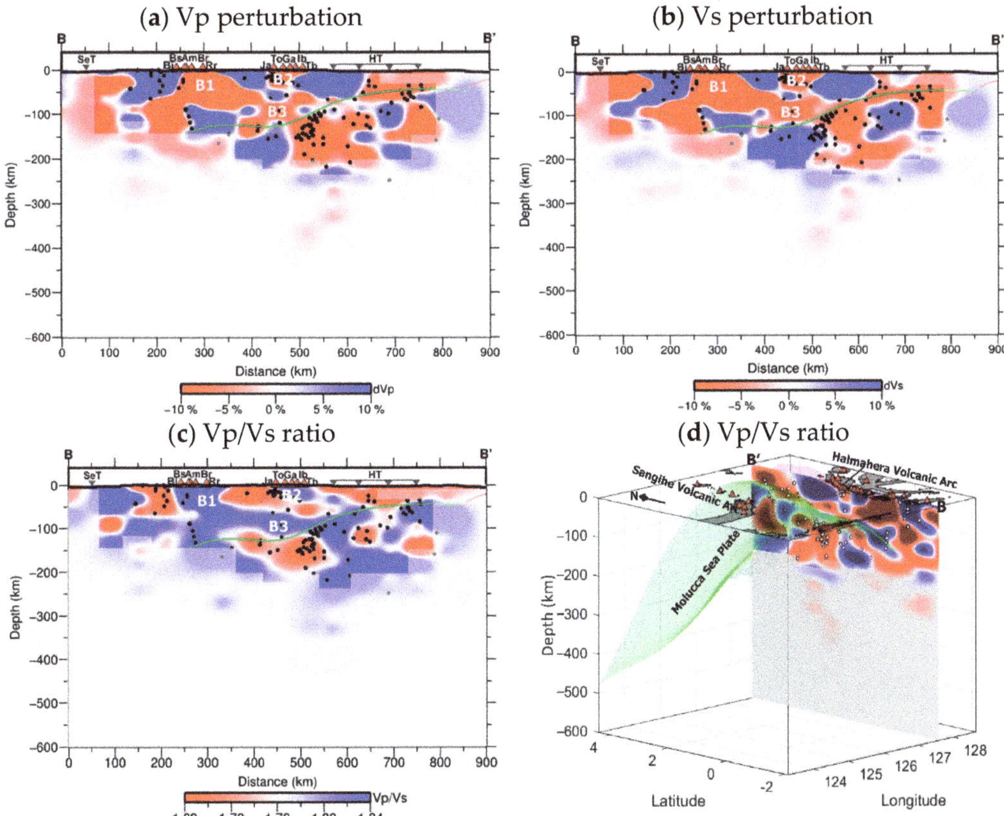

Figure 7. Vertical cross-section of B-B′ (south–north) crossing beneath the Halmahera Volcanic Arc. The red color indicates low velocity and low Vp/Vs ratio. The blue color indicates high velocity and high Vp/Vs ratio. The B1, B2, and B3 are areas for the interpretation of tomographic inversion results. The green line and the green three-dimensional surface plot are the slab2 models of the Molucca Sea Plate [45]. The black dots signify earthquakes within a radius of 30 km ≤ B-B′ ≤ 30 km. SeT: Seram Trench; HT: Halmahera Trench; Bi: Bibinoi Volcano; Bs: Batusibela Volcano; Am: Amasing Volcano; Br: Buku-Rica Volcano; Rr: Rogi-Rogi Volcano; Ja: Jailolo Volcano; To: Todoko-Ranu Volcano; Ga: Gamkonora Volcano; Ib: Ibu Volcano; Tb: Tobaru Volcano. The blurred area is an area that is not well-resolved.

Tomographic inversion results of cross-section C-C′, as shown in Figure 8 as follows:

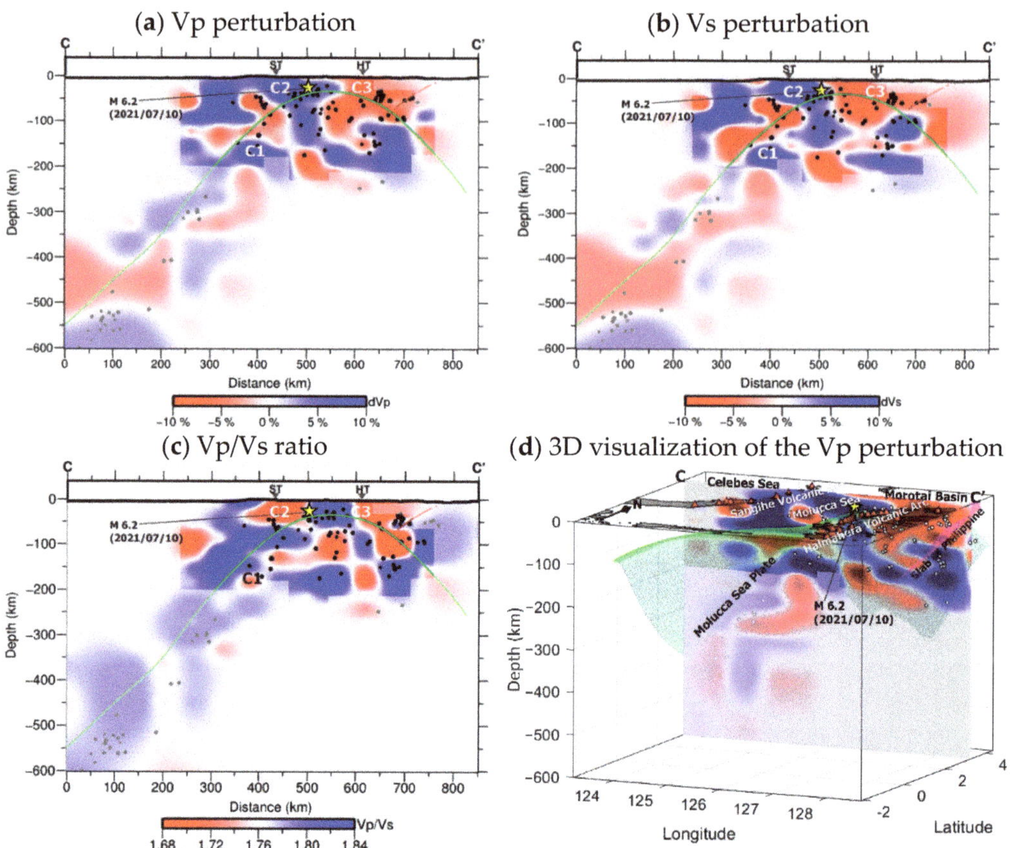

Figure 8. Vertical cross-section of C-C′ (west–east) crossing the Celebes Sea, Sangihe Island, Molucca Sea, and Morotai Basin. The red color indicates low velocity and low Vp/Vs ratio. The blue color indicates high velocity and high Vp/Vs ratio. The C1, C2, and C3 are areas for the interpretation of tomographic inversion results. The black dots signify earthquakes within a radius of 30 km $\leq$ C-C′ $\leq$ 30 km, and the yellow star is a significant earthquake. The green line and the green three-dimensional surface plot are the slab2 models of the Molucca Sea Plate, and the brown line is the slab2 model of the Philippines [45]. ST: Sangihe Trench; HT: Halmahera Trench; the blurred area is an area that is not well-resolved.

Tomographic inversion results of cross-section D-D′, as shown in Figure 9 as follows:

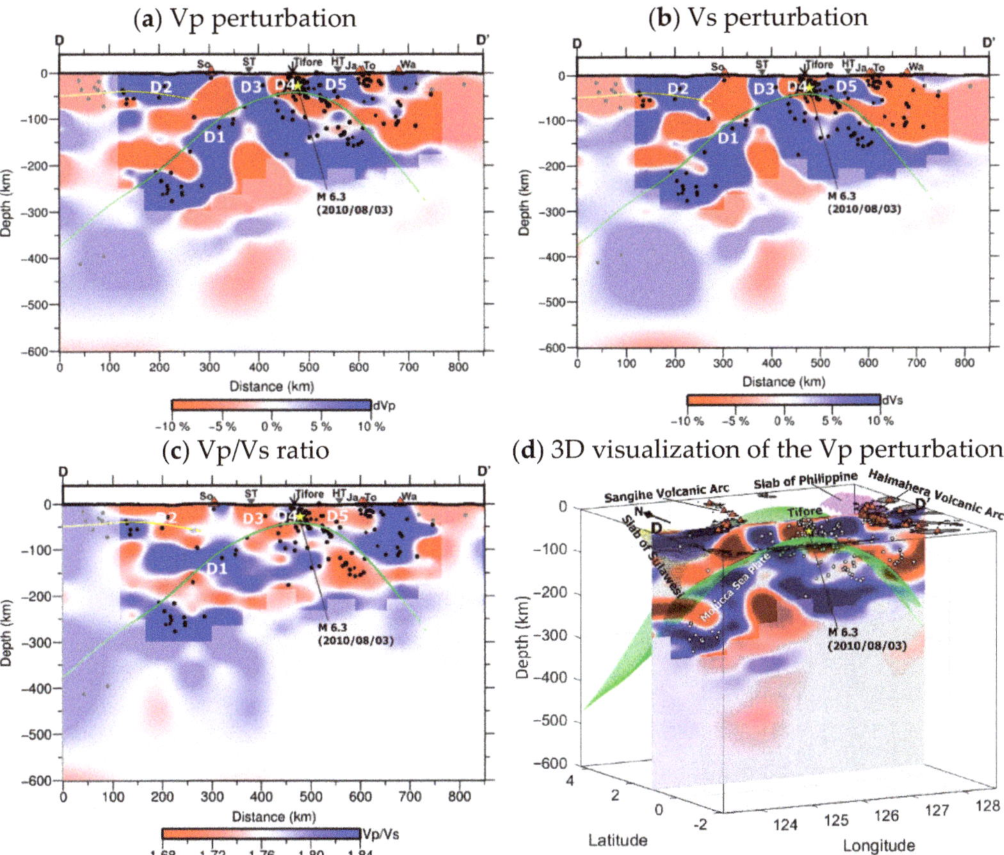

Figure 9. Vertical cross-section of D-D′ (west–east) crossing Sulawesi's north arm, the Molucca Sea, and Halmahera Island. The red color indicates low-velocity areas and low Vp/Vs ratio, and the blue color indicates high-velocity areas and high Vp/Vs ratio. The D1, D2, D3, D4, and D5 are areas for the interpretation of tomographic inversion results. The black dots signify earthquakes within a radius of 30 km ≤ D-D′ ≤ 30 km, and the yellow star is the magnitude of a significant earthquake. The green line and the green three-dimensional surface plot are the slab2 models of the Molucca Sea Plate; the yellow line is the slab2 model of Sulawesi [45]. Thus, So: Soputan Volcano; ST: Sangihe Trench; the reserve arrow is Tifore island; HT: Halmahera Trench; Ja: Jailolo Volcano; To: Todoko-Ranu Volcano; Wa: Wato-Wato Volcano; the blurred area is an area that is not well-resolved.

Three-dimensional tomographic inversion results combine cross-section profile A-A′ and profile D-D′, as shown in Figure 10 as follows:

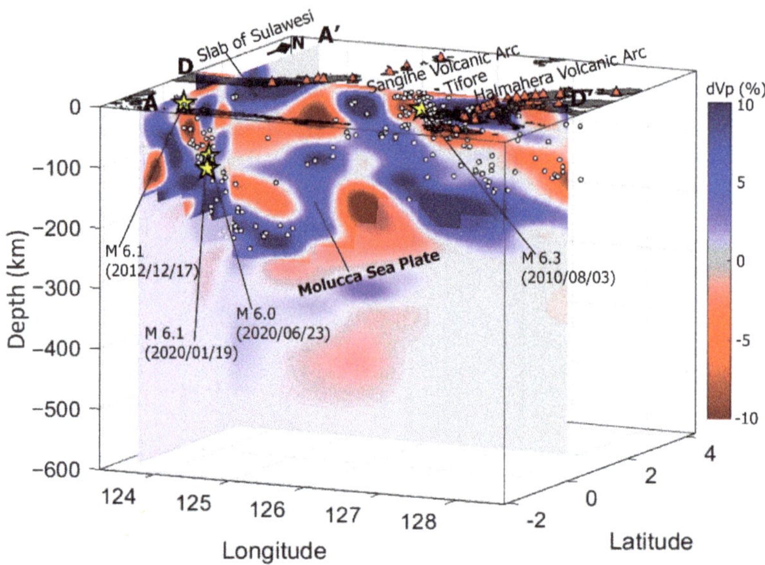

Figure 10. The 3D visualization of the Molucca Sea Plate, combining vertical cross-section profile A-A′ (south–north) and profile D-D′ (west–east) from the P-wave tomography inversion results. The earthquake events follow the high Vp anomaly pattern. The red color indicates low-velocity areas and low Vp/Vs ratio, and the blue color indicates high velocity and high Vp/Vs ratio. The white dots signify earthquakes; the yellow stars signify earthquakes with significant magnitudes. The red triangle is a volcanic arc located in Sangihe and Halmahera.

4. Discussion

4.1. Subduction Zone Beneath Sulawesi's North Arm

Subducting slabs are usually indicated by high-velocity anomalies, which are caused by denser material that has a lower temperature than its surroundings [25,46,47], along with the presence of intermediate and deep earthquakes [48,49]. Based on the results of the tomographic inversion, the section A-A′ (south–north) shows a velocity anomaly of high Vp, high Vs, and low Vp/Vs at the A1 area, down to a depth of ~210 km (Figure 6). These features are interpreted as double subduction of the Molucca Sea Plate, which subducts westward beneath Sulawesi's north arm and the Sangihe Islands with an angle of subduction from south to north, an approximate angle of ~40°. These results are consistent with previous research performed in different locations [29,30,46,50], which show that high Vp, high Vs, and low Vp/Vs can be identified as subducted slabs. It also has positive agreement the slab model of Hayes et al. [45]. In addition, most earthquakes are concentrated in a high-velocity area, these continued in the subduction zone as far down as a depth of ~500 km due to the very active movement of the subduction plate.

Three major earthquakes with magnitude ≥ 6 are observed in this area between January 2010 and June 2021. Two of these earthquakes (23 June 2020 and 19 January 2020 events) had a depth of ~100 km and were located in the South Bolaang Mongondow Regency around the Minahasa Peninsula of Sulawesi. These earthquakes are associated with the double subduction activity of the Molucca Sea Plate, which subducts to the west. Meanwhile, one shallow earthquake with a magnitude of 6.1 is located near the Gorontalo Basin and is associated with an active fault in the sea. The results of tomography inversion and hypocenters horizontally for each depth section are shown in Figure S16.

The A2 area in Figure 6 and the D2 area in Figure 9 are located in the northern part of Sulawesi's north arm, precisely in the North Sulawesi Trench (NST) Zone. This area reveals an anomalous pattern of high Vp, high Vs, and low Vp/Vs that can be interpreted as a slab of Sulawesi that subducts to the south with the structure above being the Molucca Sea Plate. This result is consistent with previous studies conducted by Chou et al. [50], Zhao [46], and Zenonos et al. [29,30], suggesting that high Vp, high Vs, and low Vp/Vs are associated with the presence of subducted slabs. It is also consistent with Fan and Zhao [27] which identified the slab of Sulawesi from P-wave tomography and is similar to the subduction zone geometry model (yellow line) described by Hayes et al. [45]. The existence of this slab is also followed by earthquake events, even though the level of seismic activity in this cross-section is less than the earthquake activity that occurs in the A1 area of the Molucca Sea Plate zone.

4.2. Subduction Zone Beneath Halmahera Volcano Arc

The subduction zone beneath the Halmahera Volcano Arc is shown in the vertical cross-section of B-B′ (south–north) in Figure 7. The B1 area is located under three large islands in the southern part of Halmahera: Bacan, Kasiruta, and Mandioli Islands. This area is also crossed by the Bacan–Sorong Fault, which is a branch, or splay, of the Sorong Fault that passes to the east [51], and is responsible for the seismicity in this region. Based on historical records, there was a strong and destructive earthquake, magnitude 7.1, on 16 April 1963 with an intensity scale of VIII Modified Mercalli Intensity (MMI). The most recent earthquake occurred on 26 February 2020, magnitude 5.2. The earthquake shocks were felt to be quite strong and were estimated on the V MMI intensity scale. There are several volcanoes on Bacan Island, consisting of the Bibinoi Volcano in the south, and the Batusibela, Amasing, and Buku Rica (Meng) Volcanoes in the north; all of these are stratovolcanoes. The tomographic inversion results show low-Vp, low-Vs, and high-Vp/Vs anomalies, which can be interpreted as the possible source of this magmatic arc [25,52,53] as also represented by Huang et al. [25] in the same region. These volcanoes have remained dormant until the present and without any historical eruptions. However, these volcanoes may possibly erupt in the future. This possibility is supported by the flow of fluids released from partial melting (B3 area) at a depth of ~100 km above the Molucca Sea Plate that subducts to the east under Halmahera Island [54].

The B2 area shown in Figure 7 is located beneath the northern arm of Halmahera Island and belongs to the West Halmahera Regency. The vertical cross-section passes through several active volcanoes: Jailolo, Todoku-Ranu, Gamkonora, Ibu, and Tobaru. Jailolo is a stratovolcano [54] that has experienced several earthquake swarms [55,56], but whose last eruption is unknown. Young lava flows have been found on the east side of the volcano, indicating that there has been an eruption. There is also no information about the last eruption of the Todoko–Ranu Volcano, which is a caldera volcano. Gamkonora, a stratovolcano, experienced its last eruption in 2013. The Ibu Volcano, also a stratovolcano, had its last eruption in 2012. The latest update was on 16 June 2021, in the form of a volcanic ash advisory. The Tobaru Volcano, also known as the Lolodai Volcano, is a stratovolcano and its last eruption is unknown [54].

The tomographic inversion results show low-Vp, low-Vs, and high-Vp/Vs anomalies, which can be interpreted as possible sources of a magmatic arc [25,52,53]. This interpretation is also supported by the ascending flow of fluid that originated from partial melting in the B3 area at depths of ~50–70 km. The partial melting and the subducted Molucca Sea Plate (the green line in Figure 7) are shallower in this area in comparison with the B1 area. We suggest that the difference in fluid depth is probably caused by the higher activity of volcanoes in the north of Halmahera compared with volcanoes in the south. However, further and more detailed research is necessary to ascertain this.

The B3 area is located beneath the Halmahera Volcanic Arc at a depth of ~100 km and above the Molucca Sea Plate that subducts to the east. The tomographic inversion results show low-Vp, low-Vs, and high-Vp/Vs anomalies, indicating the presence of fluids that promotes partial melting. Its good agreement with the previous studies by Nakajima et al. [57] and Zhao [53], the low Vp, low Vs, and high Vp/Vs are interpreted as partial melting materials. It is observed that the upwelling process is connected with zone B1 beneath the volcanic group located on Bacan Island, and zone B2 located beneath the volcanic group on the northern arm of Halmahera. High Vp/Vs above the subduction zone can be interpreted as slab dehydration, as shown by Chou et al. [50] who stated that Vp/Vs is an effective indicator of hydration, or melting, on the mantle wedge. In studies by Lin et al. [58] and Maruyama et al. [59], hydration is indicated by low Vp, Vs, and high Vp/Vs.

4.3. Double Subduction of the Molucca Sea Plate

The double subduction zone of the Molucca Sea Plate is shown in the vertical cross-section of C-C′ (west–east), crossing the Celebes Sea, Sangihe Island, Molucca Sea, and the Morotai Basin (Figure 8); and the vertical cross-section D-D′ (west–east), crossing Sulawesi's north arm, the Molucca Sea, and Halmahera Island (Figure 9). Based on tomographic inversion results in the C1 area, the high-Vp, high-Vs anomalies subduct to the west and reach the asthenosphere layer at a depth of ~300 km. Based on the resolution test (Figures S6–S8), the subduction pattern is shown to reach a depth of ~400 km (CRT result: not good resolved areas), while a low Vp/Vs is seen in the partially molten asthenosphere layer at a depth of ~210 km. High-Vp, high-Vs, and low-Vp/Vs anomalies subducting eastward to a depth of ~200 km are also observed. This feature can be interpreted as a double subduction of the Molucca Sea Plate, which subducts westward beneath the Sangihe Islands and eastward beneath the Morotai Basin. This is also followed by the presence of intermediate and deep earthquakes in the upper mantle layer that forms a double subduction slab pattern, as indicated by high seismic velocity [53]. These results are consistent with those of Chou et al. [50], Zhao [46], and Zenonos et al. [29,30] who suggest that high Vp, high Vs, and low Vp/Vs can be associated with the feature of subducted slab. Previous studies of P-wave tomography also identified a double subduction of the Molucca Sea Plate [9,22–24,27,60] in which the plate subducting to the west can reach the mantle transition layer at a depth of ~600 km. According to Fan and Zhao [27], the Molucca Sea Plate is submerged deeper in the north than in the south. These studies also indicate that the arch of double subduction of the Molucca Sea Plate has submerged into the upper mantle at a depth of ~150 km. Based on the geometry model of slab2 (green line) from Hayes et al. [45], the arch of the double subduction is closer to the surface in the uppermost mantle layer.

The C2 area is located beneath the Sangihe Trench (ST), which is a collision zone of the Sangihe continental crust that moves eastward with an overriding plate structure [26] and the Molucca Sea Plate that subducts to the west. The tomographic inversion results show that high-Vp, high-Vs, and low-Vp/Vs anomalies in the mantle crust can be interpreted as rock material that experienced micro-cracks that have since closed due to high pressure [30,61–63]. The C3 area is located beneath the Halmahera Trench (HT), which is a collision zone of the Halmahera continental crust that moves westward, having an overriding plate structure [26] with the Molucca Sea Plate subducting to the east. The tomographic inversion results show that low-Vp, low-Vs, and high-Vp/Vs anomalies in the crust and mantle can be interpreted as highly fractured, water-saturated rock material [63]. Our results obtained beneath the Central Ridge of Molucca Sea, indicate a clear presence of boundary between ST which has high-velocity anomalies and low Vp/Vs and HT, which has low-velocity anomalies and high Vp/Vs. This boundary also has a signature of high seismicity area. An earthquake of magnitude 6.2 occurred on 10 July 2021 on this boundary that can be interpreted as a collision zone between fault planes or fractures [10]. The boundary may represents material with different densities between the western and eastern parts of the Central Ridge of the Molucca Sea which have upraised ophiolite bodies [17,18].

In the vertical cross-section D-D′ of area D1 (Figure 9), the results of the inversion tomography show the presence of a double subduction of the Molucca Sea Plate indicated by high-Vp, high-Vs anomalies. It subducts westward and eastward, reaching a solid asthenosphere layer at a depth of ~300 and ~250 km, respectively. The Molucca Sea Plate is not well-imaged as far as the Vp/Vs ratio, and there is a discontinuity in the slab that subducts westward at depths of ~100–150 km. On the other hand, the slab that subducts to the east is well-imaged, as indicated by low Vp/Vs. The seismic pattern observed at intermediate and deep depths had a high seismic velocity, which can be associated with a double subduction slab [53]. This finding is consistent with Chou et al. [50], Zhao [46] and Zenonos et al. [29,30]; i.e., high Vp, high Vs, and low Vp/Vs are interpreted as a slab subduction. Previous studies of P-wave tomography have also identified a double subduction of the Molucca Sea Plate to depths of ~600–650 km subducting to the west, and depths of ~250–400 km subducting to the east [9,22–24,27,60]. The arch of double subduction in this area is in the uppermost mantle layer near the surface, which is consistent with the geometry of the slab2 model from Hayes et al. [45]. Fan and Zhao [27] showed that the arch of the Molucca Sea Plate submerged shallower in the south than in the north.

The D3 and D5 areas are located in ST and HT which is a collision zone between the continental crust and the denser, heavier oceanic crust. The ST area is the collision zone between the continental crust of Sulawesi's north arm located above the Molucca Sea Plate that subducts to the west. The HT area is the collision zone between the continental crust of Halmahera above the Molucca Sea Plate and subducting to the east [26]. According to Puspito et al. [22], low velocity dominates the Molucca Sea collision zones and volcanic zones. The results of tomography inversion in these two areas show high-Vp, high-Vs, and low-Vp/Vs anomalies in the crust and mantle. This feature can be interpreted as rock material that experienced micro-cracks that have since closed due to high pressure [30,61–63]. The D4 area, located beneath the island of Tifore in the Central Molucca Sea may correspond to upraised ophiolite body [17,18] which has an active thrust fault [11,12]. The tectonic activity in this area causes many earthquakes of both small and large magnitudes. Tomographic inversion results in the Central Molucca Sea show low-Vp, low-Vs anomalies, and high Vp/Vs in the crust and mantle which indicate the presence of highly fractured brittle material and water-saturated rock [63]. In the west and east (ST and HT), we suggest that the rock material is probably not water-saturated because the micro-cracks have been covered by high pressure. However, further north (Figure 8, cross-section C-C′), the cracked and water-saturated rock materials shift to the east (there is a clear boundary in the Central Molucca Sea) and extend to HT.

Figure 10 shows a 3D visualization of the double subduction of the Molucca Sea Plate beneath Sulawesi's north arm, the Molucca Sea, and Halmahera Island based on P-wave tomography inversion results. These results show that the Molucca Sea Plate subducts to the west beneath Sulawesi's north arm. In perspective from south to north, the state of the slab field also dips north from the Minahasa Peninsula to the Celebes Sea, having numerous earthquake events that follow the high-Vp anomaly pattern and is interpreted as a double subduction of the Molucca Sea Plate. This is consistent with the green three-dimensional surface plot of the slab2 models of the Molucca Sea Plate [45] in Figure 6.

5. Conclusions

In this research, we performed travel time tomographic inversion using an initial 3D velocity model. These are obtained by manual re-picking of the same number of P- and S-arrivals and simultaneous determination of the 3D structure for the Vp, Vs, and Vp/Vs ratio, as well as relocating the hypocenter location. The results show a high Vp, high Vs, and low-Vp/Vs anomaly beneath the Minahasa Peninsula, Sulawesi's north arm, with the Celebes Sea (cross-section A-A′, south–north) dipping to the north at a ~40° angle. Beneath the Celebes Sea, Sangihe Island, Molucca Sea, the Morotai Basin (cross-section C-C′, west–east), and Sulawesi's north arm, the Molucca Sea, and Halmahera Island (cross-section D-D′, west–east) can be interpreted as a double subduction of the Molucca Sea Plate. The tomographic image beneath the southern part of the Halmahera volcanic arc (Bacan Island) shows low Vp, low Vs, and high Vp/Vs, which indicates a possible magma source; although the volcano has been dormant until the present. The tomographic image beneath the group of volcanoes in the northern arm of Halmahera also shows low Vp, low Vs, and high Vp/Vs, which can be interpreted as a possible magma source. Differing from volcanoes in the south, however, some volcanoes in this area have erupted and are emitting volcanic ash. Both the volcanic groups on Bacan Island and the northern arm of Halmahera are connected by partial melting, as indicated by low Vp, low Vs, high Vp/Vs above the Molucca Sea Plate, which subducts to the east.

The upper curvature of the Molucca Sea Plate in the northern part is submerged deeper into the upper mantle layer at a depth of ~150 km compared with the southern part. It was also found that beneath the Central Ridge of the Molucca Sea, at latitude 1.10 °N, the low Vp, low Vs, and high Vp/Vs in the crust and mantle can be interpreted as highly fractured and water-saturated rock material; while high Vp, high Vs, and low Vp/Vs are observed in the western (beneath the ST) and eastern parts (beneath the HT), which can be interpreted as rock material that experienced micro-cracks that have since closed due to high pressure. Further north, the low Vp, low Vs, and high Vp/Vs anomalies shift eastward and extend to HT with a clear boundary between low and high velocity below the Central Ridge of the Molucca Sea.

Supplementary Materials: The following supporting information can be downloaded at: https://www.mdpi.com/article/10.3390/app122010520/s1, Figure S1: Molucca Sea Plate profile overlaid with seismic profile from 1959–1966 which forms a double subduction pattern was modified from [7]. NW: Northwest, SE: Southeast. Sign ?: the length of the plate subduction to the west and east is not yet known. The black dot denotes earthquakes.; Figure S2: Histogram of the number of phases per station; Figure S3: Horizontal and vertical ray path coverage (West-East: below and North-South: right). The brown line is the ray path connecting hypocenters locations with BMKG stations; Figure S4: Wadati diagram for all events used in this study. Tp is arrival times of the P-wave and Ts−Tp is the difference between S- and P-waves arrival times. The Gray dot is the picks of phase time from each station at a certain event, and the solid line is the slope to determine the Vp/Vs value which is 1.76; Figure S5: Trade-off curve between data variance and model variance. We chose damping parameters Vp and Vp/Vs to be 150 and 130, respectively. We use this damping value to achieve a good compromise between data variance and model variance for the inversion, both the real data and the checkerboard test; Figure S6: Vertical cross-section of Derivative Weight Sum (DWS) on profile South-North: A-A′, B-B′ and profile West-East: C-C′, D-D′. The DWS is used to find the number of ray paths that pass through a grid node. The first column is the Vp value, and the second column is Vp/Vs value (logarithmic scale). The black color is the high-resolution zone, and the red triangle is the volcanoes; Figure S7: Vertical cross-section of Ray Hit Count (RHC) on profile South-North: A-A′, B-B′ and profile West-East: C-C′, D-D′ for making a thorough cut in a particular DWS. The first column is the Vp value, and the second column is Vp/Vs value (logarithmic scale). The black color is the high-resolution zone, and the red triangle is the volcanoes; Figure S8: Vertical cross-section of Diagonal Resolution Element (DRE) on profile South-North: A-A′, B-B′ and profile West-East: C-C′, D-D′ which concerning matrix resolution of R (damped least-squares problems), with an R-value between 0 and 1. The first column is the Vp value, and the second column is Vp/Vs value. The black color is the high-resolution zone, and the red triangle is the volcanoes; Figure S9: The distribution frequency estimated error location. (A–C) are the x, y, and z directions; Figure S10: The distribution between the number of events and the azimuthal gap; Figure S11: The distribution between P- and S-travel times and epicentral distance; Figure S12: Histogram of travel time residual (black and yellow bars color are initial and final travel time residual, respectively). The mean travel time residual decreased from −1.111 s to 0.156 s. The Binwidth histogram is 0.2 s; Figure S13: The curve between the number of iterations and weighted RMS residual. Tomography inversion became convergent after 15 iterations, and the weighted RMS residual decreased by approximately 47%; Figure S14: Histogram of RMS Residual (initial: yellow bars color and final: black bars color). The mean of RMS residual decreased by approximately 33%. The Binwidth histogram is 0.1 s; Figure S15: The horizontal section of CRT profiles from a depth of 0 to 330 km. The purple dashed line is an area that is good well-resolved. The blue color indicates positive perturbations, and the red color indicates negative perturbations relative to the initial model. The first column is Vp perturbation, and the second column is Vp/Vs perturbation; Figure S16: Horizontal tomographic inversion results for Vp, Vs, dan Vp/Vs ratio from a depth of 0 to 330 km. The bright area is an area that good well-resolved, and the blurred area is an area that was not well-resolved; Table S1: The highest value of Vp and Vp/Vs for the DWS, RHC, and DRE resolution tests.

Author Contributions: Conceptualization, G.R., B.J.S., A.D.N. and S.R. (Supriyanto Rohadi); methodology, G.R., A.D.N., B.J.S. and S.R. (Supriyanto Rohadi); software, G.R., S.R. (Shindy Rosalia), P.S., M.R., A.A., F.M. and H.A.; validation, G.R., A.D.N., S.R. (Shindy Rosalia) and F.M.; formal analysis, G.R., A.D.N., B.J.S. and S.R. (Supriyanto Rohadi); investigation, A.D.N., B.J.S., S.R. (Supriyanto Rohadi) and S.R. (Shindy Rosalia); resources, G.R., A.D.N., B.J.S., S.R. (Supriyanto Rohadi), S.S. and S.R. (Shindy Rosalia); data curation, G.R., A.D.N. and S.R. (Supriyanto Rohadi); writing—original draft preparation, G.R. and F.M.; writing—review and editing, A.D.N., S.R. (Shindy Rosalia), Z.Z., D.P.S. and S.S.; visualization, G.R., F.M., A.A. and H.A.; supervision, A.D.N., S.R. (Shindy Rosalia), Z.Z., D.P.S. and S.S.; project administration, G.R. All authors have read and agreed to the published version of the manuscript.

Funding: This research was funded by BUDI-DN Ministry of Research, Technology and Higher Education (Kemristekdikti) in collaboration with the Indonesian Endowment Fund for Education (LPDP) Number: PRJ-5440/LPDP.3/2016; Addendum Number: PRJ-193/LPDP.4/2019.

Institutional Review Board Statement: Not applicable.

Informed Consent Statement: Not applicable.

Data Availability Statement: The waveform data used in this research were obtained from BMKG. The data supporting the findings of this study are available from the corresponding author upon request.

Acknowledgments: Our grateful appreciation goes to the BUDI-DN Ministry of Research, Technology, and Higher Education (Kemristekdikti) in collaboration with the Indonesian Endowment Fund for Education (LPDP), which provides scholarship and research funds. We would also like to thank the Volcanology and Geothermal Laboratory (Geophysical Engineering, Faculty of Mining and Petroleum Engineering, Institut Teknologi Bandung) for providing facilities for conducting research. The Agency for Meteorology, Climatology, and Geophysics (BMKG) provided waveform data. Anthony Lomax created and developed the NLLoc and Seisgram2Kv.7.0 software used in this study. All 2D images are plotted using the GMT tool, and 3D images using Matlab 2018b.

Conflicts of Interest: The authors declare no conflict of interest.

References

1. Hall, R.; Nichols, G.; Ballantyne, P.; Charlton, T.; Ali, J. The Character and Significance of Basement Rocks of the Southern Molucca Sea Region. *J. Southeast Asian Earth Sci.* **1991**, *6*, 249–258. [CrossRef]
2. McCaffrey, R.; Silver, E.A.; Raitt, R.W. Crustal Structure of the Molucca Sea Collision Zone, Indonesia. In *The Tectonic and Geologic Evolution of Southeast Asian Seas and Islands*; American Geophysical Union (AGU): Washington, DC, USA, 1980; pp. 161–177.
3. Moore, G.F.; Kadarisman, D.; Evans, C.A.; Hawkins, J.W. Geology of the Talaud Islands, Molucca Sea Collision Zone, Northeast Indonesia. *J. Struct. Geol.* **1981**, *3*, 467–475. [CrossRef]
4. Morris, J.D.; Jezek, P.A.; Hart, S.R.; Hill, J.B. The Halmahera Island Arc, Molucca Sea Collision Zone, Indonesia: A Geochemical Survey. In *The Tectonic and Geologic Evolution of Southeast Asian Seas and Islands: Part 2*; Geophysical Monograph Series; American Geophysical Union (AGU): Washington, DC, USA, 1983; Volume 27, pp. 373–387.
5. Silver, E.A.; Moore, J.C. The Molucca Sea Collision Zone, Indonesia. *J. Geophys. Res. Solid Earth* **1978**, *83*, 1681–1691. [CrossRef]
6. Hall, R.; Audley-Charles, M.G.; Banner, F.T.; Hidayat, S.; Tobing, S.L. Basement Rocks of the Halmahera Region, Eastern Indonesia: A Late Cretaceous–Early Tertiary Arc and Fore-Arc. *J. Geol. Soc. London.* **1988**, *145*, 65–84. [CrossRef]
7. Hamilton, W.B. *Tectonics of the Indonesian Region*; U.S. Govt. Print. Off.: Washington, DC, USA, 1979.
8. Hatherton, T.; Dickinson, W.R. The Relationship between Andesitic Volcanism and Seismicity in Indonesia, the Lesser Antilles, and Other Island Arcs. *J. Geophys. Res.* **1969**, *74*, 5301–5310. [CrossRef]
9. Hafkenscheid, E.; Buiter, S.J.; Wortel, M.J.; Spakman, W.; Bijwaard, H. Modelling the Seismic Velocity Structure beneath Indonesia: A Comparison with Tomography. *Tectonophysics* **2001**, *333*, 35–46. [CrossRef]
10. Cardwell, R.K.; Isaacks, B.L.; Karig, D.E. The Spatial Distribution of Earthquakes, Focal Mechanism Solutions, and Subducted Lithosphere in the Philippine and Northeastern Indonesian Islands. In *The Tectonic and Geologic Evolution of Southeast Asian Seas and Islands*; Geophys. Monogr., Ser.; Hayes, D.E., Ed.; AGU: Washington, DC, USA, 1980; Volume 23, pp. 1–35.
11. McCaffrey, R. Lithospheric Deformation within the Molucca Sea Arc-Arc Collision: Evidence from Shallow and Intermediate Earthquake Activity. *J. Geophys. Res.* **1982**, *87*, 3663. [CrossRef]
12. Hutchings, S.J.; Mooney, W.D. The Seismicity of Indonesia and Tectonic Implications. *Geochem. Geophys. Geosyst.* **2021**, *22*, e2021GC009812. [CrossRef]
13. Hall, R. Plate Boundary Evolution in the Halmahera Region, Indonesia. *Tectonophysics* **1987**, *144*, 337–352. [CrossRef]
14. Milsom, J.; Masson, D.; Nicols, G. Three Trench Endings in Eastern Indonesia. *Mar. Geol.* **1992**, *104*, 227–241. [CrossRef]

15. Sufni Hakim, A.; Hall, R. Tertiary Volcanic Rocks from the Halmahera Arc, Eastern Indonesia. *J. Southeast Asian Earth Sci.* **1991**, *6*, 271–287. [CrossRef]
16. Moore, G.F.; Silver, E.A. Collision Processes in the Northern Molucca Sea. In *The Tectonic and Geologic Evolution of Southeast Asian Seas and Islands: Part 2*; Geophys. Monogr., Ser.; Hayes, D.E., Ed.; AGU: Washington, DC, USA, 1983; Volume 27, pp. 360–372.
17. Widiwijayanti, C.; Tiberi, C.; Deplus, C.; Diament, M.; Mikhailov, V.; Louat, R. Geodynamic Evolution of the Northern Molucca Sea Area (Eastern Indonesia) Constrained by 3-D Gravity Field Inversion. *Tectonophysics* **2004**, *386*, 203–222. [CrossRef]
18. Widiwijayanti, C.; Mikhailov, V.; Diament, M.; Deplus, C.; Louat, R.; Tikhotsky, S.; Gvishiani, A. Structure and Evolution of the Molucca Sea Area: Constraints Based on Interpretation of a Combined Sea-Surface and Satellite Gravity Dataset. *Earth Planet. Sci. Lett.* **2003**, *215*, 135–150. [CrossRef]
19. Hédervári, P.; Papp, Z. Seismicity Maps of the Indonesian Region. *Tectonophysics* **1981**, *76*, 131–148. [CrossRef]
20. Hall, R.; Spakman, W. Mantle Structure and Tectonic History of SE Asia. *Tectonophysics* **2015**, *658*, 14–45. [CrossRef]
21. Chen, P.F.; Chien, M.; Bina, C.R.; Yen, H.Y.; Antonio Olavere, E. Evidence of an East-Dipping Slab beneath the Southern End of the Philippine Trench (1°N–6°N) as Revealed by ISC-EHB. *J. Asian Earth Sci. X* **2020**, *4*, 100034. [CrossRef]
22. Puspito, N.T.; Yamanaka, Y.; Miyatake, T.; Shimazaki, K.; Hirahara, K. Three-Dimensional P-Wave Velocity Structure beneath the Indonesian Region. *Tectonophysics* **1993**, *220*, 175–192. [CrossRef]
23. Widiyantoro, S.; Hilst, R. Mantle Structure beneath Indonesia Inferred from High-Resolution Tomographic Imaging. *Geophys. J. Int.* **1997**, *130*, 167–182. [CrossRef]
24. Widiyantoro, S. Complex Morphology of Subducted Lithosphere in the Mantle below the Molucca Collision Zone from Non-Linear Seismic Tomography. *ITB J. Eng. Sci.* **2003**, *35*, 1–10. [CrossRef]
25. Huang, Z.; Zhao, D.; Wang, L. P Wave Tomography and Anisotropy beneath Southeast Asia: Insight into Mantle Dynamics. *J. Geophys. Res. Solid Earth* **2015**, *120*, 5154–5174. [CrossRef]
26. Zhang, Q.; Guo, F.; Zhao, L.; Wu, Y. Geodynamics of Divergent Double Subduction: 3-D Numerical Modeling of a Cenozoic Example in the Molucca Sea Region, Indonesia. *J. Geophys. Res. Solid Earth* **2017**, *122*, 3977–3998. [CrossRef]
27. Fan, J.; Zhao, D. Evolution of the Southern Segment of the Philippine Trench: Constraints from Seismic Tomography. *Geochem. Geophys. Geosyst.* **2018**, *19*, 4612–4627. [CrossRef]
28. Fan, J.; Zhao, D. P-Wave Anisotropic Tomography of the Central and Southern Philippines. *Phys. Earth Planet. Inter.* **2019**, *286*, 154–164. [CrossRef]
29. Zenonos, A.; De Siena, L.; Widiyantoro, S.; Rawlinson, N. P and S Wave Travel Time Tomography of the SE Asia-Australia Collision Zone. *Phys. Earth Planet. Inter.* **2019**, *293*, 106267. [CrossRef]
30. Zenonos, A.; De Siena, L.; Widiyantoro, S.; Rawlinson, N. Direct Inversion of S-P Differential Arrival Times for Ratio in SE Asia. *J. Geophys. Res. Solid Earth* **2020**, *125*, e2019JB019152. [CrossRef]
31. Cao, L.; He, X.; Zhao, L.; Lü, C.; Hao, T.; Zhao, M.; Qiu, X. Mantle Flow Patterns Beneath the Junction of Multiple Subduction Systems Between the Pacific and Tethys Domains, SE Asia: Constraints From SKS-Wave Splitting Measurements. *Geochem. Geophys. Geosyst.* **2021**, *22*, e2021GC009700. [CrossRef]
32. Lomax, A.; Michelini, A. Mwpd: A Duration-Amplitude Procedure for Rapid Determination of Earthquake Magnitude and Tsunamigenic Potential from P Waveforms. *Geophys. J. Int.* **2009**, *176*, 200–214. [CrossRef]
33. Lomax, A.; Virieux, J.; Volant, P.; Berge-Thierry, C. *Probabilistic Earthquake Location in 3D and Layered Models*; Springer: Dordrecht, The Netherlands, 2000; pp. 101–134.
34. Tarantola, A.; Valette, B. Inverse Problems = Quest for Information. *J. Geophys.* **1982**, *50*, 159–170.
35. Moser, T.J.; van Eck, T.; Nolet, G. Hypocenter Determination in Strongly Heterogeneous Earth Models Using the Shortest Path Method. *J. Geophys. Res.* **1992**, *97*, 6563. [CrossRef]
36. Wittlinger, G.; Herquel, G.; Nakache, T. Earthquake Location in Strongly Heterogeneous Media. *Geophys. J. Int.* **1993**, *115*, 759–777. [CrossRef]
37. Kennett, B.L.N.; Engdahl, E.R. Traveltimes for Global Earthquake Location and Phase Identification. *Geophys. J. Int.* **1991**, *105*, 429–465. [CrossRef]
38. Kennett, B.L.N.; Engdahl, E.R.; Buland, R. Constraints on Seismic Velocities in the Earth from Traveltimes. *Geophys. J. Int.* **1995**, *122*, 108–124. [CrossRef]
39. Evans, J.R.; Eberhart-Phillips, D.; Thurber, C.H. *User's Manual for SIMULPS12 for Imaging vp and vp/vs; a Derivative of the "Thurber" Tomographic Inversion SIMUL3 for Local Earthquakes and Explosions*; Open-File Rep. Open-File Report 94-431; U.S. Geological Survey: Reston, VI, USA, 1994; p. 101. [CrossRef]
40. Um, J.; Thurber, C. A Fast Algorithm for Two-Point Seismic Ray Tracing. *Bull. Seismol. Soc. Am.* **1987**, *77*, 972–986. [CrossRef]
41. Eberhart-Phillips, D. Three-Dimensional Velocity Structure in Northern California Coast Ranges from Inversion of Local Earthquake Arrival Times. *Bull. Seismol. Soc. Am.* **1986**, *76*, 1025–1052. [CrossRef]
42. Thurber, C.H. *Local Earthquake Tomography: Velocities and Vp/Vs-Theory*; Chapman and Hall: London, UK, 1993; ISBN 0412371901.
43. Nugraha, A. *Tomografi Seismik*, 1st ed.; ITB Press: Bandung, Indonesia, 2017; ISBN 978-602-5417-48-1.
44. Rawlinson, N.; Sambridge, M. *Seismic Traveltime Tomography of The Crust and Lithosphere*; Elsevier: Amsterdam, The Netherlands, 2003; pp. 81–198.
45. Hayes, G.P.; Moore, G.L.; Portner, D.E.; Hearne, M.; Flamme, H.; Furtney, M.; Smoczyk, G.M. Slab2, a Comprehensive Subduction Zone Geometry Model. *Science* **2018**, *362*, 58–61. [CrossRef] [PubMed]

46. Zhao, D. *Multiscale Seismic Tomography*; Springer: Japan, Tokyo, 2015; ISBN 978-4-431-55359-5.
47. Zhao, D.; Yamashita, K.; Toyokuni, G. Tomography of the 2016 Kumamoto Earthquake Area and the Beppu-Shimabara Graben. *Sci. Rep.* **2018**, *8*, 15488. [CrossRef]
48. Hasegawa, A.; Umino, N.; Takagi, A. Double-Planed Structure of the Deep Seismic Zone in the Northeastern Japan Arc. *Tectonophysics* **1978**, *47*, 43–58. [CrossRef]
49. Hasegawa, A.; Umino, N.; Takagi, A. Double-Planed Deep Seismic Zone and Upper-Mantle Structure in the Northeastern Japan Arc. *Geophys. J. Int.* **1978**, *54*, 281–296. [CrossRef]
50. Chou, H.-C.; Kuo, B.-Y.; Chiao, L.-Y.; Zhao, D.; Hung, S.-H. Tomography of the Westernmost Ryukyu Subduction Zone and the Serpentinization of the Fore-Arc Mantle. *J. Geophys. Res. Solid Earth* **2009**, *114*, 12301. [CrossRef]
51. Hall, R.; Ali, J.R.; Anderson, C.D.; Hall, R.; Ali, J.R.; Anderson, C.D. Cenozoic Motion of the Philippine Sea Plate: Palaeomagnetic Evidence from Eastern Indonesia. *Tecto* **1995**, *14*, 1117–1132. [CrossRef]
52. Kim, K.-H.; Chiu, J.-M.; Pujol, J.; Chen, K.-C.; Huang, B.-S.; Yeh, Y.-H.; Shen, P. Three-Dimensional VP and VS Structural Models Associated with the Active Subduction and Collision Tectonics in the Taiwan Region. *Geophys. J. Int.* **2005**, *162*, 204–220. [CrossRef]
53. Zhao, D. Multiscale Seismic Tomography and Mantle Dynamics. *Gondwana Res.* **2009**, *15*, 297–323. [CrossRef]
54. Siebert, L.; Simkin, T.; Kimberly, P. *Volcanoes of the World*; Smithsonian Institution: Washington, DC, USA, 2010; ISBN 0520268776.
55. Nugraha, A.D.; Shiddiqi, H.A.; Widiyantoro, S.; Puspito, N.T.; Triyoso, W.; Wiyono, S.; Daryono; Wandono; Rosalia, S. Hypocenter Relocation of Earthquake Swarm in West Halmahera, North Molucca Region, Indonesia by Using Double-Difference Method and 3D Seismic Velocity Structure. *IOP Conf. Ser. Earth Environ. Sci.* **2017**, *62*, 012053. [CrossRef]
56. Nugraha, A.D.; Supendi, P.; Widiyantoro, S.; Daryono; Wiyono, S. Hypocenter Relocation of Earthquake Swarm around Jailolo Volcano, North Molucca, Indonesia Using the BMKG Network Data: Time Periods of 27 September–10 October 2017. *AIP Conf. Proc.* **2018**, *1987*, 020093. [CrossRef]
57. Nakajima, J.; Matsuzawa, T.; Hasegawa, A.; Zhao, D. Three-Dimensional Structure of Vp, Vs, and Vp/Vs beneath Northeastern Japan: Implications for Arc Magmatism and Fluids. *J. Geophys. Res. Solid Earth* **2001**, *106*, 21843–21857. [CrossRef]
58. Lin, J.Y.; Hsu, S.K.; Sibuet, J.C. Melting Features along the Ryukyu Slab Tear, beneath the Southwestern Okinawa Trough. *Geophys. Res. Lett.* **2004**, *31*, 20862. [CrossRef]
59. Maruyama, S.; Hasegawa, A.; Santosh, M.; Kogiso, T.; Omori, S.; Nakamura, H.; Kawai, K.; Zhao, D. The Dynamics of Big Mantle Wedge, Magma Factory, and Metamorphic-Metasomatic Factory in Subduction Zones. *Gondwana Res.* **2009**, *16*, 414–430. [CrossRef]
60. Amaru, M.L. Global Travel Time Tomography with 3-D Reference Models. *Geol. Ultraiectina* **2007**, *274*, 81–88.
61. Wagner, L.S.; Anderson, M.L.; Jackson, J.M.; Beck, S.L.; Zandt, G. Seismic Evidence for Orthopyroxene Enrichment in the Continental Lithosphere. *Geology* **2008**, *36*, 935–938. [CrossRef]
62. Zheng, Y.; Lay, T. Low Vp/Vs Ratios in the Crust and Upper Mantle beneath the Sea of Okhotsk Inferred from Teleseismic PMP, SMP, and SMS Underside Reflections from the Moho. *J. Geophys. Res. Solid Earth* **2006**, *111*, 1305. [CrossRef]
63. Bariş, S.; Nakajima, J.; Hasegawa, A.; Honkura, Y.; Ito, A.; Balamir, S. Three-Dimensional Structure of Vp, Vs and Vp/Vs in the Upper Crust of the Marmara Region, NW Turkey. *Earth Planets Sp.* **2005**, *57*, 1019–1038. [CrossRef]

Article

Estimation of Litho-Fluid Facies Distribution from Zero-Offset Acoustic and Shear Impedances

Mohammed Fathy Gouda [1,*], Abdul Halim Abdul Latiff [1] and Seyed Yasser Moussavi Alashloo [2]

[1] Department of Geosciences, Universiti Teknologi Petronas, Seri Iskander 32610, Perak, Malaysia; abdulhalim.alatiff@utp.edu.my
[2] The Institute of Digital Signal Processing, University of Duisburg-Essen, Forsthausweg 2, 47057 Duisburg, Germany; y.alashloo@gmail.com
* Correspondence: mohammed_22000724@utp.edu.my

Abstract: Seismic data are considered crucial sources of data that help identify the litho-fluid facies distributions in reservoir rocks. However, different facies mostly have similar responses to seismic attributes. In addition, seismic anisotropy negatively affects the facies predictors extracted from seismic data. Accordingly, this study aims at estimating zero-offset acoustic and shear impedances based on partial-stack inversion by two methods: statistical modeling and a multilayer feed-forward neural network (MLFN). The resulting impedance volumes are compared to those obtained from isotropic simultaneous inversion by using impedance logs. The best impedance volumes are applied to Thomsen's anisotropy equations to solve for the anisotropy parameters Epsilon and Delta. Finally, the shear and acoustic impedances are transformed into elastic properties from which the facies and fluid distributions are predicted by using the logistic regression and decision tree algorithms. The results obtained from the MLFN show better matching with the impedance and facies logs compared to those obtained from isotropic inversion and statistical modeling.

Keywords: anisotropy; rock physics; inversion; parameter estimation

Citation: Gouda, M.F.; Abdul Latiff, A.H.; Moussavi Alashloo, S.Y. Estimation of Litho-Fluid Facies Distribution from Zero-Offset Acoustic and Shear Impedances. *Appl. Sci.* **2022**, *12*, 7754. https://doi.org/10.3390/app12157754

Academic Editors: Guofeng Liu, Xiaohong Meng and Zhifu Zhang

Received: 8 June 2022
Accepted: 5 July 2022
Published: 1 August 2022

1. Introduction

Seismic data have proved to be important quantitative tools that can bring valuable information about rock properties, especially when being integrated with statistical tools. For example, multiattribute transforms were used to predict log properties from seismic data [1]. Another common approach is to invert seismic data into elastic properties and then into petrophysical properties by statistical relationships obtained from logging data [2]. Porosity, lithology, and fluid properties were also estimated with the resampling of rock physics constraints [3–5]. Russel used multivariate statistics and neural networks to predict reservoir parameters from seismic attributes [6]. In addition, Bachrach inverted porosity and water saturation by stochastic rock physics modeling [7].

Other methodologies consider reservoir parameter inversion based on Bayesian classification theory [8,9]. Moreover, a geostatistical inversion method was used to predict reservoir properties in a higher certainty than that of deterministic methods [10]. In addition, some studies derived elastic properties from prestack seismic inversion to define facies [11,12]. Another straightforward methodology is to combine petroelastic modeling and stochastic inversion to predict hydrocarbon zones based on Bayesian theory [13]. A rock physics model was postulated to derive the elastic rock properties from mineral parameters and structure information [14]. Recently, rock properties were estimated from angle stack seismic data by a Bayesian inversion based on the Gibbs sampling algorithm [15].

Machine learning (ML) tools have been widely used in reservoir characterization [16]. For instance, formation porosity and permeability were estimated from logging and core data by using an artificial neural network (ANN) [17–19]. Other studies also used neural networks to predict essential rock properties such as permeability [20–22] and shear wave velocity [23–26]. Neural networks can be combined with seismic attributes to predict rock physics properties from seismic data [27–30]. In addition, facies distributions were estimated by both supervised and unsupervised ML algorithms such as ANN [31,32], convolutional neural network (CNN) [33–35], support vector machine (SVM) [36–41], bagged tree (BT) [42,43], relevance vector machine (RVM) [44], seed region growing (SRG) [45], self-organizing map (SOM) [46,47], principal component analysis (PCA) [48–50], and generative topographic mapping (GTM) [51]. Some challenges are common in ML facies models such as over-fitting and over-parametrization. Moreover, such models should consider the limitations of seismic data to eliminate error as much as possible. One of the problems that contributes to the uncertainty of seismic-derived methods is seismic anisotropy.

This study aimed at forecasting zero-offset P-impedance (Z_P) and S-impedance (Z_S) by using statistical modeling and MLFN after applying a partial-log-constrained inversion to the near, mid, and far angle stacks. The outputs of this objective were zero-offset impedances that were more accurate than those obtained from isotropic simultaneous inversion. Another objective was to compare the abilities of both statistical modeling and neural networks to turn the randomness of the data into identified patterns. The next step was to apply the best Z_P and Z_S volumes to Thomsen's anisotropy equations to solve for the anisotropy parameters, Epsilon and Delta, assuming a vertical transverse isotropic (VTI) medium. Next, the zero-offset Z_P and Z_S were used as inputs to a facies model, which predicted the distribution of three classes: gas sand, wet sand, and shale. The idea of the facies model is to forecast the sand and gas probabilities by logistic regression [52] and then use them as inputs for a decision tree to predict the distribution of each facies class.

The three offset stacks were converted into angle stacks, which were the inputs of the partial-stack inversion, and angle gathers, which were the inputs of the simultaneous isotropic inversion. The ranges of the near, mid, and far stacks were from 5° to 15°, 15° to 25°, and 25° to 40°. The partial-stack inversion resulted in the Z_P and Z_S volumes at the central angles, 10°, 20°, and 32.5°, from which the zero-offset properties were forecasted. The target zone consisted of two gas-bearing shaly sand reservoirs in the Malay basin: A and B. The training data of the Bayesian and MLFN models were gathered from the wells ($\times$1) and ($\times$2); however, the facies log was only obtained at the well ($\times$1) due to the availability of the petrophysical cut-offs and oil-mud resistivity imaging data. Another challenge was that the Z_S log was only available at the well ($\times$1), so it was estimated from other log properties by the MLFN neural network at the well ($\times$2).

The resulting Z_P and Z_S were compared to those obtained from isotropic inversion by using the impedance logs. The MLFN resulted in the most accurate zero-offset impedances. Therefore, the Z_P and Z_S volumes were applied to Thomsen's anisotropy equations to solve for the anisotropy parameters Epsilon and Delta. The Z_P and Z_S of MLFN and isotropic inversion were transformed into the near and far elastic impedances, which were the lithology predictors, and then into the Mu–Rho (MR) ratio, Lambda–Rho/Mu–Rho (LR/MR) ratio, and Poisson's ratio (PR), which are the fluid predictors. The litho-fluid model was applied to the inverted elastic volumes to forecast the litho-fluid facies distribution. The facies log was used to validate the predicted facies obtained from the isotropic and anisotropic approaches.

2. Seismic Anisotropy

Seismic anisotropy is the dependence of the seismic velocity on the incident angle [53–55]. Wave propagation in anisotropic media has been discussed by various studies [53,56–58]. To understand the seismic anisotropy of the Earth's subsurface, the relationship between the internal forces (stress) and deformation occurring in the medium (strain) should be defined. When an elastic wave propagates through the subsurface, the stress–strain relationship can be obtained according to Hooke's law as shown below [59]:

$$\sigma_{ij} = c_{ijkl}e_{kl} \tag{1}$$

where σ_{ij} is the stress tensor, e_{kl} is the strain tensor, and c_{ijkl} is the elastic (stiffness) tensor, which is a fourth-order tensor with 81 independent components. Both the strain and stress tensors are symmetric [60], resulting in:

$$c_{ijkl} = c_{jikl} = c_{ijlk} \tag{2}$$

The 81 independent components are accordingly reduced to 36 due to the symmetry in the stress and strain tensors. Therefore, the elastic tensor (c_{ijlk}) is replaced with a 6×6 matrix ($c_{\alpha\beta}$), where [61]:

$$\alpha = ij = ji, \beta = kl = lk \tag{3}$$

where α and β are the new indices of the elastic tensor. The resulting (6×6) notation matrix is expressed as shown below:

$$\begin{bmatrix} C_{11} & C_{12} & C_{13} & C_{14} & C_{15} & C_{16} \\ C_{12} & C_{22} & C_{23} & C_{24} & C_{25} & C_{26} \\ C_{13} & C_{23} & C_{33} & C_{34} & C_{35} & C_{36} \\ C_{14} & C_{24} & C_{34} & C_{44} & C_{45} & C_{46} \\ C_{15} & C_{25} & C_{35} & C_{45} & C_{55} & C_{56} \\ C_{16} & C_{26} & C_{36} & C_{46} & C_{56} & C_{66} \end{bmatrix} \tag{4}$$

There are several types of symmetry for the independent components that determine the degree of anisotropy. The maximum anisotropy degree occurs at the lowest possible symmetry, which is called "triclinic" symmetry and occurs when the ($C_{\alpha\beta}$) matrix has 21 different nonzero components, which is considered the maximum number of elements required to describe an anisotropic medium [58]. On the other hand, the isotropic case occurs at the highest symmetry case with a smaller amount of independent components.

The most realistic symmetry case of the Earth's subsurface is polar symmetry, which has a single pole of rotational symmetry (Z-axis) that is different from the other two axes (X and Y) [61]. This anisotropy case is called transverse isotropy (TI), at which rock properties are assumed invariant about an axis of symmetry. If the symmetry axis is vertical, this means that the layers are considered horizontal and is hence called vertical transverse isotropy (VTI). Another case is called tilted transverse isotropy (TTI), which is a more realistic case because it follows the variations of the dip angles of layers. Polar symmetry can be described by a matrix of five different independent components, as shown below:

$$\begin{bmatrix} C_{11} & C_{11} - 2C_{66} & C_{13} & 0 & 0 & 0 \\ C_{11} - 2C_{66} & C_{11} & C_{13} & 0 & 0 & 0 \\ C_{13} & C_{13} & C_{33} & 0 & 0 & 0 \\ 0 & 0 & 0 & C_{44} & 0 & 0 \\ 0 & 0 & 0 & 0 & C_{44} & 0 \\ 0 & 0 & 0 & 0 & 0 & C_{66} \end{bmatrix} \tag{5}$$

There are three anisotropy parameters that can define any anisotropy degree in terms of the stiffness components of a TI medium [53]. The anisotropy parameters are called Epsilon (ϵ), Delta (δ), and Gamma (γ). The weaker the anisotropy degree is, the closer these parameters are to zero. The anisotropy parameters in the weak anisotropy case have values less than 1.

Epsilon is the fractional difference between the vertical and horizontal P-wave velocities. It generally has a positive sign because rocks are less compressible along bedding than across bedding. It can be obtained by the following equation:

$$\epsilon = \frac{V_{PH} - V_{PV}}{V_{PV}} = \frac{C_{11} - C_{33}}{2C_{33}} \tag{6}$$

where ϵ is the anisotropy parameter Epsilon, V_{PH} is the horizontal P-wave velocity, V_{PV} is the vertical P-wave velocity, and C_{11} and C_{33} are the stiffness components of the elastic tensor.

The Delta parameter controls the near-vertical anisotropy and does not relate to the horizontal velocity, so it can be positive or negative. Assuming weak anisotropy, the Delta parameter refers to the stiffness components by the following equation:

$$\delta = \frac{(C_{13} + C_{44})^2 - (C_{33} - C_{44})^2}{2C_{33}(C_{33} - C_{44})} \tag{7}$$

where δ is the anisotropy parameter Delta, while C_{13}, C_{44}, and C_{33} are the stiffness components of the elastic tensor.

The Gamma parameter is the fractional difference between the vertical and horizontal S-wave velocities and can be given as follows:

$$\gamma = \frac{V_{SH} - V_{SV}}{V_{SV}} = \frac{C_{66} - C_{44}}{2C_{44}} \tag{8}$$

where γ is the anisotropy parameter Gamma, V_{SH} is the horizontal S-wave velocity, V_{SV} is the vertical S-wave velocity, and C_{66} and C_{44} are the stiffness components of the elastic tensor.

The P- and S-wave velocities can be defined in terms of the anisotropy parameters based on the weak anisotropy assumption [53,61], as shown below:

$$V_P(\theta) = V_{P0}\left[1 + \delta sin^2(\theta)cos^2(\theta) + \epsilon sin^4(\theta)\right] \tag{9}$$

$$V_{SV}(\theta) = V_{S0}\left[1 + \left(\frac{V_{P0}}{V_{S0}}\right)^2(\epsilon - \delta)sin^2(\theta)cos^2(\theta)\right] \tag{10}$$

$$V_{SH}(\theta) = V_{S0}(1 + \gamma sin^2(\theta)) \tag{11}$$

where $V_P(\theta)$ and $V_{SV}(\theta)$ are the vertical P-wave and S-wave velocities, respectively, $V_{SH}(\theta)$ is the horizontal S-wave velocity, V_{P0} and V_{S0} are the zero-offset P-wave and S-wave velocities, respectively, at $\theta = 0$, and θ is the incident angle.

3. The Effect of Seismic Anisotropy on Seismic Data

Seismic anisotropy causes a misinterpretation of seismic data, which affects the final reservoir model. For example, if seismic anisotropy is neglected, a target may be expected at the wrong depth. Another problem is that seismic anisotropy affects the angle-dependent reflectivity calculation, which is further used for prestack seismic applications such as amplitude versus offset (AVO) and elastic impedance (EI). Figure 1 shows a comparison between the isotropic and anisotropic assumptions, which yield significantly different reflectivities at the top of the C-sand reservoir in the Sawan gas field [62]. In addition to that, seismic anisotropy adds uncertainty to prestack seismic inversion, which results in inaccurate rock properties.

A common workflow of anisotropic seismic inversion is to obtain the anisotropy parameters and then process seismic data before the inversion process. The anisotropy parameters can be obtained from various types of data. However, the most confident sources of anisotropy quantification are core data. Traditional laboratory ultrasonic measurements of intrinsic anisotropy [63] depend on a limited number of orientations, resulting in inaccurate anisotropy measurements. On the other hand, more effective methods [64–67] depend on many ray paths having various inclination angles to the bedding plane.

A numerical study was applied to ultrasonic measurements according to Markov chain Monte Carlo simulation (MCMC) [68]. The idea of this study was to obtain the probability distribution curves of the anisotropy parameters based on the layering effects of five lithofacies: dry sandstone, wet sandstone, dry carbonate, wet carbonate, and shale. Accordingly, the anisotropy parameters were simulated for each facies class. The Backus averaging method helped calculate the harmonic means of the anisotropy parameters, assuming a horizontally layered inhomogeneous medium [69].

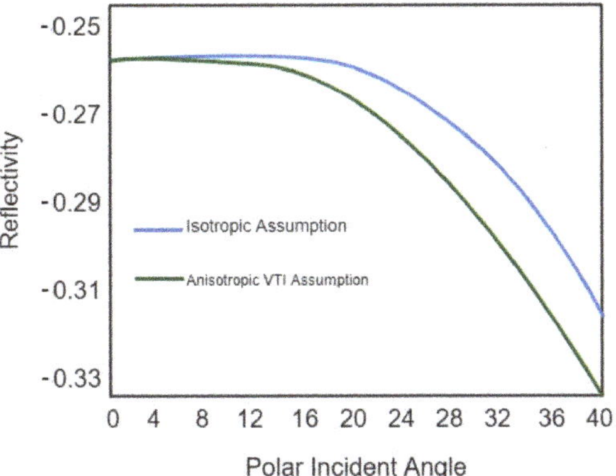

Figure 1. A plot of the reflection coefficient on the Y-axis versus the angle of incidence on the X-axis. The blue and green lines represent the reflectivities obtained based on the isotropic and anisotropic assumptions, respectively.

Another source of anisotropy information is the vertical seismic profile (VSP). For instance, one study aimed at obtaining the horizontal slowness (Sx) as well as the vertical slowness (Sz) and then combining them in the vector way to yield the scalar slowness [70,71]. This method results in accurate arrival time and depth values, but it assumes that layers are homogeneous and horizontal, which is not realistic [61].

The Delta parameter was estimated by using both prestack time data and sonic logs [72]. Moreover, a method, called The Empirical Relationship relates the anisotropy parameters to the shale volume (Vsh) that causes the intrinsic anisotropy [73–75]. The volume fraction and degree of compaction are assumed to be the controlling factors of the orientation of clay minerals and hence the anisotropy parameters. This method is convenient for heterogeneous media because it solves for the intrinsic anisotropy components. However, the problem of obtaining an accurate zero-offset velocity away from wells still exists. Another study considered the intrinsic anisotropy in shale based on an orientation distribution function [76]. This study aimed at modeling anisotropic elastic properties based on two approaches. The first approach was to apply an anisotropic differential effective medium model (DEM) based on the textural information about the kerogen network in kerogen-rich shale. The second approach was to use the DEM model with a compaction-dependent orientation distribution function to study the anisotropy in laminated shaly sand rocks.

The anisotropic elastic constants were estimated as functions of the fluid-filled porosity and the aspect ratio of the inclusions [77]. Another study introduced an anisotropic dual porosity (ADP) model to forecast the elastic properties of shaly sand rocks based on a combination of the anisotropic formulations of self-consistent approximation (SCA), DEM, and anisotropic Gassmann theory [78].

Kelter also estimated both Epsilon and Delta from synthetic modeling through some algorithms such as neural networks and regridding inversion [56]. The idea of this method is to model synthetics by using a PP and PS survey such that layer boundaries and material properties are predefined, and then the survey is simulated by the ray tracing method to generate a synthetic model from which the anisotropy parameters are forecasted. The best parameter estimates are selected according to the comparison between the synthetic model and the real seismic data. This method gives reliable results but only at well locations.

Another method was set by Liner and Fe [79], who estimated Thomsen's parameters from routine well logging data, following a theoretical framework which uses sonic logs to provide a fine-layered isotropic elastic model that can be used to calculate anisotropy parameters as functions of depth and seismic wavelength [80]. Thomsen's parameters can be also estimated from the average velocity, obtained from sonic data, and the mean inclination angles of deviated wells [81]. However, the results of that study were inaccurate because most of the wells' inclination angles were below 5°, showing no anisotropy effect on seismic data. The residual moveout analysis of diving waves has been used to estimate anisotropy parameters by applying different parametrization scenarios [82]. This method is reasonably accurate even for large values of velocity gradients, but it assumes that the velocity increases linearly with depth, while the anisotropy parameters do not. However, this assumption can only apply to homogeneous media.

The previous methods can be used to obtain the anisotropy parameters and then process seismic data to obtain zero-offset images, which can be inverted into accurate rock properties. However, a different approach is to transform the layer properties, obtained from isotropic inversion, into zero-offset parameters based on Ruger's reflectivity function in horizontal transverse isotropy (HTI) media [83]. Another anisotropic prestack seismic inversion [84] is based on the Markov random field (MRF), which can establish dependencies between the spacial wave propagation field's nodes based on prior constraints [85]. The weights of those constraints have been optimized by the anisotropic diffusion method to remove the anisotropy effect in three different models: the symmetrical, vertical asymmetrical, and asymmetrical models. This method has produced encouraging results for Vp, Vs, and density.

Based on the elastic wave equation, the anisotropic elastic waveform inversion was applied to invert for the P- and S-wave velocities, density, and Thomsen's parameters [86]. The method is based on the idea of eliminating the error between the synthetic and field seismic waveform data according to a misfit function [87]. The anisotropic inversion outputs were successfully used to reduce the ground-roll noise in 2D seismic data.

Another methodology depends on multiscale phase inversion, assuming an anisotropic medium [88]. The advantage of this method is that it avoids the nonlinearity of the misfit functions [89] by adding wave-number details to the velocity model by a skeletonized inversion. This method assumes a VTI medium with Delta equal to zero, while the zero-offset P-wave velocity and Epsilon are inverted, resulting in more accurate results compared to full-wave inversion.

It can be concluded that most of the previous methods have quantified seismic anisotropy only at well locations. The interpolation of the anisotropy parameters between wells is challenging and should be further linked to the effect of the facies distribution. Therefore, the current study has improved a previous anisotropic inversion approach [90] which solves for the zero-offset impedances and anisotropy parameters, Epsilon and Delta, based on partial-stack inversion and Thomsen's anisotropy equations [91]. On the other hand, our study estimates Z_P and Z_S based on statistical modeling and MLFN rather than calculating them empirically.

The advantage of ML algorithms is that they reduce the random error of partial-stack inversion and result in accurate impedances that best match the logs. Another contribution is that the facies model uses logistic regression to model lithologies and fluids and then combines them using the decision tree algorithm. The integration of two ML algorithms adds stability to the model and raises the chance of being applied to other fields. Moreover, the model includes the depth variable, which acts as a trend factor that highly affects the facies distribution.

4. Methodology

The study workflow, as shown in Figure 2, begins with inverting the angle gathers into Z_P, Z_S, and density volumes and inverting the angle stacks into Z_P and Z_S volumes at the near, mid, and far incident angles. Next, the partially inverted impedances are used to forecast the zero-offset impedances by the statistical and MLFN models. The impedance volumes are then compared to each other by using the impedance logs. The best Z_P and Z_S volumes are applied to Thomsen's anisotropy equations to solve for the anisotropy parameters, Epsilon and Delta. The next step is to transform the impedance and density volumes into the elastic volumes, from which the sand and gas probabilities are estimated by using the logistic regression model. These probabilities are then inserted into a decision tree model to forecast the distribution of the facies classes: gas sand, wet sand, and shale. The inversion processes in our study were carried out by using the Hampson–Russel software provided by the CGG company.

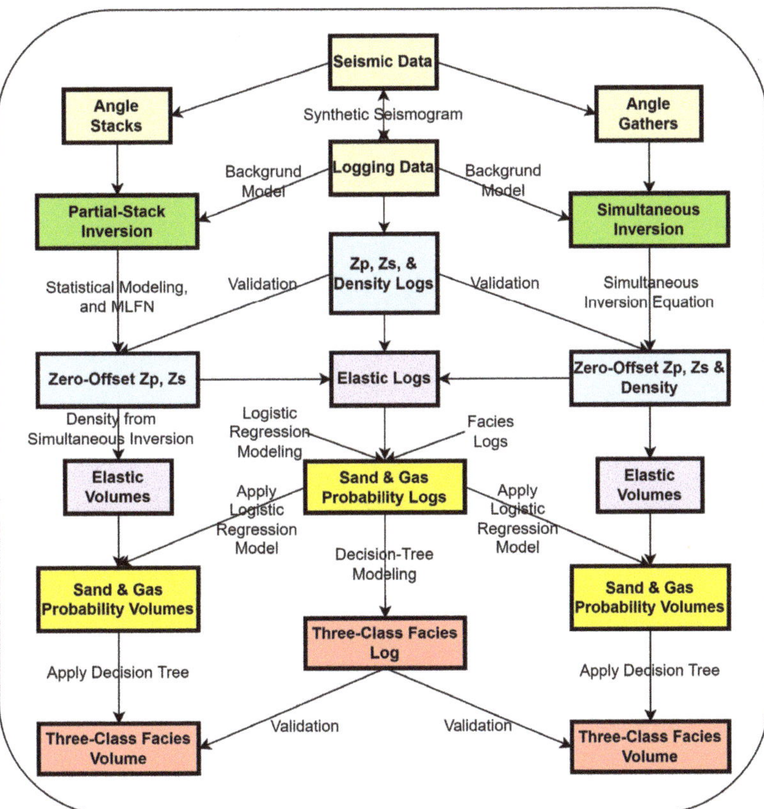

Figure 2. A flow diagram showing the methodology of estimating the facies distribution from the zero-offset acoustic and shear impedances, which are obtained from isotropic inversion, statistical modeling, and MLFN.

This study is not applicable to poststack data. The impedance model's accuracy depends on the seismic data quality as well as the angle gathers' range. The wider the angle range, the more enhanced signal-to-noise ratio and hence the more accurate the model. Moreover, both the P- and S-impedances should be available in order to solve for the anisotropy parameters: Epsilon and Delta. The study assumes a VTI medium which neglects some anisotropy-related factors such as the layers' dip angles and fractures; however, impedance modeling by MLFNs accurately solves for the zero-offset impedances by normalizing the error and mitigating the anisotropy effect simultaneously. The study's facies model is only applicable to the three facies classes: gas sand, wet sand, and shale. However, the model can be extended to include more classes. Moreover, the applicability of the model to other fields can be enhanced if there are more facies data available from a large number of wells.

4.1. Isotropic Simultaneous Inversion

Simultaneous inversion is a prestack seismic inversion that simultaneously inverts the angle gathers into the reflectivities of the Z_P, Z_S, and density based on the linear relationships between the Z_P and each of the Z_S and density. The methodology used in this study [92] directly inverted for the Z_P, Z_S, and density [93,94].

The Aki–Richards equation [95] is an approximation of the relationship between the P-wave reflectivity between two media and the incident angle as follows:

$$R(\theta) = \frac{1}{2}(\Delta V/V + \Delta\rho/\rho) - 2(W/V)^2(2\Delta W/W + \Delta\rho/\rho)sin^2\theta + \frac{1}{2}(\Delta V/V)tan^2\theta \quad (12)$$

where $R(\theta)$ is the P-wave reflectivity at an incident angle (θ), while V, W, and ρ are the average P-wave velocity, S-wave velocity, and density between the two media, respectively. The idea of Fatti's equation [93] is to re-express the Aki–Richards equation in (12) in terms of the acoustic and shear impedances as follows:

$$R(\theta) = \frac{1}{2}\frac{\Delta I}{I}(1+tan^2\theta) - 4\left(\frac{W}{V}\right)^2\left(\frac{\Delta J}{J}\right)sin^2\theta - \left[\frac{1}{2}\left(\frac{\Delta\rho}{\rho}\right)tan^2\theta - 2\left(\frac{W}{V}\right)^2\left(\frac{\Delta\rho}{\rho}\right)sin^2\theta\right] \quad (13)$$

where I is the acoustic impedance, J is the shear impedance, the term ($\frac{1}{2}\frac{\Delta I}{I}$) is the zero-offset P-wave reflectivity, and the term ($\frac{1}{2}\frac{\Delta J}{J}$) is the zero-offset S-wave reflectivity, such that:

$$\frac{1}{2}\frac{\Delta J}{J} = \frac{1}{2}(\Delta V/V + \Delta\rho/\rho), \frac{1}{2}\frac{\Delta J}{J} = \frac{1}{2}(\Delta W/W + \Delta\rho/\rho) \quad (14)$$

Fatti's equation could be re-expressed by replacing the reflectivities by the rock properties as follows:

$$s(\theta) = \frac{1}{2}C_1 W_\theta Dln(Z_P) + \frac{1}{2}C_2 W_\theta Dln(Z_S) + C_3 W_\theta Dln(\rho) \quad (15)$$

where $s(\theta)$ is the angle trace, $W(\theta)$ is the angle-dependent wavelet, D is a derivative operator, Z_P is the acoustic impedance, Z_S is the shear impedance, ρ is the density, and θ is the incident angle. The terms C_1, C_2, and C_3 are given by:

$$C_1 = 1 + tan^2\theta, C_2 = -8(\frac{V_S}{V_P})tan^2\theta, C_3 = -0.5tan^2\theta + 2(\frac{V_S}{V_P})^2 sin^2\theta \quad (16)$$

where C_1, C_2, and C_3 are constants, θ is the incident angle, V_S is the S-wave velocity, and V_P is the P-wave velocity.

Next, the terms $ln(Z_P)$, $ln(Z_S)$, and $ln(\rho)$ were coupled by the following equations:

$$ln(Z_s) = kln(Z_p) + k_c + \Delta ln(Z_s) \quad (17)$$

$$ln(\rho) = mln(Z_p) + m_c + \Delta ln(\rho) \quad (18)$$

where k and m are the intercepts of the linear relations and k_c and m_c are constants, while $\Delta ln(Z_S)$ and $\Delta ln(\rho)$ are controlled by the deviations away from the best-fit lines of the linear relationships. ΔL_D and ΔL_S are the displacements of $ln(\rho)$ and $ln(Z_S)$ due to the deviation from the best-fit lines.

The next step was to apply the derivative operator (D) to Equations (3) and (4):

$$D\ln Z_S = kDlnZ_P + D\Delta lnZ_S \quad (19)$$

$$D\ln\rho = mDlnZ_P + D\Delta ln\rho \quad (20)$$

where D is the derivative operator, while k and m are the intecepts of the linear relations. Eventually, the algorithm was applied to an initial guess and then iterated until it reached a solution at which the initial model matched the real data with the least-squared error.

4.2. Partial-Log-Constrained Inversion

The separate inversion of angle stacks is based on a partial inversion approach that obtains the zero-offset P- and S-wave velocities from the inverted P- and S-waves at different angular ranges [90]. The idea is based on Thomsen's equations in (9) and (10), which relate the vertical P-wave (V_P) and S-wave (V_S) velocities to the anisotropy parameters, Epsilon and Delta, in VTI media.

Partial-log-constrained inversion requires a wavelet to be extracted from each angle stack so that the wavelet is centered at the average angle of the angle stack [90]. An initial model was created from logging data to guide the inversion. This method required S-wave velocity logs in the initial model to constrain the inversion. Each partial seismic volume was inverted into a velocity volume at the central angle of its angle range. Next, the velocity-domain Thomsen's equations in (9) and (10) were used to solve for the zero-offset P- and S-wave velocities as well as the Epsilon and Delta parameters, as shown in the following equations:

$$\begin{bmatrix} 1 & sin^2(\theta_1)cos^2(\theta_1) & sin^4(\theta_1) \\ 1 & sin^2(\theta_2)cos^2(\theta_2) & sin^4(\theta_2) \\ \vdots & \vdots & \vdots \\ 1 & sin^2(\theta_n)cos^2(\theta_n) & sin^4(\theta_n) \end{bmatrix} \begin{bmatrix} V_{P0} \\ V_{P0}\delta \\ V_{P0}\epsilon \end{bmatrix} = \begin{bmatrix} V_{qP}(\theta_1) \\ V_{qP}(\theta_2) \\ \vdots \\ V_{qP}(\theta_n) \end{bmatrix} \tag{21}$$

where V_{P0} is the zero-offset P-wave velocity, ϵ and δ are the anisotropy parameters, V_{qP} is the quasivertical P-wave velocity, θ is the incident angle, and n is the number of partial stacks.

$$\begin{bmatrix} 1 & sin^2(\theta_1)cos^2(\theta_1) \\ 1 & sin^2(\theta_2)cos^2(\theta_2) \\ \vdots & \vdots \\ 1 & sin^2(\theta_n)cos^2(\theta_n) \end{bmatrix} \begin{bmatrix} V_{s0} \\ \frac{V_{P0}^2(\epsilon-\delta)}{V_{s0}} \end{bmatrix} = \begin{bmatrix} V_{qSV}(\theta_1) \\ V_{qSV}(\theta_2) \\ \vdots \\ V_{qSV}(\theta_n) \end{bmatrix} \tag{22}$$

where V_{S0} is the zero-offset P-wave velocity, ϵ and δ are the anisotropy parameters, V_{qSV} is the quasivertical S-wave velocity, θ is the incident angle, and n is the number of the partial stacks.

The log-constrained inversion of partial stacks was adopted from the nonlinear optimization inversion [90], such that the objective function is obtained by:

$$f(V) = \parallel S - D \parallel \rightarrow min \tag{23}$$

where V is the inverted P- or S-wave velocity, D is the angle-domain seismic data, and S is the corresponding prediction, which can be expressed according to the convolution model [96]. The reflectivity function is based on the Aki and Richard approximation of the angle-dependent reflectivity [97], as shown below:

$$R(\theta) = R(0) + Asin^2(\theta) + Btan^2(\theta) \tag{24}$$

where $R(\theta)$ is the reflectivity at an incident angle θ, while $R(0)$ is the zero-offset reflectivity, which is given by:

$$R(0) = \frac{1}{2}\left(\frac{\Delta V_P}{V_P} + \frac{\Delta\rho}{\rho}\right) \tag{25}$$

$$A = -2\left(\frac{V_S}{V_P}\right)^2\left(\frac{2\Delta V_S}{V_S} + \frac{\Delta\rho}{\rho}\right) \tag{26}$$

$$B = \frac{1}{2}\left(\frac{\Delta V_P}{V_P}\right) \tag{27}$$

where V_P, V_S, and ρ are, respectively, the average P-wave velocity, S-wave velocity, and density of the underlying and overlying layers, while ΔV_P, ΔV_S, and $\Delta\rho$ are the differences between these properties at the underlying and overlying strata.

The term $\frac{\Delta\rho}{\rho}$ can be obtained by Gardner's equation as follows [98]:

$$\frac{\Delta\rho}{\rho} = \frac{1}{4}\left(\frac{\Delta V_P}{V_P}\right) \tag{28}$$

According to Equations (24)–(28), the reflectivity will be as follows:

$$R(\theta) = \left(\frac{5}{8} - \frac{1}{2}\frac{V_S^2}{V_P^2}sin^2\theta + \frac{1}{2}tan^2\theta\right)\frac{\Delta V_P}{V_P} - 4\frac{V_S^2}{V_P^2}sin^2\theta\frac{\Delta V_S}{V_S} \tag{29}$$

After inverting the partial stacks separately into V_P and V_S volumes at the near, mid, and far angles, the zero-offset velocities Epsilon and Delta were solved by Equations (21) and (22).

The current study applied the same methodology to Z_P and Z_S by replacing the velocities of Equations (9) and (10) by impedances, as shown below:

$$Z_P(\theta) = Z_{P0}\left[1 + \delta sin^2(\theta)cos^2(\theta) + \epsilon sin^4(\theta)\right] \tag{30}$$

$$Z_S(\theta) = Z_{S0}\left[1 + \left(\frac{Z_{P0}}{Z_{S0}}\right)^2(\epsilon - \delta)sin^2(\theta)cos^2(\theta)\right] \tag{31}$$

Partial inversion is sensitive to the quality of seismic data because the magnitude of the noise may be more than that of the anisotropy effect. Hence, the zero-offset impedances were estimated from the impedances at the near, mid, and far incident angles by using statistical modeling and MLFN. Seismic data consisted of three 3D volumes of angle ranges, from 5° to 15°, 15° to 25°, and 25° to 40°. Accordingly, three partial stacks were created at the near, mid, and far central incident angles: 10°, 20°, and 32.5°, respectively. The advantage of impedance modeling is that it normalizes the error and reduces the anisotropy effect simultaneously.

An important step prior to the modeling process is to calculate Epsilon and Delta at the well locations in order to have an idea about the degree of anisotropy within the study area and to specify the model's assumptions accordingly. The anisotropy parameters were calculated by the methodology of Liner and Fei [79], who calculated the layer-induced stiffness parameters from the averaged properties of the isotropic elastic layers and then obtained Epsilon and Delta as functions of those parameters. The stiffness parameters were obtained from the following equations:

$$a = \langle\frac{\lambda}{\lambda + 2\mu}\rangle^2\langle\frac{1}{\lambda + 2\mu}\rangle^{-1} + 4\langle\frac{\mu(\lambda + \mu)}{\lambda + 2\mu}\rangle \tag{32}$$

$$c = \langle\frac{1}{\lambda + 2\mu}\rangle^{-1} \tag{33}$$

$$f = \langle\frac{1}{\lambda + 2\mu}\rangle\langle\frac{1}{\lambda + 2\mu}\rangle^{-1} \tag{34}$$

$$l = \langle\frac{1}{\mu}\rangle^{-1} \tag{35}$$

where $\langle\rangle$ is the Backus averaging symbol, μ is the shear modulus, and λ is Lame's constant, while a, c, f, and l are the stiffness parameters. Next, Epsilon and Delta were obtained from the following equations:

$$\epsilon = \frac{a - c}{2c} \tag{36}$$

$$\delta = \frac{(f + l)^2 - (c - l)^2}{2c(c - l)} \quad (37)$$

Another premodeling step is to explore the relations between the zero-offset impedances, which are the Z_P and Z_S logs, and the angle impedances, which are the composite traces obtained from the partial-stack inversion. Figure 3 shows a fair linear relationship based on the data of wells (×1) and (×2), which consist of 1292 observations for Z_P 1582 observations for Z_S.

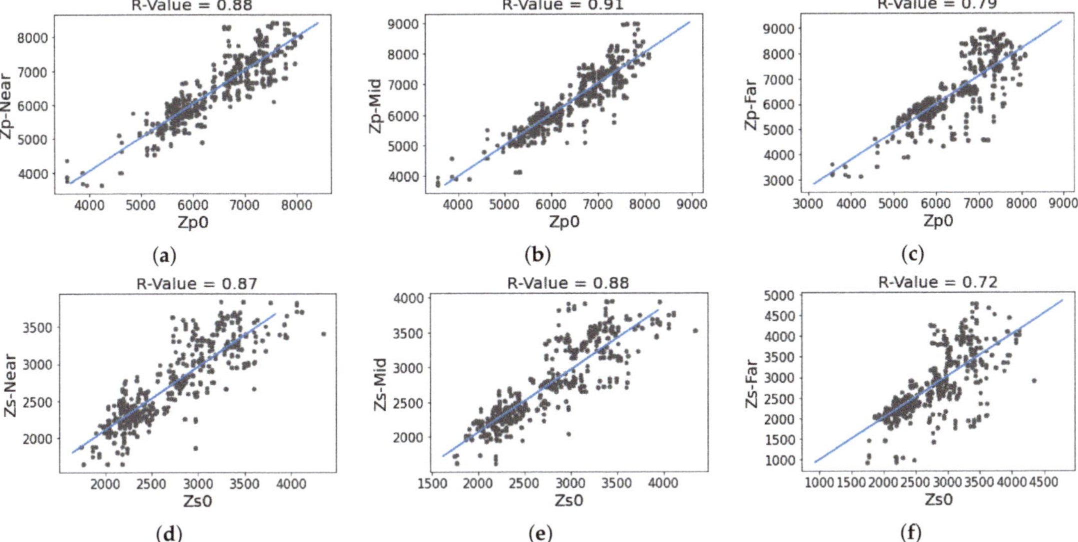

Figure 3. Plots of the zero-offset P-impedance (Z_{P0}) versus each of the partially inverted P-impedances (**a**) Z_p-Near, (**b**) Z_p-Mid, and (**c**) Z_p-Far, while the zero-offset S-impedance (Z_{S0}) is plotted versus each of the partially inverted S-impedances (**d**) Z_s-Near, (**e**) Z_s-Mid, and (**f**) Z_s-Far.

The linearity between variables is proven visually through the best-fit lines (in blue). In addition, the Pearson correlation coefficient (R-value) was calculated in Python using NumPy. The R-values show strong positive correlations at the near and mid angles, while the correlation is moderate at the far angle. This is matching with Thomsen's equations in (9) and (10), which consider a linear relationship between the zero-offset velocity and the velocity at an angle theta.

Accordingly, a linear regression model was firstly created to forecast the zero-offset impedance from the partially inverted impedances based on the MCMC simulation. As the R-value between each of the P- and S-impedances and their far-angle impedances was moderate due to the anisotropy effect, the MLFN was created to better estimate the zero-offset impedances and turn the randomness in the data points into recognized patterns.

4.3. Zero-Offset Impedance Modeling

4.3.1. Statistical Modeling

The idea of the model is to fit a linear relationship between the zero-offset impedance, which is the response variable, and the partially inverted impedances, which are the explanatory variables. The output of the model consists of the posterior means of the coefficients of the three partially inverted impedances. The general form of the model is:

$$Imp_0 = b_1 Imp_{\theta 1} + b_2 Imp_{\theta 2} + b_3 Imp_{\theta 3} \quad (38)$$

where Imp_0 is the zero-offset Z_P or Z_S, while b_1, b_2, and b_3 are the model's coefficients and $Imp_{\theta 1}$, $Imp_{\theta 2}$, and $Imp_{\theta 3}$ are the impedances at the near, mid, and far angles, respectively. The model's parameters are estimated based on Bayesian theory and MCMC simulation. The Bayesian model consists of three components: the likelihood, priors, and posterior. The priors of the model were the parameters of the proposed distribution of the model coefficients. The posterior component was considered the probability distribution function of all realizations of the simulated coefficients.

The linear expression of the Z_{P0} model was obtained from Thomsen's equation in (9), which can be rewritten as follows:

$$Z_{P0} = bZ_P(\theta) \tag{39}$$

where:

$$b = \frac{1}{1 + \delta sin^2(\theta)cos^2(\theta) + \epsilon sin^4(\theta)} \tag{40}$$

By substituting the near, mid, and far P-impedances, Z_{P1}, Z_{P2}, and Z_{P3}, respectively, to Equation (39), three equations were obtained as follows:

$$Z_{P0} = b_1 Z_{P1} \tag{41}$$

$$Z_{P0} = b_2 Z_{P2} \tag{42}$$

$$Z_{P0} = b_3 Z_{P3} \tag{43}$$

By adding the three Equations (41)–(43), the linear expression will be as follows:

$$Z_{P0} = 0.33b_1 Z_{P1} + 0.33b_2 Z_{P2} + 0.33b_3 Z_{P3} \tag{44}$$

Similarly, the linear expression of the Z_{S0} model was obtained based on Thomsen's equation in (10), which can be rewritten as shown below:

$$Z_{S0} = b.Z_S(\theta) \tag{45}$$

where

$$b = \frac{1}{1 + \left(\frac{Z_{P0}}{Z_{S0}}\right)^2 (\epsilon - \delta) sin^2(\theta) cos^2(\theta)} \tag{46}$$

After adding the three equations of the near, mid, and far angles, the linear expression will be as follows:

$$Z_{S0} = 0.33b_1 Z_{S1} + 0.33b_2 Z_{S2} + 0.33b_3 Z_{S3} \tag{47}$$

The likelihood function of the Z_P or Z_S models can be written as shown below:

$$Imp_0|b_j, (Imp_\theta)_i, \sigma^2 \sim N\left(\left[b_1(Imp_1)_i + b_2(Imp_2)_i + b_3(Imp_3)_i\right], \sigma^2\right) \tag{48}$$

Given the coefficients vector b, the explanatory variables vector Imp_θ, and the variance σ^2, the zero-offset impedance Imp_0 follows a normal distribution with a mean equal to the linear expression of the covariates and coefficients, where (i) is the number of observations.

The next layer of the model is the coefficients vector b_j, which consists of b_1, b_2, and b_3 such that the three coefficients follow a common normal distribution. The mean and variance of this distribution were obtained from Equation (40) for the Z_P model and Equation (46) for the Z_S model. Based on the weak anisotropy assumption, the mean of the anisotropy parameters was considered zero and the standard deviation was 0.2. Moreover, the squared ratio of Z_P and Z_S was set, according to the wells' data, to have a mean equal to 6 and a standard deviation (SD) equal to 3. The prior function for the Z_P and Z_S models can be written as follows:

$$bj \sim N(\mu, \tau), j = \{1, 2, 3\} \tag{49}$$

where μ and τ are, respectively, the mean and variance of the normal distribution of the coefficients vector b, which has an index (j) from 1 to 3.

The variance of the model was assumed to follow an inverse gamma distribution, which depends on the shape parameter α and the scale parameter β, which act as priors of the parameter σ^2. The distribution function of the σ^2 is shown below:

$$\sigma^2 \sim IG(\alpha, \beta) \tag{50}$$

The idea of the model is to calculate the the zero-offset impedance by substituting the posterior means of the coefficients to Equation (38). The MCMC was used to draw multiple random realizations from the proposed normal distribution. Based on the law of large numbers, the average of a random sample from a distribution converges to the true mean of that distribution [99]. In other words, the Markov chains would converge to the posterior means of the model's coefficients. The advantage of the MCMC is that it can solve for complicated posterior distributions which are difficult to be calculated mathematically.

Finally, the posterior means of the Z_P and Z_S models were substituted into Equation (38) to solve for the zero-offset impedances. The final graphical representation of the impedance model is shown in Figure 4, where Imp_0 is the zero-offset impedance, Imp_1, Imp_2, and Imp_3 are the impedances at the near, mid, and far angles, respectively, i is the observation index that ranges from 1 to n, bj is the common distribution of the coefficients which depends on the mean μ and variance τ, and σ^2 is the variance of the model which depends on shape parameter α and scale parameter β.

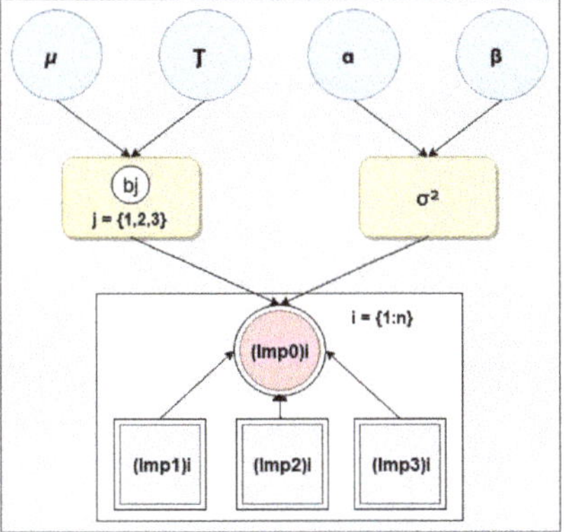

Figure 4. The graphical representation of the statistical Z_P or Z_S models.

4.3.2. Multilayer Feed-Forward Neural Network (MLFN)

Due to the availability of vast amounts of data in recent years, ANNs have become the most common prediction methods [100]. One of the widely used ANNs is the multilayer feed-forward neural network (MLFN) [101,102]. The MLFN consists of an input layer, one or more hidden layers, and an output layer [103]. An activation function is used to map the summation of the weighted inputs into the output layer [104]. The number of hidden layers and number of neurons in each of them should be carefully determined because this step controls the model's accuracy [105].

The selection of appropriate activation functions has a high impact on a model's performance. There are various types of activation functions such as the sigmoid, or logistic, function, which is commonly used in classification models. The current study considers a linear activation function to solve for the zero-offset impedance in a regression model. The choice of linear function is based on the moderate-to-strong correlation between the input and output variables as discussed earlier.

The connection between any two neurons is controlled by a random weight. A model's random weights are adjusted until reaching the least mean square error and the best match between the predicted and actual output values. The backpropagation method is commonly used to obtain the optimal set of weights and calculate the error in a repetitive way until reaching the best training accuracy [106,107]. The backpropagation process involves an optimization algorithm such as stochastic gradient descent (SGD) [108] and adaptive moment estimation (Adam) [109]. The algorithm used in our model is the Levenberg–Marquardt (LMA) [110] due to its fast performance and ability to train MLFN models [111].

A model's error can be represented by various loss functions such as cross-entropy [112], weighted binary cross-entropy (WCE) [113], and dice loss [114]. The loss function used in our study is the mean squared error (MSE), which is one of the best unbiased estimators used in regression models [115]. The MSE can be defined by the following equation:

$$MSE = \frac{1}{N}\sum_{i=1}^{N}(y_i - \hat{y}_i)^2 \quad (51)$$

where N is the number of samples and y_i is the output variable at an observation (i).

The structure of the MLFN, in this study, consists of three layers: an input, a hidden, and an output layer. Figure 5 shows the graphical representation of the MLFN models for both Z_P and Z_S, where the input layer consists of three neurons, which are the near-, mid-, and far-angle impedances, while the output layer consists of one neuron, which is the zero-offset impedance. Several trials were performed in MATLAB to determine the best number of neurons for the hidden layer. The best performance was observed at 20 neurons for both the Z_P and Z_S models.

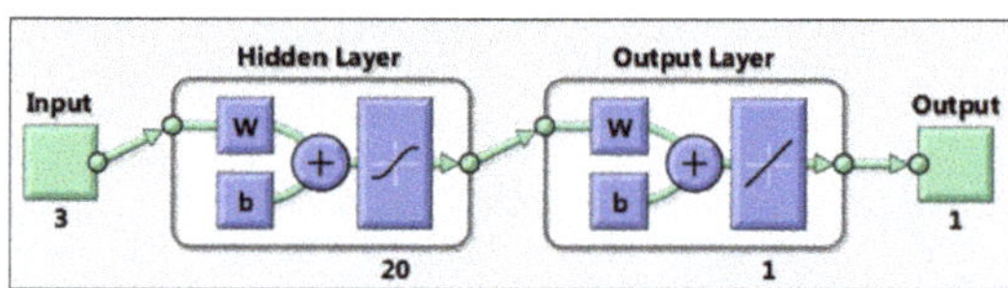

Figure 5. A graphical representation of the MLFN for both the Z_P and Z_S models. The input layer consists of three nodes, which are the near, mid, and far impedances, the hidden layer consists of 20 nodes, and the output layer consists of one node, which is the zero-offset impedance.

4.4. Calculation of the Anisotropy Parameters: Epsilon and Delta

After comparing and validating Z_P and Z_S, obtained from the isotropic inversion, statistical modeling, and MLFN, the best P- and S-impedance volumes were applied to Thomsen's anisotropy equations to solve for the parameters Epsilon and Delta. Both the partially inverted impedances, zero-offset impedances, and central angles of the angle stacks (10, 20, and 32.5 degrees) were applied to the impedance-domain Thomsen's equations in (30) and (31), yielding six equations, as shown below:

$$Z_{P1} = Z_{P0}\left[1 + \delta sin^2(10)cos^2(10) + \epsilon sin^4(10)\right] \quad (52)$$

$$Z_{P2} = Z_{P0}\left[1 + \delta sin^2(20)cos^2(20) + \epsilon sin^4(20)\right] \quad (53)$$

$$Z_{P3} = Z_{P0}\left[1 + \delta sin^2(32.5)cos^2(32.5) + \epsilon sin^4(32.5)\right] \quad (54)$$

$$Z_{S1} = Z_{S0}\left[1 + \left(\frac{Z_{P0}}{Z_{S0}}\right)^2(\epsilon - \delta)sin^2(10)cos^2(10)\right] \quad (55)$$

$$Z_{S2} = Z_{S0}\left[1 + \left(\frac{Z_{P0}}{Z_{S0}}\right)^2 (\epsilon - \delta)sin^2(20)cos^2(20)\right] \tag{56}$$

$$Z_{S3} = Z_{S0}\left[1 + \left(\frac{Z_{P0}}{Z_{S0}}\right)^2 (\epsilon - \delta)sin^2(32.5)cos^2(32.5)\right] \tag{57}$$

where Z_{P1}, Z_{P2}, and Z_{P3} are the P-impedance volumes at the near, mid, and far angles, respectively, while Z_{S1}, Z_{S2}, and Z_{S3} are the S-impedance volumes at the near, mid, and far angles, respectively. Z_{P0} and Z_{S0} are the zero-offset P- and S-impedances, respectively, and ϵ and δ are the anisotropy parameters.

By adding Equations (52)–(54) to each other and Equations (55)–(57) to each other, two equations were obtained such that one of them related Epsilon and Delta to the Z_P volumes and the other one related Epsilon and Delta to the Z_S volumes. Finally, the two resulting equations were solved by MATLAB for the anisotropy parameters Epsilon and Delta, as shown below:

$$\text{Epsilon} = 2.3(Z_{P0}Z_{P1} + Z_{P0}Z_{P2} + Z_{P0}Z_{P3} + Z_{S0}Z_{S1} + Z_{S0}Z_{S2} + Z_{S0}Z_{S3} - 3Z_{P0}^2 - 3Z_{S0}^2)/(Z_{P0}^2) \tag{58}$$

$$\text{Delta} = 2.3(Z_{P0}Z_{P1} + Z_{P0}Z_{P2} + Z_{P0}Z_{P3} - 0.3Z_{S0}Z_{S1} - 0.3Z_{S0}Z_{S2} - 0.3Z_{S0}Z_{S3} - 3Z_{P0}^2 + 0.9Z_{S0}^2)/(Z_{P0}^2) \tag{59}$$

4.5. Estimation of Facies Distribution

After obtaining the zero-offset P- and S-impedances, they were transformed into elastic properties in order to forecast the lithology and fluid probabilities separately by two logistic regression models. The lithology and fluid models were merged together to estimate the distribution of gas sand, wet sand, and shale by using the decision tree algorithm. Z_P and Z_S were firstly transformed into the lithology predictors: near and far elastic impedances [116]. Next, the fluid predictors, Mu–Rho, Lambda–Mu–Rho, and Poisson's ratios, were calculated based on the LMR theory [117]. The idea of the facies model is based on the combination of the logistic regression [52] and decision tree algorithms.

Logistic regression is one of the linear regression models that deal with binary response variables. The idea of logistic regression is to provide model estimates which lie between zero and one [118]. These estimates can be considered the probabilities of occurrence of both zero and one. Therefore, the likelihood of a logistic regression model follows a Bernoulli distribution which takes on the value 1 with a probability ϕ and 0 with a probability $(1 - \phi)$, as shown in the following equation:

$$Y_i|\phi_i \sim Bernoulli(\phi_i) \tag{60}$$

where Y_i is a binary event at an observation (i), while ϕ_i is the probability of the success of this event at the observation (i). Instead of directly relating the expected value of Y_i to the model's variables, it can be assigned to the value of ϕ_i, which can be related to the model's variables and coefficients by using a link function, as shown in the following equation:

$$logit(\phi_i) = \log\left(\frac{\phi_i}{1 - \phi_i}\right) = \beta_0 + \beta_1(X_1)_i + \ldots \beta_n(X_n)_i \tag{61}$$

where $logit(\phi_i)$ is the logarithmic function of the odds of ϕ_i, β_0 is the intercept, β_1 to β_n are the model's coefficients, and $(X_1)_i$ to $(X_n)_i$ are the model's explanatory variables at an observation (i). After simplifying the previous equation in (61), the ϕ_i can be obtained as shown below:

$$E(Y_i) = \phi_i = (1/1 + e^{-(\beta_0 + \beta_1(X_1)_i + \ldots \beta_n(X_n)_i)}) \tag{62}$$

where $E(Y_i)$ is the expected value of Y_i. Assuming a three-class medium, which consists of gas sand, wet sand, and shale, the lithology model aims at predicting the sand probability from the lithology predictors, while the fluid model aims at estimating the gas probability from the fluid predictors. Given the observations of the elastic attributes and the facies log, the model's coefficients can be obtained based on Bayes's theorem [119].

The lithology model aims at predicting the sand probability from the near-EI and far-EI attributes. On the other hand, the fluid model aims at estimating the gas probability from the Mu–Rho, Lambda–Rho/Mu–Rho, and Poisson's ratios. The posterior means of the coefficients of the two models were obtained by MCMC simulation. The facies and fluid models were merged together by using the sand and gas probabilities, along with depth values, as inputs to a decision tree, which turned the probabilities into facies estimates.

A decision tree is a classification method that generates questions over training samples and answers them using a set of statistical criteria for data classification [42]. There are two types of decision trees, which are regression and classification trees. As the response variable in this study was categorical, a decision tree was considered.

A decision tree begins with a root node where the samples are classified into two branches such that if the answer to the question is "Yes", the sample goes under the "Yes" branch, while if the answer is "NO", the sample goes under the "No" branch. A distinct series of questions and branches are set in the internal nodes until satisfying the stopping rule in the last terminal of the tree, which is called the leaf node. According to the classification rule [120], each leaf in a tree represents a class (A or B).

A popular method used to construct decision trees is called classification and regression trees (CART) [121]. The CART method splits observations into two parts by minimizing the relative sum of squared errors [122] between the two partitions, according to the principle of binary recursive partitioning (BRP) [123]. The advantages of this method are that it is nonparametric, free of significance tests, and has low sensitivity to outliers [124].

The process of tree shortening is called pruning. Avoiding the large size of a tree is crucial to eschew over-fitting. The size of a decision tree is controlled by many factors, such as the minimum number of samples that should be left in a node to achieve splitting [125]. A pruned tree mostly leads to results which are close to those of the original tree. However, the pruned tree may be more accurate than the initial one in some cases [126].

This study used the "rpart" package in R to model the three-class facies variable based on the CART method. The inputs of the tree were the sand and gas probabilities that had been estimated by logistic regression as well as the depth variable as trend factors. The output of the tree was a categorical variable having three values, 1, 2, and 3, which corresponded to the three litho-fluid classes: gas sand, wet sand, and shale, respectively. The tree was assessed by calculating the correct classification rate (CCR), which is the number of correctly classified observations over the total number of samples.

5. Results and Discussion

The wells were tied to seismic data by check shots of wells $\times 1$ and $\times 2$. A constant wavelet was extracted from the wells to achieve the tie between the synthetic logs and seismic traces. Next, three horizons were picked: A, B, and C, which appear as troughs in seismic data, as shown in Figure 6.

Both the conventional well logs and oil-mud resistivity imager (OMRI) were interpreted at the well ($\times 1$) such that the zone of interest extended from 1300 m to 1480 m. Firstly, the sand and shale could be preliminarily discriminated by using the shale volume at the cut-off 0.3. The gas sand and wet sand were differentiated by V_P/V_S and water saturation at the cut-offs 2.8 and 0.45, respectively. The thicknesses of the gas sand zones were validated by the net pay thickness mentioned in the well report.

Figure 7 shows the OMRI-derived facies against the well-log-derived facies obtained from the petrophysical cut-offs at the well ($\times 1$). Both facies logs confirm that the dominant facies is a massive shale that corresponds to the high shale volume and water saturation zones. The massive and planar-laminated sandstone layers correspond to the gas intervals of the two reservoirs: A and B. Those gas zones appear as fining-upward cycles in the shale volume (Vsh) log where V_P/V_S and water saturation are low. Minor siltstone zones appear in track (a) and coincide with some of the wet sand zones in track (b). The layers become thinner as the depth increases, coinciding with abrupt changes in the well logs.

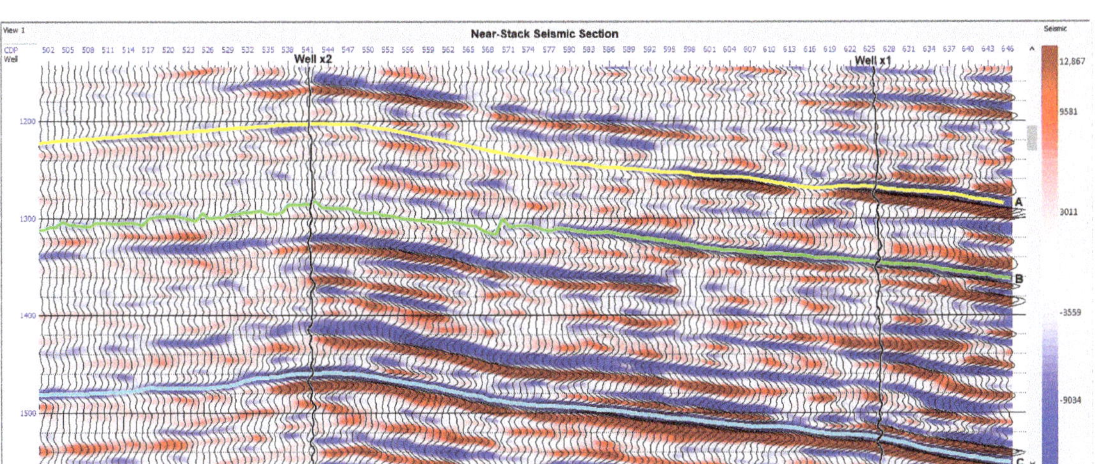

Figure 6. A seismic section extracted from the near-stack volume and tied to the wells ×1, ×2, and ×3, where the horizons' tops A, B, and C are in yellow, green, and cyan colors, respectively.

The top of the target zone (1300–1316 m) consists of massive shale, which can be interpreted as a delta plain. The next interval extends from depth 1316 m to 1330 m, where the A reservoir appears as a distributary channel that consists of fining-upward sediments. The top of this interval is considered a fine-grained sandstone, which appears as wet sand in track (b). The rest of the A reservoir consists of equal amounts of the planar and cross-bedded gas-bearing sandstones, where V_P/V_S, Vsh, and water saturation are low. A sharp contact is clear at the base of the channel at depth 1330 m due to the change from sandstone to shale.

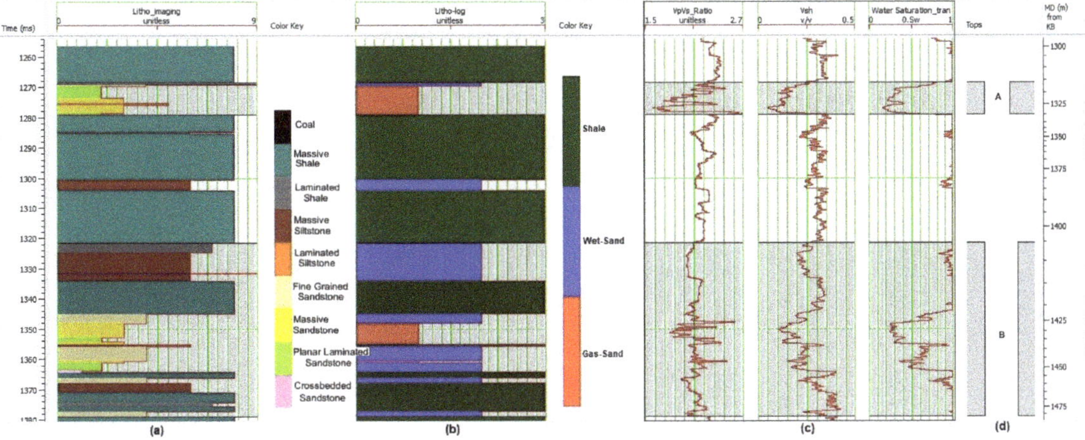

Figure 7. Well (×) facies and conventional logs: (**a**) OMRI-derived facies, (**b**) well-log-derived facies, (**c**) well logs, and (**d**) well tops.

Another depositional cycle begins at depth 1330 m, where the massive and laminated shales are interbedded with minor siltstone. The OMRI images show broad and spiral bands in darker clay-rich portions. The shale extends until depth 1422 m, where another distributary channel appears within the B reservoir. This unit is dominated by massive-to-weakly stratified porous sands. A sharp contact is defined at depth 1438 m, below which the lower interval consists of laminated sandstone with low porosity. The depth interval from 1452 m to 1481 m consists of interbedded sandstone, siltstone, and shale. The reservoir quality in this zone is poor, with high water-saturation and shale-volume. This is matching with the facies log in track (b), where the sandstone is wet.

The OMRI images are crucial for data filtering due to their high resolution relative to the conventional well logs. For example, there is a possible streak of coal at depth 1316.8 m which could not be detected by the conventional logs, as shown in Figure 7. Accordingly, all thin layers and coal streaks have been excluded from the facies model. This data preparation step is essential to reduce misclassification. In other words, if the OMRI images are not available, the collected training data cannot be accurate due to the presence of coal streaks and thin layers which are below logging data resolution.

A group wavelet was extracted from the angle gathers such that it consisted of three wavelets at the three central angles: 10°, 20°, and 32.5°. The initial model of the simultaneous inversion was created from wells ×1 and ×2 such that the target zone ranged from two-way time 1000 to 1800 ms with a two-millisecond (ms) sample rate. A smoother was applied to the modeled trace in the output domain after lateral interpolation. The inverted properties of the inversion were the P-impedance, S-impedance, and density. Similarly, an initial model was created for each angle stack to invert for P- and S-impedances at the near, mid, and far incident angles.

Figures 8 and 9 show the change in the P- and S-impedance values by changing the incident angle for each facies class at the well (×1). The anisotropy effect appears in the shale-dominated zones, as shown in Figure 8d. The P-impedance seems directly proportional to the incident angle, especially in the black circles, where the values gradually increase from the near to the far impedance sections. On the other hand, the gas sands of the A and B horizons appear in green color and almost have the same impedance values at the near, mid, and far angles.

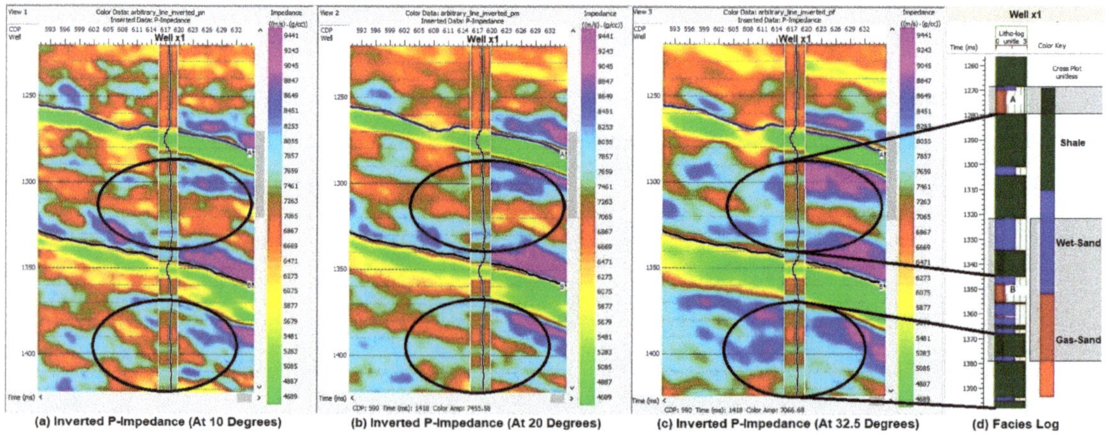

Figure 8. The partially inverted Z_P sections at the near, mid, and far incident angles (from (**a**–**c**), respectively) against the facies log at well ×1 (**d**).

These results seem consistent with a previous study that calculates the P-wave velocity for the Floyd shale by three different models, as shown in Figure 10 [127]. The first and second methods, T_45 and T_60, are based on Thomsen's model [53], where the Delta parameter is obtained using the C_{13}-component on the 45° and 60° plugs, respectively. The third model, Vp_b, is based on the strong anisotropy assumption [128]. The three models show that the P-wave velocity increases by increasing the incident angle due to the intrinsic anisotropy of shale. The results are confirmed by the ultrasonic measurements applied to the core sample.

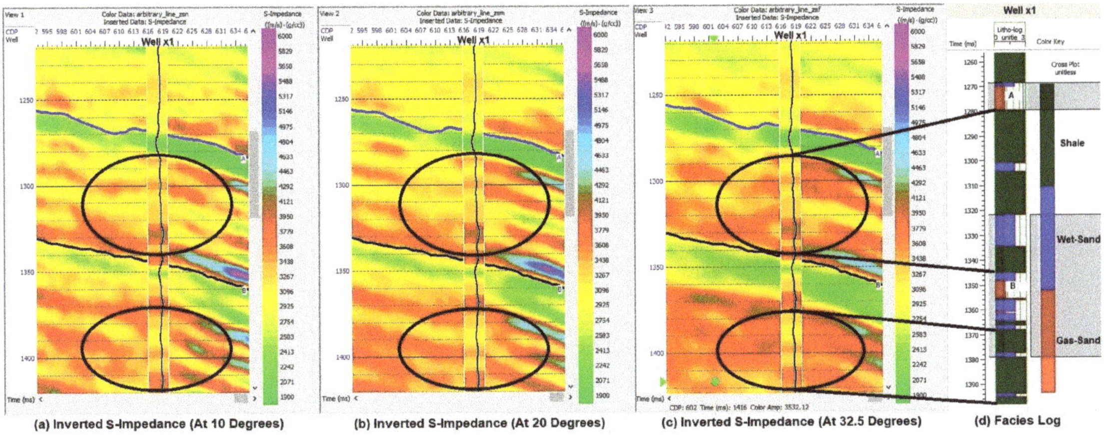

(a) Inverted S-Impedance (At 10 Degrees) (b) Inverted S-Impedance (At 20 Degrees) (c) Inverted S-Impedance (At 32.5 Degrees) (d) Facies Log

Figure 9. The partially inverted Z_S sections at the near, mid, and far incident angles (from (**a**–**c**), respectively) against the facies log at well ×1 (**d**).

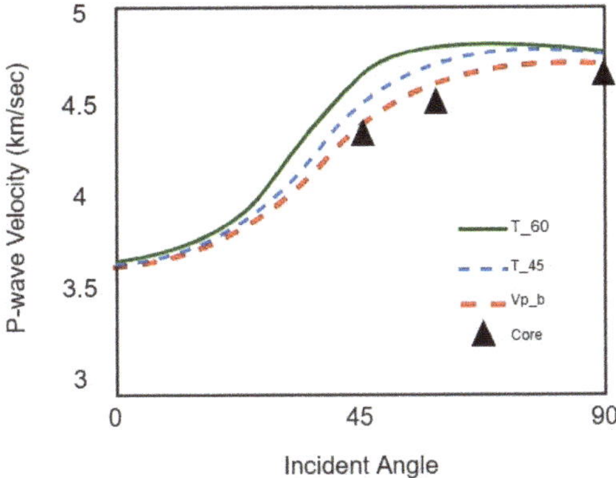

Figure 10. A plot of the P-wave velocities vs. the incident angles for the Floyd shale.

A previous study shows the measurements of the P- and S-wave velocities in shale core samples which indicate moderate intrinsic anisotropy, where the Epsilon (ϵ) is 0.16, Delta (δ) is 0.08, and Gamma (γ) is 0.29 [129]. Figure 11 shows a plot of velocity versus angle, where the quasivertical P-wave velocity (qP), quasivertical S-wave velocity (qSV), and horizontal S-wave velocity (SH) increase by increasing the incident angle at the angle range up to approximately 35°. These results are consistent with the current study that claims that the P- and S-impedances increase by increasing angles, as shown in the black circles of Figures 8 and 9.

An important note about seismic data is that the velocity change with angle may be due to other reasons rather than seismic anisotropy, such as tuning effect and noise. Accordingly, the next step is to estimate the zero-offset impedance from the partially inverted volumes to normalize the random error and anisotropy effect simultaneously.

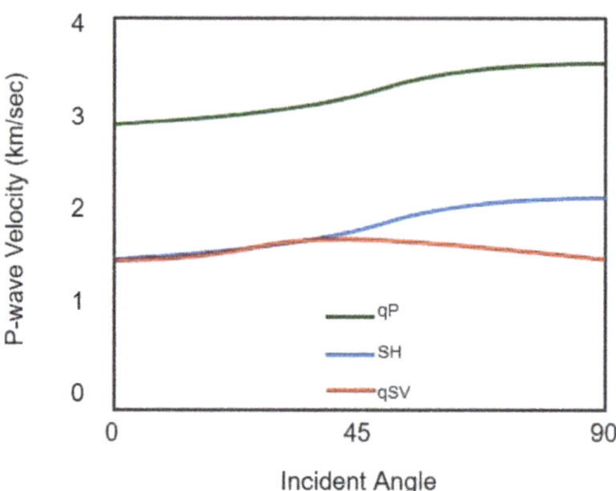

Figure 11. A plot of the velocity versus incident angle, obtained from ultrasonic measurements on shale cores.

The total data points collected for the Z_{P0} and Z_{S0} models are 1292 and 1582, respectively. These data points were gathered from the Backus-averaged impedance logs of wells ×1 and ×2 and the depth-converted composite traces of the partially inverted impedances at the wells' locations. The data points were randomly divided into three parts such that 70% of them were used for training, 15% for testing, and 15% for validation.

Figures 12 and 13 show the trace plots of the Markov chains, which aim at estimating the coefficients for both the Z_P and Z_S models, respectively. The number of iterations is shown on the X-axis and the value of each coefficient is shown on the Y-axis. Each coefficient was simulated in R by three chains, which appear in green, black, and red colors. For both the Z_P and Z_S models, the chains reached their stationary distribution in less than 5000 iterations. However, the chains of the Z_P model seem more converged than those of the Z_S model due to the high random error of the Z_S inversion.

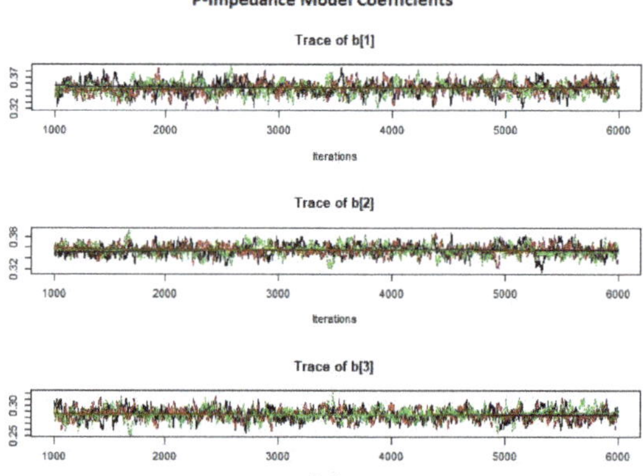

Figure 12. The trace plots of the Markov chains that simulate the coefficients of the variables for the Z_P statistical model: b1, b2, and b3.

The posterior means and standard deviations of the coefficients were finally obtained for each model, as shown in Table 1. It is clear from the posterior means of the coefficients that the rate of change of the zero-offset impedance with the angle of incidence is almost constant at the near and mid incident angles (10° and 20°), while the rate of change decreases at the far angle (32.5°). In other words, the zero-offset Z_P changes with the far Z_P at a lower rate compared to the near- and mid-angle impedances.

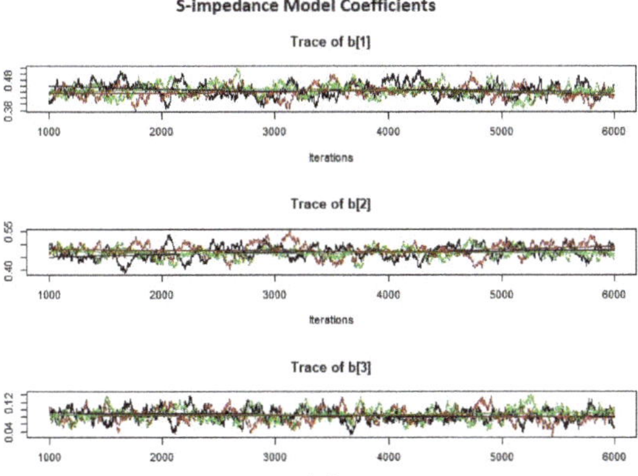

Figure 13. The trace plots of the Markov chains that simulate the coefficients of the variables for the Z_S statistical model: b1, b2, and b3.

Table 1. The statistical properties, mean (Mu) and standard deviation (SD), of the coefficients of the zero-offset Z_P and Z_S statistical models.

Coefficient	b_1		b_2		b_3	
Property	**Mu**	**SD**	**Mu**	**SD**	**Mu**	**SD**
Z_P Model	0.3532	0.01	0.3555	0.01	0.286	0.01
Z_S Model	0.3672	0.01	0.364	0.01	0.266	0.01

After applying the posterior means of the coefficients to Equation (38), the Z_P and Z_S equations will be:

$$Z_{P0} = 0.3532Z_{P1} + 0.3555Z_{P2} + 0.286Z_{P3} \quad (63)$$

$$Z_s0 = 0.3672Z_{S1} + 0.364Z_{S2} + 0.266Z_{S3} \quad (64)$$

The next step is to estimate the zero-offset impedances by the MLFN. The training, testing, and validation data were the same as those used for the statistical model. Two networks were trained for the Z_P and Z_S models by the Levenberg–Marquardt algorithm such that the best performance was reached at 21 iterations for the Z_P model and 58 for the Z_S model, as shown in Figure 14, where the number of iterations is on the X-axis and the mean square error is on the Y-axis. The error reduction trends of the training, validation, and testing data appear in blue, green, and red colors, respectively. The three trends are close to each other, which means that there is no over-fitting.

The networks reached the least mean square error (MSE) error values at 92,840 for the Z_P model and 36,893 for the Z_S model. In addition, the error values were plotted in a histogram, as shown in Figure 15, where the errors of the training, validation, and testing data appear in blue, green, and red colors, respectively. The error distribution for the Z_P and Z_S models is centered around the zero-error (yellow) line.

The quality check of the resulting zero-offset Z_P and Z_S starts with color matching with the impedance logs, as shown in Figures 16 and 17, respectively. It is clear that the MLFN's impedances (Figures 16c and 17c) better match the impedance logs at the wells ($\times$1) and ($\times$2) than those of simultaneous inversion (Figures 16a and 17a) and statistical modeling (Figures 16b and 17b). In addition, there are extreme impedance values shown in isotropic inversion compared to the anisotropic methods that have more normalized inversion error values.

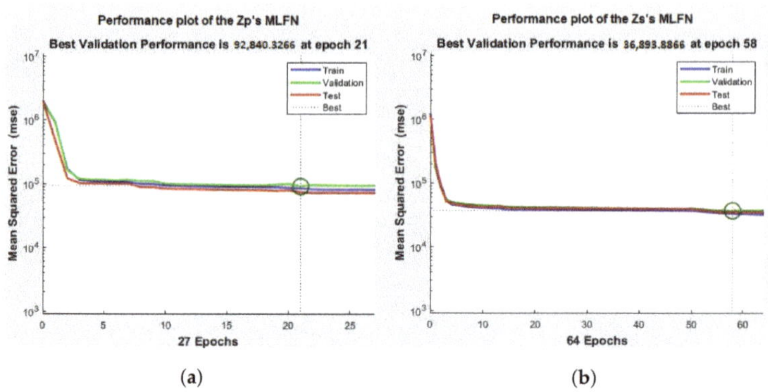

(**a**) (**b**)

Figure 14. The performance plots of the (**a**) Z_P and (**b**) Z_S MLFN models, where the number of iterations is on the X-axis and the mean square error is on the Y-axis. The error reduction trends of the training, validation, and testing data appear in blue, green, and red colors, respectively.

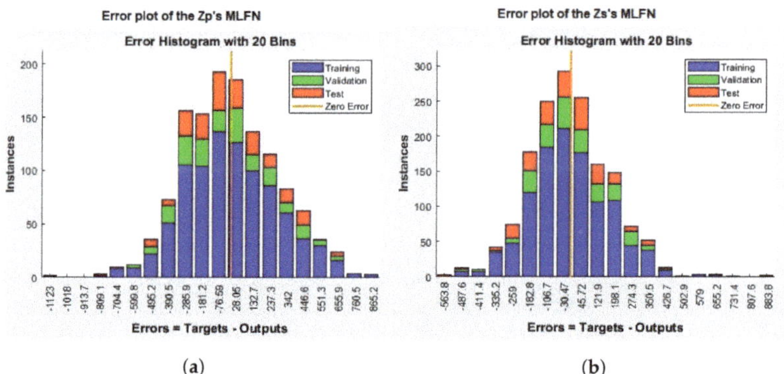

(**a**) (**b**)

Figure 15. The error histograms of the (**a**) Z_P and (**b**) Z_S MLFN models, where the error value is on the X-axis and the error frequency (instances) is on the Y-axis. The errors of the training, validation, and testing data appear in blue, green, and red colors, respectively.

Another visual comparison is made by the histogram plots of the inverted Z_P for the three models, as shown in Figure 18, where the Z_P, obtained from the simultaneous inversion, has extreme low and high values of impedance with relatively high standard deviation (1445) and low mean (6195). The extreme values of simultaneous inversion appear in purple and green colors in Figure 16a. On the other hand, the least standard deviation (1068) belongs to the Z_P of the MLFN. Moreover, the means of the anisotropic models are higher than those of simultaneous inversion, which show extremely low Z_P values down to 2000 ((m/s)·(g/cc)).

Similarly, Figure 19 shows the histogram plots extracted from the the S-impedance sections, where the isotropic method has extremely low Z_S values, down to 1000 ((m/s)·(g/cc)), high values, up to 9000 ((m/s)·(g/cc)), and relatively high standard deviation (1671). On the other hand, the statistical model and MLFN have lower standard deviations, which are 821 and 886, respectively.

Another comparison is to evaluate the residuals' plots of all models by calculating the difference between the impedance logs and the predicted impedance for each model. Figure 20 shows the plots of the calculated residuals versus the observation indexes for the Z_P models (Figure 20a–c) and Z_S models (Figure 20d–f). It can be noticed that the residuals of the MLFN model are centered around zero and have low magnitude compared to those of the simultaneous inversion and statistical modeling. In order to determine the best Z_P model, the R-value was calculated for each model, as shown in the Table 2.

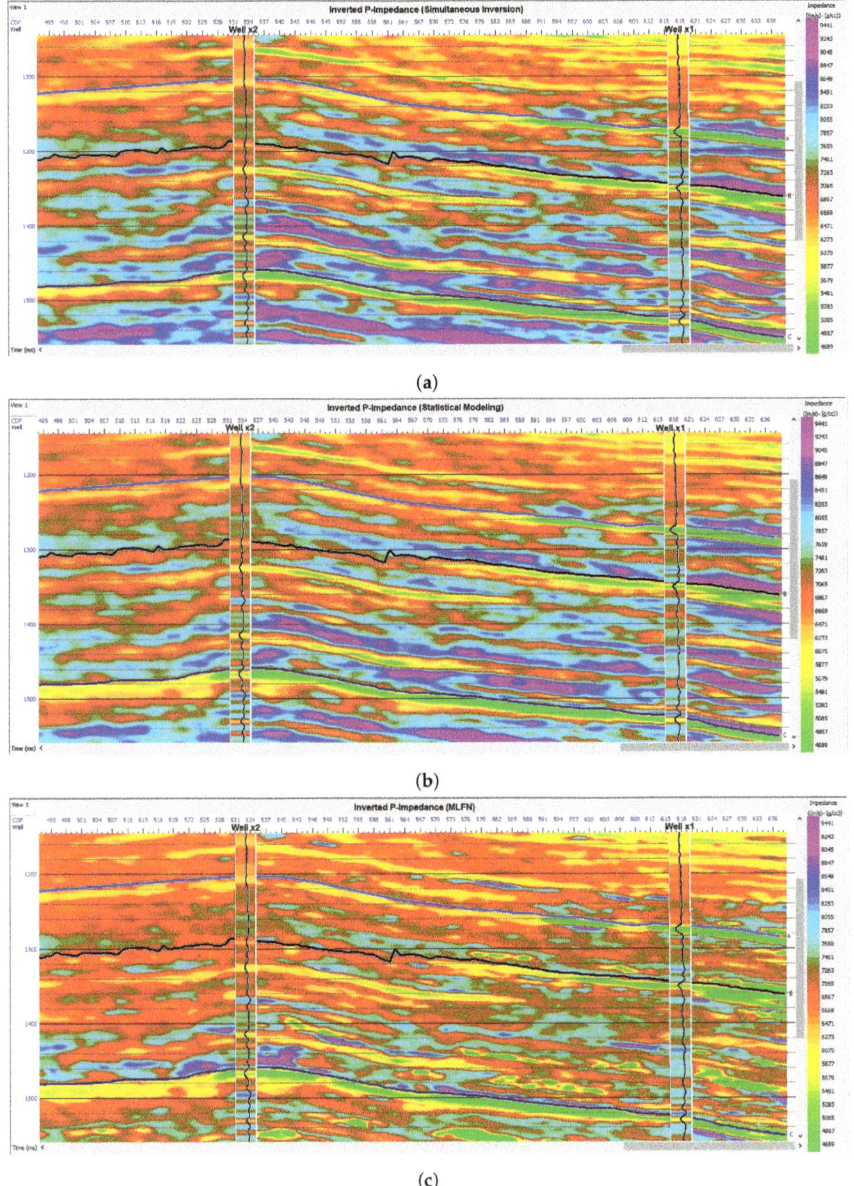

Figure 16. Arbitrary sections extracted from the zero-offset acoustic impedance volumes obtained from (**a**) isotropic inversion, (**b**) statistical modeling, and (**c**) MLFN. The inserted curves are the acoustic impedance logs of the wells $\times 1$ and $\times 2$.

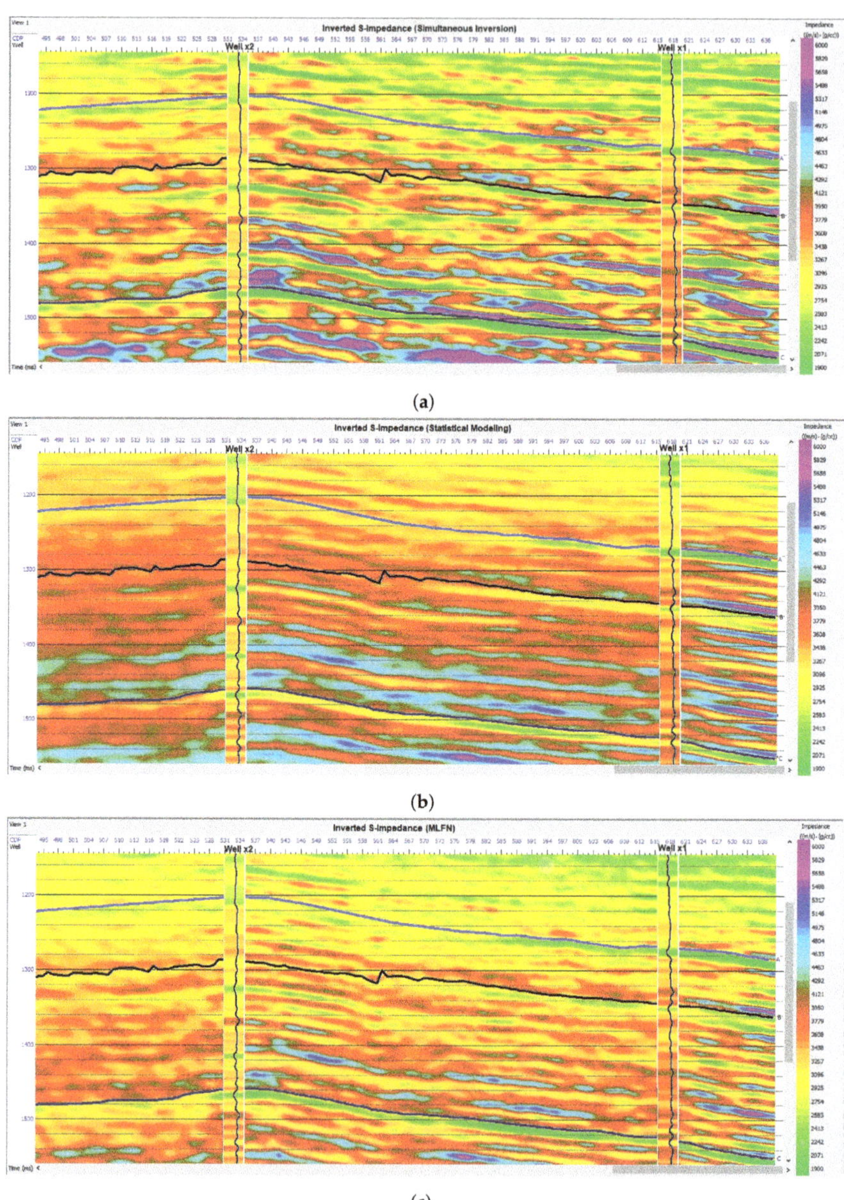

(a)

(b)

(c)

Figure 17. Arbitrary sections extracted from the zero-offset shear impedance volumes obtained from (**a**) isotropic inversion, (**b**) statistical modeling, and (**c**) MLFN. The inserted curves are the shear impedance logs of the wells ×1 and ×2.

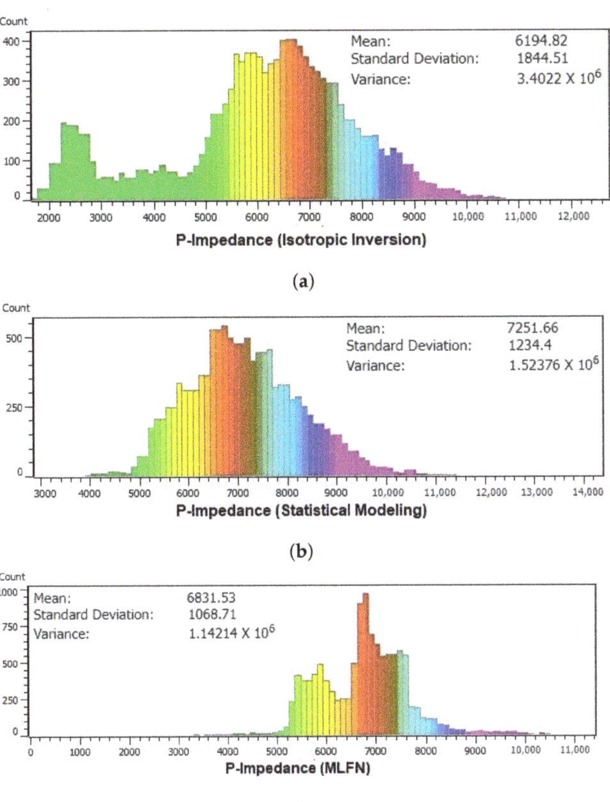

Figure 18. The histogram plots of the zero-offset P-impedance obtained from (**a**) isotropic inversion, (**b**) statistical modeling, and (**c**) MLFN.

Table 2. The correlation coefficients of the predicted Z_P and Z_S obtained from isotropic inversion, statistical modeling, and MLFN at the wells (×1) and (×2).

Property	Isotropic Inversion	Statistical Modeling	MLFN
Z_P	84%	88%	94%
Z_S	73%	85%	92%

The simultaneous (isotropic) inversion shows low R-values for Z_P and Z_S (84% and 73%, respectively), which indicates high inversion errors. The statistical model has enhanced the correlations to be 88% and 85% for Z_P and Z_S, respectively. However, the statistical models' results are still close to the isotropic case because the random error has been reduced by using only one value of each coefficient for the entire volume. Therefore, the error normalization is not as much as desired. On the other hand, the MLFN method has the highest R-values, 94% and 92% for Z_P and Z_S, respectively. This shows how ML can reduce the randomness in the data and turn it into recognized patterns.

It can be concluded that the statistical model has resulted in more accurate impedance volumes than the simultaneous inversion; however, the statistical model has only one value for each coefficient, which is not enough to normalize the random error as desired. The model can be more hierarchical if there are more facies data in several wells. On the other hand, the MLFN model could estimate the zero-offset impedance efficiently by turning the randomness in data into identified patterns.

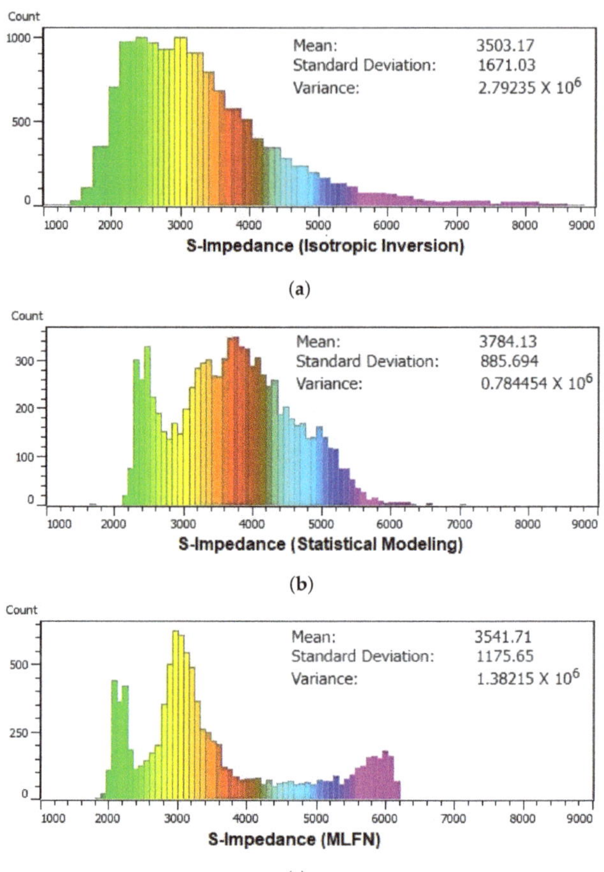

Figure 19. The histogram plots of the zero-offset S-impedance obtained from (**a**) isotropic inversion, (**b**) statistical modeling, and (**c**) MLFN.

A previous method [90] aimed at calculating the zero-offset velocities and anisotropy parameters empirically by inverting the partial stacks into velocities at different incident angles and solving Thomsen's anisotropy equations [61]. However, this method is applicable to high-resolution seismic data. On the other hand, the current study uses statistical modeling and neural networks (MLFN) to reduce the anisotropy effect and normalize the error simultaneously.

Ma estimated the P- and S-impedances from the P- and S-wave reflectivities by using a joint inversion [130]. The author reduced the nonuniqueness of the inversion output by adding the V_S/V_P ratio as a lithology constraint. Compared to simultaneous inversion, joint inversion follows the background lithology trend and hence yields more accurate impedances. The current study also follows a lithology constraint by estimating the zero-offset impedances from the impedances at different incident angles. In other words, the impedances at non-normal angles add facies information to the model and yield a unique solution for the zero-offset impedance.

The Z_P and Z_S volumes, obtained from the MLFN method, were applied to Equations (58) and (59) in order to solve for the anisotropy parameters Epsilon and Delta. Figure 21 shows the resulting Epsilon and Delta sections plotted against those obtained from the methodology of Liner and Fei [79] at the well ($\times 1$).

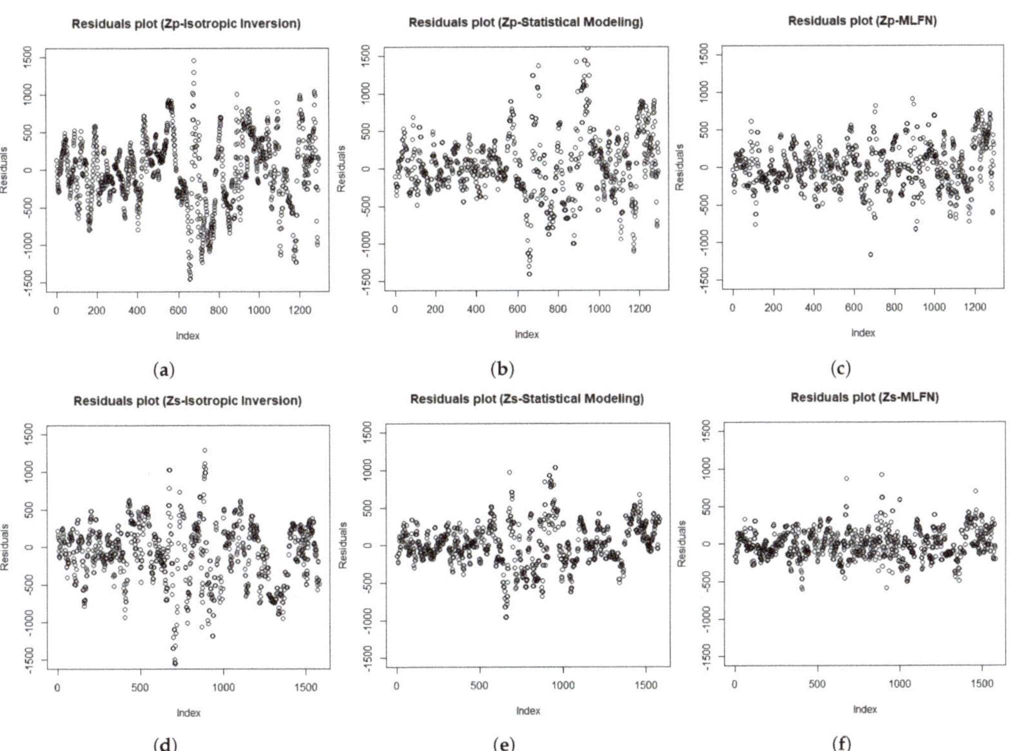

Figure 20. The residual plots of the isotropic inversion, statistical modeling, and MLFN for each of the zero-offset Z_P (**a–c**) and Z_S (**d–f**), where the number of samples of each model is on the X-axis and residuals are on the Y-axis.

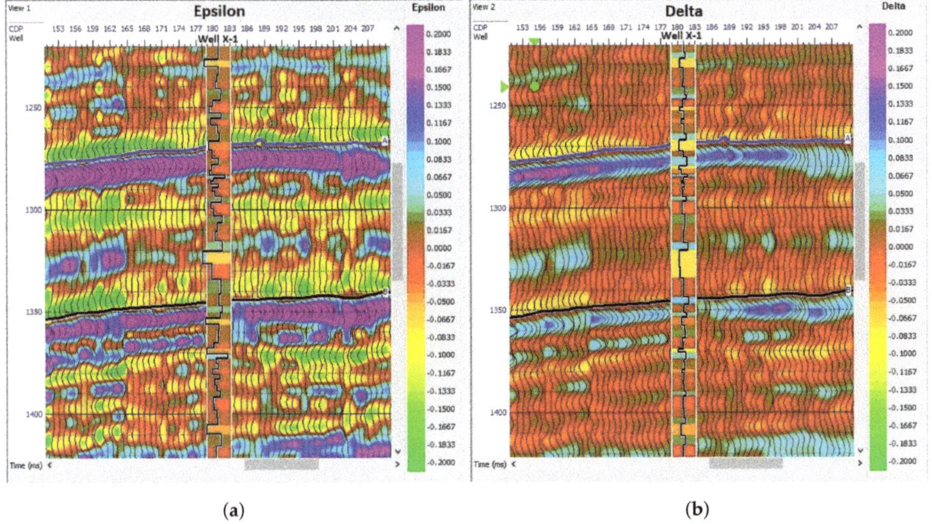

Figure 21. Cross-sections extracted from the anisotropy parameters' volumes Epsilon (**a**) and Delta (**b**) calculated by applying the MLFN-based Z_P and Z_S to the impedance-domain Thomsen's anisotropy equations. The inserted curves are the Epsilon and Delta calculated at the well (×1).

Figure 22 shows the histograms of Epsilon and Delta sections, where the values of Epsilon range from −0.2 to 0.2 with a mean value of 0.002, while the Delta ranges from −0.075 to +0.075 with a mean value of 0.001. These values are within the weak-anisotropy ranges, according to Thomsen's model [53].

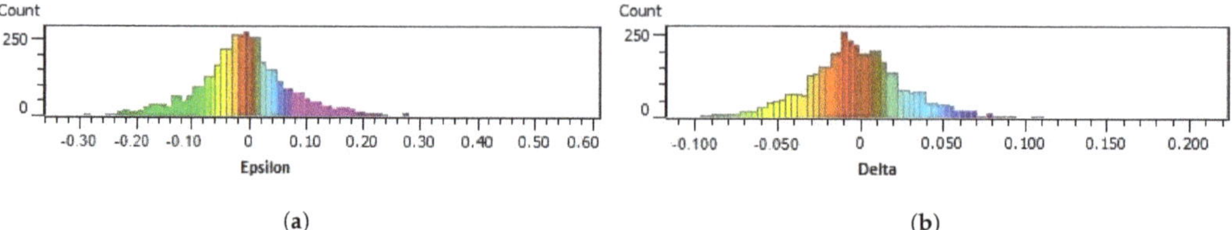

(**a**) (**b**)

Figure 22. The histograms of the anisotropy parameters (**a**) Epsilon and (**b**) Delta sections, which were calculated by applying the MLFN-based Z_P and Z_S to the impedance-domain Thomsen's anisotropy equations.

The composite traces of the resulting Epsilon and Delta were extracted from their volumes and plotted against the facies log at the well (×1), as shown in Figure 23. The tops of the gas horizons, A and B, show high positive Epsilon values (blue rectangles) due to the reduction in the P-wave velocity. The positive sign is reasonable because the vertical velocity decreases relative to the horizontal velocity, which is in the same direction as the bedding.

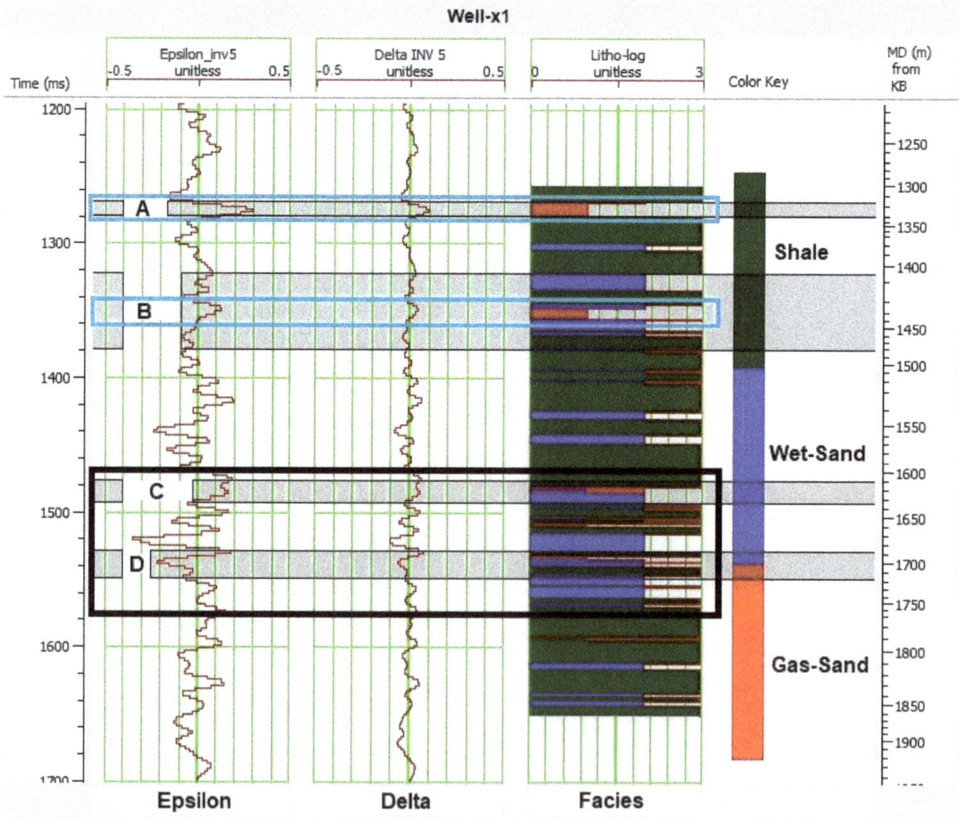

Figure 23. The anisotropy parameters, Epsilon and Delta, against the facies log at the well (×1), where A, B, C, and D are sand zones.

Another observation is that the anisotropy degree is directly proportional to the contrast of the elastic properties between layers. This phenomenon appears in the abrupt change in facies, as shown in Figure 23, where the black rectangle shows high contrast boundaries between layers coinciding with significant fluctuations in the Epsilon parameter. The same conclusion is mentioned by Bandyopadhyay [76], who plotted the anisotropy parameters, Epsilon and Delta, versus the depth for a laminated shaly sand succession, as shown in Figure 24. The magnitude of the anisotropy is low for the water-saturated sand (Figure 24a), which has Lame's parameters that are similar to those of shale. On the other hand, the Lame's parameters of the gas-saturated sand (Figure 24b) and shale are significantly different, resulting in a high magnitude of the anisotropy parameters.

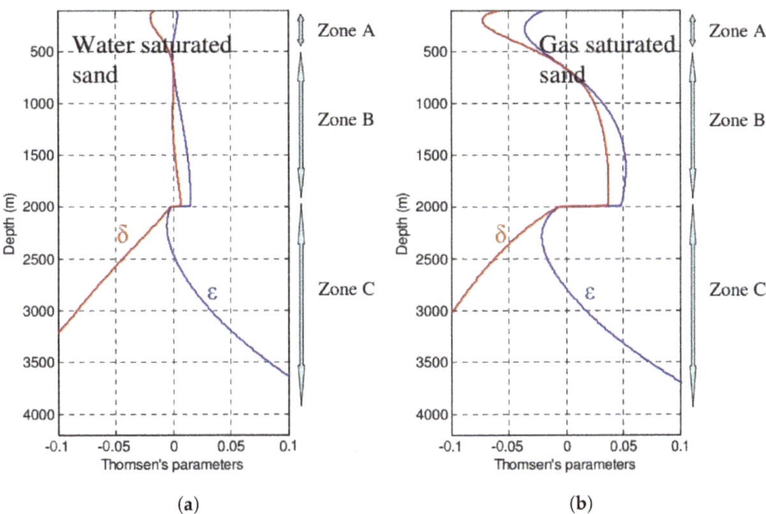

(a) (b)

Figure 24. Plots of the anisotropy parameters, Epsilon (blue curve) and Delta (red curve), on the X-axes versus the depth on the Y-axes, for both the water- and gas-saturated sand (**a** and **b**, respectively). The depth range is divided into three zones, (A), (B), and (C), which are the shallow, compaction, and diagenesis regimes, respectively [76].

Most of the previous studies stated that Epsilon is always positive or zero. However, the resulting Epsilon, as shown in Figure 21a, has a negative sign in some zones. This is matching with Bandyopadhyay's study [76], which stated that Epsilon can be negative in laminated shaly gas sand layers, as shown in the anisotropy–depth plot (Figure 24b), where the depth interval is divided into three zones, (A), (B), and (C), which are the shallow, compaction, and diagenesis regimes, respectively. Epsilon is significantly fluctuating around zero in the gas-saturated sand such that it becomes negative at the shallow and diagenesis zones.

The Epsilon and Delta sections are plotted in Figure 25, showing a strong positive correlation between them. The color legend represents the two-way time value for each observation. The best-fit line gives the following linear relationship between Epsilon and Delta:

$$\delta = 0.31\epsilon + 0.004 \tag{65}$$

Interestingly, the observed relationship is mostly similar to that obtained by Li [73], who implemented the following equation:

$$\delta = 0.32\epsilon \tag{66}$$

The next step is to transform the Z_P and Z_S into the elastic properties from which the lithology and fluid distributions should be forecasted. The facies integrated model was firstly applied to logging data at the well (×1) for training and then used to predict the facies distribution from the elastic volumes. The lithology model was postulated by using R, based on logistic regression, to forecast the sand probability from the near and far elastic impedances. Similarly, the fluid model was created to predict the gas probability from the Mu–Rho, Lambda–Rho/Mu–Rho, and Poisson's ratios [52]. Finally, the two models were merged by a decision tree to obtain the distribution of each facies class.

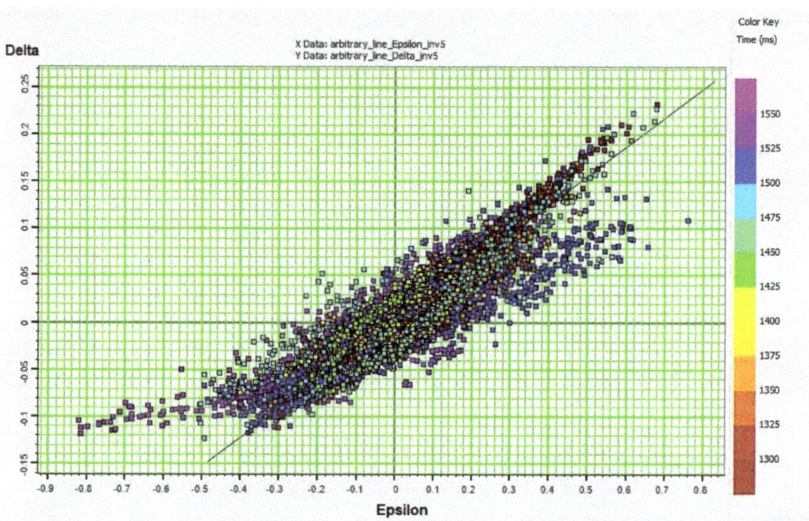

Figure 25. A linear regression plot between Epsilon, on the X-axis, and Delta, on the Y-axis. The data points were obtained from the Epsilon and Delta volumes, which were calculated by applying the MLFN-based Z_P and Z_S to the impedance-domain Thomsen's anisotropy equations.

The predicted sand and gas probabilities (Phi-sand and Phi-gas-sand, respectively) of the training data were plotted versus the lithology and fluid variables, respectively, as shown in Figure 26. As expected, the sand probability values are high where the lithology variable is closer to 1, indicating sandstone, and low where the lithology variable is closer to 0, indicating shale (Figure 26a). Like the sand probability, the gas probability is high where the fluid variable is 1, indicating gas sand, and generally low where the fluid variable equals 0, indicating wet sand (Figure 26b). The best cut-off values were selected to differentiate between the sand and shale observations in the lithology model as well as the gas sand and wet sand in the fluid model.

A decision tree was created to estimate the litho-fluid classes from the sand probability, gas probability, and depth. The total data points are 1084, starting from depth 1300 m to 1851 m. According to the facies log of the well (×1), the data consist of 75 gas sand, 286 wet sand, and 723 shale data points. The data were divided into three parts for training (80%), testing, (10%), and validation (10%). The best minimum split value has been observed to be 10, which leads to the best performance of the tree. Figure 27 shows the graphical representation of the decision tree, which classifies the training data into gas sand in red, wet sand in blue, and shale in green. The input variables are the gas probability (Pg), sand probability (Ps), and depth (d).

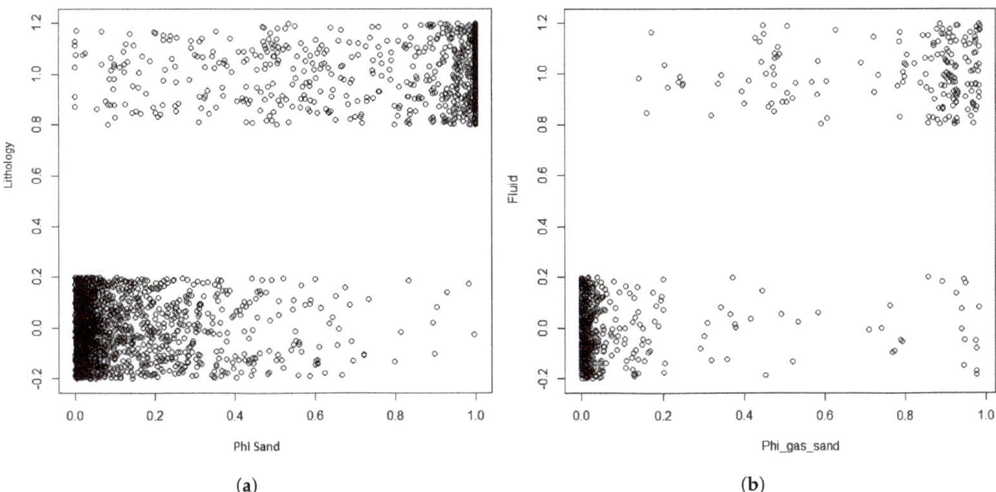

Figure 26. (**a**) Plot of the sand probability (Phi-sand) versus the lithology (facies) variable. (**b**) Plot of the gas probability (Phi-gas-sand) versus the fluid variable.

The accuracy of the tree was calculated for the training, testing, and validation data to be 91%, 89%, and 90%, respectively, and the average accuracy is 91%. These results show that the tree is accurate enough to be applied to the depth-converted seismic data. A function was created in R-Studio to transform the zero-offset Z_P and Z_S volumes, obtained from the isotropic inversion and MLFN, into elastic properties and then into sand and gas probabilities based on the facies model.

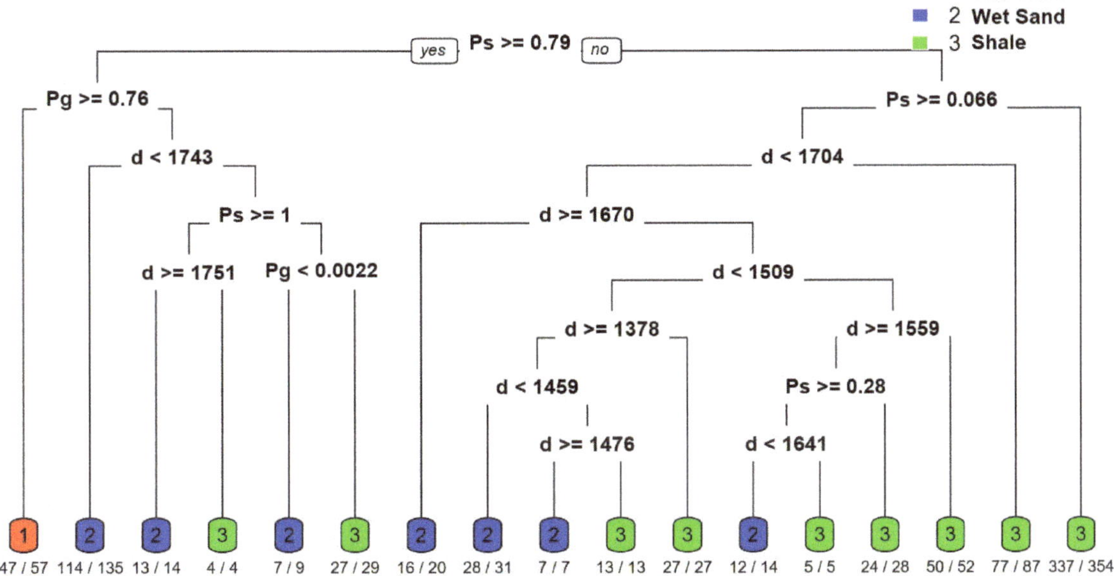

Figure 27. The graphical representation of the decision tree model, which classifies the training data of the well (×1) into gas sand in red, wet sand in blue, and shale in green. The input variables are the gas probability (Pg), sand probability (Ps), and depth (d).

Eventually, the decision tree was applied to predict the facies classes from the probabilities and depth. The composite traces of the facies volumes were extracted at the well (×1) to compare the isotropic simultaneous inversion and the anisotropic MLFN outputs, as shown in Figure 28, where the gas sand, wet sand, and shale are represented in the red, blue, and green colors, respectively. The CCR was calculated for each litho-fluid class for both the isotropic and anisotropic cases, as shown in Table 3.

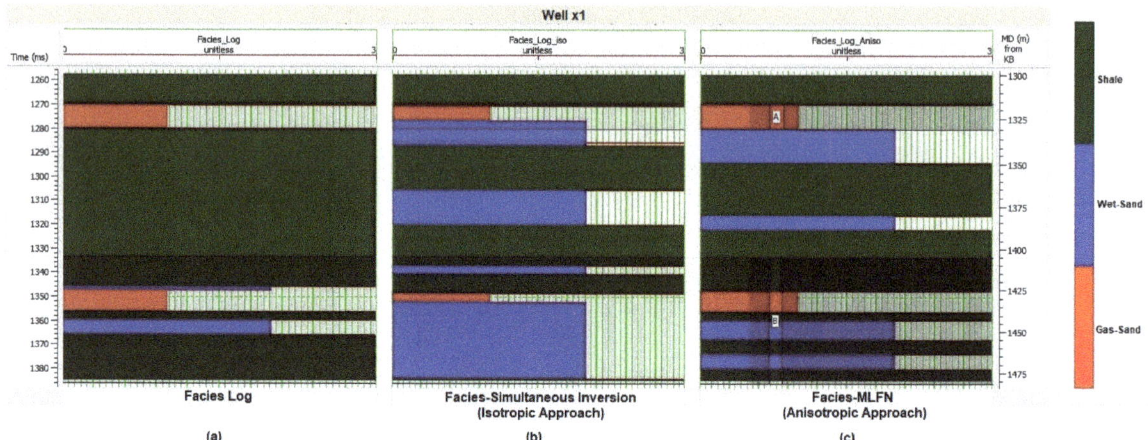

Figure 28. The litho-fluid facies at the well (×1): (**a**) log facies, (**b**) simultaneous inversion facies, and (**c**) MLFN facies.

Table 3. The correct classification rates of the litho-fluid classes for isotropic inversion and MLFN.

Method	Gas Sand	Wet Sand	Shale	All Facies
Isotropic Inversion	51%	82%	55%	56%
MLFN	97%	83%	73%	77%

The CCR values of the gas sand are 51% and 97% for the isotropic and anisotropic cases, respectively. This improvement is evident in Figure 28, where the gas sand is correctly detected at the anisotropic case. Similarly, the CCR values of the shale are 55% and 73% for the isotropic and anisotropic cases, respectively. The CCR values of the wet sand are almost the same for the two cases, 82% and 83% for the isotropic and anisotropic cases, respectively.

The average CCR value is 56% for the isotropic facies and 77% for the anisotropic case. This means that the accuracy of litho-fluid detection has been enhanced by about 21%. This improvement is not only due to the reduction in the anisotropy effect but also because the MLFN has normalized the inversion errors due to the tuning effect.

These results seem comparable to those of another case study, from the Okoli field in the Sava depression, which aimed at predicting clastic facies using neural networks [107]. Although the lithology was estimated based on various logs, the results showed correlation levels ranging from 78.3% to 82.1%, which are close to the results of this study.

A recent study set various ML algorithms to estimate four facies classes from seismic attributes [131]. The logistic regression showed encouraging success rates, up to 0.98, compared to other algorithms, such as the Gaussian process, linear discriminant analysis, and stochastic gradient descent. However, a high success rate does not indicate the model's stability, especially in heterogeneous fields. The advantage of the current study's facies model is that the litho-fluid probabilities as well as the depth variable are inserted into a decision tree which turns the probabilities into facies estimates. The integration of two different algorithms guarantees the model's stability and reduces the chance of over-fitting.

6. Conclusions

The zero-offset Z_P and Z_S volumes were obtained by an isotropic simultaneous inversion and then by two anisotropic methods: statistical modeling and MLFN. The validation of the P-impedance volumes shows that the MLFN model had near-zero residuals and a high R-value (0.94) compared to simultaneous inversion and statistical modeling, which had low R-squared values (0.84 and 0.88, respectively). Similarly, the MLFN resulted in a high correlation for the S-impedance (0.92), while simultaneous inversion and statistical modeling had lesser R-values (0.73 and 0.85, respectively).

The Z_P and Z_S volumes, obtained from the MLFN models, were applied to the impedance-domain Thomsen's anisotropy equations to solve for the anisotropy parameters: Epsilon and Delta. The anisotropy magnitude was observed to be low in the shaly wet sand successions, where the contrast of the elastic properties between layers was low, while it showed higher magnitude in the shaly gas sand successions, which had high-contrast fluctuations in the elastic properties.

The Z_P and Z_S volumes, obtained from the isotropic inversion and MLFN, were converted into elastic properties from which the sand and gas probabilities were forecasted by logistic regression. The predicted probabilities were inserted into a decision tree to estimate the distribution of gas sand, wet sand, and shale classes within the target zone. The predicted facies distributions were validated by the facies log at the well ($\times 1$), where the MLFN showed a high average success rate (0.77) compared to the simultaneous inversion, which had a low success rate (0.56).

Conventional inversion algorithms have proved to be misleading if seismic anisotropy is neglected. On the other hand, statistical methods and ML are more efficient tools for obtaining more accurate rock properties, especially when considering the dependence of these properties on incident angles.

The decision tree of the facies model shows encouraging results due to the contribution of the depth constraint. This proves that the trend factor is crucial for accurate reservoir characterization. Therefore, the availability of geochemical data would be an added value that links other properties, such as elastic properties and anisotropy parameters, to the pressure regime and compaction profile.

The study considers that the subsurface follows the VTI assumption, which applies to horizontal layers. However, the methodology can be extended to include more realistic cases, such as the TTI and HTI assumptions, which can enhance the accuracy of zero-offset parameter modeling. In addition, core data are needed to validate anisotropy measurement and yield confident models.

Author Contributions: Conceptualization, M.F.G.; methodology, M.F.G. and S.Y.M.A.; software, M.F.G.; validation, M.F.G., A.H.A.L. and S.Y.M.A.; formal analysis, M.F.G.; investigation, M.F.G.; resources, A.H.A.L.; data curation, M.F.G. and A.H.A.L.; writing—original draft preparation, M.F.G.; writing—review and editing, M.F.G., A.H.A.L. and S.Y.M.A.; visualization, M.F.G.; supervision, A.H.A.L.; project administration, A.H.A.L.; funding acquisition, A.H.A.L. All authors have read and agreed to the published version of the manuscript

Funding: This research was funded by YUTP-PRG OF Grant Cost Center: 015PBC-021.

Institutional Review Board Statement: Not applicable.

Informed Consent Statement: Not applicable.

Data Availability Statement: The data that support the findings of this study are available from Petronas. Restrictions apply to the availability of these data, which were used under license for this study.

Acknowledgments: We would like to thank the Department of Petroleum Geosciences at Universiti Teknologi Petronas. In addition, we appreciate all the efforts and support of our colleagues in the Center of Seismic Imaging. A special thanks to Amir Abbas for his valuable help and support.

Conflicts of Interest: The authors declare no conflict of interest.

Abbreviations

The following abbreviations are used in this manuscript:

MLFN	Multi Layer Feed Forward
ML	Machine Learning
ANN	Artificial Neural Network
CNN	Convolutional Neural Network
SVM	Support Vector Machine
BT	Bagged Tree
RVM	Relevance Vector Machine
SRG	Seed Region Growing
SOM	Self-Organizing Map
PCA	Principal Component Analysis
GTM	Generative Topographic Mapping
EI	Elastic Impedance
MR	Mu-Rho
LR	Lambda-Rho
PR	Poisson's Ratio
VTI	Vertical Transverse Isotropy
TI	Transverse Isotropy
TTI	Tilted Transverse Isotropy
AVO	Amplitude Versus Offset
MCMC	Markov Chain Monte Carlo
VSP	Vertical Seismic Profile
Vsh	Shale Volume
DEM	Differential Effective Medium
ADP	Anisotropic Dual Porosity
SCA	Self-Consistent Approximation
HTI	Horizontal Transverse Isotropy
MRF	Markov-Random Field
SD	Standard Deviation
CCR	Correct Classification Rate
OMRI	Oil-Mud Resistivity Imager

References

1. Hampson, D.; Todorov, T.; Russell, B. Using multi-attribute transforms to predict log properties from seismic data. *Explor. Geophys.* **2000**, *31*, 481–487. [CrossRef]
2. Dubucq, D.; Busman, S.; Van Riel, P. Turbidite reservoir characterization: Multi-offset stack inversion for reservoir delineation and porosity estimation; A Golf of Guinea example. In *SEG Technical Program Expanded Abstracts 2001*; Society of Exploration Geophysicists: Houston, TX, USA, 2001; pp. 609–612.
3. Johansen, T.A.; Spikes, K.; Dvorkin, J. Strategy for estimation of lithology and reservoir properties from seismic velocities and density. In *SEG Technical Program Expanded Abstracts 2004*; Society of Exploration Geophysicists: Houston, TX, USA, 2004; pp. 1726–1729.
4. Ganguli, S.S.; Vedanti, N.; Dimri, V. 4D reservoir characterization using well log data for feasible CO2-enhanced oil recovery at Ankleshwar, Cambay Basin-A rock physics diagnostic and modeling approach. *J. Appl. Geophys.* **2016**, *135*, 111–121. [CrossRef]
5. Ganguli, S.S. *Integrated Reservoir Studies for CO2-Enhanced Oil Recovery and Sequestration: Application to an Indian Mature Oil Field*; Springer: Berlin/Heidelberg, Germany, 2017.
6. Russell, B.H. The Application of Multivariate Statistics and Neural Networks to the Prediction of Reservoir Parameters Using Seismic Attributes. Ph.D. Thesis, University of Calgary, Calgary, AB, Canada, 2004.
7. Bachrach, R. Joint estimation of porosity and saturation using stochastic rock-physics modeling. *Geophysics* **2006**, *71*, O53–O63. [CrossRef]
8. Doyen, P. *Seismic Reservoir Characterization: An Earth Modelling Perspective*; EAGE Publications: Houten, The Netherlands, 2007; Volume 2.

9. Hu, H.; Yin, X.; Wu, G. Joint inversion of petrophysical parameters based on Bayesian classification. *Geophys. Prospect. Pet.* **2012**, *51*, 225–232.
10. Yin, X.Y.; Sun, R.Y.; Wang, B.L.; Zhang, G.Z. Simultaneous inversion of petrophysical parameters based on geostatistical a priori information. *Appl. Geophys.* **2014**, *11*, 311–320. [CrossRef]
11. Di Luca, M.; Salinas, T.; Arminio, J.F.; Alvarez, G.; Alvarez, P.; Bolivar, F.; Marín, W. Seismic inversion and AVO analysis applied to predictive-modeling gas-condensate sands for exploration and early production in the Lower Magdalena Basin, Colombia. *Lead. Edge* **2014**, *33*, 746–756. [CrossRef]
12. Yoong, A.A.; Lubis, L.A.; Ghosh, D.P. Application of Simultaneous Inversion Method to Predict the Lithology and Fluid Distribution in "X" Field, Malay Basin. *Proc. IOP Conf. Ser. Earth Environ. Sci.* **2016**, *38*, 012007. [CrossRef]
13. Babasafari, A.A.; Ghosh, D.P.; Salim, A.M.A.; Ratnam, T.; Sambo, C.; Rezaee, S. Petro-elastic modeling for enhancement of hydrocarbon prediction: Case study in Southeast Asia. In *SEG Technical Program Expanded Abstracts 2018*; Society of Exploration Geophysicists: Houston, TX, USA, 2018; pp. 3141–3145.
14. Babasafari, A.A.; Ghosh, D.; Salim, A.M.A.; Alashloo, S.Y.M. Rock Physics Modeling Assisted Reservoir Properties Prediction: Case Study in Malay Basin. *Int. J. Eng. Technol.* **2018**, *7*, 24–28. [CrossRef]
15. de Figueiredo, L.P.; Grana, D.; Bordignon, F.L.; Santos, M.; Roisenberg, M.; Rodrigues, B.B. Joint Bayesian inversion based on rock-physics prior modeling for the estimation of spatially correlated reservoir properties. *Geophysics* **2018**, *83*, M49–M61. [CrossRef]
16. Saikia, P.; Baruah, R.D.; Singh, S.K.; Chaudhuri, P.K. Artificial Neural Networks in the Domain of Reservoir Characterization: A Review From Shallow to Deep Models. *Comput. Geosci.* **2020**, *135*, 104357. [CrossRef]
17. Verma, A.K.; Cheadle, B.A.; Routray, A.; Mohanty, W.K.; Mansinha, L. Porosity and Permeability Estimation Using Neural Network Approach From Well Log Data. In Proceedings of the GeoConvention Vision Conference, Calgary, AB, Canada, 14–18 May 2012.
18. Cao, J.; Yang, J.; Wang, Y.; Wang, D.; Shi, Y. Extreme Learning Machine for Reservoir Parameter Estimation in Heterogeneous Sandstone Reservoir. *Math. Probl. Eng.* **2015**, *2015*. [CrossRef]
19. Okon, A.N.; Adewole, S.E.; Uguma, E.M. Artificial Neural Network Model for Reservoir Petrophysical Properties: Porosity, Permeability and Water Saturation Prediction. *Model. Earth Syst. Environ.* **2021**, *7*, 2373–2390. [CrossRef]
20. Ahmadi, M.A.; Zendehboudi, S.; Lohi, A.; Elkamel, A.; Chatzis, I. Reservoir permeability prediction by neural networks combined with hybrid genetic algorithm and particle swarm optimization. *Geophys. Prospect.* **2013**, *61*, 582–598. [CrossRef]
21. Gholami, R.; Moradzadeh, A.; Maleki, S.; Amiri, S.; Hanachi, J. Applications of Artificial Intelligence Methods in Prediction of Permeability in Hydrocarbon Reservoirs. *J. Pet. Sci. Eng.* **2014**, *122*, 643–656. [CrossRef]
22. Liu, S.; Zolfaghari, A.; Sattarin, S.; Dahaghi, A.K.; Negahban, S. Application of Neural Networks in Multiphase Flow Through Porous Media: Predicting Capillary Pressure and Relative Permeability Curves. *J. Pet. Sci. Eng.* **2019**, *180*, 445–455. [CrossRef]
23. Asoodeh, M.; Bagheripour, P. Prediction of Compressional, Shear, and Stoneley Wave Velocities From Conventional Well Log Data Using a Committee Machine With Intelligent Systems. *Rock Mech. Rock Eng.* **2012**, *45*, 45–63. [CrossRef]
24. Akhundi, H.; Ghafoori, M.; Lashkaripour, G.R. Prediction of Shear Wave Velocity Using Artificial Neural Network Technique, Multiple Regression and Petrophysical Data: A Case Study in Asmari Reservoir (SW Iran). *Open J. Geol.* **2014**, *2014*. [CrossRef]
25. Anemangely, M.; Ramezanzadeh, A.; Tokhmechi, B. Shear Wave Travel Time Estimation from Petrophysical Logs Using ANFIS-PSO Algorithm: A Case Study From Ab-Teymour Oilfield. *J. Nat. Gas Sci. Eng.* **2017**, *38*, 373–387. [CrossRef]
26. Anemangely, M.; Ramezanzadeh, A.; Amiri, H.; Hoseinpour, S.A. Machine Learning Technique for the Prediction of Shear Wave Velocity Using Petrophysical Logs. *J. Pet. Sci. Eng.* **2019**, *174*, 306–327. [CrossRef]
27. Li, H.; Lin, J.; Wu, B.; Gao, J.; Liu, N. Elastic Properties Estimation From Prestack Seismic Data Using GGCNNs and Application on Tight Sandstone Reservoir Characterization. *IEEE Trans. Geosci. Remote Sens.* **2021**, *60*, 1–21. [CrossRef]
28. Validov, M.F.; Nurgaliev, D.K.; Sudakov, V.A.; Murtazin, T.A.; Golod, K.A.; Galimova, A.R.; Shamsiev, R.R.; Lutfullin, A.A.; Amerhanov, M.I.; Aslyamov, N.A. The Use of Neural Network Technologies in Prediction the Reservoir Properties of Unconsolidated Reservoir Rocks of Shallow Bitumen Deposits. In Proceedings of the SPE Annual Caspian Technical Conference, OnePetro, Virtual Event, 1–5 October 2021.
29. Weinzierl, W.; Wiese, B. Deep Learning a Poroelastic Rock-Physics Model For Pressure and Saturation Discrimination Deep Learning a Rock-Physics Model. *Geophysics* **2021**, *86*, MR53–MR66. [CrossRef]
30. Banerjee, A.; Chatterjee, R. Mapping of Reservoir Properties Using Model-Based Seismic Inversion and Neural Network Architecture in Raniganj Basin, India. *J. Geol. Soc. India* **2022**, *98*, 479–486. [CrossRef]
31. Kapur, L.; Lake, L.W.; Sepehrnoori, K.; Herrick, D.C.; Kalkomey, C.T.; et al. Facies Prediction From Core and Log Data Using Artificial Neural Network Technology. In Proceedings of the SPWLA 39th annual logging symposium. Society of Petrophysicists and Well-Log Analysts, Keystone, CO, USA, 26–28 May 1998.
32. Wang, G.; Carr, T.R. Organic-rich Marcellus Shale Lithofacies Modeling and Distribution Pattern Analysis in the Appalachian Basin. *AAPG Bull.* **2013**, *97*, 2173–2205. [CrossRef]
33. Zhang, G.; Wang, Z.; Chen, Y. Deep learning for seismic lithology prediction. *Geophys. J. Int.* **2018**, *215*, 1368–1387. [CrossRef]
34. Li, F.; Zhou, H.; Wang, Z.; Wu, X. ADDCNN: An attention-based deep dilated convolutional neural network for seismic facies analysis with interpretable spatial–spectral maps. *IEEE Trans. Geosci. Remote Sens.* **2020**, *59*, 1733–1744. [CrossRef]

35. Feng, R.; Balling, N.; Grana, D.; Dramsch, J.S.; Hansen, T.M. Bayesian Convolutional Neural Networks for Seismic Facies Classification. *IEEE Trans. Geosci. Remote Sens.* **2021**, *59*, 8933–8940. [CrossRef]
36. Al-Anazi, A.; Gates, I. A support vector machine algorithm to classify lithofacies and model permeability in heterogeneous reservoirs. *Eng. Geol.* **2010**, *114*, 267–277. [CrossRef]
37. Saraswat, P.; Sen, M.K. Artificial immune-based self-organizing maps for seismic-facies analysis. *Geophysics* **2012**, *77*, O45–O53. [CrossRef]
38. Roy, A. *Latent Space Classification of Seismic Facies*; The University of Oklahoma: Norman, OK, USA, 2013.
39. Torres, A.; Reveron, J.; Infante, J. Lithofacies discrimination using support vector machines, rock physics and simultaneous seismic inversion in clastic reservoirs in the Orinoco Oil Belt, Venezuela. In Proceedings of the 2013 SEG Annual Meeting, OnePetro, Houston, TX, USA, 1–3 October 2013.
40. Wang, G.; Carr, T.R.; Ju, Y.; Li, C. Identifying organic-rich Marcellus Shale lithofacies by support vector machine classifier in the Appalachian basin. *Comput. Geosci.* **2014**, *64*, 52–60. [CrossRef]
41. Hall, B. Facies classification using machine learning. *Lead. Edge* **2016**, *35*, 906–909. [CrossRef]
42. Amraee, T.; Ranjbar, S. Transient instability prediction using decision tree technique. *IEEE Trans. Power Syst.* **2013**, *28*, 3028–3037. [CrossRef]
43. Keynejad, S.; Sbar, M.L.; Johnson, R.A. Assessment of machine-learning techniques in predicting lithofluid facies logs in hydrocarbon wells. *Interpretation* **2019**, *7*, SF1–SF13. [CrossRef]
44. Liu, X.; Chen, X.; Li, J.; Zhou, X.; Chen, Y. Facies identification based on multikernel relevance vector machine. *IEEE Trans. Geosci. Remote Sens.* **2020**, *58*, 7269–7282. [CrossRef]
45. Liu, J.; Dai, X.; Gan, L.; Liu, L.; Lu, W. Supervised seismic facies analysis based on image segmentation. *Geophysics* **2018**, *83*, O25–O30. [CrossRef]
46. Liu, R.; Zhang, B.; Wang, X. Patterns classification in assisting seismic-facies analysis. In *SEG Technical Program Expanded Abstracts 2017*; Society of Exploration Geophysicists: Houston, TX, USA, 2017; pp. 2127–2131.
47. Zhao, T.; Li, F.; Marfurt, K.J. Constraining self-organizing map facies analysis with stratigraphy: An approach to increase the credibility in automatic seismic facies classification. *Interpretation* **2017**, *5*, T163–T171. [CrossRef]
48. Guo, H.; Marfurt, K.J.; Liu, J. Principal component spectral analysis. *Geophysics* **2009**, *74*, P35–P43. [CrossRef]
49. Haykin et al., S. *Neural Networks and Learning Machines*, 3rd ed.; Pearson Prentice Hall: New York, NY, USA, 2009.
50. Roden, R.; Smith, T.; Sacrey, D. Geologic pattern recognition from seismic attributes: Principal component analysis and self-organizing maps. *Interpretation* **2015**, *3*, SAE59–SAE83. [CrossRef]
51. Roy, A.; Romero-Peláez, A.S.; Kwiatkowski, T.J.; Marfurt, K.J. Generative topographic mapping for seismic facies estimation of a carbonate wash, Veracruz Basin, southern Mexico. *Interpretation* **2014**, *2*, SA31–SA47. [CrossRef]
52. Gouda, M.F.; Salim, A. Litho-Fluid Facies Modeling by Using Logistic Regression. *Pet. Coal* **2020**, *62*.
53. Thomsen, L. Weak elastic anisotropy. *Geophysics* **1986**, *51*, 1954–1966. [CrossRef]
54. Thomsen, L.; Anderson, D.L. Weak elastic anisotropy in global seismology. *Geol. Soc. Am. Spec. Pap.* **2015**, *514*, 39–50.
55. Zhang, Y.; Eisner, L.; Barker, W.; Mueller, M.C.; Smith, K.L. Effective anisotropic velocity model from surface monitoring of microseismic events. *Geophys. Prospect.* **2013**, *61*, 919–930. [CrossRef]
56. Kelter, A. Estimation of Thomsen's anisotropy parameters from compressional and converted wave surface seismic travel-time data using NMO equations, neural networks and regridding inversion. In *SEG Technical Program Expanded Abstracts 2005*; Society of Exploration Geophysicists: Houston, TX, USA, 2005; pp. 120–122.
57. Mainprice, D. Seismic anisotropy of the deep Earth from a mineral and rock physics perspective. In *Treatise on Geophysics*, 2nd ed.; Elsevier: Oxford, UK, 2015; pp. 487–538.
58. Almqvist, B.S.G.; Mainprice, D. Seismic properties and anisotropy of the continental crust: predictions based on mineral texture and rock microstructure. *Rev. Geophys.* **2017**, *55*, 367–433. [CrossRef]
59. Love, A.E.H. *A Treatise on the Mathematical Theory of Elasticity*; Cambridge University Press: Cambridge, UK, 2013.
60. Bergstrom, J.S. *Mechanics of Solid Polymers: Theory and Computational Modeling*; William Andrew: Norwich, NY, USA, 2015.
61. Thomsen, L. *Understanding Seismic Anisotropy in Exploration and Exploitation*; Distinguished Instructor Series; Society of Exploration Geophysicists: Houston, TX, USA, 2002; pp. 3–27.
62. Anwer, H.M.; Alves, T.M.; Ali, A. Effects of sand-shale anisotropy on amplitude variation with angle (AVA) modelling: The Sawan gas field, Pakistan, as a key case-study for South Asia's sedimentary Basins. *J. Asian Earth Sci.* **2017**, *147*, 516–531. [CrossRef]
63. Dewhurst, D.N.; Siggins, A.F.; Sarout, J.; Raven, M.D.; Nordgård-Bolås, H.M. Geomechanical and ultrasonic characterization of a Norwegian Sea shale. *Geophysics* **2011**, *76*, WA101–WA111. [CrossRef]
64. Nadri, D.; Sarout, J.; Bóna, A.; Dewhurst, D. Estimation of the anisotropy parameters of transversely isotropic shales with a tilted symmetry axis. *Geophys. J. Int.* **2012**, *190*, 1197–1203. [CrossRef]
65. Sarout, J.; Esteban, L.; Delle Piane, C.; Maney, B.; Dewhurst, D.N. Elastic anisotropy of Opalinus Clay under variable saturation and triaxial stress. *Geophys. J. Int.* **2014**, *198*, 1662–1682. [CrossRef]
66. Sarout, J.; Delle Piane, C.; Nadri, D.; Esteban, L.; Dewhurst, D.N. A robust experimental determination of Thomsen's δ parameter. *Geophysics* **2015**, *80*, A19–A24. [CrossRef]
67. Yan, F.; Han, D.H.; Chen, X.L. Practical and robust experimental determination of c13 and Thomsen parameter δ. *Geophys. Prospect.* **2018**, *66*, 354–365. [CrossRef]

68. Yan, F.; Han, D.H.; Sil, S.; Chen, X.L. Analysis of seismic anisotropy parameters for sedimentary strata. *Geophysics* **2016**, *81*, D495–D502. [CrossRef]
69. Backus, G.E. Long-wave elastic anisotropy produced by horizontal layering. *J. Geophys. Res.* **1962**, *67*, 4427–4440. [CrossRef]
70. Miller, D.E.; Spencer, C. An exact inversion for anisotropic moduli from phase slowness data. *J. Geophys. Res. Solid Earth* **1994**, *99*, 21651–21657. [CrossRef]
71. Miller, D.E.; Leaney, S.; Borland, W.H. An in situ estimation of anisotropic elastic moduli for a submarine shale. *J. Geophys. Res. Solid Earth* **1994**, *99*, 21659–21665. [CrossRef]
72. Alkhalifah, T.; Tsvankin, I. Velocity analysis for transversely isotropic media. *Geophysics* **1995**, *60*, 1550–1566. [CrossRef]
73. Li, Y. Anisotropic well logs and their applications in seismic analysis. In *SEG Technical Program Expanded Abstracts 2002*; Society of Exploration Geophysicists: Houston, TX, USA, 2002; pp. 2459–2462.
74. Li, Y. An empirical method for estimation of anisotropic parameters in clastic rocks. *Lead. Edge* **2006**, *25*, 706–711. [CrossRef]
75. Ehirim, C.; Chikezie, N. The effect of anisotropy on amplitude versus offset (AVO) synthetic modelling in Derby field southeastern Niger delta. *J. Pet. Explor. Prod. Technol.* **2017**, *7*, 667–672. [CrossRef]
76. Bandyopadhyay, K. *Seismic Anisotropy: Geological Causes and its Implications to Reservoir Geophysics*; Stanford University: Standford, CA, USA, 2009.
77. Hornby, B.E.; Schwartz, L.M.; Hudson, J.A. Anisotropic effective-medium modeling of the elastic properties of shales. *Geophysics* **1994**, *59*, 1570–1583. [CrossRef]
78. Sun, J.; Jiang, L.; Liu, X. Anisotropic effective-medium modeling of the elastic properties of shaly sandstones. In *SEG Technical Program Expanded Abstracts 2010*; Society of Exploration Geophysicists: Houston, TX, USA, 2010; pp. 212–216.
79. Liner, C.L.; Fei, T.W. Layer-induced seismic anisotropy from full-wave sonic logs: Theory, application, and validation. *Geophysics* **2006**, *71*, D183–D190. [CrossRef]
80. Berryman, J.G.; Grechka, V.Y.; Berge, P.A. Analysis of Thomsen parameters for finely layered VTI media. *Geophys. Prospect.* **1999**, *47*, 959–978. [CrossRef]
81. Haktorson, H. Estimation of Anisotropy Parameters and AVO modeling of the Troll field, North Sea. Master's Thesis, Institutt for petroleumsteknologi og anvendt geofysikk, Trondheim , Norway, 2012.
82. Xu, S.; Stovas, A.; Alkhalifah, T. Estimation of the anisotropy parameters from imaging moveout of diving wave in a factorized anisotropic medium. *Geophysics* **2016**, *81*, C139–C150. [CrossRef]
83. Mesdag, P. A new approach to quantitative azimuthal inversion for stress and fracture detection. In *SEG Technical Program Expanded Abstracts 2016*; Society of Exploration Geophysicists: Houston, TX, USA, 2016; pp. 357–361.
84. Guo, Q.; Zhang, H.; Han, F.; Shang, Z. Prestack seismic inversion based on anisotropic Markov random field. *IEEE Trans. Geosci. Remote Sens.* **2017**, *56*, 1069–1079. [CrossRef]
85. Benedek, C.; Shadaydeh, M.; Kato, Z.; Szirányi, T.; Zerubia, J. Multilayer Markov random field models for change detection in optical remote sensing images. *ISPRS J. Photogramm. Remote Sens.* **2015**, *107*, 22–37. [CrossRef]
86. Huang, L.; Gao, K.; Huang, Y.; Cladouhos, T.T. Anisotropic seismic imaging and inversion for subsurface characterization at the Blue Mountain geothermal field in Nevada. In Proceedings of the 43rd Workshop on Geothermal Reservoir Engineering, Stanford, CA, USA, 12–14 February 2018; pp. 12–14.
87. Chi, B.; Dong, L.; Liu, Y. Correlation-based reflection full-waveform inversion. *Geophysics* **2015**, *80*, R189–R202. [CrossRef]
88. Fu, L.; Guo, B.; Schuster, G.T. Multiscale phase inversion of seismic data. *Geophysics* **2018**, *83*, R159–R171. [CrossRef]
89. Fu, L.; Symes, W.W. A discrepancy-based penalty method for extended waveform inversion. *Geophysics* **2017**, *82*, R287–R298. [CrossRef]
90. Wei, X.; Jiang, X.; Booth, D.; Liu, Y. The inversion of seismic velocity using a partial-offset stack with well-log constraints. *J. Geophys. Eng.* **2006**, *3*, 50–58. [CrossRef]
91. Gouda, M.; Salim, A.; Hamada, G. Estimation of Anisotropy-Free Acoustic Impedance from Partial-Stack Seismic Inversion: A Case Study From Inas field, Malay Basin. In Proceedings of the EAGE-GSM 2nd Asia Pacific Meeting on Near Surface Geoscience and Engineering, Kuala Lumpur, Malaysia, 24–25 April, 2019.
92. Hampson, D.P.; Russell, B.H.; Bankhead, B. Simultaneous inversion of pre-stack seismic data. In *SEG Technical Program Expanded Abstracts 2005*; Society of Exploration Geophysicists: Houston, TX, USA, 2005; pp. 1633–1637.
93. Fatti, J.L.; Smith, G.C.; Vail, P.J.; Strauss, P.J.; Levitt, P.R. Detection of gas in sandstone reservoirs using AVO analysis: A 3-D seismic case history using the Geostack technique. *Geophysics* **1994**, *59*, 1362–1376. [CrossRef]
94. Simmons Jr, J.L.; Backus, M.M. Waveform-based AVO inversion and AVO prediction-error. *Geophysics* **1996**, *61*, 1575–1588. [CrossRef]
95. Aki, K. Quantitative seismology. *Theory and Method*; Freeman: San Francisco, CA, USA, 1980; pp. 304–308.
96. Oldenburg, D.; Scheuer, T.; Levy, S. Recovery of the acoustic impedance from reflection seismograms. *Geophysics* **1983**, *48*, 1318–1337. [CrossRef]
97. Richards, P.G.; Aki, K. *Quantitative Seismology: Theory and Methods*; Freeman: Stuttgart, Germany 1980.
98. Gardner, G.; Gardner, L.; Gregory, A. Formation velocity and density—The diagnostic basics for stratigraphic traps. *Geophysics* **1974**, *39*, 770–780. [CrossRef]
99. Gao, R.; Sheng, Y. Law of large numbers for uncertain random variables with different chance distributions. *J. Intell. Fuzzy Syst.* **2016**, *31*, 1227–1234. [CrossRef]

100. Kumar, N. Deep Learning: Feedforward Neural Networks Explained. 2019. Available online: https://medium.com/hackernoon/deep-learning-feedforward-neural-networks-explained-c34ae3f084f1 (accessed on 4 July 2022)
101. Rosenblatt, F. The perceptron: A probabilistic model for information storage and organization in the brain. *Psychol. Rev.* **1958**, *65*, 386. [CrossRef] [PubMed]
102. Jain, A.K.; Mao, J.; Mohiuddin, K.M. Artificial Neural Networks: A tutorial. *Computer* **1996**, *29*, 31–44. [CrossRef]
103. Nikravesh, M.; Zadeh, L.A.; Aminzadeh, F. *Soft Computing and Intelligent Data Analysis in Oil Exploration*; Elsevier: Amsterdam, The Netherlands, 2003.
104. Zhang, X.; Liu, X.; Wang, X.; Band, S.S.; Bagherzadeh, S.A.; Taherifar, S.; Abdollahi, A.; Bahrami, M.; Karimipour, A.; Chau, K.W.; et al. Energetic Thermo-Physical Analysis of MLP-RBF Feed-Forward Neural Network Compared With RLS Fuzzy to Predict CuO/liquid Paraffin Mixture Properties. *Eng. Appl. Comput. Fluid Mech.* **2022**, *16*, 764–779. [CrossRef]
105. Rizk-Allah, R.M.; Hassanien, A.E. COVID-19 forecasting based on an improved interior search algorithm and multilayer feed-forward neural network. In *Medical Informatics and Bioimaging Using Artificial Intelligence*; Springer: Berlin/Heidelberg, Germany, 2022; pp. 129–152.
106. Werbos, P.J. *The Roots of Backpropagation: From Ordered Derivatives to Neural Networks and Political Forecasting*; John Wiley & Sons: Hoboken, NJ, USA, 1994; Volume 1.
107. Malvić, T. Clastic facies prediction using neural network (case study from Okoli field). *Nafta* **2006**, *57*, 415–431.
108. Robbins, H.; Monro, S. A stochastic approximation method. *Ann. Math. Stat.* **1951**, 400 –407. [CrossRef]
109. Kingma, D.P.; Ba, J. Adam: A method for stochastic optimization. *arXiv* **2014**, arXiv:1412.6980.
110. Moré, J.J. The Levenberg-Marquardt algorithm: Implementation and theory. In *Numerical Analysis*; Springer: Berlin/Heidelberg, Germany, 1978; pp. 105–116.
111. Jozanikohan, G.; Norouzi, G.H.; Sahabi, F.; Memarian, H.; Moshiri, B. The application of multilayer perceptron neural network in volume of clay estimation: Case study of Shurijeh gas reservoir, Northeastern Iran. *J. Nat. Gas Sci. Eng.* **2015**, *22*, 119–131. [CrossRef]
112. Yi-de, M.; Qing, L.; Zhi-Bai, Q. Automated image segmentation using improved PCNN model based on cross-entropy. In Proceedings of 2004 International Symposium on Intelligent Multimedia, Video and Speech Processing, Hong Kong, China, 20–22 October 2004; pp. 743–746.
113. Pihur, V.; Datta, S.; Datta, S. Weighted rank aggregation of cluster validation measures: A monte carlo cross-entropy approach. *Bioinformatics* **2007**, *23*, 1607–1615. [CrossRef]
114. Sudre, C.H.; Li, W.; Vercauteren, T.; Ourselin, S.; Jorge Cardoso, M. Generalised dice overlap as a deep learning loss function for highly unbalanced segmentations. In *Deep Learning in Medical Image Analysis and Multimodal Learning for Clinical Decision Support*; Springer: Berlin/Heidelberg, Germany, 2017; pp. 240–248.
115. James, W.; Stein, C. Estimation with quadratic loss. In *Breakthroughs in Statistics*; Springer: Berlin/Heidelberg, Germany, 1992; pp. 443–460.
116. Veeken, P.C.; Rauch-Davies, M. AVO attribute analysis and seismic reservoir characterization. *First Break* **2006**, *24*, 41–52. [CrossRef]
117. Goodway, B.; Chen, T.; Downton, J. Improved AVO fluid detection and lithology discrimination using Lamé petrophysical parameters;"$\lambda\rho$","$\mu\rho$", & "λ/μ fluid stack", from P and S inversions. In *SEG Technical Program Expanded Abstracts 1997*; Society of Exploration Geophysicists: Houston, TX, USA 1997; pp. 183–186.
118. Kleinbaum, D.G.; Klein, M. Analysis of matched data using logistic regression. In *Logistic Regression: A Self-Learning Text*; Springer: Berlin/Heidelberg, Germany, 2002; pp. 227–265.
119. Stuart, A. Kendall's advanced theory of statistics. *Distrib. Theory* **1994**, *1*, 128.
120. Quinlan, J.R. Learning decision tree classifiers. *ACM Comput. Surv. (CSUR)* **1996**, *28*, 71–72. [CrossRef]
121. Breiman, L.; Friedman, J.; Olshen, R.; Stone, C. *Classification and Regression Trees*; Wadsworth International Group: Belmont, CA, USA, 1984; Volume 37, pp. 237–251.
122. Venkatasubramaniam, A.; Wolfson, J.; Mitchell, N.; Barnes, T.; JaKa, M.; French, S. Decision trees in epidemiological research. *Emerg. Themes Epidemiol.* **2017**, *14*, 11. [CrossRef] [PubMed]
123. Strobl, C.; Malley, J.; Tutz, G. An introduction to recursive partitioning: Rationale, application, and characteristics of classification and regression trees, bagging, and random forests. *Psychol. Methods* **2009**, *14*, 323. [CrossRef]
124. Merkle, E.C.; Shaffer, V.A. Binary recursive partitioning: Background, methods, and application to psychology. *Br. J. Math. Stat. Psychol.* **2011**, *64*, 161–181. [CrossRef]
125. Therneau, T.; Atkinson, B.; Ripley, B.; Ripley, M.B. *Package 'rpart'*. The Comprehensive R Archive Network (CRAN). 2019. Available online: https://cran.microsoft.com/snapshot/2020-04-20/web/packages/rpart/rpart.pdf (accessed on 7 June 2022).
126. Almuallim, H. An efficient algorithm for optimal pruning of decision trees. *Artif. Intell.* **1996**, *83*, 347–362. [CrossRef]
127. Sondergeld, C.H.; Rai, C.S. Elastic anisotropy of shales. *Lead. Edge* **2011**, *30*, 324–331. [CrossRef]
128. Berryman, J.G. Exact seismic velocities for transversely isotropic media and extended Thomsen formulas for stronger anisotropies. *Geophysics* **2008**, *73*, D1–D10. [CrossRef]
129. Vernik, L.; Fisher, D.; Bahret, S. Estimation of net-to-gross from P and S impedance in deepwater turbidites. *Lead. Edge* **2002**, *21*, 380–387. [CrossRef]

130. Ma, X.Q. Global joint inversion for the estimation of acoustic and shear impedances from AVO derived P-and S-wave reflectivity data. *First Break* **2001**, *19*, 557–566.
131. Wrona, T.; Pan, I.; Gawthorpe, R.L.; Fossen, H. Seismic facies analysis using machine learning. *Geophysics* **2018**, *83*, O83–O95. [CrossRef]

Article

Application of an Automatic Noise or Signal Removal Algorithm Based on Synchrosqueezed Continuous Wavelet Transform of Passive Surface Wave Imaging: A Case Study in Sichuan, China

Jie Fang, Guofeng Liu * and Yu Liu

School of Geophysics and Information Technology, China University of Geosciences, Beijing 100083, China; fangjie@cugb.edu.cn (J.F.); 3010200013@cugb.edu.cn (Y.L.)
* Correspondence: liugf@cugb.edu.cn

Abstract: Passive surface wave imaging based on noise cross-correlation has been a research hotspot in recent years. However, because randomness of noise is difficult to achieve in reality, prominent noise sources will inevitably affect the dispersion measurement. Additionally, in order to recover high-fidelity surface waves, the time series input during cross-correlation calculation is usually very long, which greatly limits the efficiency of passive surface wave imaging. With an automatic noise or signal removal algorithm based on synchrosqueezed continuous wavelet transform (SS-CWT), these problems can be alleviated. We applied this method to 1-h passive datasets acquired in Sichuan province, China; separated the prominent noise events in the raw field data, and enhanced the cross-correlation reconstructed surface waves, effectively improving the accuracy of the dispersion measurement. Then, using the conventional surface wave inversion method, the shear wave velocity profile of the underground structure in this area was obtained.

Keywords: passive seismic; noise separation; surface wave; dispersion curve

Citation: Fang, J.; Liu, G.; Liu, Y. Application of an Automatic Noise or Signal Removal Algorithm Based on Synchrosqueezed Continuous Wavelet Transform of Passive Surface Wave Imaging: A Case Study in Sichuan, China. *Appl. Sci.* **2021**, *11*, 11718. https://doi.org/10.3390/app112411718

Academic Editor: Amerigo Capria

Received: 6 October 2021
Accepted: 7 December 2021
Published: 9 December 2021

1. Introduction

Passive seismic methods do not require the excitation of explosives, vibroseis, etc.; they simply require the placement of node geophones according to the active source acquisition line and point spacing, and the data can be collected by continuously receiving ambient noises. Then, by using seismic interferometry (SI), the responses of virtual sources can be retrieved through the correlated responses on the two receivers [1–4]. The term "interferometry" generally refers to the study of interference between signals in order to obtain information through the differences between them. In the field of SI, it is used to study the interference of seismic-related signals. Its basic operation steps are very simple, mainly divided into two steps: the first step is cross-correlation calculation, which can be understood as detecting the travel time difference of waves recorded between two receivers; the second step is stacking—that is, integrating all actual sources.

Although the theory of SI is not limited to either body or surface waves [5,6], extracting surface waves from the cross-correlation of ambient noise is quite robust and has become a mature technology [7]. In contrast, it is much more difficult to extract body waves [8,9]. However, some studies have also shown its possibility at various scales [10–13]. Even so, in most SI studies, retrieving surface waves is still one of the major applications. Surface waves with different periods can be used to investigate the earth structure at different depths [14]. Generally speaking, most studies focus on frequencies lower than 1 Hz to provide images of the structure of the crust and upper mantle [15,16]. However, some studies have proven that structures tens to hundreds of meters underground can be imaged with frequencies higher than 1 Hz, such as by using traffic noise to detect urban underground space [17,18], which has attracted the attention of more and more engineering seismologists.

Green's function reconstruction of diffuse wavefield (i.e., waves arrive from all angles with equal strength) observations in an open configuration is the basis for SI with passive data [19]. However, it is not realistic to obtain a pure diffuse wavefield from observations. Non-ideal source distributions such as limited azimuthal distribution or near-source effects [20] can significantly affect the reliability and interpretation of observations due to the violation of the stationarity assumptions required by passive survey methods [21]. In order to improve the accuracy of the dispersion measurement, some studies have begun to consider how to attenuate the influence of noise sources with prominent directivity. Aki [5] proposed a classical spatial autocorrelation method which used the mathematical transformation of a symmetric receiving array (e.g., circle) to eliminate the azimuth factor of the noise source. Park et al. [22] introduced a novel strategy for imaging dispersion of passive surface waves with an active scheme based on phase-shift measurement, called roadside passive multichannel analysis of surface waves. Scanning processes were implemented along potential azimuthal directions of noise. Feuvre et al. [23] proposed a cross-correlation-based beamforming average dispersion analysis method which effectively restrained the directionality of noise and reduced the aliasing effect. Cheng et al. [24] proposed a multichannel analysis of the passive surface (MAPS) wave method based on long noise sequence cross-correlation to handle the azimuthal effects for directional noise sources. Distinct from the above methods, Mousavi et al. [25] proposed a new and fast algorithm for accurate noise removal/signal removal based on higher-order statistics (HOS), general cross-validation (GCV), and wavelet hard thresholding (WHT) in synchrosqueezed domains and tested the performance of the proposed algorithm with synthetic and real seismic data. In addition, he indicated that it can also be an effective procedure in ambient noise studies.

In this paper, we applied the noise or signal removal algorithm based on SS-CWT proposed by Mousavi et al. [25] to a 1-h passive seismic dataset acquired in Sichuan, China, and verified the effectiveness of this method in enhancing passive surface waves from short time series. First, we used this method to preprocess the 1-hour passive seismic data in Sichuan to remove the prominent noise events and used cross-correlation to produce virtual shot gathers from the preprocessed noise records. Then, after cross-correlation, the method was used as a post-processing step to denoise the virtual shot gathers, improving the signal-to-noise ratio (S/N) of the virtual shot gathers and further improving the reconstructed surface waves. Next, after preprocessing and post-processing, virtual shot gathers were used for dispersion analysis with an active scheme based on the phase-shift measurement, and inversion was conducted to obtain the underground shear wave velocity profile in this area. Finally, the applicability and shortcomings of this method are discussed.

2. Methodology

The methodology used in this paper was proposed by Mousavi et al. [25]. We summarized and sorted out their methods, mainly including the following three processing steps.

2.1. Preprocessing

For the passive seismic data y collected in the field, first, we carried out CWT and calculated its coefficient W_y, which represents the finite energy of the data in a concentrated time–frequency picture. CWT is a multiresolution transform method given by [26], as shown in Equation (1):

$$W_y(a,\tau) = \langle y, \psi_{a,\tau}\rangle = \int_{-\infty}^{+\infty} y(t) a^{-\frac{1}{2}} \psi^*\left(\frac{t-\tau}{a}\right) dt, \tag{1}$$

where a and τ denote the scale and time shift, respectively; $*$ is the complex conjugate; y, ψ denotes the inner product; and ψ is the mother wavelet.

Then, we used HOS and the kurtosis criterion to detect the scale of and remove the pure Gaussian noise correlation coefficient from the time–frequency analysis (TFR), leaving the combination of noise and signal. The *Kurt* of N observation coefficients W_y can be calculated by Equation (2):

$$Kurt_y = \frac{\sum_{n=1}^{N}\left(W_{y_n} - \mu_{W_y}\right)^4}{N\sigma_{W_y}^4} - 3, \tag{2}$$

where σ_{W_y} and μ_{W_y} are the estimated standard deviation and mean value of wavelet coefficient W_y, respectively, and N denotes the sample sequence. HOS measure the degree of Gaussianity, so they have been used as detectors of non-Gaussian signals in Gaussian noise [27]. Equation (3) defines the HOS criterion, which is retained for the Gaussianity measure for distinguishing Gaussian distributions from non-Gaussian distributions:

$$\left|Kurt_y\right| \leq \frac{\sqrt{24/N}}{\sqrt{1-\alpha}}, \tag{3}$$

where α is the level of confidence. Ravier et al. [28] estimated that the optimal value of α is 90%.

2.2. Thresholding

After obtaining the processed coefficient W_y through the preprocessing steps, the synchrosqueezed transform was performed according to Equation (4) to obtain the synchrosqueezed-CWT (SS-CWT) coefficients T_y.

$$T_y(\omega_\ell, \tau) = \Delta\omega^{-1} \sum_{a_k \mid \omega(a_k,\tau) - \omega_\ell \mid \leq \frac{\Delta\omega}{2}} W_y(a_k, \tau) a_k^{-\frac{3}{2}} \Delta a_k, \tag{4}$$

where ω is the discrete frequency and ω_ℓ represents the ℓth discrete frequency, a_k represents the kth scale, τ is the time shift, and Δ represents the increment symbol, where $\Delta \mathrm{a} = a_k - a_{k-1}$.

Then, we used the hard threshold scheme (Equation (5)) proposed by Donoho et al. [29,30] to provide a threshold for the SS-CWT coefficients T_y.

$$\eta_\lambda^h(W_y, \lambda) = \begin{cases} W_y & , \left|W_y\right| > \lambda \\ 0 & , \left|W_y\right| < \lambda \end{cases}, \tag{5}$$

where λ is the selected appropriate threshold; η is the selected threshold rule, which can be divided into a hard threshold and a soft threshold; η_λ^h represents the hard threshold rule; and W_y represents the wavelet coefficients of observation y.

The optimal threshold λ was automatically determined by the GCV method, which was proposed by Nason et al. [31]. The GCV function is defined as follows:

$$\mathrm{GCV}(\lambda) = \frac{1}{N} \frac{\|T_y - \widetilde{T}_\lambda\|^2}{\|\frac{N_0}{N}\|^2}, \tag{6}$$

where $\widetilde{T_\lambda}$ is the threshold coefficient using the threshold λ, and N_0 is the number of coefficients using the threshold λ to return to zero. This function simulated the error between the estimated signal and the real signal, so its minimum value was used to select an optimal threshold. By selecting thresholds for the main components of the signal, the initial estimation of the signal was obtained using Equation (7):

$$y_k(t) = 2C_\psi^{-1} Re\left(\sum_{l \in L_k(t)} T_y(\omega_\ell, t)\right), \tag{7}$$

where $L_{k(t)}$ denotes a small frequency band around the ridge of the kth component in the SS-CWT, the symbol Re represents the real part, and the constant C_ψ is given by Thakur et al. [32]:

$$C_\psi = \int_0^\infty \xi^{-1}\hat{\psi}^*(\xi)d\xi, \quad (8)$$

where $*$ represents the complex conjugate; ψ is the mother wavelet, which is a square integrateable function; ξ is the angular frequency; and $\hat{\psi}$ is the Fourier transform of ψ.

2.3. Postprocessing

In the third step, a simple level-dependent wavelet threshold was applied to the signal obtained in the thresholding steps; the initial estimation of the seismic signal was subjected to CWT, shown in Equation (1); and thresholds of the coefficients of all scales were selected step by step using the hard threshold rule again, as shown in Equation (5). However, the threshold λ estimation used here was given by Donoho et al. [29], which was calculated using Equation (9):

$$\lambda = \sigma_n\sqrt{2\ln N}, \quad (9)$$

where $\sigma_n = median(|W_y|)/0.6745$. Finally, the signal in the original data was extracted by applying an inverse CWT, as shown in Equation (10):

$$y(t) = \frac{1}{C_\psi}\iint W_y(a,\tau)d\tau\frac{da}{a^2}. \quad (10)$$

For the passive seismic raw records, the extracted signal $y(t)$ represented the records that contained unnecessary prominent noise events, and the uniform background noise could be extracted by subtracting $y(t)$ from the original data y.

3. The Field Data

The passive seismic data acquisition area is located in a mining area in Sichuan (as shown in the blue square in Figure 1a), where Triassic feldspar quartz sandstone, Jurassic marl-dolomite-feldspar quartz sandstone, Cretaceous purplish red mudstone-siltstone-sandstone, and Quaternary strata are exposed. Figure 1b shows a diagram of the geological structure of the mining area. The main structural direction of this area is north–northeast, followed by nearly east–west. Magmatic rocks in the mining area are mainly gabbro and diabase from the late Jinning period, which are exposed as rock masses and dikes of different sizes. The wall rock alteration is developed, among which the wall rock alteration related to mineralization mainly includes chloritization, silicification, and carbonization.

We arranged a two-dimensional passive seismic survey line, P18-P18″, at the periphery of the mining area (as shown by the red dotted line in Figure 1b). The passive seismic data were continuously recorded for 4 days with a 1-millisecond sample interval using 120 node-type single-component geophone I-Nodal at 1-240 Hz natural frequency. The distance between each geophone was 12.5 m, and the total length of the survey line was 1.5 km. In order to avoid surface wind noise, the geophones were buried at depths of about 20 cm.

The maximum elevation difference of the terrain in the test area was 100 m, and the slope was gentle. The surrounding areas were sparsely populated and underdeveloped, with one or two village-level roads. There were many rivers nearby, and there were mining activities about 4 km northwest. Comprehensive analysis showed that the noise in this area mainly came from rivers, traffic, and mining activities, and mining activities may be the main source of prominent noise events. Figure 2 shows the daily noise data recorded by a geophone.

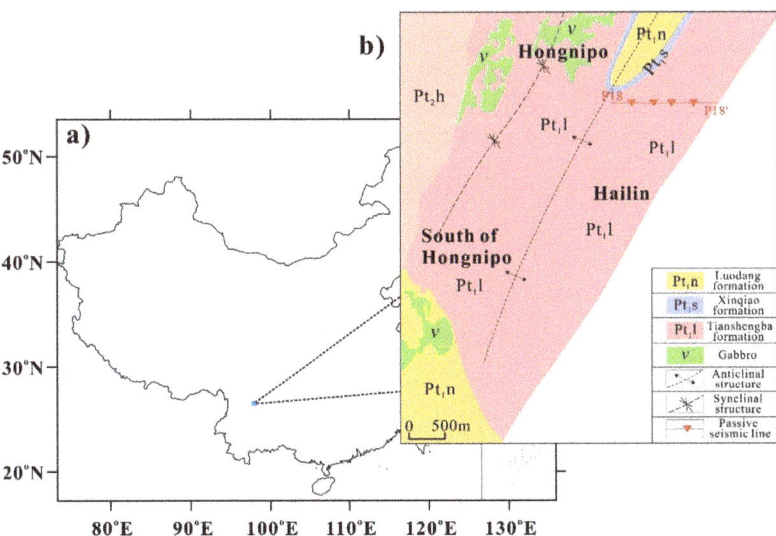

Figure 1. Location of the passive seismic survey line. (**a**) Location of mining area where data were collected (blue square); (**b**) geological structure of the mining area. The red dotted line P18-P18′ indicates the passive seismic survey line.

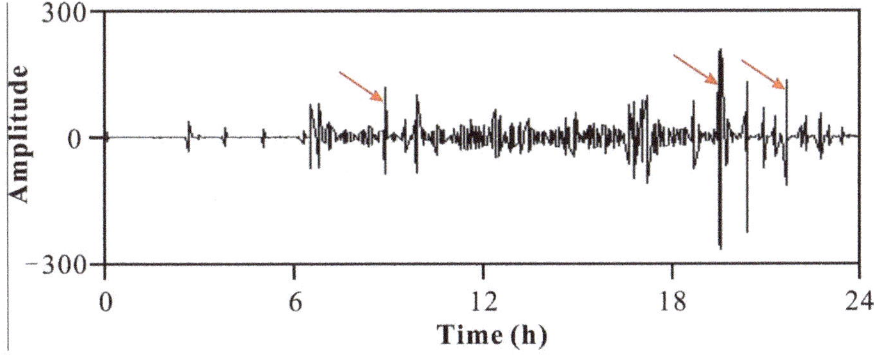

Figure 2. Typical ambient noise data recorded by a geophone in the survey area.

There was very little noise from night to morning (0–6 h), with relatively low amplitude and stable change, whereas during the day, due to activities such as surface mining, the noise amplitude was relatively high and changed drastically, including some prominent high-amplitude noise events (red arrows). Therefore, in order to satisfy the hypothesis of a theoretical wavefield as much as possible, the original field data were processed.

4. Field Data Processing

4.1. Preprocessing for Raw Field Data

Raw field data will inevitably be affected by prominent noise events (e.g., mechanical construction), which makes it impossible to satisfy the assumption of a diffuse wavefield. Therefore, in order to reconstruct passive surface waves better, we first used the noise or signal removal algorithm based on SS-CWT proposed by Mousavi et al. [25] to preprocess the field data and separate the prominent noise events. Figure 3a shows an original field record of 90 s on which several relatively prominent noise events (red arrows) are distributed. After noise and signal separation, the amplitude of the separated noise records was relatively evenly distributed (as shown in Figure 3b). Figure 3c shows the removed records containing prominent noise events (red arrows).

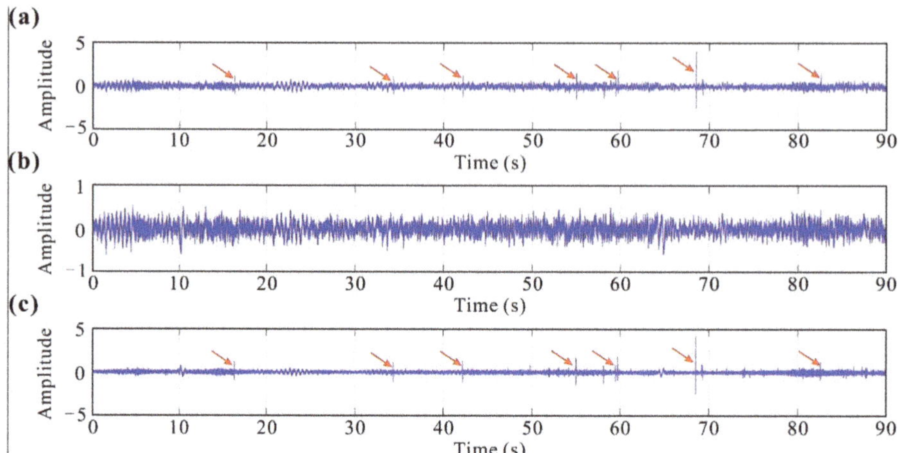

Figure 3. (**a**) Raw field data of 90 s. (**b**) Separated noise records. (**c**) Removed records. The red arrows indicate prominent noise events.

Power spectral density (PSD) plots are useful for visualizing variations in the frequency content of ambient data over time [33]. In order to further show the application effect of the noise separation algorithm in preprocessing the field data, we also calculated the PSD plots (as shown in Figure 4) corresponding to the three noise records in Figure 3.

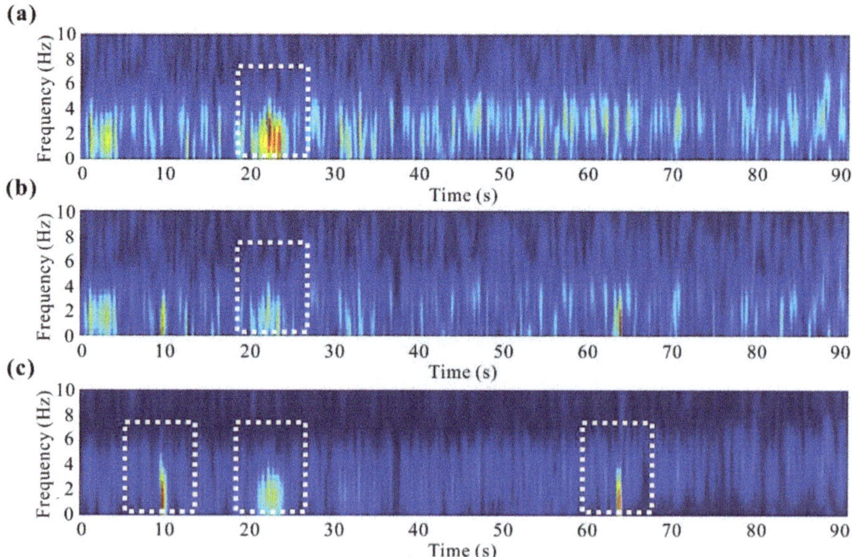

Figure 4. PSD plots. (**a**) Raw field data. (**b**) Separated noise records. (**c**) Removed records. The white dotted boxes indicate prominent noise events.

Figure 4a shows the PSD plots of the original field data, on which a prominent high-energy event (white dotted box) can be seen. After noise separation, as shown in Figure 4b, the high-energy event in the white dotted box was attenuated, and the PSD of the whole record was evenly distributed. Figure 4c shows the PSD of the removed noise record, which was occupied by three prominent high-energy events (white dashed boxes). According to the results of Figures 3 and 4, the noise or signal removal algorithm based on SS-CWT could effectively remove the prominent noise events in the original passive seismic data.

4.2. Reconstruction of Virtual Shot Gathers by Cross-Correlation

After preprocessing the raw field data, one geophone served as a virtual source for waves recorded by other receivers when using the cross-correlation calculation; thus, we could obtain all virtual shot gathers from every receiver without using an active source. In this paper, we showed the virtual shot gathered with the first receiver as the virtual source. In addition, in order to better characterize the intensity of the surface waves on the reconstructed virtual shot gathers, we used the dispersion curve image obtained by the phase-shift measurement method as the visualization tool.

Figure 5a shows the virtual shot gather formed by the cross-correlation of raw passive seismic data acquired in Sichuan for a certain hour. The surface waves (indicated by red arrows) can be seen in the red box on the virtual shot gather but are concealed by the energy of a strong transverse axis. The corresponding dispersion curve image is shown in Figure 5b, and the dispersion curve (red part) is disordered and discontinuous.

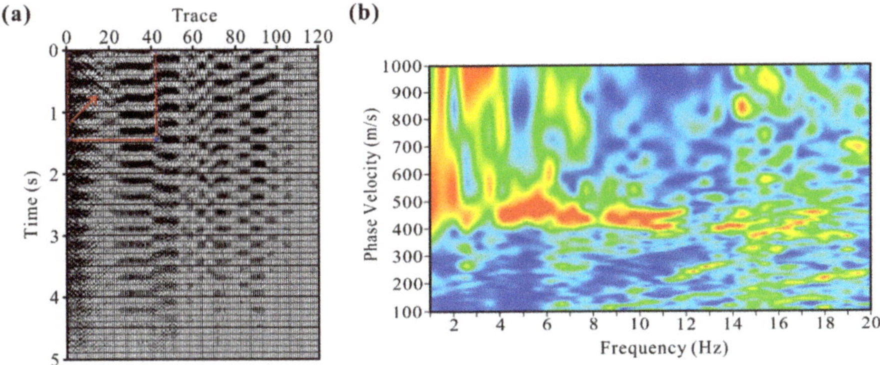

Figure 5. (**a**) Virtual shot gather formed by cross-correlation of raw one-hour field data and (**b**) the corresponding dispersion curve image.

After preprocessing the raw field data, the virtual shot gather formed by the separated noise records is shown in Figure 6a. Compared with Figure 5a, the in-phase axis (indicated by the red arrow) of the internal surface waves in the red box for the processed virtual shot gather is clearer, but it is still concealed by the energy of the strong horizontal axis. The corresponding dispersion curve image is shown in Figure 6b, where the curve (red part) is clearer and more continuous than in Figure 5b.

By comparing the virtual shot gathers and their corresponding dispersion curve images in Figures 5 and 6, we saw that the virtual shot gather formed by the separated noises had clearer surface wave in-phase axes and dispersion curves, which proved that the noise or signal removal algorithm based on SS-CWT could effectively enhance the reconstructed surface waves after preprocessing the raw field data.

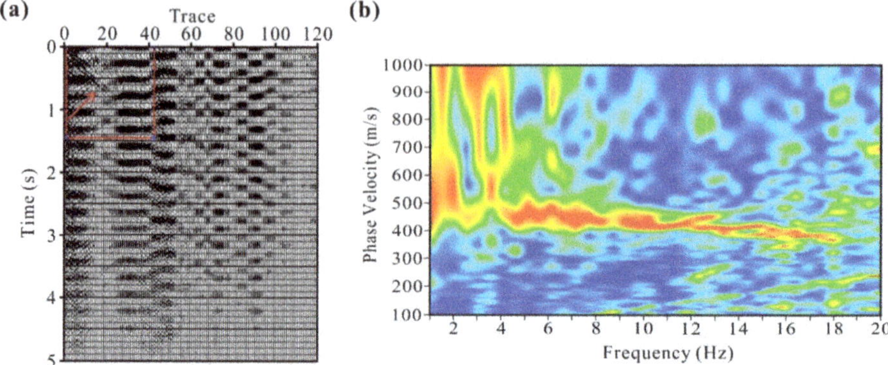

Figure 6. (**a**) Virtual shot gather formed by cross-correlation of preprocessed separated noises and (**b**) the corresponding dispersion curve image.

4.3. Post-Processing for Virtual Shot Gathers

Although preprocessing the raw field data before cross-correlation could effectively enhance the surface waves, the surface waves on the virtual shot gather were still concealed by the strong transverse axis energy. Determining how to eliminate the strong transverse axis energy, improve the S/N of the virtual shot gather, and further enhance the surface waves is of great significance for subsequent inversion and imaging.

Baig et al. [34] proposed several methods to improve the fidelity of noise cross-correlation based on the discrete orthogonal S transform, which showed that time–frequency denoising of correlograms (i.e., virtual shot gathers) can alleviate this problem. The noise or signal removal algorithm based on SS-CWT proposed by Mousavi et al. [25] can also be applied to the virtual shot gather after cross-correlation to denoise correlograms and improve the S/N. Figure 7 shows the virtual shot gather (Figure 7a) and its dispersion curve image (Figure 7b) after denoising the correlogram of Figure 5a using the noise or signal removal algorithm based on SS-CWT. Comparing Figure 7a with Figure 5a, the strong transverse axis energy on the correlogram was effectively eliminated, and the surface waves were highlighted. The dispersion curve in Figure 7b (red part) is more concentrated and continuous than that in Figure 5b.

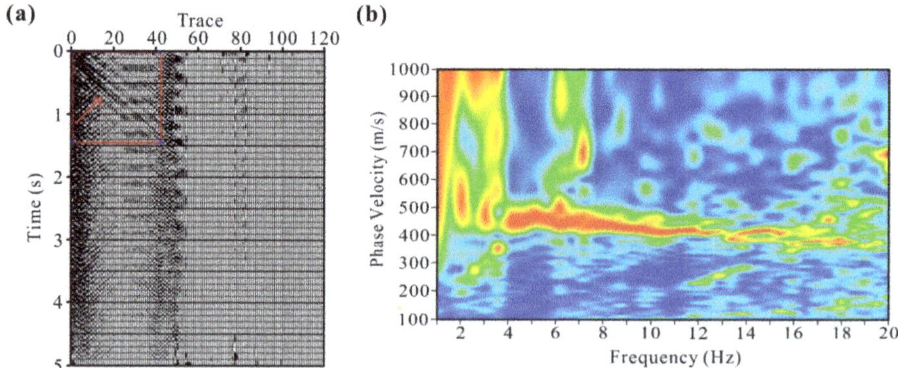

Figure 7. (**a**) Virtual shot gather after denoising the correlogram of Figure 5a and (**b**) the corresponding dispersion curve image.

The above research showed that the noise or signal removal algorithm based on SS-CWT could effectively enhance the surface waves after preprocessing the raw field data. At the same time, the method could also denoise the correlogram to further enhance the surface waves. Therefore, for the passive seismic datasets acquired in Sichuan, we first preprocessed the raw field data, then used the separated noise records to perform cross-correlation calculations to form the virtual shot gathers, and finally post-processed the correlograms, thus obtaining the final virtual shot gather, shown in Figure 8a, and its corresponding dispersion curve image (Figure 8b).

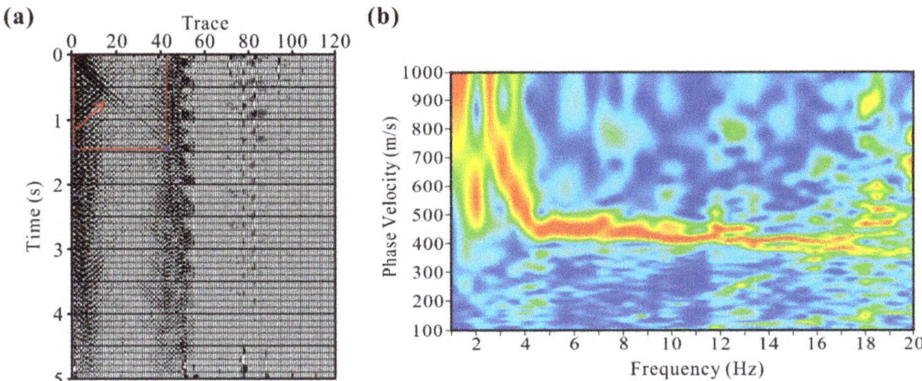

Figure 8. (**a**) Virtual shot gather after preprocessing and post-processing and (**b**) the corresponding dispersion curve image.

Comparing Figure 8 with Figures 5–7, we observed that the strong transverse axis energy in the virtual shot gather (Figure 8a) was eliminated, and the surface waves were richer and clearer; the information of the low-frequency part in the dispersion map shown in Figure 8b is prominent, and the dispersion curve is clear and continuous.

4.4. Shear Wave Velocity Profile

According to the pre- and post-processing steps mentioned above, we obtained virtual shot gathers with each receiver as the virtual shot point. We used phase-shift measurement for dispersion analysis and then used the genetic algorithm [35] to invert the underground shear wave velocity structure.

The genetic algorithm is an efficient, parallel, random global optimization search algorithm based on biological genetic mechanisms and natural selection. It searches for the optimal solution by giving a hypothetical solution and simulating the process of natural evolution. This method has been widely promoted and successfully applied in many fields, and it is also a commonly used Rayleigh wave dispersion curve inversion method. The initial model parameters (i.e., hypothetical solution) and the inversion parameters are shown in Table 1.

Table 1. Initial model parameters and the parameters used for inversion by genetic algorithm.

	Parameters Used to Constrain the Initial Model			Parameters Used for Inversion		
Layer Number	**Lower Limit of Shear Wave Velocity (m/s)**	**Upper Limit of Shear Wave Velocity (m/s)**	**Layer Thickness (m)**	**Vp/Vs**	**Density (kg/m^3)**	**Number of Iterations**
1	200	400	10			
2	400	500	20			
3	500	600	30	2	2	10,000
4	600	800	50			
5	800	1000	90			

In Figure 9, we show the inversion result of the dispersion curve in Figure 8. Figure 9a shows the initial picked dispersion curve (red dotted line) and the inverted dispersion curve (blue line), and Figure 9b shows the given upper and lower limits of the initial shear wave velocity model (red dotted line) and the inverted underground shear wave velocity structure (blue line). The high-frequency component of the surface wave mainly reflects the shallow underground information, while the low-frequency component mainly reflects the deep underground information. As there were few or no dispersion data recorded by the passive seismic data in the low-frequency range (1–2.5 Hz), the inversion results of the shallow part (0–200 m in Figure 9b) of the shear wave velocity structure are more accurate, while the inversion results of the deep part (200–300 m in Figure 9b) are not necessarily the same. Therefore, the depth range of the finally inverted shear wave velocity profile is 0–200 m underground.

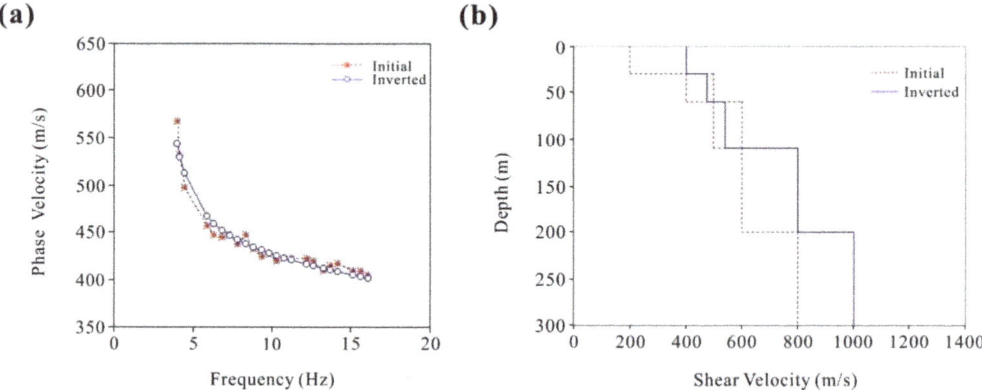

Figure 9. The inversion result of the dispersion curve in Figure 8. (**a**) The initial picked dispersion curve (red dotted line) and the inverted dispersion curve (blue line). (**b**) The given upper and lower limits of the initial shear wave velocity model (red dotted line) and inversed shear wave velocity structure (blue line).

We then selected the virtual shot gathers at 30 receivers, used phase-shift measurement for dispersion analysis, and obtained the inverse shear wave velocity structures of those 30 receivers. Finally, the shear wave velocity structure profile in this area was obtained by Kriging interpolation [36], as shown in Figure 10; the red inverted triangle indicates the position of the inverted shear wave velocity structure, and the section length of the final inversion is 1300 m. As the data quality of the following (1300–1500 m) is poor, it cannot avoid bringing errors to the whole inversion, so it was abandoned. Its underground structure is clear, and there is a fault, F, at approximately 900 m.

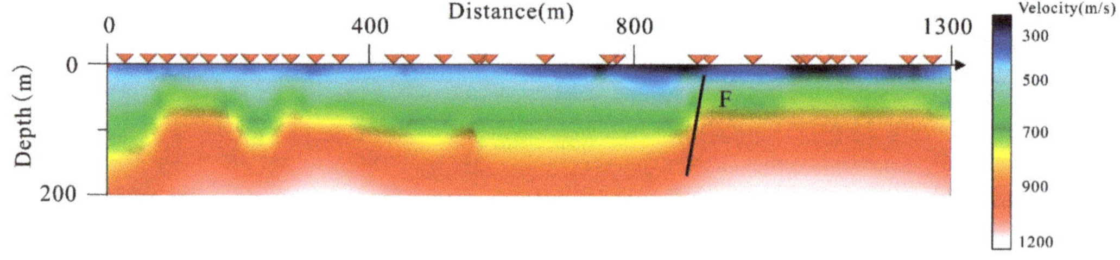

Figure 10. Shallow shear wave velocity profile from passive surface wave inversion.

5. Discussion

Although the noise or signal removal algorithm based on SS-CWT proposed by Mousavi et al. [25] has achieved outstanding results in this case study, there are still some problems that need further discussion. In this section, we highlight two issues based on our experience of processing the Sichuan passive seismic dataset: (1) the applicability of this method in other passive seismic datasets, and (2) the efficiency of this method in processing passive seismic datasets.

5.1. The Applicability of This Method in Other Passive Seismic Datasets

Due to differences in collection environment and conditions, there are often great differences between different passive seismic datasets. Therefore, in addition to testing the passive seismic dataset in Sichuan, we also attempted to apply this method to a passive seismic dataset acquired in Inner Mongolia [37] to enhance the surface waves but did not achieve very obvious results. As the survey line arranged in the research area of Inner Mongolia was located near a village, and it was the busy farming season, when farmers used agricultural machinery to harvest corn in the field, the human activities were complex and the interference was serious, so this method might not be able to completely eliminate these complex noises. However, the study area in this paper is far away from the village, with little human interference, and only the mining activities about 4 km northwest of the survey line are prominent. Therefore, although this method had a good effect in processing the Sichuan dataset, it may not have obvious effects on the datasets collected in different environments (especially in areas where human activities are prominent).

5.2. The Efficiency of This Method in Processing Passive Seismic Datasets

The amount of passive seismic data was relatively large. Taking the Sichuan dataset as an example, the amount of data collected by 120 geophones for 1 ms was approximately 40 GB a day. When we used this method to process the 1-h data selected by each receiver, it took more than 20 min to preprocess each trace, and it took 40 h to complete the whole 1-h dataset of 120 receivers; thus, the calculation efficiency was relatively low. Therefore, further improvement of this method for calculation efficiency is required before applying it to larger passive seismic datasets.

6. Conclusions

We introduced an accurate noise or signal removal algorithm based on SS-CWT. The performance of this algorithm was tested with actual passive seismic data acquired in a certain area of Sichuan, China. The results show that this algorithm can not only eliminate unnecessary prominent coherent noise events in the original passive seismic data but can also denoise the correlogram formed after cross-correlation calculation, improve the S/N of the restored Green function, and enable the reconstruction of a high-fidelity Green function from short time series. However, the computational efficiency of this method is low, and it may not be able to achieve good application results on datasets that are greatly disturbed by human activities.

Author Contributions: Methodology, J.F. and G.L.; validation, J.F., G.L. and Y.L.; writing—original draft preparation, J.F.; writing—review and editing, J.F. and Y.L.; supervision, G.L. All authors have read and agreed to the published version of the manuscript.

Funding: This work was supported by the National Natural Science Foundation of China (grant no. 42074131).

Institutional Review Board Statement: Not applicable.

Informed Consent Statement: Not applicable.

Data Availability Statement: The data presented in this study are available on request from the corresponding author.

Acknowledgments: We thank the editors and four anonymous reviewers for their careful reviews and constructive comments.

Conflicts of Interest: The authors declare no conflict of interest.

References

1. Wapenaar, K. Retrieving the elastodynamic Green's function of an arbitrary inhomogeneous medium by crosscorrelation. *Phys. Rev. Lett.* **2004**, *93*, 254–301. [CrossRef]
2. Schuster, G.T.; Yu, J.; Sheng, J.; Rickett, J. Interferometric/daylight seismic imaging. *Geophys. J. Int.* **2004**, *157*, 838–852. [CrossRef]
3. Weaver, R.L. Information from Seismic Noise. *Science* **2005**, *307*, 1568–1569. [CrossRef]
4. Snieder, R.; Şafak, E. Extracting the building response using seismic interferometry: Theory and application to the Millikan library in Pasadena, California. *Bull. Seismol. Soc. Am.* **2006**, *96*, 586–598. [CrossRef]
5. Aki, K. Space and time spectra of stationary stochastic waves, with special reference to microtremors. *Bull. Earthq. Res. Inst.* **1957**, *35*, 415–456.
6. Claerbout, J.F. Synthesis of a layered medium from its acoustic transmission response. *Geophysics* **1968**, *33*, 264–269. [CrossRef]
7. Nakata, N.; Snieder, R.; Tsuji, T.; Larner, K.; Matsuoka, T. Shear-wave imaging from traffic noise using seismic interferometry by cross-coherence. *Geophysics* **2011**, *76*, SA97–SA106. [CrossRef]
8. Roux, P.; Sabra, K.G.; Gerstoft, P.; Kuperman, W.A. P-waves from cross-correlation of seismic noise. *Geophys. Res. Lett.* **2005**, *32*, 1944–8007. [CrossRef]
9. Forghani, F.; Snieder, R. Underestimation of body waves and feasibility of surface-wave reconstruction by seismic interferometry. *Lead. Edge* **2010**, *29*, 790–794. [CrossRef]
10. Draganov, D.; Wapenaar, K.; Mulder, W.; Singer, J.; Verdel, A. Retrieval of reflections from seismic background-noise measurements. *Geophys. Res. Lett.* **2007**, *34*, L04305. [CrossRef]
11. Draganov, D.; Campman, X.; Thorbecke, J.; Verdel, A.; Wapenaar, K. Reflection images from ambient seismic noise. *Geophysics* **2009**, *74*, A63–A67. [CrossRef]
12. Nakata, N.; Boué, P.; Brenguier, F.; Roux, P.; Ferrazzini, V.; Campillo, M. Body and surface wave reconstruction from seismic noise correlations between arrays at Piton de la Fournaise volcano. *Geophys. Res. Lett.* **2016**, *43*, 1047–1054. [CrossRef]
13. Chamarczuk, M.; Malinowski, M.; Draganov, D. 2D body-wave seismic interferometry as a tool for reconnaissance studies and optimization of passive reflection seismic surveys in hardrock environments. *J. Appl. Geophys.* **2021**, *187*, 104288. [CrossRef]
14. Cheng, F.; Xia, J.H.; Xu, Y.X.; Xu, Z.B.; Pan, Y.D. A new passive seismic method based on seismic interferometry and multichannel analysis of surface waves. *J. Appl. Geophys.* **2015**, *117*, 126–135. [CrossRef]
15. Yang, Y.J.; Ritzwoller, M.H.; Lin, F.C.; Moschetti, M.P.; Shapiro, N.M. Structure of the crust and uppermost mantle beneath the western United States revealed by ambient noise and earthquake tomography. *J. Geophys. Res. Solid Earth* **2008**, *113*, B12310. [CrossRef]
16. Guo, Z.; Chen, Y.S.; Yin, W.W. Three-dimensional crustal model of Shanxi graben from 3D joint inversion of ambient noise surface wave and Bouguer gravity anomalies. *Chin. J. Geophys.* **2015**, *58*, 821–831. [CrossRef]
17. Behm, M.; Leahy, M.; Snieder, R. Retrieval of local surface wave velocities from traffic noise: An example from the La Barge basin (Wyoming). *Geophys. Prospect.* **2014**, *62*, 223–243. [CrossRef]
18. Cheng, F.; Xia, J.H.; Zhang, K.; Zhou, C.J.; Ajo-Franklin, J.B. Phase-weighted slant stacking for surface wave dispersionmeasurement. *Geophys. J. Int.* **2021**, *226*, 256–269. [CrossRef]
19. Wapenaar, K.; Draganov, D.; Robertsson, J.O.A. *Seismic Interferometry: History and Present Status*; Society of Exploration Geophysicists: Tulsa, OK, USA, 2008; pp. 99–101.
20. Roberts, J.; Asten, M. A study of near source effects in arraybased (SPAC) microtremor surveys. *Geophys. J. Int.* **2008**, *174*, 159–177. [CrossRef]
21. Claprood, M.; Asten, M.W. Statistical validity control on SPAC microtremor observations recorded with a restricted number of sensors. *Bull. Seismol. Soc. Am.* **2010**, *100*, 776–791. [CrossRef]
22. Park, C.B.; Miller, R.D.; Laflen, D.; Neb, C.; Ivanov, J.; Bennett, B.; Huggins, R. Imaging dispersion curves of passive surface waves. In Proceedings of the SEG 74th Annual International Meeting, Denver, CO, USA, 10–15 October 2004; Society of Exploration Geophysicists: Tulsa, OK, USA, 2004; pp. 1357–1360. [CrossRef]
23. Feuvre, M.; Joubert, A.; Leparoux, D.; Côte, P. Passive multi-channel analysis of surface waves with cross-correlations and beamforming: Application to a sea dike. *J. Appl. Geophys.* **2015**, *114*, 36–51. [CrossRef]
24. Cheng, F.; Xia, J.H.; Luo, Y.H.; Xu, Z.B.; Wang, L.M.; Shen, C.; Liu, R.F.; Pan, Y.D.; Mi, B.B.; Hu, Y. Multichannel analysis of passive surface waves based on crosscorrelations. *Geophysics* **2016**, *81*, EN57–EN66. [CrossRef]
25. Mousavi, S.M.; Langston, C.A. Automatic noise-removal/signal-removal based on general cross-validation thresholding in synchrosqueezed domain and its application on earthquake data. *Geophysics* **2017**, *82*, V211–V227. [CrossRef]
26. Mallat, S. *A Wavelet Tour of Signal Processing, in Wavelet Analysis and Its Applications*, 2nd ed.; Academic Press: Salt Lake City, UT, USA, 1999; p. 620.
27. Swami, A.; Giannakis, G.B. Higher-Order statistics. *Signal Process.* **1996**, *53*, 89–91. [CrossRef]
28. Ravier, P.; Amblard, P.O. Wavelet packets and de-noising based on higher-order-statistics for transient detection. *Signal Process.* **2001**, *81*, 1909–1926. [CrossRef]
29. Donoho, D.L.; Johnstone, I.M. Ideal spatial adaption by wavelet shrinkage. *Biometrika* **1994**, *81*, 425–455. [CrossRef]
30. Donoho, D.L.; Johnstone, I.M. Adapting to unknown smoothness via wavelet shrinkage. *J. Am. Stat. Assoc.* **1995**, *90*, 1200–1224. [CrossRef]
31. Nason, G.P. Wavelet shrinkage using cross-validation. *J. R. Stat. Soc. Ser. B-Stat. Methodol.* **1996**, *58*, 463–479. [CrossRef]

32. Thakur, G.; Brevdo, E.; Fučkar, N.S.; Wu, H.T. The Synchrosqueezing algorithm for time-varying spectral analysis: Robustness properties and new paleoclimate applications. *Signal Process.* **2013**, *93*, 1079–1094. [CrossRef]
33. Draganov, D.; Campman, X.; Thorbecke, J.; Verdel, A.; Wapenaar, K. Seismic exploration-scale velocities and structure from ambient seismic noise (>1 Hz). *J. Geophys. Res. Solid Earth* **2013**, *118*, 4345–4360. [CrossRef]
34. Baig, A.; Campillo, M.; Brenguier, F. Denoising seismic noise cross correlations. *J. Geophys. Res.* **2009**, *114*, B08310. [CrossRef]
35. John, H. Genetic Algorithms and the Optimal Allocation of Trials. *SIAM J. Comput.* **1973**, *2*, 88–105. [CrossRef]
36. Hansen, R.O. Interpretive gridding by anisotropic kriging. *Geophysics* **1993**, *58*, 1491–1497. [CrossRef]
37. Liu, G.F.; Liu, Y.; Meng, X.H.; Liu, L.B.; Su, W.J.; Wang, Y.Z.; Zhang, Z.F. Surface wave and body wave imaging of passive seismic exploration in shallow coverage area application of Inner Mongolia. *Chin. J. Geophys.* **2021**, *64*, 937–948. [CrossRef]

MDPI

Article

Three-Dimensional Processing of Reflections for Passive-Source Seismology Based on Geometric Design

Yu Liu and Guofeng Liu *

School of Geophysics and Information Technology, China University of Geosciences (Beijing), Beijing 100083, China; 3010200013@cugb.edu.cn
* Correspondence: liugf@cugb.edu.cn

Abstract: Passive-source exploration is a method of seismic exploration that has loose requirements on the conditions of the surface, is cheap, and does not require excitation by an active source. The ambient seismic signals collected from the field over an extended period of time can be used to generate virtual-shot seismic records similar to those obtained from the seismic exploration of an active source based on the relevant correlations, and this can in turn yield information on the underground structure through a series of conventional methods of processing seismic data. Three-dimensional (3D) processing can mitigate the influence of the azimuth of random noise to yield a more representative underground structure, but requires intensive computation. In this paper, we propose a 3D method to compute reflections of a passive source based on the geometry of seismic exploration. Assuming a high quality of imaging, we use information on the predesigned geometry to choose and correlate noisy synthetic data on the reflections by a seismic body to create virtual shot data, and subsequently capture images of the 3D data on passive reflection. The use of the predesigned geometry ensures that only the important and useful parts of the dataset are used for correlation and imaging, where this reduces the cost of computation. The proposed method can thus efficiently generate high-quality 3D synthetic data.

Keywords: correlation; direct migration method; generation of virtual shot records; geometric design; passive seismic exploration

Citation: Liu, Y.; Liu, G. Three-Dimensional Processing of Reflections for Passive-Source Seismology Based on Geometric Design. *Appl. Sci.* **2023**, *13*, 6126. https://doi.org/10.3390/app13106126

Academic Editor: José A. Peláez

Received: 18 March 2023
Revised: 15 May 2023
Accepted: 15 May 2023
Published: 17 May 2023

1. Introduction

Passive-source seismic exploration, which is based on seismic wave interferometry, was developed by Schuster [1] and Bakulin and Calvert [2]. This method is different from active-source seismic exploration in that it does not require excitation by an artificial source or knowledge of the number and locations of the sources. Correlations can be used to convert environmental seismic noise or micro-earthquakes into a determined seismic response. The theory of seismic interferometry involves "perform[ing] correlations on the seismic records received by receivers at two points on the surface, and the results can be regarded as the records of the wave field of the active source, with one point as the receiver and the other point as the source" [3,4]. Applying this calculation to all random signals of the receiver can yield a group of virtual shot records of the passive source that are similar to forward records of the active source in seismic exploration.

Methods of passive-source seismic exploration can be divided into techniques for investigating passive-source surface waves [5–7] and passive-source body waves [3,8–11], according to the types of seismic waves. Because the Earth absorbs body waves to a much greater extent than surface waves, it is much more difficult to capture images of the former than the latter [12]. Some researchers have retrieved reflections of waves from ambient noise by using illumination-based diagnosis [13,14], while others have sought to improve the quality of data on reflections of the passive source [15–19].

Methods of exploring reflections of the passive source have been applied in many areas. Masatoshi et al. [20] calculated the velocities of the P-wave and the S-wave, the bundle factor of the P-wave, and other structural parameters based on seismic noise in field data obtained from observation wells in Cold Lake, Alberta, Canada. Seismic wave interference technology was used to investigate the presence of metal ores by Cheraghi et al. [21]. Cheraghi et al. [22] applied passive source exploration to the four-dimensional dynamic monitoring of carbon dioxide storage sites at an Aquistore facility. Eric et al. [23] used data on underground activity in a mine and vibrations of the ventilation pipes in the context of exploring mining resources, and Brenguier et al. [24] used the noise in waves generated by a railway to monitor the shallow crust. Brenguier et al. [25] also used a receiver array to monitor long-term seismic velocity and anticipate geological events.

According to the theory of passive seismic exploration, sources of random noise are needed to evenly identify geological targets from all directions and obtain accurate kinematical results. Artifacts may be obtained when the source is not located along the direction of the receiver line [14]. Three-dimensional (3D) passive-source seismic exploration can be used to obtain more uniform information and images of better quality than two-dimensional (2D) passive seismic exploration. Draganov et al. [12] and Nakata et al. [26] obtained the underground velocity from environmental noise by using seismic wave interference technology and generated a corresponding 3D seismic profile. Chamarczuk et al. [27] used the full-scale 3D seismic method of the virtual source survey to explore the Kylylahti polymetallic mine in Finland and obtained geological information that could not be obtained through active seismic exploration.

While 3D passive seismic reflections have delivered promising results, they require intensive computations as well as a large storage space for the data obtained. Given these issues, we propose a method to calculate images of 3D reflections based on the geometric design of the area. We use information on the area, the surface, and the underground structure to design a reasonable geometry for it, subsequently use that to screen for effective data to obtain correlations, and finally generate an imaging profile with uniform folds and azimuths. The passive-source direct migration method is another means to improve computational efficiency. It was proposed by Artman [28] and has been applied to passive seismic exploration [29–32]. The passive-source direct migration method does not generate correlations to obtain virtual shot records, but directly performs migration imaging to obtain records of random noise and improve computational efficiency.

2. Extracting 3D Virtual Shot Records Based on Geometry

2.1. Principle of Reflection of Waves from Passive Sources

Interferometry is the basis of calculations for passive seismic exploration and can be derived from the equation of the acoustic wave according to the reciprocity theorem [31,32]:

$$2R\{\,G(x_{A,},x_{B,}\omega)\}\,S(\omega) = \frac{2}{v\rho}\,\langle u(x_A,\omega)u(x_B,\omega)\rangle \tag{1}$$

where $G(x_{A,},x_{B,}\omega)$ is Green's function of the observation points x_A and x_B in the frequency domain, ω is the angular frequency, $S(\omega)$ is a random source, and v and ρ are the velocity and the density of medium, respectively. $u(x_A,\omega)$ and $u(x_B,\omega)$ are random signals received by the receivers, $\langle * \rangle$ represents the superposition of the results of the mutual interferometry of different windows, and $R\{*\}$ is calculated by taking the real part.

Figure 1 shows that the above formula can be used to obtain virtual shot records similar to those due to excitation by an active source, with one point as the source and the other as the receiver, by performing interferometry on two points of the passive source. If point A does not move, and point B is transformed, we can obtain virtual shot records with point A as the source.

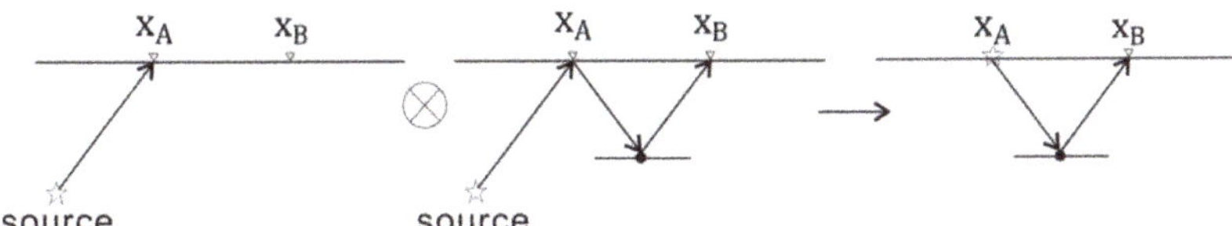

Figure 1. Schematic diagram of reflection interferometry.

We use a 2D forward model of a passive source to illustrate the process of extraction of the virtual shot records of the reflections by the passive source. We established a model of velocity as shown in Figure 2a, with a length of 3000 m, a depth of 800 m, and a grid spacing of dx = dz = 10 m. The lower, left, and right boundaries of the model were the boundaries of absorption, and we set 31 absorption layers. The upper boundary was a free boundary. The model contained three geological horizons with velocities of 1000, 2500, and 4000 m/s from top to bottom. A receiver was arranged after every 10 m on the surface, with a total of 301 receivers. There were 2000 sources of random noise (Figure 2b), and 60 s of random noise records were obtained through forward modeling (Figure 3). The source was made to continuously shake during acquisition, and its amplitudes and phases were completely random. Each of the channels of random noise was related to the other 300 channels. That is, one channel of the seismic record of a virtual source was chosen as the source, and the others were used as receivers (Figure 4a). The signal-to-noise ratios (SNRs) of the near-offset data in the red box in the figure were high. The profile of migration was obtained, as shown in Figure 4b, by calculating all 301 respective channels, using the conventional method of processing seismic data.

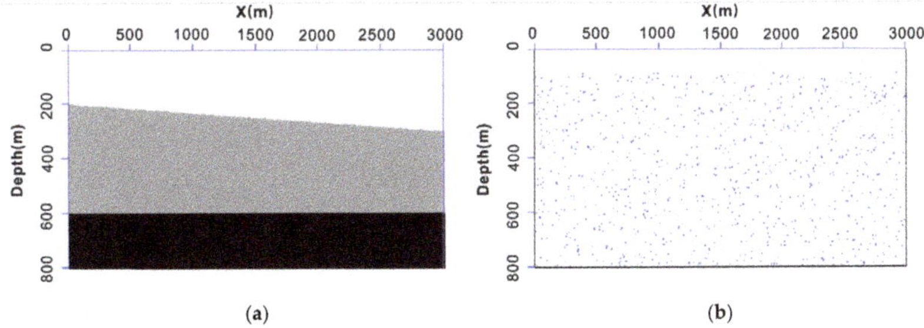

Figure 2. (**a**) Model of velocity. (**b**) Distribution of the sources of random noise.

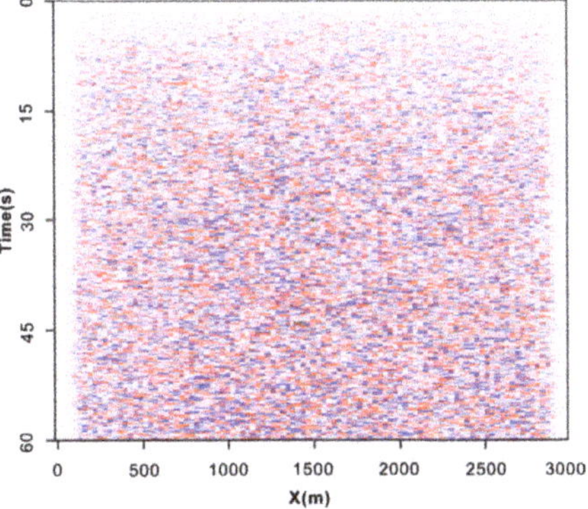

Figure 3. Random noise records lasting 60 s.

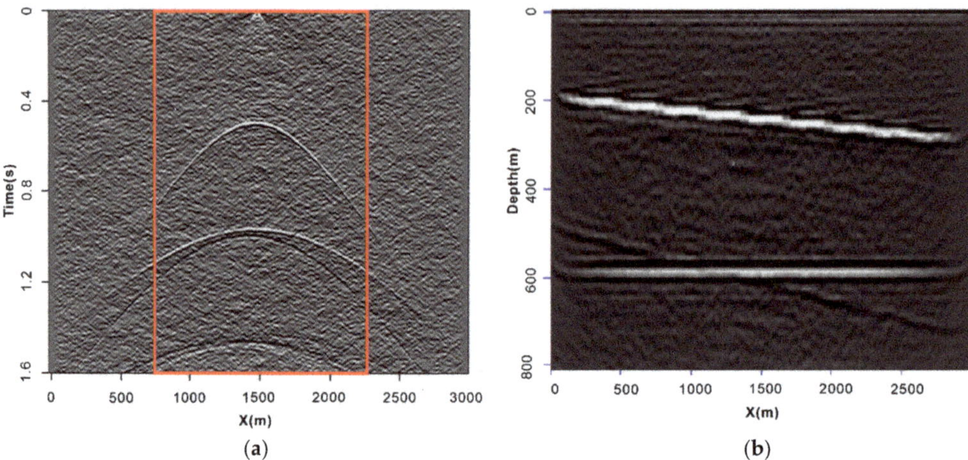

Figure 4. (**a**) Virtual shot records (the source point x = 1500 m). (**b**) Migration profile obtained by conventional processing.

2.2. Calculating 3D Virtual Shot Records of Passive Source Based on Geometry

The core of passive seismic reflection exploration is to obtain virtual shot records based on correlations between a channel of ambient noise records and other channels. When the length of the correlation is given, the ambient seismic records received over a long time need to be divided into several windows according to this length, and can then be correlated and stacked. To reconstruct a high-fidelity Green function, the time series of mutual correlation among the seismic data is usually very long. This makes it necessary to increase the number of folds to improve the quality of the Green function, which in turn reduces the computational efficiency of imaging of the passive source. Therefore, the number of virtual shot records, length of the ambient seismic records, length of the correlation window, and number of seismic channels involved in the calculation all influence the time needed to generate the virtual shot records.

Passive seismic exploration using reflections of waves from the source is based on correlations between all channels of random signals, and the generated virtual shot records are similar to those of an active source. They should also comply with the corresponding requirements of geometric design in theory. An appropriate geometry can not only improve the quality of imaging, but can also reduce the computational expense. Many seismic channels are involved in correlation and stacking in the calculation of virtual shot records, and the long time needed to record the ambient noise incurs a high computational intensity. The range of seismic channels needed to obtain correlations in 2D calculations is usually selected by setting the spacing and range of offset of points on the source. However, a rule is required in case of 3D calculations to ensure the quality of the generated virtual shot records. The data processing features uniform folds, wide azimuths, and other requirements that are similar to those for active seismic exploration. The geometric design has been extensively studied in the context of conventional seismic exploration, and exploration based on reflections from a passive source can directly learn from the experience accumulated by this research. We designed a method to compute the virtual shot records of reflections from a 3D passive source based on the design of the geometry for active-source seismic data to ensure a sufficient number of folds and quality of imaging while reducing the computational cost.

Before acquiring 3D data on the active source for seismic exploration, the geometry required to determine how to position the sources and receivers can be designed according to the known surface and the underground geological conditions of the given area. One can then specify the location of each source and receiver and the corresponding relationships between them. Following this, we must ensure that the collected data contain uniform folds, and then set the range of the offset distance and the azimuths. The geometric design usually considers the geological target elements and cost of acquisition. The file containing these geometric data is usually called the Shell Processing Support (SPS) Format for 3D Land Survey, and is a standard file for recording information on the source and the receiver as well as their relationship. The SPS file is mainly composed of a source-information file (S file, Table 1), a receiver-information file (R file, Table 2), and a source–receiver relationship file (X file, Table 3). The SPS file determines the unique physical location of the sources and receivers in the working area as well as their relationships. In the context of correlations, the SPS.S file stores information on the sources of all virtual shot records, and the SPS.X file contains the range of the receivers in the SPS.R file that need to be correlated with these sources.

Table 1. Key header words in the SPS.S (source) file.

Parameter	Line	Point	Easting	Northing	Elevation
Byte range	2–11	12–21	30–37	47–55	66–71
Number of bytes	10	10	8	8	6

Table 2. Key header words in the SPS.R (receiver) file.

Parameter	Line	Point	Easting	Northing	Elevation
Byte range	2–11	12–21	30–37	47–55	66–71
Number of bytes	10	10	8	8	6

Table 3. Key header words in the SPS.X (relation) file.

Parameter	Field File Number	Source Line	Source Point	From	To	Receiver Line	From	To
Byte range	8–15	18–27	28–37	39–43	44–48	50–59	60–69	70–79
Number of bytes	8	10	10	4	5	10	10	1

The difference between the geometry of the reflections from a passive source and that used in conventional seismic exploration should be considered to formulate the geometric design. Because each point representing a receiver in passive seismic exploration can become a point representing a source through correlation, the density of the points representing the source in passive seismic exploration may be much higher than that in active seismic exploration. On the premise of satisfying the requirements related to the number of folds and the SNR of the data, it is thus necessary to set an appropriate number of passive sources and receivers to significantly reduce the cost of processing.

Based on the above requirements, we propose a method to generate virtual shot records of 3D reflections from a passive source based on the above-mentioned geometry file, as shown in Figure 5. The procedure is as follows: (1) Having collected random noise records from the field, a geometry with a sufficient number of folds and uniform azimuths should be designed by using a design software according to the known surface, information on the underground structure, and geological targets, and should be stored in the SPS file. (2) The information is obtained on the source and receivers from the SPS.X file. The length of the correlation should be determined according to the conditions of the working area. After completion of the generation of virtual data for a receiver from its related shots, the initial data of this receiver are obtained for one correlation time. Then, we slide the time window by a correlation time, and repeat the above process until the end of the random noise records. By vertically stacking all of the correlation results, we can obtain the seismic record of this receiver. Following this, the remaining receivers corresponding to the source are calculated to obtain one shot of the virtual seismic records. (3) The entire SPS.X file is read to obtain all of the virtual shot records according to the previous step, and the imaging section is then obtained by processing seismic records of reflections of waves from the passive source according to the conventional method of processing.

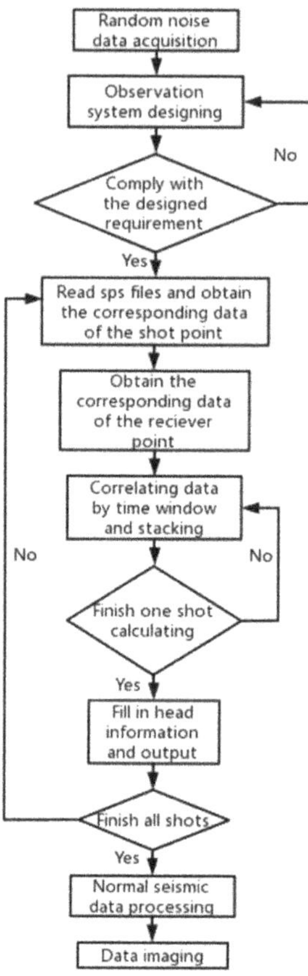

Figure 5. Correlation processing using the SPS file.

2.3. Model Test

We designed a forward model to verify the theory of 3D passive seismic exploration. The size of the model was 3000 m × 3000 m × 2100 m, and the grid spacing dx = dy = dz = 30 m. The model consisted of four geological horizons with velocities of 1000, 1600, 2500, and 4000 m/s from top to bottom, as shown in Figure 6a. As we sought to study reflections from the source in deep layers, we used the acoustic finite-difference algorithm to generate data on ambient noise, and randomly set 160,000 sources in the range of 200–1800 m from the model, as shown in Figure 6b. A survey line was arranged every 30 m on the surface, and 100 receivers were arranged for each survey line. A total of 10,000 receivers were thus set, that is, the distance between the receivers in both the x and the y directions was 30 m. The seismic waves were stimulated at randomly distributed sources, and a total of 200 s of seismic records were collected by the receivers on the surface (shown in Figure 7). During acquisition, the source was made to continuously shake at random amplitudes and phases. This was done to simulate the collection of noise records of reflections from a passive source by the surface receivers.

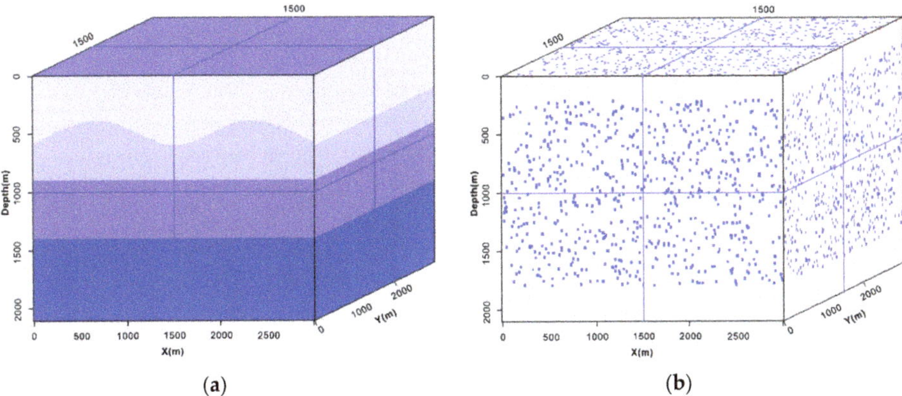

Figure 6. (**a**) A 3D model of velocity. (**b**) Distribution of sources of random noise in the model (blue dots).

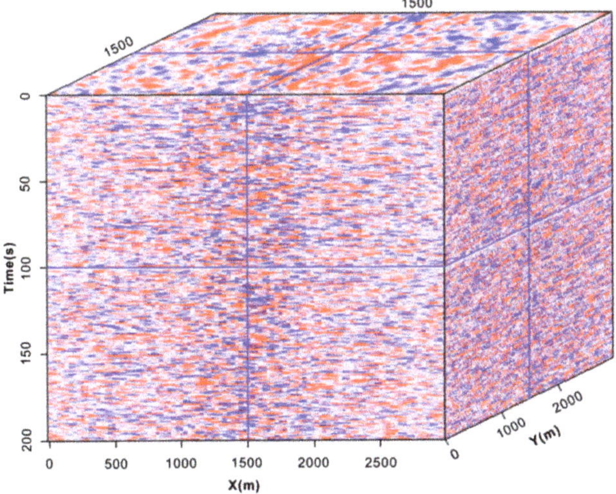

Figure 7. Random noise records (200 s).

2.3.1. Conventional Forward Model

All data from each receiver are successively used according to the time window as the source to obtain the correlations and stack them with data from the other receivers to generate a group of virtual shot records in the conventional method. The proposed geometry is shown in Figure 8. It is clear from the map of the folds that the maximum number of folds of the virtual shot records was 10,000, and gradually decreased from the center of the working area to the surrounding area. The 5050th source and receiver lines 50–52 of the seismic records were extracted. We can clearly see the non-phase axial morphology of the reflections and its conformance to the model as well as the low SNR of the far-offset part. Figure 9 shows the results of processing one shot of the virtual shot records. It yielded the imaging section that was stacked by using conventional seismic processing, with 40 s as the time window of correlation.

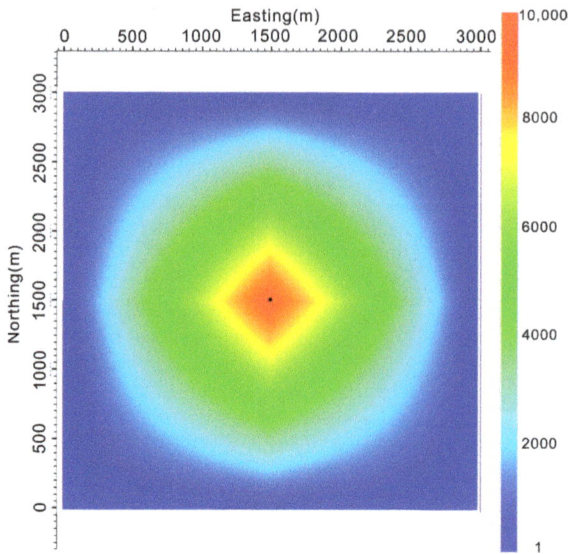

Figure 8. Folds of results of conventional correlation.

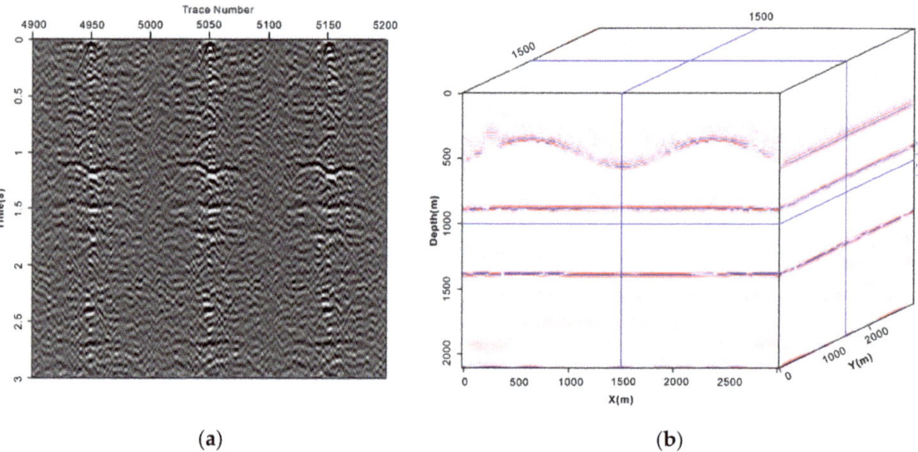

Figure 9. (**a**) Three shots of the conventional virtual shot records. (**b**) Imaging section of the conventional correlation.

2.3.2. Forward Model with a Narrow Azimuth

According to the proposed scheme and the method for active seismic exploration, we designed a geometry with a narrow azimuth: The maximum offset along the inline direction was 1500 m, and that along the crossline direction was 300 m, that is, each shot was recorded by 51 × 11 receivers at most, and the aspect ratio was 0.2. Figure 10a shows the source point and the range of its corresponding receiver point, and Figure 10b shows the folds of the geometry designed on the basis of this scenario. Figure 11a shows the records collected by using this geometry at the center of the model, Figure 11b shows the part of the shot records in the red box in Figure 11a, and Figure 12 shows the results of migration imaging and stacking, performed by calculating shot records by extracting one line of every four, with 40 s as the time window of correlation. The velocities required for migration and nmo were obtained from the model of velocity.

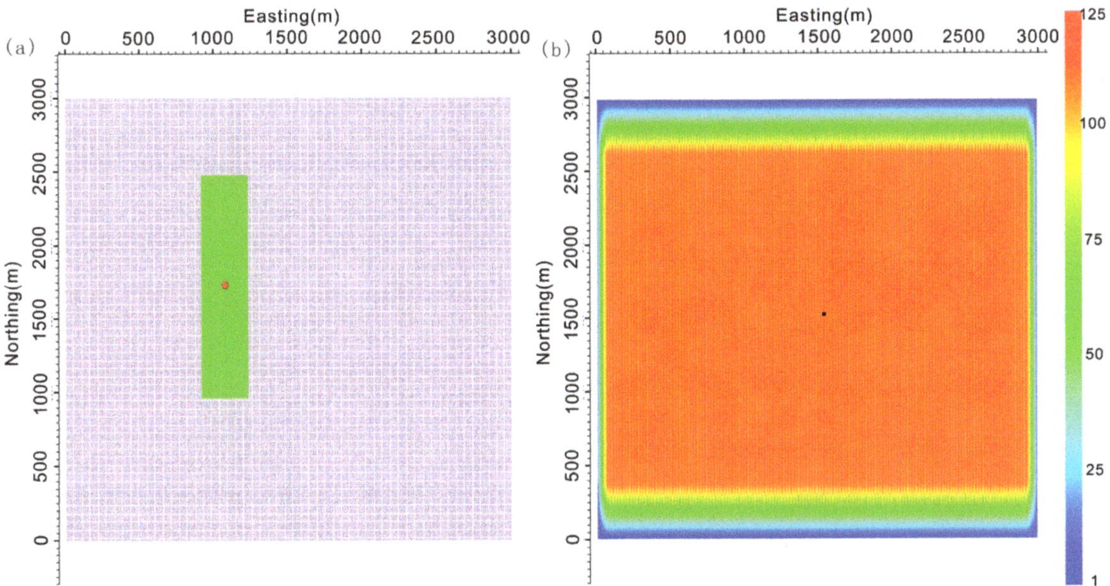

Figure 10. (**a**) Coverage of the folds of a source in the geometry with a narrow azimuth (the red point is the position of the source, the green rectangle is the range of the receiver corresponding to the source, east is the inline direction, and north is the crossline direction). (**b**) Folds of this geometry.

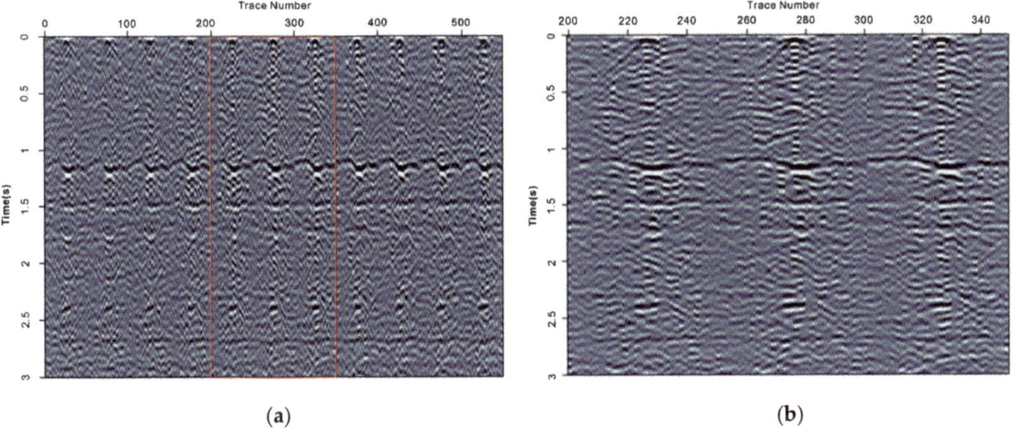

Figure 11. (**a**) Three-dimensional (3D) virtual shot records at the 5050th source in the case of a geometry with a narrow azimuth (source at line 51, source point 5050); (**b**) 3D virtual shot records at the 5050th source and receiver lines 50–52 in the case of a geometry with a narrow azimuth.

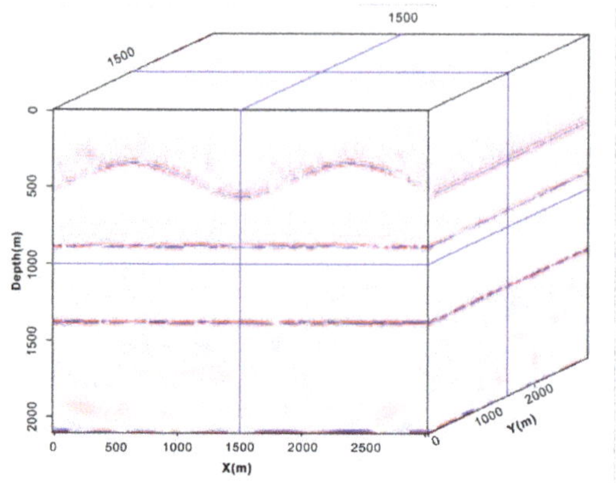

Figure 12. Migration profile of 3D virtual shot records in the case of a geometry with a narrow azimuth.

2.3.3. Forward Model with a Wide Azimuth

We also designed a geometry with a wide azimuth: The maximum offsets along the direction of the line and the direction vertical to it were both 1500 m, that is, each source received data from 51 × 51 receivers at most, and the aspect ratio was one. Figure 13a shows the source point and the range of its corresponding receiver point, and Figure 13b shows folds of the geometry designed on the basis of this scenario. Figure 14a shows the shot records collected by using this geometry at the center of the model, and Figure 14b shows shot records depicted by the red box in Figure 14a. Figure 15 shows the results of migration imaging and stacking, performed by calculating shot records by extracting one line from every four with a time window of correlation of 40 s. The velocities required for the migration and nmo were obtained from the model of velocity.

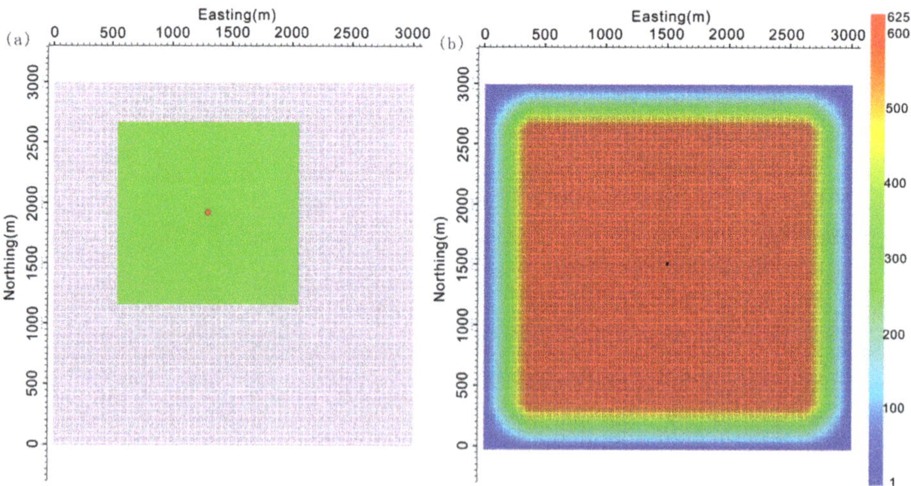

Figure 13. (**a**) Coverage of folds of a source in the geometry with a wide azimuth (the red point is the source point, the green rectangle is the range of the receiver corresponding to the source point, east is the inline direction, and north is the crossline direction). (**b**) Folds of this geometry.

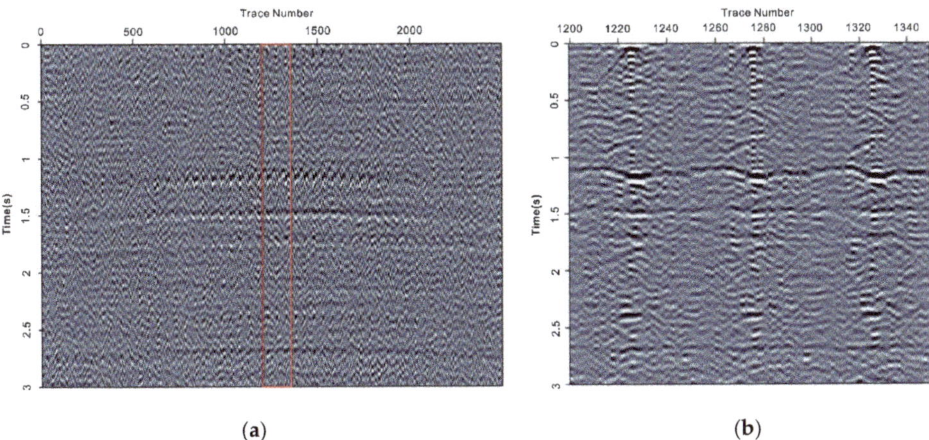

Figure 14. (**a**) Three-dimensional (3D) virtual shot records of the 5050th source in the geometry with a wide azimuth (source at line 51; source point 5050); (**b**) 3D virtual shot records of the 5050th source at receiver lines 50–52 in the geometry with a wide azimuth.

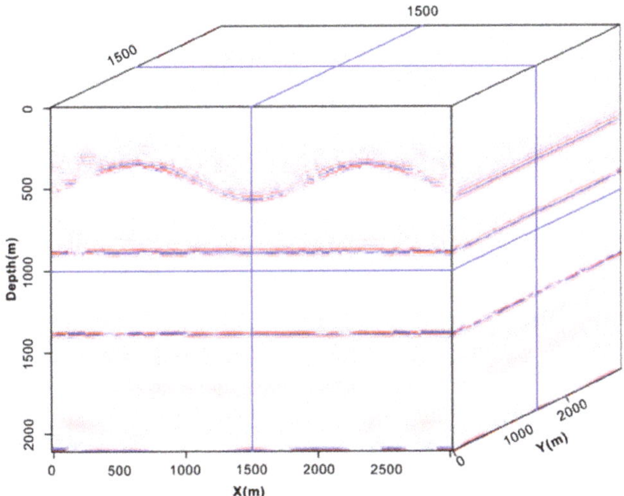

Figure 15. Migration profile of 3D virtual shot records in the geometry with a wide azimuth.

A comparison of the complete model (Figure 9b), narrow azimuth model (Figure 12) and wide azimuth model (Figure 15) shows that the three methods generated similar migration profiles by using a model-compliant geometry and the processing workflow of conventional seismic exploration, and their results were consistent with the 3D model of velocity in Figure 6a. This shows that the proposed method of generating virtual shot records of a 3D passive source based on SPS files is accurate. Due to the geometric design, data based on narrow and wide azimuths from all receiver points are not needed to participate in the calculation of each source, which significantly reduces the computation time (narrow azimuth model and wide azimuth model took about 1/20 and 1/4, respectively, of the time taken to calculate the virtual shot records by using all of the receiver points) and ensures uniform folds.

3. Direct Migration of Reflections of 3D Waves from Passive Source Based on Geometry

3.1. Principle of Calculation of Direct Migration

The principle of calculating random noise by direct migration can be expressed as follows:

$$I(x) = \sum_{k=1}^{N_s} \int_0^T R_K^A(x,t) R_K^B(x,t) dt \quad (2)$$

$R_K^A(x,t)$ and $R_K^B(x,t)$ are the wave fields from the kth (k = 1, 2, . . . , N_s) source at a certain time, and are received by the receivers and on the surface. The two are extended downward, and the underground image of the reflection $I(x)$ can be obtained through correlation (Figure 16). The final result of imaging is the sum of the results of correlation of a series of sources and receivers in a series of time windows. Artman [27] summarized the process of calculation of direct migration of data from a passive source as the extrapolation, correlation, and summation of the wave field. All algorithms for migration imaging that involve the extrapolation of the wave field and are conducive for use in the relevant imaging conditions can apply the direct migration method

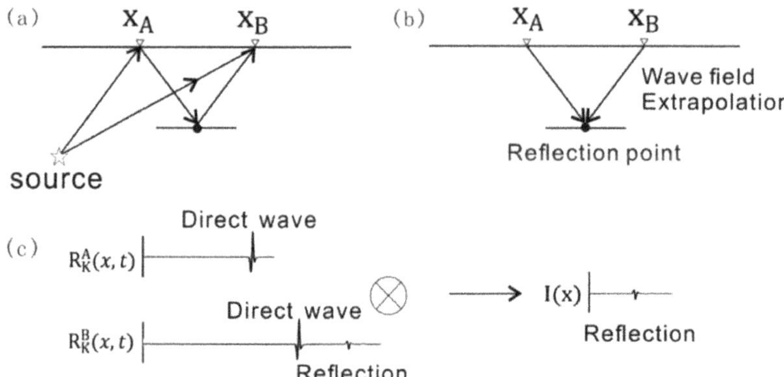

Figure 16. (**a**) Receivers A and B receive a single vibration from underground at a certain time. (**b**) The image of the underground reflection is obtained by downward continuation. (**c**) Diagram of correlation.

We used the one-way wave pre-stack depth migration method to process the original random noise records. It consisted of two steps: recursion of the source and receiver points along the direction of the depth, and imaging [33] to satisfy the requirements of the direct migration method.

The forward calculation of the 2D passive source by using direct migration was carried out by using the models of velocity and random noise records described in Section 2, and shown in Figures 2 and 3, to illustrate the process of imaging of the reflections by the passive source. One of the random noise records was set as the source point, with the remaining records used for one-way wave migration. All 301 data sources were calculated and stacked, and the migration profile was obtained by using conventional seismic processing, as shown in Figure 17.

One channel of the random noise records was set as the source point, and the other 300 channels were set as receiver points, that is, the range of geometry included all of the data. Performing one-way wave migration and stacking for all 301 "sources" and models of velocity yielded a direct migration profile (Figure 17b). The results were consistent with those of migration imaging based on the conventional processing of the 2D passive source in Figure 17a. That is, the direct migration imaging of reflections by a passive source was accurate, and the proposed method is thus feasible.

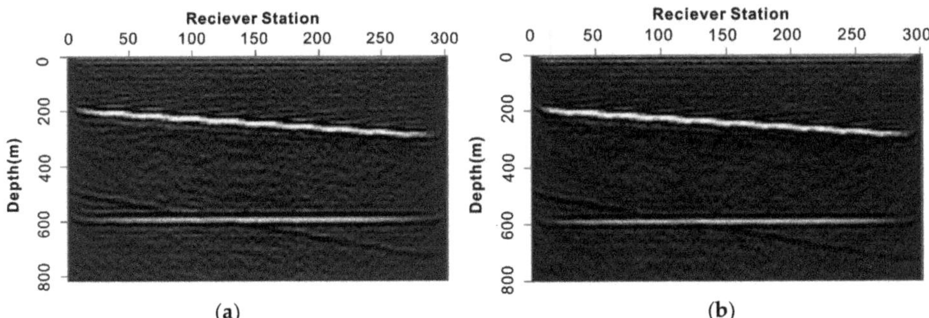

Figure 17. Comparison of (**a**) results of migration imaging of a 2D passive source using conventional processing and (**b**) direct migration imaging.

3.2. Calculation of 3D Reflections from a Passive Source Using Direct Migration Based on Geometry

The process of calculating the 3D reflection from a passive source by using direct migration based on the geometry is similar to the method of generating virtual shot records of reflections from a 3D passive source based on the relevant geometry file, as shown in Figure 18:

1. Having collected random noise records from the field, the geometry that satisfies the given requirements is designed by using a design software according to the known surface and information on the underground structure as well as the geological targets, and is stored in the SPS file.
2. A random noise record is set at a specified location as the noise record of the source point according to the content of the SPS.S file. The range of the corresponding receiver point is obtained from the SPS.X file according to the source point. The noise records of all receiver points are read, the one-way wave offset is calculated, stacking according to the time window is performed, and the information stored in the SPS.S file and SPS.R files is inserted into the header.
3. All of the results of the migration of virtual shot records are calculated, and the imaging section is obtained after processing the stack.

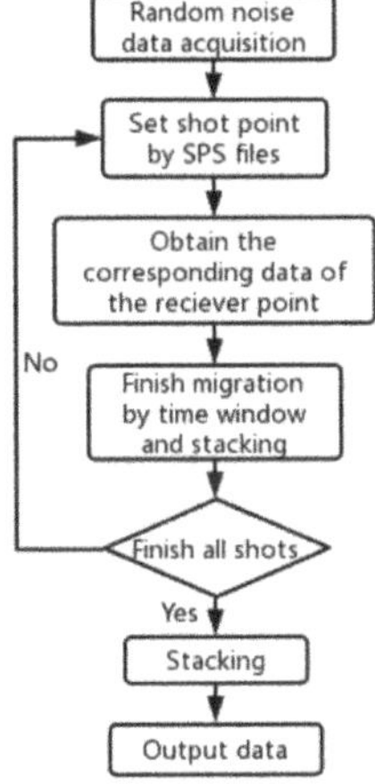

Figure 18. Processing flow of calculations of the direct migration method for 3D passive seismic exploration.

3.3. Testing the Model of Direct Migration

The model of velocity (Figure 6a) and the random noise records (Figure 7) described in the previous section were used to conduct an experiment on the direct migration of random reflection-induced noise. The process of calculation is shown in Figure 18. A geometry with a wide azimuth was used, as shown in Figure 14. The maximum offset along the inline direction and the crossline direction was 1500 m, that is, the signal from each source was received by 51 × 51 receivers at most, and the aspect ratio was one. One line was extracted from every four for the calculations. After reading the SPS.X file, the source points of the data on the offset noise were placed in the channels described in the file, and the corresponding seismic channels of the receiver point in the SPS.X file were extracted to calculate the one-way wave migration with a time window of 20 s. The final migration profile obtained by stacking all of the results of migration is shown in Figure 19b. The velocities required for the migration and nmo were taken from the model of velocity. The results obtained by the conventional method of passive seismic exploration were consistent with those based on calculating the direct migration of reflections. The direct migration method required less computational and storage-related resources, and was accurate.

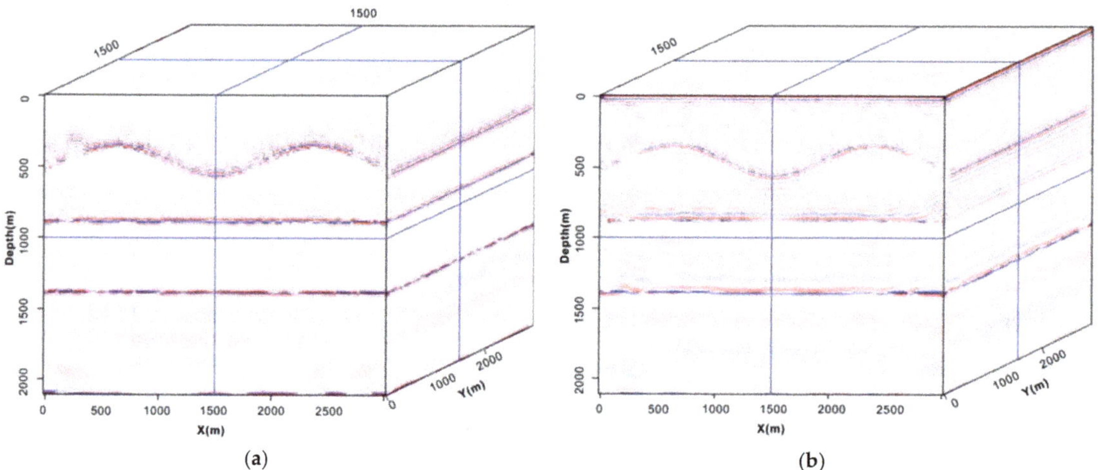

Figure 19. (**a**) Migration profile of 3D virtual shot records obtained by using a geometry with a wide azimuth. (**b**) Direct migration profile of reflection from a 3D passive source using a geometry with a wide azimuth.

When the forward migration-based imaging is carried out by using a geometry with a wide azimuth, there is no need for all sources or receivers to participate in the calculation; this significantly reduces the time needed for computations and yields uniform folds. By applying this geometry, we used only 1/4 of the source lines, and each source used only 1/4 of the near-offset receivers (the computational time was only about 1/16 of that incurred when using the entire dataset). As shown in Figure 18, the 3D seismic records obtained by the wide-azimuth geometry were processed by the direct migration method and yielded results of imaging similar to those of the virtual shot records of reflections from a passive source by using the migration method. This indicates that the proposed method can obtain accurate migration profiles by omitting the correlation in the virtual shot records, and thus requires less time and computational resources.

4. Conclusions

Seismic exploration based on reflections from a passive source uses the imaging of background noise, is cheap, and can be used as an effective supplement to methods of exploring deep active seismic sources. However, its computational cost and storage-related requirements constitute constraints on its further development.

Using a geometry based on the given geological structure can solve the above problem in practice. A set of geometric parameters that are suitable for the given area can reduce the cost of acquisition and the difficulty of data processing while improving their resolution. In this paper, we used experience in processing data on seismic exploration based on reflections from a 3D active source to design a reasonable and efficient geometry to obtain data for a 3D passive source. On the premise of ensuring a high signal-to-noise ratio, uniform folds, and reasonable azimuth, we used SPS files to guide the relevant correlations and obtain the virtual shot records; this improved the computational efficiency. At the same time, we obtained data on reflections of waves from the 3D passive source through forward modeling, and generated a clear seismic profile by calculating according to the above process.

In addition, we tested the feasibility of imaging based on direct migration using random reflection-induced noise. The method of generating virtual shot records is similar to the migration imaging algorithm based on correlation-based imaging conditions, that is, this method of migration can omit the correlations in the virtual shot records, thus reducing the requisite computational resources and storage space and improving efficiency. After intercepting multiple, random time windows of noise for multiple folds, the results also yielded images of high quality. The proposed method is thus efficient in terms of the seismic exploration of a passive source based on reflections.

The proposed method can accelerate the calculation of reflections from a passive source, and its many virtues can help promote the use of such technology on a large scale.

Author Contributions: Conceptualization, Y.L.; Software, Y.L.; Validation, Y.L.; Writing–original draft, Y.L.; Project administration, G.L.; Funding acquisition, G.L. All authors have read and agreed to the published version of the manuscript.

Funding: This study was supported by the NSFC (Grant No. 42074131), and the National Key R&D Program of China (Grant Nos. 2021YFC2801404 and 2022YFC280016803).

Institutional Review Board Statement: Not applicable.

Informed Consent Statement: Not applicable.

Data Availability Statement: Not applicable.

Conflicts of Interest: The authors declare no conflict of interest.

References

1. Schuster, G.T. Theory of daylight/ interferometric imaging: Tutorial. In Proceedings of the 63rd EAGE Conference & Exhibition, Amsterdam, The Netherlands, 11–15 June 2001; p. A32, Extended Abstracts.
2. Bakulin, A.; Calvert, R. Virtual source: New method for imaging and 4D below complex overburden. In Proceedings of the 74th Annual International Meeting, SEG, Denver, Colorado, 10–15 October 2004; pp. 2477–2480, Expanded Abstracts.
3. Claerbout, J.F. Synthesis of a layered medium from its acoustic transmission response. *Geophysics* **1968**, *33*, 264–269. [CrossRef]
4. Rickett, J.; Claerbout, J. Acoustic daylight imaging via spectral factorization: Helioseismology and reservoir monitoring. *Lead. Edge* **1999**, *18*, 957–960. [CrossRef]
5. Campillo, M.; Paul, A. Long-range correlations in the diffuse seismic coda. *Science* **2003**, *299*, 547–549. [CrossRef] [PubMed]
6. Shapiro, N.M.; Campillo, M. Emergence of broadband Rayleigh waves from correlations of the ambient seismic noise. *Geophys. Res. Lett.* **2004**, *31*, L07614-1–L07614-4. [CrossRef]
7. Sabra, K.G.; Gerstoft, P.; Roux, P.; Kuperman, W.A.; Fehler, M.C. Extracting time-domain Green's function estimates from ambient seismic noise. *Geophys. Res. Lett.* **2005**, *32*, L03310-1–L03310-5. [CrossRef]
8. Scherbaum, F. Seismic imaging of the site response using microearthquake recordings. Part I. Method. *Bull. Seismol. Soc. Ofamerica* **1987**, *77*, 1905–1923. [CrossRef]
9. Scherbaum, F. Seismic imaging of the site response using microearthquake recordings. Part II. Application to the Swabian Jura, southwest Germany, seismic network. *Bull. Seismol. Soc. Am.* **1987**, *77*, 1924–1944. [CrossRef]
10. Draganov, D.; Wapenaar, K.; Mulder, W.; Singer, J.; Verdel, A. Retrieval of reflections from seismic background-noise measurements. *Geophys. Res. Lett.* **2007**, *34*, L04305. [CrossRef]

11. Draganov, D.; Campman, X.; Thorbecke, J.; Verdel, A.; Wapenaar, K. Reflection images from ambient seismic noise. *Geophysics* **2009**, *74*, A63–A67. [CrossRef]
12. Draganov, D.; Campman, X.; Thorbecke, J. Seismic exploration-scale velocities and structure from ambient seismic noise (>1 Hz). *J. geophys. Res.* **2013**, *118*, 4345–4360. [CrossRef]
13. Vidal, C.A.; Draganov, D.; Van der Neut, J.; Drijkoningen, G.; Wapenaar, K. Retrieval of reflections from ambient noise using illumination diagnosis. *Geophys. J. Int.* **2014**, *198*, 1572–1584. [CrossRef]
14. Chamarczuk, M.; Malinowski, M.; Draganov, D. 2D body-wave seismic interferometry as a tool for reconnaissance studies and optimization of passive reflection seismic surveys in hardrock environments. *J. Appl. Geophys.* **2021**, *187*, 104288. [CrossRef]
15. Larose, E.; Derode, A.; Campillo, M. Imaging from one-bit correlations of wide band diffuse wave fields. *J. Appl. Phys.* **2004**, *95*, 8393–8399. [CrossRef]
16. Roux, P.; Sabra, K.G.; Gerstoft, P.; Kuperman, W.A.; Fehler, M.C. P-wavesfrom cross-correlation of seismic noise. *Geophys. Res.* **2005**, *32*, L19303.
17. Landès, M.; Hubans, F.; Shapiro, N.M.; Paul, A.; Campillo, M. Origin of deep ocean microseisms by using teleseismic body waves. *J. Geophys. Res.* **2010**, *115*, B05302. [CrossRef]
18. Ruigrok, E.; Campman, X.; Draganov, D.; Wapenaar, K. High-resolution lithospheric imaging with seismic interferometry. *Geophys. J. Int.* **2010**, *183*, 339–357. [CrossRef]
19. Girard, A.J.; Shragge, J. Automated processing strategies for ambient seismic data. *Geophys. Prospect.* **2020**, *68*, 293–312. [CrossRef]
20. Miyazawa, M.; Snieder, R.; Venkataraman, A. Application of seismic interferometry to extract P- and S-wave propagation and observation of shear-wave splitting from noise data at Cold Lake, Alberta, Canada. *Geophysics* **2008**, *73*, 35. [CrossRef]
21. Cheraghi, S.; Craven, J.A.; Bellefleur, G. Feasibility of virtual source reflection seismology using interferometry for mineral exploration: A test study in the Lalor Lake VMS mining area, Manitoba, Canada. *Geophys. Prospect.* **2015**, *63*, 833–848. [CrossRef]
22. Cheraghi, S.; White, D.J.; Draganov, D.; Bellefleur, G.; Craven, J.A.; Roberts, B. Passsive seismic reflection interferometry:A case study from the Aquistore CO_2 storage site, Saskatchewan, Canada. *Geophysics* **2017**, *82*, B79–B93. [CrossRef]
23. Roots, E.; Calvert, A.J.; Craven, J. Interferometric seismic imaging around the active Lalor mine in the Flin Flon greenstone belt, Canada. *Tectonophysics* **2017**, *718*, 92–104. [CrossRef]
24. Brenguier, F.; Boué, P.; Ben-Zion, Y.; Vernon, F.; Johnson, C.W.; Mordret, A.; Coutant, O.; Share, P.E.; Beaucé, E.; Hollis, D.; et al. Train traffic as a powerful noise source for monitoring active faults with seismic interferometry. *Geophys. Res. Lett.* **2019**, *46*, 9529–9536. [CrossRef] [PubMed]
25. Brenguier, F.; Courbis, R.; Mordret, A.; Campman, X.; Boué, P.; Chmiel, M.; Takano, T.; Lecocq, T.; Van Der Veen, W.; Postif, S.; et al. Noise-based ballistic wave passive seismic monitoring. Part 1: Body-waves. *Geophys. J. Int.* **2020**, *221*, 683–691. [CrossRef]
26. Nakata, N.; Chang, J.P.; Lawrence, J.F.; Boué, P. Body wave extraction and tomography at Long Beach, California, with ambient-noise interferometry. *J. Geophys. Res. Solid Earth* **2015**, *120*, 1159–1173. [CrossRef]
27. Chamarczuk, M.; Draganov, D.; Malinowski, M.; Koivisto, E.; Heinonen, S.; Rötsä, S. Reflection Image Beyond the Known Extent of the Prospective Zone Provided by 3D Virtual-Source Methodology. In Proceedings of the 83rd EAGE Annual Conference & Exhibition, Madrid, Spain, 6–9 June 2022; pp. 1–5. [CrossRef]
28. Artman, B. Imaging passive seismic data. *Geophysics* **2006**, *71*, SI177–SI187. [CrossRef]
29. Girard, A.J.; Shragge, J. Direct migration of ambient seismic data. *Geophys. Prospect.* **2020**, *68*, 270–292. [CrossRef]
30. Shiraishi, K.; Watanabe, T. Passive seismic reflection imaging based on acoustic and elastic reverse time migration without source information: Theory and numerical simulations. theory and numerical simulations. *Explor. Geophys.* **2022**, *53*, 198–210. [CrossRef]
31. Wapenaar, K.; Draganov, D.; Snieder, R.; Campman, X.; Verdel, A. Tutorial on seismic interferometry: Part 1—Basic principles and applications. *Geophysics* **2010**, *75*, 75A195–75A209. [CrossRef]
32. Wapenaar, K.; Slob, E.; Snieder, R.; Curtis, A. Tutorial on seismic interferometry: Part 2—Underlying theory and new advances. *Geophysics* **2010**, *75*, A211–A275. [CrossRef]
33. Liu, G.-F.; Meng, X.-H.; Liu, H. Accelerating finite difference wavefi eld-continuation depth migration by GPU. *Appl. Geophys.* **2012**, *9*, 41–48. [CrossRef]

Article

Simulation of Elastic Wave Propagation Based on Meshless Generalized Finite Difference Method with Uniform Random Nodes and Damping Boundary Condition

Siqin Liu [1,2], Zhusheng Zhou [2,*] and Weizu Zeng [3]

1 Key Laboratory of Metallogenic Prediction of Nonferrous Metals and Geological Environment Monitoring, Ministry of Education, Changsha 410083, China
2 School of Geosciences and Info-Physics, Central South University, Changsha 410017, China
3 Sichuan Earthquake Administration, Chengdu 610041, China
* Correspondence: geophys@126.com

Abstract: When the grid-based finite difference (FD) method is used for elastic wavefield forward modeling, it is inevitable that the grid divisions will be inconsistent with the actual velocity interface, resulting in problems related to the stepped grid diffraction and inaccurate travel time of reflected waves. The generalized finite difference method (GFDM), which is based on the Taylor series expansion and weighted least square fitting, solves these problems. The partial derivative of the unknown parameters in the differential equation is represented by the linear combination of the function values of adjacent nodes. In this study, the Poisson disk node generation algorithm and the centroid Voronoi node adjustment algorithm were combined to obtain an even and random node distribution. The generated nodes fit the internal boundary more accurately for model discretization, without the presence of diffracted waves caused by the stepped grid. To avoid the instability caused by the introduction of boundary conditions, a Cerjan damping boundary condition was proposed for boundary reflection processing. The test results generated by the different models showed that the generalized finite difference method can effectively solve the problems related to inaccurate travel time of reflection waves and stepped grid diffraction.

Keywords: generalized finite difference method (GFDM); elastic wave modeling; centroid Voronoi; Cerjan damping boundary condition

Citation: Liu, S.; Zhou, Z.; Zeng, W. Simulation of Elastic Wave Propagation Based on Meshless Generalized Finite Difference Method with Uniform Random Nodes and Damping Boundary Condition. *Appl. Sci.* **2023**, *13*, 1312. https://doi.org/10.3390/app13031312

Academic Editor: Emanuel Guariglia

Received: 8 December 2022
Revised: 10 January 2023
Accepted: 13 January 2023
Published: 18 January 2023

1. Introduction

Elastic wave numerical simulation is an important means by which to study the law of seismic wave propagation within the ground, and it plays a vital role in the whole process of seismic exploration, including data acquisition, processing, and interpretation. Many numerical simulation methods, including the finite element [1,2], finite difference [3–5], spectral element [6], and pseudo spectral [7] methods have been successfully applied to elastic wave forward modeling. The staggered grid finite difference method (SGFD) is widely used in the forward modeling of elastic wave equations owing to its high calculation efficiency, high accuracy, and convenient implementation process. However, there are two problems with SGFD. One is that the fixed grid step may discretize the interface to the wrong depth, resulting in an inaccurate travel time for the reflected waves. The second is that a stepped interface will appear when the undulating interface is discretized with a regular grid, which will generate false diffracted waves. To eliminate grid diffraction, many scholars have carried out extensive research, including using variable grid algorithms [8], vertical grid mapping based on coordinate transformation [9], body fitted grids [10], and other methods. These methods effectively suppress stepped diffraction to a certain extent, but they also have problems. For example, the variable mesh algorithm essentially uses a smaller mesh step size to discretize the undulating interface, so it cannot fundamentally eliminate the stepped diffraction; the difference method based on coordinate transformation requires a corresponding transformation of the wave equation; and the implementation process is relatively complex.

In recent years, the meshless method, which avoids mesh generation, discretizes the solution area into a series of independent nodes and constructs approximate functions on these discrete nodes to obtain linear equations [11]. Since the nodes change flexibly with the velocity model, the meshless method can effectively avoid grid diffraction [12]. The generalized finite difference method is a meshless method with a simple principle, flexible node arrangement, and high calculation accuracy, and it has been widely used to solve a variety of mathematical and engineering problems [13–15]. Jensen (1972) first used the Taylor series expansion of several adjacent nodes around a center point, based on a distance function, to solve differential equations [16]. Later, Benito et al. (2001) developed the explicit formula of the generalized finite difference method [17]. Based on the Taylor series expansion and weighted least square fitting, the partial derivative of the unknown quantity in the differential equation is expressed as a linear combination of the function values of several adjacent nodes. Thus, the basic theory of the generalized finite difference method was formed. Since then, many scholars have improved the theory of this method and its implementation technology. For example, Benito and Ureña systematically analyzed various factors that affect the accuracy of the GFDM calculations, such as node generation [18,19], star of nodes shape [20], and weight function [17], and they pointed out that as the discretization (cloud of nodes) becomes more regular, the results obtained become more stable [21].

Ureña et al. (2011) first applied the generalized finite difference method to the forward modeling of elastic wave equations, and proposed an adaptive method to minimize the effect of the irregularity of node distributions on dispersion [22]; on this basis, Ureña et al. (2012) discussed the dispersion and stability of elastic wave forward modeling using the generalized finite difference method under regular and irregular node conditions [23]. Benito et al. (2013) further studied the generalized finite difference method for solving the problem of determining seismic wave propagation in homogeneous isotropic media [24]. Benito et al. (2015) discussed the influence of node settings on simulation accuracy using circular hole models with Dirichlet uniform boundary conditions and square hole model scatterers with free boundary conditions [25]. Benito et al. (2017) applied the generalized finite difference method (GFDM) to solve the problem of elastic wave propagation, and analyzed the influence of the type of node clouds (regular and irregular) on discretization [17]. Salete et al. (2017) put forward a generalized finite difference scheme to solve the two-dimensional seismic wave propagation problem with a perfectly matched layer absorption boundary, and discussed the stability of PML. They also pointed out that the stability condition at the boundary of PML is stricter than that in the internal computational domain [26].

In this study, the elastic wave equation was solved using the meshless generalized finite difference method. This method first discretizes the solution area into a series of independent nodes, and then constructs an approximate function for these discrete nodes based on the Taylor series expansion and weighted least squares fitting. Finally, it solves the linear system to obtain an approximate solution for the boundary value of the elastic wave equation. An algorithm combining the Poisson disk node generation algorithm and the centroid Voronoi node adjustment algorithm is suggested to improve the stability of the node "star". The Cerjan damping boundary condition is introduced to avoid the instability caused by the absorbing boundary conditions. In some worked examples, the GFDM is compared with SGFD, and analytical solutions based on homogeneous, horizontal layered, undulating interface, and fault models are also tested.

2. Methodology

2.1. Elastic Wave Equation

The two-dimensional elastic wave equation can be expressed as:

$$\begin{cases} \rho\frac{\partial^2 u_x}{\partial t^2} = (\lambda+2\mu)\frac{\partial^2 u_x}{\partial x^2} + (\lambda+\mu)\frac{\partial^2 u_z}{\partial x \partial z} + \mu\frac{\partial^2 u_x}{\partial^2 z} + f_x \\ \rho\frac{\partial^2 u_z}{\partial t^2} = \mu\frac{\partial^2 u_z}{\partial x^2} + (\lambda+\mu)\frac{\partial^2 u_x}{\partial x \partial z} + (\lambda+2\mu)\frac{\partial^2 u_z}{\partial z^2} + f_z \end{cases} \tag{1}$$

where t is the time, ρ is the density of the medium, λ and μ are Lame constants, $\lambda = \rho\left(v_p^2 - 2v_s^2\right)$, $\mu = \rho v_s^2$, v_p, and v_s are the velocities of compressional and shear waves, u_x and u_z represent displacement in the x and z directions, and f_x and f_z represent source terms in the x and z directions.

2.2. Central Difference for Time Partial Derivative Approximation

The displacement at time $t+\Delta t$ and $t-\Delta t$ can be expanded at time t via the Taylor series:

$$u(t+\Delta t) = u(t) + \Delta t\frac{\partial u(t)}{\partial t} + \frac{(\Delta t)^2}{2!}\frac{\partial^2 u(t)}{\partial t^2} + \frac{(\Delta t)^3}{3!}\frac{\partial^3 u(t)}{\partial t^3} + \ldots \tag{2}$$

$$u(t-\Delta t) = u(t) - \Delta t\frac{\partial u(t)}{\partial t} + \frac{(\Delta t)^2}{2!}\frac{\partial^2 u(t)}{\partial t^2} - \frac{(\Delta t)^3}{3!}\frac{\partial^3 u(t)}{\partial t^3} + \ldots \tag{3}$$

Truncating the 2M-order obtains the central difference expression:

$$u(t+\Delta t) = 2u(t) - u(t-\Delta t) + 2\sum_{m=1}^{M}\frac{\Delta t^{2m}}{2m!}\frac{\partial^{2m}u(t)}{\partial t^{2m}} + O\left(\Delta t^M\right) \tag{4}$$

When $M = 1$, then:

$$u(t+\Delta t) = 2u(t) - u(t-\Delta t) + \Delta t^2\frac{\partial^2 u(t)}{\partial t^2} + O\left(\Delta t^2\right) \tag{5}$$

The time partial derivatives $\partial^2 u_x(t)/\partial t^2$ and $\partial^2 u_z(t)/\partial t^2$ in Equation (1) can be converted using Equation (5) into the following form:

$$\begin{cases} u_x(t+\Delta t) = 2u_x(t) - u_x(t-\Delta t) \\ \quad +\Delta t^2\left[\frac{\lambda+2\mu}{\rho}\frac{\partial^2 u_x}{\partial x^2} + \frac{\mu}{\rho}\frac{\partial^2 u_x}{\partial^2 z} + \frac{\lambda+\mu}{\rho}\frac{\partial^2 u_z}{\partial x \partial z}\right] + f_x(t+\Delta t) \\ u_z(t+\Delta t) = 2u_z(t) - u_z(t-\Delta t) \\ \quad +\Delta t^2\left[\frac{\lambda+2\mu}{\rho}\frac{\partial^2 u_z}{\partial x^2} + \frac{\mu}{\rho}\frac{\partial^2 u_z}{\partial^2 z} + \frac{\lambda+\mu}{\rho}\frac{\partial^2 u_x}{\partial x \partial z}\right] + f_z(t+\Delta t) \end{cases} \tag{6}$$

2.3. GFDM for Spatial Partial Derivative Approximation

In the GFDM, through utilizing the Taylor series expansion and weighted least squares fitting, the derivatives of unknown variables can be approximated by a linear combination of function values of some neighboring nodes, which are located inside a star. First, a regular or irregular cloud of points is generated in the computation domain and along the boundary. For a given node i, which is denoted as a central node, the m nearest nodes surrounding the node i will be found. The concept of the star refers to a group of established nodes in relation to the central node, as shown in the black circle in Figure 1. Each node is assigned an associated star.

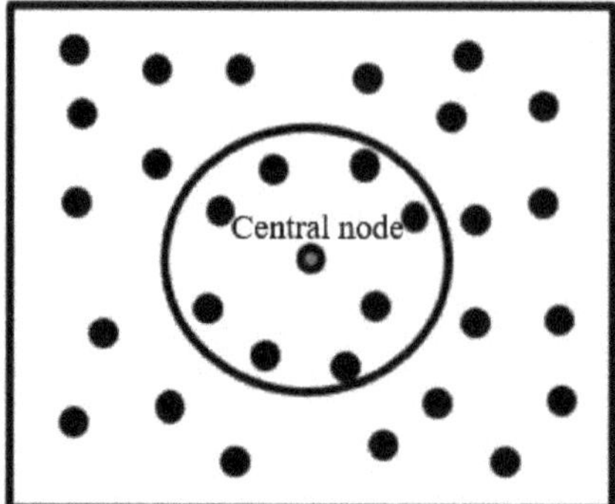

Figure 1. An irregular cloud of points and the selection of star via distance criterion.

If u_0 is the function value at the central node x_0 of the star and $u_i(i = 1, 2, \cdots, m)$ is the function value at one of the rest of the nodes, then the Taylor series expansion around the central node can be expressed in the following form:

$$u_i = u_0 + h_i \frac{\partial u_0}{\partial x} + l_i \frac{\partial u_0}{\partial z} + \frac{h_i^2}{2}\frac{\partial^2 u_0}{\partial x^2} + \frac{l_i^2}{2}\frac{\partial^2 u_0}{\partial z^2} + h_i l_i \frac{\partial^2 u_0}{\partial x \partial z} + \cdots \quad (7)$$

where the coefficients $h_i = x_i - x_0$, $l_i = z_i - z_0$, and (x_0, z_0) are the coordinates of the central node, and (x_i, z_i) are the coordinates of the node i in the star. By truncating the Taylor series with the second derivative, the residual function $B(u)$ can be defined by the following equation:

$$B(u)_{2^{nd}} = \sum_{i=1}^{m}\left[\left(u_0 - u_i + h_i \frac{\partial u_0}{\partial x} + l_i \frac{\partial u_0}{\partial z} + \frac{h_i^2}{2}\frac{\partial^2 u_0}{\partial x^2} + \frac{l_i^2}{2}\frac{\partial^2 u_0}{\partial z^2} + h_i l_i \frac{\partial^2 u_0}{\partial x \partial z}\right) w_i\right]^2 \quad (8)$$

By truncating the Taylor series with the fourth derivative, the residual function $B(u)$ can be defined by the following:

$$\begin{aligned} B(u)_{4^{th}} = \sum_{i=1}^{m}\Bigg[\Bigg(& u_0 - u_i + h_i \frac{\partial u_0}{\partial x} + l_i \frac{\partial u_0}{\partial z} + \frac{h_i^2}{2}\frac{\partial^2 u_0}{\partial x^2} + \frac{l_i^2}{2}\frac{\partial^2 u_0}{\partial z^2} + h_i l_i \frac{\partial^2 u_0}{\partial x \partial z} \\ &+ \frac{h_i^3}{6}\frac{\partial^3 u_0}{\partial x^3} + \frac{l_i^3}{6}\frac{\partial^3 u_0}{\partial z^3} + \frac{h_i^2 l_i}{2}\frac{\partial^3 u_0}{\partial x^2 \partial z} + \frac{h_i l_i^2}{2}\frac{\partial^3 u_0}{\partial x \partial z^2} + \frac{h_i^4}{24}\frac{\partial^4 u_0}{\partial x^4} + \frac{l_i^4}{24}\frac{\partial^4 u_0}{\partial z^4} \\ &+ \frac{h_i^3 l_i}{6}\frac{\partial^4 u_0}{\partial x^3 \partial z} + \frac{h_i l_i^3}{6}\frac{\partial^4 u_0}{\partial x \partial z^3} + \frac{h_i^2 l_i^2}{4}\frac{\partial^4 u_0}{\partial x^2 \partial z^2}\Bigg) w_i\Bigg]^2 \end{aligned} \quad (9)$$

In Equations (8) and (9), $w_i = w(h_i, l_i)$ represents the distance weight function of the star. In all the examples considered in this paper, the weighting function was chosen as per [27]:

$$w_i = \begin{cases} 1 - 6\left(\frac{d_i}{d_{\max}}\right)^2 + 8\left(\frac{d_i}{d_{\max}}\right)^3 - 3\left(\frac{d_i}{d_{\max}}\right)^4, & d_i \le d_{\max} \\ 0, & d_i > d_{\max} \end{cases} \quad (10)$$

where d_i denotes the distance between nodes (x_i, z_i) and (x_0, z_0), and $d_{\max}$ is the distance between the central node and the farthest node in the star. Let the following terms be defined:

$$\mathbf{D_u} = \left[\frac{\partial u_0}{\partial x}, \frac{\partial u_0}{\partial z}, \frac{\partial^2 u_0}{\partial x^2}, \frac{\partial^2 u_0}{\partial z^2}, \frac{\partial^2 u_0}{\partial x \partial z}, \cdots\right] \quad (11)$$

$$\mathbf{p}_i = \left\{h_i, l_i, \frac{h_i^2}{2}, \frac{l_i^2}{2}, h_i l_i, \cdots\right\}, i = 1, 2, \cdots, m \quad (12)$$

$$\mathbf{P} = \begin{bmatrix} \mathbf{p}_1 \\ \mathbf{p}_2 \\ \vdots \\ \mathbf{p}_m \end{bmatrix} = \begin{bmatrix} h_1 & l_1 & \cdots & h_1 l_1 & \cdots \\ h_2 & l_2 & \cdots & h_2 l_2 & \cdots \\ \vdots & \vdots & \ddots & \vdots & \cdots \\ h_m & l_m & \ddots & h_m l_m & \cdots \end{bmatrix} \tag{13}$$

The residual function defined in Equations (8) and (9) can be expressed in matrix form:

$$B(u) = (\mathbf{PD_u} + \mathbf{u}_0 - \mathbf{u})^T \mathbf{W} (\mathbf{PD_u} + \mathbf{u}_0 - \mathbf{u}) \tag{14}$$

where $\mathbf{u} = [u_1, u_2, \cdots, u_m]^T$, $\mathbf{u}_0 = [u_0, u_0, \cdots, u_0]^T$, and $\mathbf{W} = diag(w_1^2, w_2^2, \cdots, w_m^2)$.

To minimize the residual function $B(u)$ with respect to the unknown derivatives at the central point (x_0, z_0):

$$\frac{\partial B(u)}{\partial \mathbf{D_u}} = 0 \tag{15}$$

yields the following linear equation system,

$$\mathbf{AD_u{=}b} \tag{16}$$

and

$$\mathbf{A} = \mathbf{P}^T \mathbf{W} \mathbf{P} \tag{17}$$

$$\mathbf{b} = \mathbf{P}^T \mathbf{W} (\mathbf{u}\text{-}\mathbf{u_0}) \tag{18}$$

According to Equations (16)–(18), the partial derivative vector $\mathbf{D_u}$ can be expressed as:

$$\mathbf{D_u} = \mathbf{A}^{-1}\mathbf{b} = \mathbf{A}^{-1}\mathbf{P}^T\mathbf{W}(\mathbf{u} - \mathbf{u}_0) = -\mathbf{a}u_0 + \sum_{i=1}^{m} \mathbf{a}_i u_i \tag{19}$$

where $\mathbf{a} = \sum_{i=1}^{m} \mathbf{a}_i$, $\mathbf{a} = [a_1, a_2, a_3, a_4, a_5]^T$, and $\sum_{i=1}^{m} \mathbf{a}_i = \left[\sum_{i=1}^{m} a_{1i}, \sum_{i=1}^{m} a_{2i}, \sum_{i=1}^{m} a_{3i}, \sum_{i=1}^{m} a_{4i}, \sum_{i=1}^{m} a_{5i} \right]^T$.
Expanding the partial derivative vector $\mathbf{D_u}$, we can obtain:

$$\begin{cases} \frac{\partial u_0}{\partial x} = -a_1 u_0 + \sum_{i=1}^{m} a_{1i} u_i \\ \frac{\partial u_0}{\partial z} = -a_2 u_0 + \sum_{i=1}^{m} a_{2i} u_i \\ \frac{\partial^2 u_0}{\partial x^2} = -a_3 u_0 + \sum_{i=1}^{m} a_{3i} u_i \\ \frac{\partial^2 u_0}{\partial z^2} = -a_4 u_0 + \sum_{i=1}^{m} a_{4i} u_i \\ \frac{\partial^2 u_0}{\partial x \partial z} = -a_5 u_0 + \sum_{i=1}^{m} a_{5i} u_i \end{cases} \tag{20}$$

Substituting Equation (20) into the Equation (6), the discrete formula for solving the elastic wave equation can be obtained:

$$\begin{cases} u_{x,0}^{n+1} = 2u_{x,0}^n - u_{x,0}^{n-1} + \Delta t^2 \left[\frac{\lambda+2\mu}{\rho} \left(-a_3 u_{x,0}^n + \sum_{i=1}^{m} a_{3i} u_{x,i}^n \right) \right. \\ \left. + \frac{\lambda+\mu}{\rho} \left(-a_5 u_{z,0}^n + \sum_{i=1}^{m} a_{5i} u_{z,i}^n \right) + \frac{\mu}{\rho} \left(-a_4 u_{x,0}^n + \sum_{i=1}^{m} a_{4i} u_{x,i}^n \right) \right] + f_x^{n+1} \\ u_{z,0}^{n+1} = 2u_{z,0}^n - u_{z,0}^{n-1} + \Delta t^2 \left[\frac{\lambda+2\mu}{\rho} \left(-a_4 u_{z,0}^n + \sum_{i=1}^{m} a_{4i} u_{z,i}^n \right) \right. \\ \left. + \frac{\lambda+\mu}{\rho} \left(-a_5 u_{x,0}^n + \sum_{i=1}^{m} a_{5i} u_{x,i}^n \right) + \frac{\mu}{\rho} \left(-a_3 u_{z,0}^n + \sum_{i=1}^{m} a_{3i} u_{z,i}^n \right) \right] + f_z^{n+1} \end{cases} \tag{21}$$

For the stability analysis, the aim is to construct a harmonic decomposition of the approximated solution at the grid points at a given time level (n). The amplification factor must be less than or equal to one in order to determine the stability limit. This has been studied by [23] and the condition for the stability of the star has been established as:

$$\Delta t \leq \sqrt{\frac{4}{\left(v_p^2 + v_s^2\right)(|a_3| + |a_4|) + \left(v_p^2 - v_s^2\right)\sqrt{(a_3 - a_4)^2 + 4a_5^2}}} \tag{22}$$

where v_p and v_s are the velocity of compressional and shear waves, respectively.

2.4. Node Generation Algorithm

Benito et al. (2017) pointed out that when the discretization (cloud of nodes) becomes more regular, the obtained results become more logical [21]. Furthermore, to avoid the diffraction caused by the difference between the node and the interface positions, it is necessary to adjust the node position at the inner part of the interface and the boundary, so that the nodes are located at the interface and the boundary. In this study, we proposed an algorithm combining the Poisson disk node generation algorithm and the centroid Voronoi adjustment algorithm, to obtain a reasonable node distribution according to the velocity model.

(1) Randomly Distributed Node Generation

The Poisson disk node generation algorithm proposed by Fornberg et al. (2015) was used [28]. In the two-dimensional case, the calculation steps for the random node generation algorithm (Figure 2) are as follows:

(a) Set the horizontal positions (potential dot positions, PDPs) of the initial nodes, and randomly set the vertical coordinates;
(b) Out of all the current PDPs, find the PDP closest to the bottom of the model, and classify this PDP as the determined dot position (DDP). The newly determined DDP is at the center of the circle, and the discrete distance set by the position model parameters is the radius (noting the radius is a function of velocity [29]);
(c) No other center may appear inside the circle, so all PDPs except the DDP within the circle are removed;
(d) Determine the two PDPs closest to the DDP at the circle center, make a circle through these two PDPs with the DDP as the center, and select multiple (we used five) new PDPs at equal angles on the arc between the two DDPs;
(e) Then, select the PDP closest to the bottom (excluding the cycled PDP and adding a new PDP) and repeat steps (b)–(d) until the cycle for all points in the calculation area is completed. At this point, uniformly and randomly distributed node coordinates will have been obtained in the computing area.

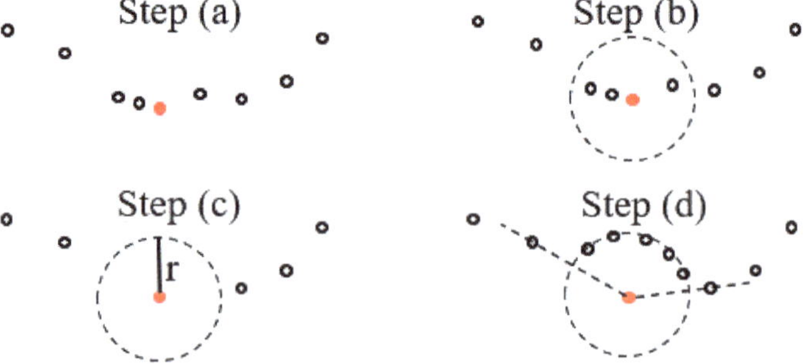

Figure 2. Generation process for the Poisson disk algorithm [28]. The red solid point is the DDP, and the black hollow points are PDPs.

(2) Node Adjustment

To adjust the nodes, overlay the random nodes obtained in the previous step with the interface and boundary nodes, and discard the nodes within a half radius of any boundary node, as shown in Figure 3b. After the redundant nodes are discarded, the nodes at the internal interface and the boundary will be unevenly distributed. Therefore, we propose a node adjustment algorithm based on the centroid Voronoi structure. This algorithm divides Voronoi polygons into multiple polygonal regions formed by the boundary and interfaces, and uses the centroid point of the Voronoi polygon as the new node location. Since the distribution of the nodes before adjustment is relatively uniform, only 10–15 iterations are needed to evenly distribute the nodes near the interface. This method has good stability and can obtain a more uniform node distribution, however, since the conventional Voronoi polygon division is borderless, the Voronoi polygon division requires computing the intersection of the Voronoi diagram and the irregular boundary, which leads to a high computation cost. To solve this problem, GPU was used to accelerate the Voronoi polygon division and the adjustment of the centroid points in the irregular boundary area.

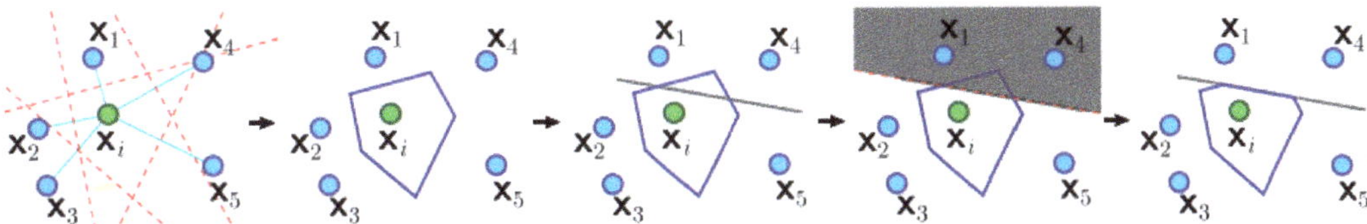

Figure 3. Generation steps for the Voronoi element under the restriction of an irregular boundary (step (a) and step (b)) [30]. X_i is the point to be adjusted, and X_1-X_5 are the five points surrounding X_i.

The algorithm takes the internal node and boundary node coordinates as its inputs, and its main steps are as follows:

(a) Calculate the coordinates of the k points closest to each node, and connect the node and the k points into a line segment Voronoi unit (see Figure 3);
(b) Calculate the intersection between each Voronoi element and the boundary polygon, and trim the Voronoi diagram (see Figure 3);
(c) The centroid of the Voronoi element is calculated and used as the new node position;
(d) Repeat the above three steps 10–15 times to get the final node distribution.

Figure 4 shows the node generation steps for the two-layer simple model. It can be seen from the figure that the algorithm first generates variable density nodes through Poisson disks in the boundary region (Figure 4a). Then, the nodes that are too close to the interface nodes are deleted, and nodes with an uneven distribution at the boundary are obtained (Figure 4c). Finally, variable density internal nodes, and a uniform distribution of boundary nodes, are obtained through the centroid Voronoi adjustment algorithm (Figure 4d). The figure shows not only that the final node distribution is accurately arranged at the interface location, but also that the overall nodes layout is evenly distributed across the different regions.

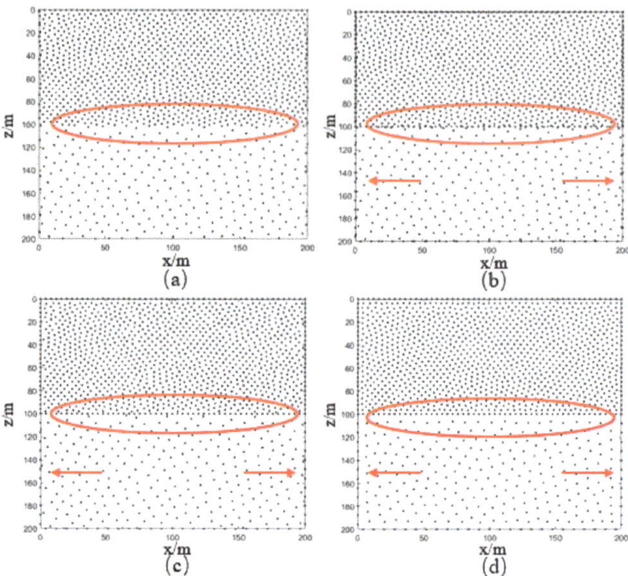

Figure 4. Node generation of two-layer model: (**a**) Variable density nodes; (**b**) Variable density nodes overlap with boundary nodes and interface nodes; (**c**) Delete the nodes in (**b**) that are too close to the interface or boundary; (**d**) Nodes after adjustment based on Voronoi centroid algorithm.

2.5. Boundary Condition

In practice, the seismic wave propagates in the underground infinite medium after being excited. When the elastic wave equation is used for seismic wave numerical simulation, it is difficult to simulate the infinite medium, so it is necessary to control the size of the model within a certain range to generate an artificial boundary. If the artificial boundary is not treated with a numerical method, the elastic wave will be reflected when it propagates to the boundary, and will interfere with the wavefield inside the model. For the generalized finite difference method in the time domain, the damping attenuation boundary condition does not involve the problem of stability. By selecting appropriate attenuation parameters, a good absorption effect can be achieved.

The damping factor proposed by Cerjan [31] is as follows:

$$damp = \exp\left[-\alpha^2 (I - i)^2\right] \tag{23}$$

where I is the grid number of the attenuation boundary thickness, $i = 1, 2, \cdots, I$, and α is the absorption coefficient. It can be seen from the formula that the selection of the absorption coefficient and the thickness of the attenuation boundary directly influence the absorption effect. The generalized finite difference method discretizes the model into nodes instead of into grids, so I represents the thickness of the attenuation boundary divided by the exclusion radius, r; and i represents the distance between the node and the inner boundary of the attenuation layer divided by the exclusion radius, r. The parameters used in this study were referenced to those used by to Li (2014) [32]. Figure 5 compares the wave field before and after adding the attenuation boundary. When there is no attenuation boundary, the energy of the wave field is covered by the boundary reflection. After adding 45 layers of attenuation boundary, the boundary reflection can be effectively absorbed.

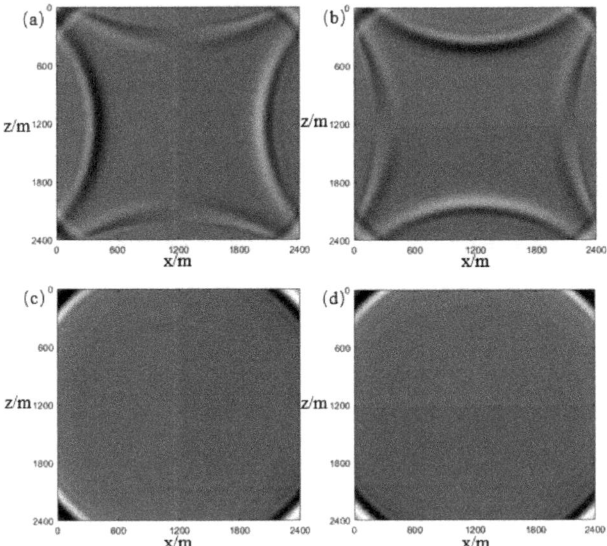

Figure 5. Wave field computed by GFDM, (**a**) X component (without attenuation boundary); (**b**) Z component (without attenuation boundary); (**c**) X component (45 layers of Cerjan boundary); (**d**) Z component (45 layers of Cerjan boundary).

2.6. Source Term

The source term in Equation (1) is the product of the seismic wavelet and the source spatial position function:

$$f(x,z;t) = s(t)\varphi(x,z) \tag{24}$$

where $s(t)$ is the seismic wavelet function, and $\varphi(x,z)$ represents the impulse function. Ricker wavelet [33] and Gaussian pulse functions are used in this paper:

$$s(t) = \left[1 - 2\pi^2 {f_m}^2 (t-t_0)^2\right] \exp\left[-\pi^2 {f_m}^2 (t-t_0)^2\right] \tag{25}$$

where f_m is the dominant frequency of the wavelet, and t_0 is the delay time Moreover:

$$\varphi(x,z) = e^{-[(x-x_0)^2+(z-z_0)^2]} \tag{26}$$

where (x_0, z_0) is the source position.

3. Examples

3.1. Homogeneous Model

To verify the correctness of the generalized finite difference method for solving the elastic wave equation, a two-dimensional homogeneous medium model was set (2400 m × 2400 m), with a P-wave velocity of 4000 m/s, an S-wave velocity of 2300 m/s and a density of 2000 kg/m^3. The dominant frequency of the Ricker wavelet was 20 Hz, and the source was located at the center of the model. For comparison, all sources were loaded onto the vertical component of the displacement. The generalized finite difference method was compared with both the analytical solution and the staggered grid finite difference method (SGFD). The node distribution was consistent with the rectangular grid of the staggered grid finite difference method, and the spacing of the grid (nodes) was 10 m.

To ensure the stability of the simulation, the time step was set at 0.5 ms, and the total computation time at 1.0 s. The forward modeling was performed using a 2nd-order precision 13 points GFDM, a 4th-order precision 21 nodes GFDM, and an 8th-order staggered grid finite difference method. Figure 6 shows a comparison of the 400 ms wavefield obtained by the different methods. The wavefront obtained by the GFDM is consistent with the analytical solution, which proves the correctness of the GFDM. The dispersion energy of the shear wave can be seen in Figure 6a,b, which was obtained through the 2nd-order 13 points GFDM. The results obtained using the 4th-order 21 nodes GFDM are almost the same as those obtained from the SGFD method, without any visible dispersion. This shows that improving the difference order can achieve a high calculation accuracy. In further comparing the shot records (Figure 7), the same conclusion can be drawn. In extracting the 80th trace from the shot record for comparison (Figure 8), little difference can be seen between the 4th-order GFDM results and the analytical solution, and the overall calculation accuracy is almost equivalent to that of the SGFD method. Therefore, it can be concluded that the time domain GFDM for elastic wave simulation is correct and effective.

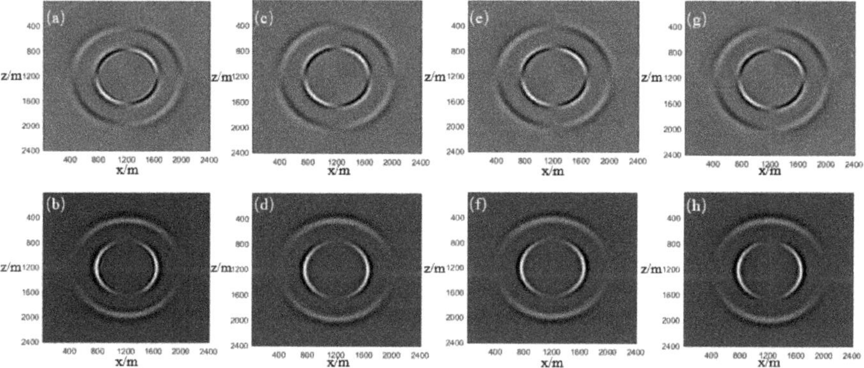

Figure 6. 400 ms wavefield of homogeneous model. (**a**) 2nd-order GFDM X component; (**b**) 2nd-order GFDM Z component; (**c**) 4th-order GFDM X component; (**d**) 4th-order GFDM Z component; (**e**) SGFD X component; (**f**) SGFD Z component; (**g**) Analytical solution for the X component; (**h**) Analytically solved Z component.

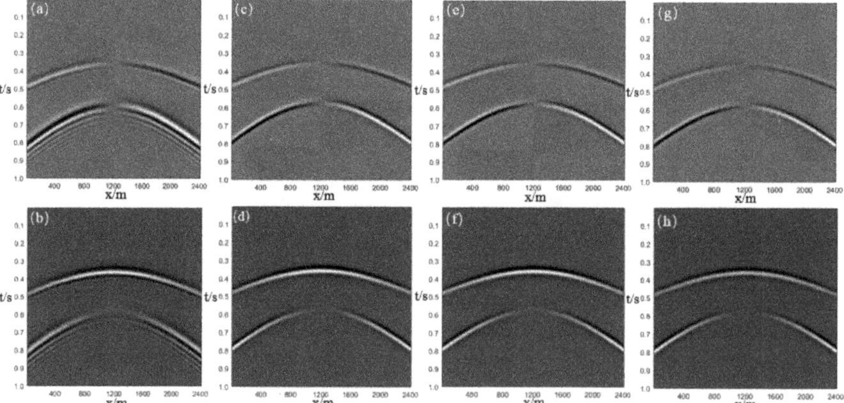

Figure 7. Shot record of homogeneous model, (**a**) 2nd-order GFDM X component; (**b**) 2nd-order GFDM Z component; (**c**) 4th-order GFDM X component;(**d**) 4th-order GFDM Z component; (**e**) SGFD X component; (**f**) SGFD Z component;(**g**) Analytical solution for the X component; (**h**) Analytically solved Z component.

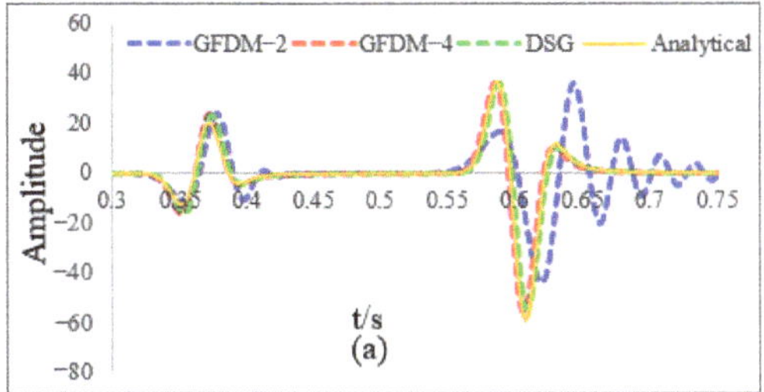

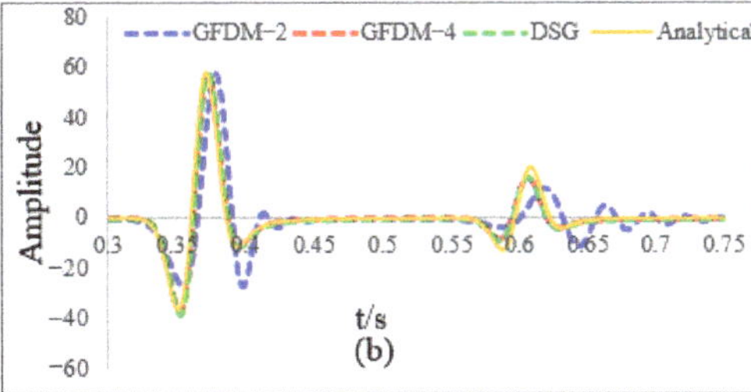

Figure 8. Comparison between the 80th trace of shot records: (**a**) X component; (**b**) Z component.

3.2. Two-Layer Model

When SGFD methodology is used for elastic wave simulation, the grid step is determined. If the interface is located at an integral multiple of the step, the grid can accurately describe the interface; otherwise, it will cause an inaccuracy in the travel time of the reflected wave. The GFDM can directly set the nodes at the interface, therefore it can accurately describe the changes in the interface and can obtain more accurate travel time simulation results. A two-layer horizontal model (2000 m $\times$ 2000 m) was designed to show the advantages of the GFDM over the SGFD model. The P-velocity, S-velocity, and density of the upper layer of the model were 4000 m/s, 2300 m/s, and 2.4 g/cm^3, respectively, and those of the lower layer were 6000 m/s, 3500 m/s, and 2.6 g/cm^3, respectively. The simulation time step was set at 0.5 ms, and the total computation time at 1.0 s. The source was located at (1000 m, 0 m), and the dominant frequency of the Ricker wavelet was 20 Hz. When the model is discretized, the SGFD model adopted a fixed grid step of 10 m. Therefore, only when the interface was located at an integral multiple of 10 (for example, when the interface was located at 1000 m), can the grid points accurately describe the interface. However, when the interface is located between 1000 m and 1010 m, the results obtained by SGFD are the same, and the travel time of the reflected wave will not change. When the 4th-order 21 nodes GFDM was used, the proposed node generation algorithm was applied to discretize the model, using a node radius of 10 m for the first layer, and 12 m for the second layer. The coordinates of the nodes change along with the real interface position, so more accurate travel time information of the reflected wave can be obtained. When the interfaces were located at 1002 m, 1005 m, and 1008 m, respectively, the GFDM and SGFD results obtained are shown in Figure 9. The black and blue dotted lines represent the seismic records obtained by SGFD when the interface depth was set as 1000 m and 1010 m, respectively. The green, red, and blue solid lines correspond to seismic records obtained by GFDM when the interfaces were located at 1002 m, 1005 m, and 1008 m, respectively. Comparing the reflected wave information for the two algorithms, the SGFD can only obtain the exact record time of interfaces located at 1000 m and 1010 m—when the interface depths are 1002 m, 1005 m, and 1008 m, the record time will never change. In contrast, for the GFDM results, the record time shifts with the interface depth, which is more consistent with reality compared to the SGFD results. This shows that the GFDM can accurately describe the changes in the interface, and can obtain more accurate travel time simulation results.

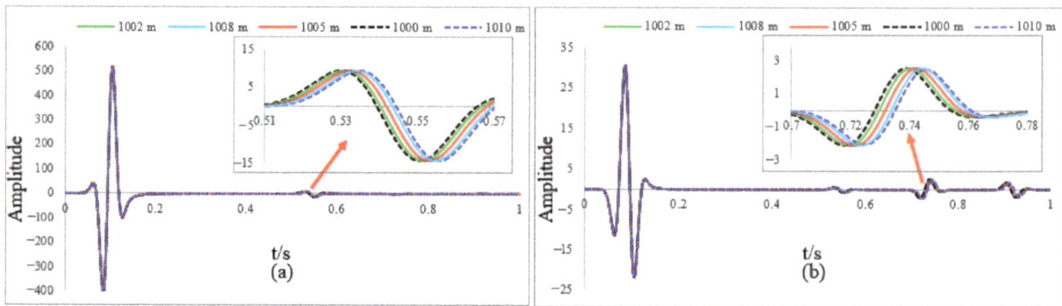

Figure 9. Seismic records at different interface depths: (**a**) X component; (**b**) Z component.

3.3. Undulating Interface Model

The SGFD model will cause false diffraction due to the stepped grid when dealing with undulating interfaces. These diffraction waves will affect the accuracy of elastic wave modeling. The GFDM can simulate an elastic wave based on nodes adapting to the change of an undulating interface, so there will be no stepped grid diffraction. When the undulating interface model, as shown in Figure 10, was used, the velocity and density of the upper layer of the model were 4000 m/s, 2300 m/s, and 2.4 g/cm^3, respectively, those of the middle layer were 5000 m/s, 2800 m/s, and 2.5 g/cm^3, and those of the lower layer were 6000 m/s, 3500 m/s, 2.6 g/cm^3. The time sampling interval was 0.5 ms, and the total computation time was 1.0 s. The source was located at (1000 m, 0 m), and the dominant frequency of the Ricker wavelet was 20 Hz. When the 4th-order 21 nodes GFDM was used, the proposed node generation algorithm was applied to discretize the model, with the node radius at 10 m, 12 m, and 15 m for the first, second, and third layers, respectively. For comparison, the SGFD model was used based on regular grids with a grid spacing of 10 m. The local grid distribution of the SGFD model is shown in Figure 11a, in which the stepped grid can be seen clearly, while the local node distribution of the GFDM is shown in Figure 11b, in which the nodes are directly and accurately distributed on the interface. It can be seen from the seismic records depicted in Figure 12a,b that there are a lot of diffracted waves behind the primary reflected wave when modeled by SGFD, while diffraction waves are practically non-existent in the GFDM record (Figure 12c,d). The results demonstrate that the forward modeling using GFDM avoids the effect of stepped grid diffraction, and is suitable for the forward modeling of a formation with undulating interfaces.

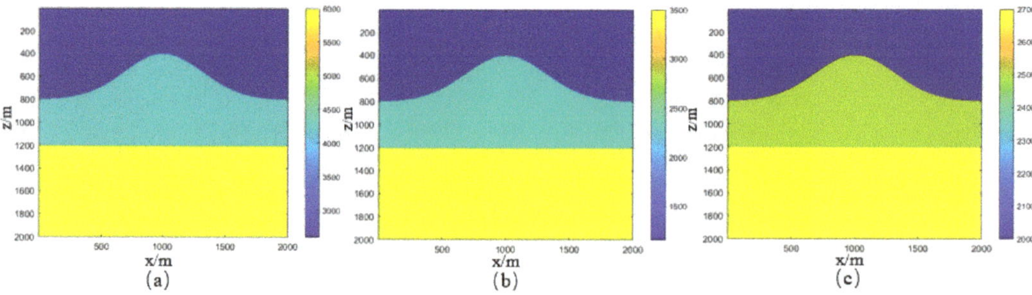

Figure 10. Velocity model of undulating interface: (**a**) longitudinal wave velocity; (**b**) shear wave velocity; (**c**) density.

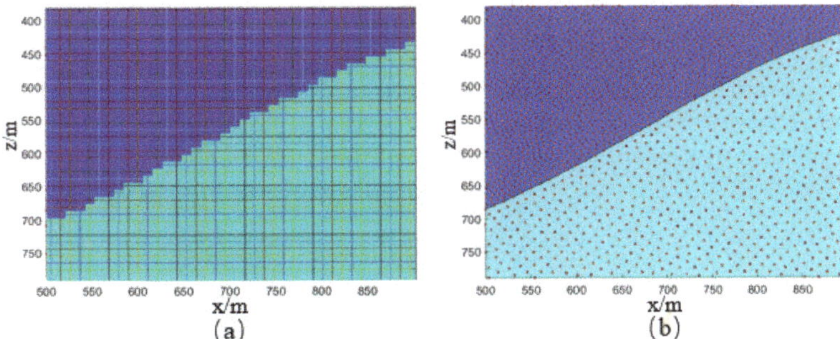

Figure 11. Local grid (node) distribution: (**a**) staggered grid finite difference method, (**b**) generalized finite difference method.

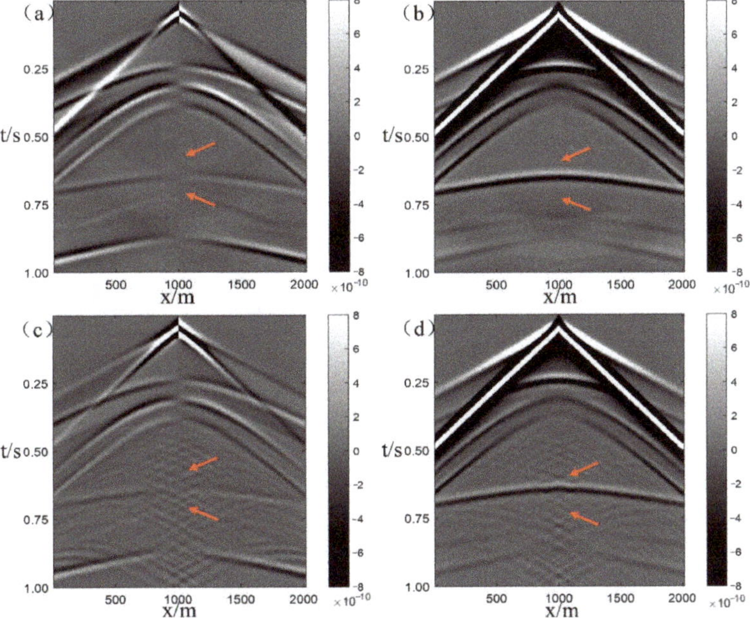

Figure 12. Shot record of undulating interface model: (**a**) SGFD X component; (**b**) SGFD Z component; (**c**) 4th-order GFDM X component; (**d**) 4th-order GFDM Z component.

3.4. Fault Model

Fault zones are closely related to the genesis of many fossil deposits. Therefore, a model with igneous rock (model parameters in gray) and fault zones, as shown in Figure 13, was designed with an irregular boundary for elastic wave numerical simulation. The time sampling interval was 0.5 ms, and the total computation time was 2.0 s. The source was located at (1200 m, 10 m), and the dominant frequency of the Ricker wavelet was 10 Hz. When the 4th-order 21 nodes GFDM was used, the proposed node generation algorithm was applied to discretize the model. For comparison, the SGFD model was used based on regular grids with a grid spacing of 10 m and with the irregular boundary extended to a rectangular boundary. It can be seen from the seismic records obtained using the two different methods that the direct and reflected wave events are essentially the same, but it is noted that there were a number of scattered waves behind the reflection wave in the SGFD records (Figure 14a,b), while there are almost no scattered waves in the GFDM records (Figure 14c,d). This shows that the method used in this study is suitable for application to such complex models without any sharp lateral changes. Moreover, the GFDM method does not require an extension of the irregular boundary to a rectangular boundary, as it can calculate the wavefield by directly setting nodes on the irregular boundary.

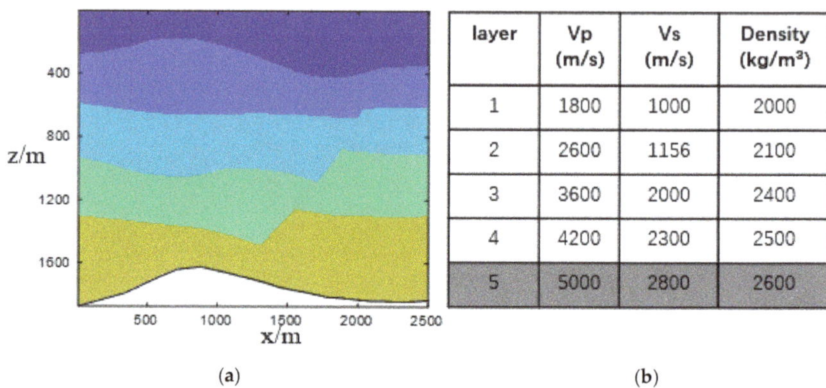

layer	Vp (m/s)	Vs (m/s)	Density (kg/m³)
1	1800	1000	2000
2	2600	1156	2100
3	3600	2000	2400
4	4200	2300	2500
5	5000	2800	2600

Figure 13. Fault model with irregular boundary: (**a**) model structure; (**b**) model parameters.

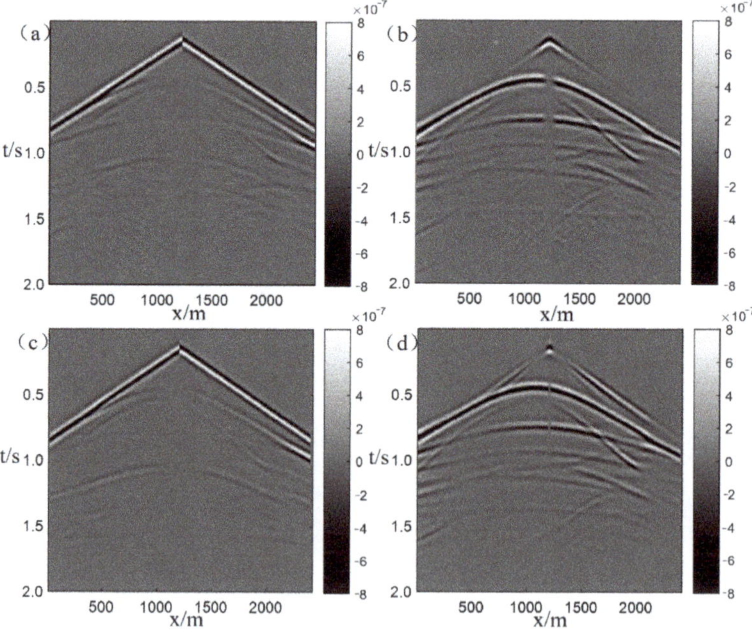

Figure 14. Shot record of fault model: (**a**) SGFD X component; (**b**) SGFD Z component; (**c**) 4th-order GFDM X component; (**d**) 4th-order GFDM Z component.

4. Conclusions

In this study, the time partial derivative of the elastic wave equation was discretized using the central difference scheme, and the spatial partial derivative was discretized using a meshless GFDM. When realizing the elastic wave numerical simulation via the GFDM in the time domain, the Poisson disk node generation algorithm and the centroid Voronoi node adjustment algorithm were combined to obtain an even and random distribution of nodes, and the Cerjan damping boundary condition was introduced. Through numerical calculation, the following understandings and conclusions were obtained:

(a) The GFDM is a meshless numerical calculation method based on scattered node approximation; it overcomes the dependence of traditional methods on grids, and can lay nodes in the simulation area in a flexible manner;
(b) When using the GFDM for forward modeling, the nodes can be rendered consistent with the real velocity interface by generating a suitable distribution of nodes, so that the velocity interface can be accurately depicted, avoiding the situation where the grid and velocity interface cannot be aligned in the conventional finite difference forward modeling; thus eliminating the diffraction problems due to a stepped grid, and the record time error caused by an inaccurate description of the interface position;
(c) One of the main problems in the GFDM forward modeling is in how to best discretize the model. The node discretization scheme used in this study is applicable to a model with gentle changes in lateral velocity. For models with sharp changes in lateral velocity, the simulation stability will be affected, so it would be necessary to explore a more stable and applicable node discretization scheme;
(d) The 2nd-order GFDM has high computational efficiency, but low accuracy. The 4th-order GFDM is unstable when using a "star" composed of fewer points (for example, 13 points), therefore it can only use a "star" composed of more calculation nodes (21 nodes were used in this study), causing the calculation efficiency to be greatly reduced. To improve computing efficiency, high-performance computing can be considered to enable better efficiency;
(e) Compared with SGFD, the GFDM requires some preprocessing before forward modeling can be applied, including calculation of the node distribution and the difference stencil, which makes GFDM more complicated in practical application.

Author Contributions: Conceptualization, S.L. and Z.Z.; methodology, S.L.; software, S.L.; validation, S.L.; formal analysis, W.Z.; review and editing, W.Z. and Z.Z.; supervision, Z.Z. All authors have read and agreed to the published version of the manuscript.

Funding: This work was supported by the Fundamental Research Funds for the Central Universities of Central South University 2019ZZTS300.

Institutional Review Board Statement: Not applicable.

Informed Consent Statement: Not applicable.

Data Availability Statement: The data that support the findings of this study are available from the corresponding author upon reasonable request.

Acknowledgments: We are grateful for technical support from the High-Performance Computing Center of Central South University. We gratefully acknowledge the editors and reviewers for providing thoughtful and useful suggestions.

Conflicts of Interest: The authors declare no competing financial interests.

References

1. Marfurt, K.J. Accuracy of finite difference and finite element modeling of the scalar and elastic wave equation. *Geophysics* **1984**, *49*, 533–549. [CrossRef]
2. Shi, R.; Wang, S.; Guo, R.; Zhao, J. Finite-element numerical modeling of elastic wave on unstructured meshes. *Oil Geophys. Prospect.* **2013**, *48*, 915–923. (In Chinese) [CrossRef]
3. Virieux, J. P-SV wave propagation in heterogeneous media; velocity-stress finite-difference method. *Geophysics* **1986**, *51*, 1933–1942. [CrossRef]
4. Wang, Y.; Liu, H.; Zhang, H.; Wang, Z.Y.; Tang, X.D. A Global Optimized Implicit Staggered-Grid Finite-Difference Scheme for Elastic Wave Modeling. *Chin. J. Geophys.* **2016**, *58*, 2508–2524. (In Chinese) [CrossRef]

5. Savovi, S.; Drljaa, B.; Djordjevich, A. A comparative study of two different finite difference methods for solving advection–diffusion reaction equation for modeling exponential traveling wave in heat and mass transfer processes. *Ric. Mat.* **2022**, *71*, 245–252. [CrossRef]
6. Komatitsch, D.; Tromp, J. Spectral-element simulations of global seismic wave propagation—I. Validation. *Geophys. J. Int.* **2002**, *149*, 390–412. [CrossRef]
7. Tang, X.P.; Bai, C.Y. Elastic wavefield simulation using separated equations through pseudo-spectral method. *Oil Geophys. Prospect.* **2012**, *47*, 10. (In Chinese) [CrossRef]
8. Li, Z.C. Variable-grid high-order finite-difference numeric simulation of first-order elastic wave equation. *Oil Geophys. Prospect.* **2008**, *43*, 711–716. (In Chinese) [CrossRef]
9. Jastram, C.; Tessmer, E. Elastic modelling on a grid with vertically varying spacing. *Geophys. Prospect.* **2010**, *42*, 357–370. [CrossRef]
10. Hao, C.; Wang, X.; Zhao, H. Rotated staggered grid and perfectly matched layer. *Chin. Ence Bull.* **2006**, *51*, 1985–1994. [CrossRef]
11. Zhang, X.; Song, K.; Lu, M. Research progress and application of meshless method. *Chin. J. Comput. Mech.* **2003**, *20*, 730–742. (In Chinese) [CrossRef]
12. Liu, L.; Duan, P.; Zhang, Y.; Tian, K.; Tan, M.; Li, Z.; Dou, J.; Li, Q. Overview of mesh-free method of seismic forward numerical simulation. *Prog. Geophys.* **2020**, *35*, 117. (In Chinese) [CrossRef]
13. Gu, Y.; Wang, L.; Chen, W.; Zhang, C.; He, X. Application of the meshless generalized finite difference method to inverse heat source problems. *Int. J. Heat Mass Transf.* **2017**, *108*, 721–729. [CrossRef]
14. Jian, C.; Yan, G.; Wang, M.; Wen, C.; Liu, L. Application of the generalized finite difference method to three-dimensional transient electromagnetic problems. *Eng. Anal. Bound. Elem.* **2017**, *92*, 257–266. [CrossRef]
15. Zhang, T.; Ren, Y.F.; Yang, Z.Q.; Fan, C.M.; Li, P.W. Application of generalized finite difference method to propagation of nonlinear water waves in numerical wave flume. *Ocean Eng.* **2016**, *7*, 38. [CrossRef]
16. Jensen, P.S. Finite difference techniques for variable grids. *Comput. Struct.* **1972**, *2*, 17–29. [CrossRef]
17. Benito, J.J.; Urena, F.; Gavete, L. Influence of several factors in the generalized finite difference method. *Appl. Math. Model.* **2001**, *25*, 1039–1053. [CrossRef]
18. Benito, J.J.; Urea, F.; Gavete, L.; Alvarez, R. An h-adaptive method in the generalized finite differences. *Comput. Methods Appl. Mech. Eng.* **2003**, *192*, 735–759. [CrossRef]
19. Gavete, L.; Gavete, M.L.; Benito, J.J. Improvements of generalized finite difference method and comparison with other meshless method. *Appl. Math. Model.* **2003**, *27*, 831–847. [CrossRef]
20. Urea, F.; Benito, J.J.; Alvarez, R.; Gavete, L. Computational Error Approximation and H-Adaptive Algorithm for the 3-D Generalized Finite Difference Method. *Int. J. Comput. Methods Eng. Sci. Mech.* **2005**, *6*, 31–39. [CrossRef]
21. Benito, J.J.; Urea, F.; Gavete, L.; Salete, E.; Urea, M. Implementations with generalized finite differences of the displacements and velocity-stress formulations of seismic wave propagation problem. *Appl. Math. Model.* **2017**, *52*, 1–14. [CrossRef]
22. Ureña, F.; Benito, J.J.; Salete, E.; Gavete, L. Seismic Wave Propagation and Perfectly Matched Layers Using a GFDM. In Proceedings of the International Conference on Computational Science and Its Applications; Springer: Berlin, Heidelberg.
23. Urea, F.; Benito, J.J.; Salete, E.; Gavete, L. A note on the application of the generalized finite difference method to seismic wave propagation in 2D. *J. Comput. Appl. Math.* **2012**, *236*, 3016–3025. [CrossRef]
24. Benito, J.J.; Urea, F.; Gavete, L.; Salete, E.; Muelas, A. A GFDM with PML for seismic wave equations in heterogeneous media. *J. Comput. Appl. Math.* **2013**, *252*, 40–51. [CrossRef]
25. Benito, J.J.; Urea, F.; Salete, E.; Muelas, A.; Gavete, L.; Galindo, R. Wave propagation in soils problems using the Generalized Finite Difference Method. *Soil Dyn. Earthq. Eng.* **2015**, *79*, 190–198. [CrossRef]
26. Salete, E.; Benito, J.J.; Urea, F.; Gavete, L.; García, A. Stability of perfectly matched layer regions in generalized finite difference method for wave problems. *J. Comput. Appl. Math.* **2017**, *312*, 231–239. [CrossRef]
27. Jia, Z.F.; Wu, G.C.; Li, Q.Y.; Yang, L.Y.; Wu, Y. Generalized finite difference forward modeling of scalar wave equation. *Oil Geophys. Prospect.* **2022**, *57*, 101–110. (In Chinese) [CrossRef]
28. Fornberg, B.; Flyer, N. Fast generation of 2-D node distributions for mesh-free PDE discretizations. *Comput. Math. Appl.* **2015**, *69*, 531–544. [CrossRef]
29. Liu, X. Absorbing Boundary Conditions in Staggered-Grid and Mesh-Free Finite-Difference Numerical Modeling for Wave Equa-tions. Master's Thesis, China University of Petroleum, Beijing, China, 2018. [CrossRef]
30. Liu, X.; Ma, L.; Guo, J.; Yan, D.M. Parallel Computation of 3D Clipped Voronoi Diagrams. *IEEE Trans. Vis. Comput. Graph.* **2020**, *1*, 2288. [CrossRef] [PubMed]
31. Cerjan, C.; Dan, K.; Kosloff, R.; Reshef, M. A Nonreflecting boundary-condition for discrete acoustic and elastic wave-equations. *Geophysics* **1985**, *50*, 705–708. [CrossRef]

32. Li, N. The Study on Numerical Simulation Method and Wave Field Characteristics of Orthorhombic Anisotropic Media. Ph.D. Thesis, China University of Petroleum, Qingdao, China, 2014.
33. Wang, Y. Frequencies of the Ricker wavelet. *Geophys. J. Soc. Explor. Geophys.* **2015**, *80*, A31–A37. [CrossRef]

applied sciences

Article

Study on Seismic Attenuation Based on Wave-Induced Pore Fluid Dissolution and Its Application

Ziqi Jin, Xuelin Zheng, Ying Shi * and Weihong Wang

School of Earth Sciences, Northeast Petroleum University, Daqing 163318, China
* Correspondence: shiying@nepu.edu.cn

Abstract: Seismic wave attenuation is affected by wave-induced pore fluid dissolution. The mechanism of wave-induced pore fluid dissolution is the mutual dissolution between different fluids caused by pore fluid pressure. Compared with the traditional WIFF (wave-induced fluid flow) mechanism, the wave-induced pore fluid dissolution mechanism can predict the attenuation of the seismic frequency band and can be used in well-to-seismic calibration. Conventional methods neglect the velocity dispersion caused by the interaction between pore fluids, which will lead to errors in attenuation prediction. In this paper, we focus on accurately predicting the velocity dispersion at low porosity and permeability, which can be used in multi-scale data matching. The stretch between the synthetic data by using logging data and seismic data needs to be calibrated for more accurate interpretation. The kernel of well-to-seismic calibration is the knowledge of the velocity dispersion between the logging frequency band and seismic frequency band. We calibrate the difference between the two kinds of data by using the rock physical model. Both the model test and field data application prove the feasibility and accuracy of the proposed strategy.

Keywords: rock physical model; upscaling model; pore fluid dissolution

Citation: Jin, Z.; Zheng, X.; Shi, Y.; Wang, W. Study on Seismic Attenuation Based on Wave-Induced Pore Fluid Dissolution and Its Application. *Appl. Sci.* **2023**, *13*, 74. https://doi.org/10.3390/app13010074

Academic Editor: Francesco Clementi

Received: 30 November 2022
Revised: 7 December 2022
Accepted: 17 December 2022
Published: 21 December 2022

1. Introduction

Wave-induced fluid flow (WIFF) is considered to be the main cause of seismic wave dispersion and attenuation in fluid-saturated porous media [1,2]. Among many theories, mesoscopic heterogeneity and microscopic heterogeneity are considered to be the main mechanisms leading to WIFF. In addition, in most rocks, the coexistence of mesoscopic heterogeneity and microscopic heterogeneity can cause a significant shift in fast p-wave velocity, which means that the effects of both mechanisms on dispersion and attenuation need to be considered simultaneously. The Squirt microscopic model mainly uses the squirt mechanism of solid/fluid interaction to estimate velocity dispersion and attenuation in fully saturated rocks [3]. Pride et al. proposed the two-pore theory to explain the corresponding seismic waves in water-bearing and gas-saturated porous media, which can explain the attenuation of magnitude $10^{-2} - 10^{-1}$ within the seismic frequency band. White's model describes the complex moduli of a partially saturated spherical gas encapsulated medium and a layered medium composed of two heterogeneous porous media [4,5]. Johnson generalized gas patches of arbitrary shape [6]. In recent years, under the condition of saturated fluid and bubble phase in rock, it is believed that an obvious attenuation phenomenon will occur at the low-frequency end of an earthquake, resulting in the corresponding mechanism of gas dissolution and dissolution induced by seismic wave (pore fluid dissolution). The dissolution of wave-induced gas exsolution–dissolution is used to explain the obvious attenuation phenomenon of seismic waves [7–9], and these models reflect the meso-loss mechanism. Chapman measured the attenuation of two Berea sandstone samples, and the results show that compared with the WIFF mechanism, the pore fluid dissolution mechanism can be closer to the measured attenuation data [10]. Moreover, as a seismic wave attenuation mechanism, pore fluid dissolution can describe micron pores. It is more suitable for shale reservoirs with low porosity and low permeability.

Well logging data and seismic data are the two most basic forms of raw data in the process of oilfield exploration and development, and they play an important role in detailed reservoir description, including reservoir prediction. The reflected reservoir information is inconsistent, resulting in multiple solutions for reservoir prediction and description. Determining how to perform well-to-seismic matching has become an indispensable and important part of predicting reservoir lithology, physical parameters and oil potential in oil and gas exploration and development research. There are three main methods for velocity matching and calibration of well-to-seismic data:

1. The Backus effective average method uses effective medium theory and a layered method to achieve high-frequency to low-frequency velocity correction. [11] deeply discussed the effect of Backus effective averaging on elastic scattering characteristics in the medium; ref. [12] used Backus effective averaging to conduct in-depth research on logging data scaling upscaling. The P and S wave velocities lead to rock physical analysis conclusions comparable to seismic scales.
2. The multi-resolution analysis method stretches the signals between adjacent events of the sonic logging synthetic seismic traces to align them with the corresponding events on the side-well traces. Ref. [13] applied multi-resolution analysis technology to the resampling of logging data and the matching of synthetic records and seismic traces, and the matching accuracy was high.
3. The rock physical model calibration method combines velocity dispersion theory and an absorption attenuation mechanism to achieve velocity calibration at different scales. Based on the resonant Q model [14], many researchers [15] measured the sonic logging velocity in different target blocks. Dispersion calibration was performed to improve the well–seismic matching effect. Ref. [16] combined the DEM model and the microscopic dispersion theoretical model to extrapolate the logging frequency band velocity to the seismic frequency band for well–seismic matching and inversion calculations and achieved good results in carbonate reservoirs.

In practical application, the Backus effective averaging method is simple and effective but does not consider the attenuation effect. Multiresolution analysis is highly automated but still requires accurate synthetic seismic record and stratigraphic correlation. The rock physical model calibration method can directly match data of different scales. In order to combine well logging and seismic data, it is necessary to select an appropriate rock physical model for dispersion calibration, thereby improving the accuracy of well–seismic matching.

Because the parameters of a rock physical model are large, difficult to obtain and have no actual physical significance, it is better to use a viscoelastic model instead of a rock physical model. The Zener model can fit the mesoscale White model [17], the Biot–Rayleigh model [18] and the dispersion and attenuation at two scales (meso and micro) [19]. Some researchers feel that the Cole–Cole model can more accurately simulate the acoustic properties of porous media than the rough approximation of Zener's mechanical model [20,21], but there are five parameters, which is relatively more, while the Power Law model only needs two parameters [22], which makes it easier to use. Ref. [10] studied the seismic attenuation in Berea sandstone saturated with bidirectional fluid, compared the attenuation curves of the WIFF model and the pore fluid dissolution model under different saturation and pressure, and used SLS model to fit, and they found that the fitting effect with pore fluid dissolution is very good. Some scholars also compared multiple viscoelastic models. Paul compared the attenuation and frequency relationship curves of Maxwell Model, Voight Solid, Standard Linear, Burgers Solid and Power Law. It was found that attenuation and elastic stiffness curves vary considerably with frequency, and each model has a different variation law [23]. Toverud et al. compared eight models using the zero-offset vertical seismic profile (VSP) dataset: the Kolsky–Futterman model, Power Law model, Kjartansson model, Muller model, Azimi second model, Azimi third model, Cole–Cole model and standard linear-solid-state model (SLS). It was found that in the same depth region, the SLS model has the best results in simulating attenuation [24].

2. Materials and Methods

Rock Physics Model Upscaling by Visco-Elastic Model

Pore fluid dissolution, as an attenuation mechanism of seismic waves, will lead to dissolution and dissolution between gas and its surrounding liquid, which will lead to the attenuation of seismic waves. In the sphere area of each bubble, the gas dissolution rate can be expressed as [7]:

$$dn/dt = -4\Pi r D_w \left(C_w - \frac{P_f + 2\gamma/r}{RTK_H} \right) \tag{1}$$

where *dn/dt* is the dissolution rate of gas, r is the bubble radius, D_w is the diffusion coefficient, C_w is the concentration, P_f is the pressure in the fluid, γ is the surface tension coefficient, R is the gas constant, T is the temperature, and K_H is the Henry's Law constant.

Rock physics models provide velocity dispersion between the seismic data and logging data for specific rock properties. Attenuation curve $Q_{rock}(f)$ can be used for generating synthetic seismic records:

$$u(t,\omega) = u_0 exp(-i\omega t) exp\left(-\frac{\omega t}{2Q_{rock}(f)} \right) exp\left(\frac{i\omega t}{\pi Q_{rock}(f)} \ln\left| \frac{\omega}{\omega_r} \right| \right) \tag{2}$$

Considering the explicit $Q(f)$ expression in rock physics model are unavailable, visco-elastic models are upscaled with non-physical meaning parameters. Attenuation properties can then be well modeled by solving the following problem by using least-squares objective function [25,26]:

$$\min \sum_f |Q_{rock}(f) - Q_{visco}(f)|^2 \tag{3}$$

SLS model can model most rock physics models [10], and the $Q_{SLS}(f)$ can be expressed by:

$$Q_{SLS}(f) = 1 + \frac{(2\pi f)^2 \tau_\sigma{}^2}{1 + \omega (2\pi f)^2 \tau_\sigma{}^2} \tau / \frac{(2\pi f) \tau_\sigma{}^2}{1 + (2\pi f)^2 \tau_\sigma{}^2} \tau \tag{4}$$

where τ_σ is stress relaxation time, and τ is strain relaxation time. Combining Equations (2) and (3), the upscaled $Q(f)$ by the SLS model can be derived:

$$\min_{\tau,\tau_\sigma,L} \sum_f \left| Q_{rock}(f) - (1 + \sum_{l=1}^{L} \frac{(2\pi f)^2 \tau_\sigma(l)^2}{1 + (2\pi f)^2 \tau_\sigma(l)^2} \tau(l)) / (\sum_{l=1}^{L} \frac{(2\pi f) \tau_\sigma(l)}{1 + (2\pi f)^2 \tau_\sigma(l)^2} \tau(l))^2 \right| \tag{5}$$

where L is relaxation mechanisms, and $\tau = \frac{\tau_\varepsilon}{\tau_\sigma} - 1$, $Q_{\text{rock}}(f)$ is the numerically modeled attenuation curve based on the rock physics model. Here, we use the wave-induced gas exsolution–dissolution (WIGED) model to calculate the $Q_{\text{rock}}(f)$, and the parameters used are listed in Table 1. The volumetric strain of the pore fluid is calculated, followed by the modulus K, and attenuation can be obtained by $Q(f)^{-1} = \frac{Im(K(f))}{Re(K(f))}$.

Table 1. Parameters used in three rock physics models: the pore fluid dissolution model, the SLS model and the Pride model.

Parameter	Symbol	Value	Unit
bulk modulus of oil	K_O	1.8 × 109	Pa
surface tension of gas	Sigma	0.012	Pa·m
gas constant	R	8.3144621	m^3·Pa/K mol
Henry's law constant of gas	KHpc	4535	Pa·m^3/mol
temperature	T	350	K
gas diffusion coefficient in water	DW	1.00×10^{-8}	m^2/s
gas solubility	S1	4535	Pa·m^3/mol
initial pressure	Pi	2.00×10^{7}	Pa
delta pressure	dP	100	Pa
water viscosity	visc_f	1.00×10^{-3}	Pa·s
CO_2 viscosity	visc_g	4.80×10^{-5}	Pa·s
water density	rho_w	840	kg/m^3
gas density	rho_g	119	kg/m^3
CH_4 viscosity	visc_g	6.00×10^{-6}	Pa·s
oil viscosity	visc_f	8.00×10^{-4}	Pa·s
high pressure	Ph	2.00×10^{7}	Pa
initial radius of bubble	r	5.00×10^{-6}	m
water saturation	Sat	1.50×10^{-3}	unitless
porosity	phi	6.00×10^{-2}	unitless
mineral content of sand	Sand	0.35	unitless
mineral content of clay	Clay	0.65	unitless
Poisson ratio	pois	0.14	unitless
bulk modulus of the gas	Kg	1.00×10^{4}	Pa
permeability	perm	1.00×10^{-12}	m^2

Well-to-seismic calibration based on rock physics model upscaling.

The misfit between the seismic data and the logging data can then be calibrated by implementing the following procedures (Figure 1):

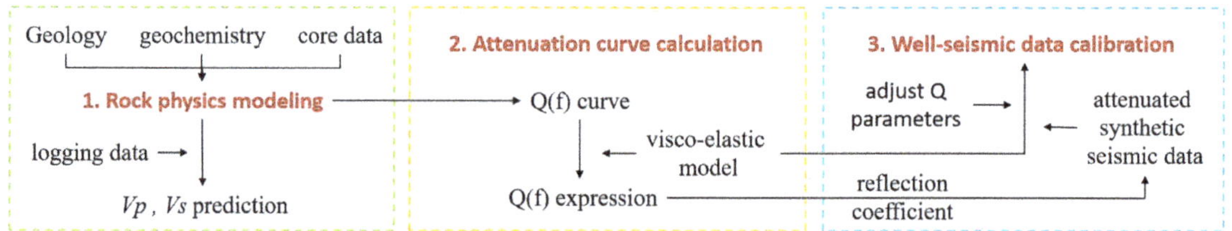

Figure 1. Procedures of the proposed well-to-seismic calibration method.

1. Rock physics modeling. Based on the geology, geochemistry and core data, a suitable rock physics model is selected, and the parameters in the model are upgraded. The WIGED model is used here for fluid, and the SCA model is used for rock matrix. Then, the saturated rock is obtained by Gassmann substitution. The key parameters in the WIGED model are the size of the bubble and diffusion length.
2. Attenuation curve calculation. *Q(f)* curve is then calculated from the rock physics model. The explicit *Q(f)* expression can be derived through upscaling by visco-elastic model. Here, we use the SLS model for upscaling, which yields best results compared with other visco-elastic models, such as the power-law model.
3. Well-to-seismic data calibration. Based on the reflection coefficient and *Q(f)*, the attenuated synthetic seismic data can be generated. By adjusting the *Q* related parameters in visco-elastic model (strain relaxation time and stress relaxation time in the SLS model), the optimal match between synthetic seismic data and field data can be found.

3. Synthetic Data Test

3.1. Rock Physics Modeling and Velocity Dispersion Analyzing

Here, three rock physics models, the pore fluid dissolution model, the SLS model and the Pride model, are applied for P-wave velocity dispersion calculation, as shown in Figure 2. Here, we consider rocks with micrometric pores, and gas micro-bubbles exist in the pores with bubble radium 1 um. Other parameters used in the models are listed in Table 1.

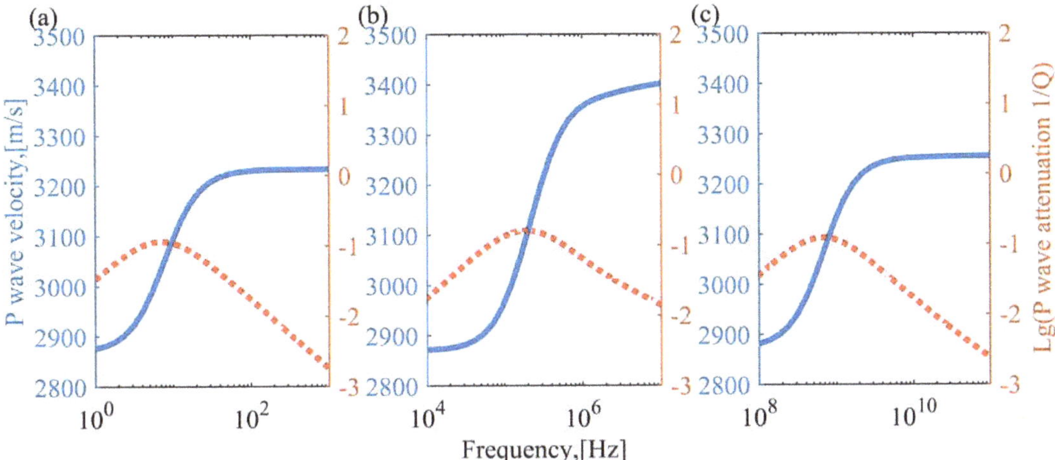

Figure 2. Comparison of velocity dispersion and attenuation calculated by three rock physics models. (**a**) Pore fluid dissolution model, (**b**) Squirt model and (**c**) Pride model. The blue solid line is the P wave velocity, and the red dash line is attenuation.

The corner frequency of the Squirt model is around 10^5 Hz, and the Pride model's around 10^9 Hz. Compared with the other two models, the pore fluid dissolution model is suitable for description of this rock's seismic frequency band, the corner frequency of which occurs between 0.1 and 100 Hz. The maximum values of the attenuation curves of the three models are similar. The Squirt model and the Pride model predict slightly larger attenuation. For the velocity dispersion, the pore fluid dissolution model and the Pride model share similar velocity range, with lower bound 2880 m/s and upper bound 3230 m/s. Although the lower bound is similar, the Squirt model yields a larger upper bound of 3400 m/s. The blue solid line is the P_wave velocity, and the red dash line is attenuation. In our model, the mobility of fluid is low due to the low porosity and low permeability, which makes it impossible for the conventional WIFF model to be used. The choice of a suitable rock physics model need to be tested by comparing the attenuation and velocity dispersion curves, and the best one is that which can model the attenuation in the seismic frequency range.

Considering that the constant Q assumption is always accepted for real data application, conventional constant Q theory is then compared for showing the velocity dispersion between logging frequency and seismic frequency. The results of velocity dispersion are shown in Figure 3. The porosity ranges from 0.04 to 0.08, and gas saturation ranges from 0.15% to 0.25%. The SLS model yields almost the same results as the pore fluid dissolution model, while the constant Q model yields a monotonous velocity curve, which contains large difference from that of the pore fluid dissolution model. As shown in Figure 3a–c, as the porosity increases, the velocity range decreases in the pore fluid dissolution model. The upper bound decreases from 3500 m/s to 3000 m/s, and the lower bound decreases from 3300 m/s to 2550 m/s. As shown in Figure 3d–f, as the gas saturation increases, the velocity range shows little difference.

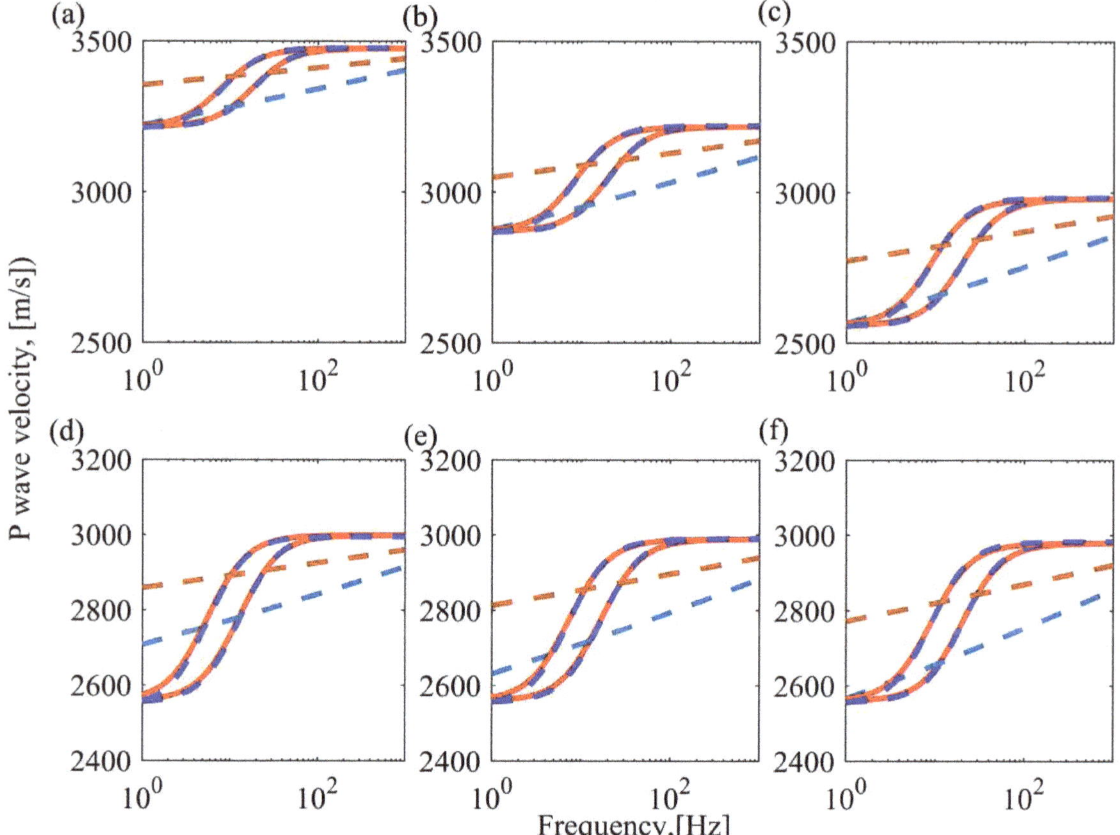

Figure 3. Comparison of velocity dispersion calculated by using the pore fluid dissolution model, the SLS model and the constant Q model. (**a**) Porosity 0.04 and gas saturation 0.25%, (**b**) Porosity 0.06 and gas saturation 0.25%, (**c**) Porosity 0.08 and gas saturation 0.25%, (**d**) Porosity 0.08 and gas saturation 0.15%, (**e**) Porosity 0.08 and gas saturation 0.2%, (**f**) Porosity 0.08 and gas saturation 0.25%. The two red solid curves are the results from the pore fluid dissolution model, the dash blue curves are from the SLS model, and the other dash curves (orange curve and cyan curve) are from the constant Q model.

The two red solid curves are the results from the pore fluid dissolution model, where the bubble radius are 4×10^{-6} and 6×10^{-6}, respectively. The dash blue curves are from the SLS model. The other dash curves (orange curve and cyan curve) are from the constant Q model, and the Q values are picked from the pore fluid dissolution model at 100 Hz.

The corresponding $1/Q$ curves are compared in Figure 4. The two curves in each subfigure represent different bubble radii. As porosity increases, the attenuation becomes larger, while the gas saturation increase influences little about the attenuation.

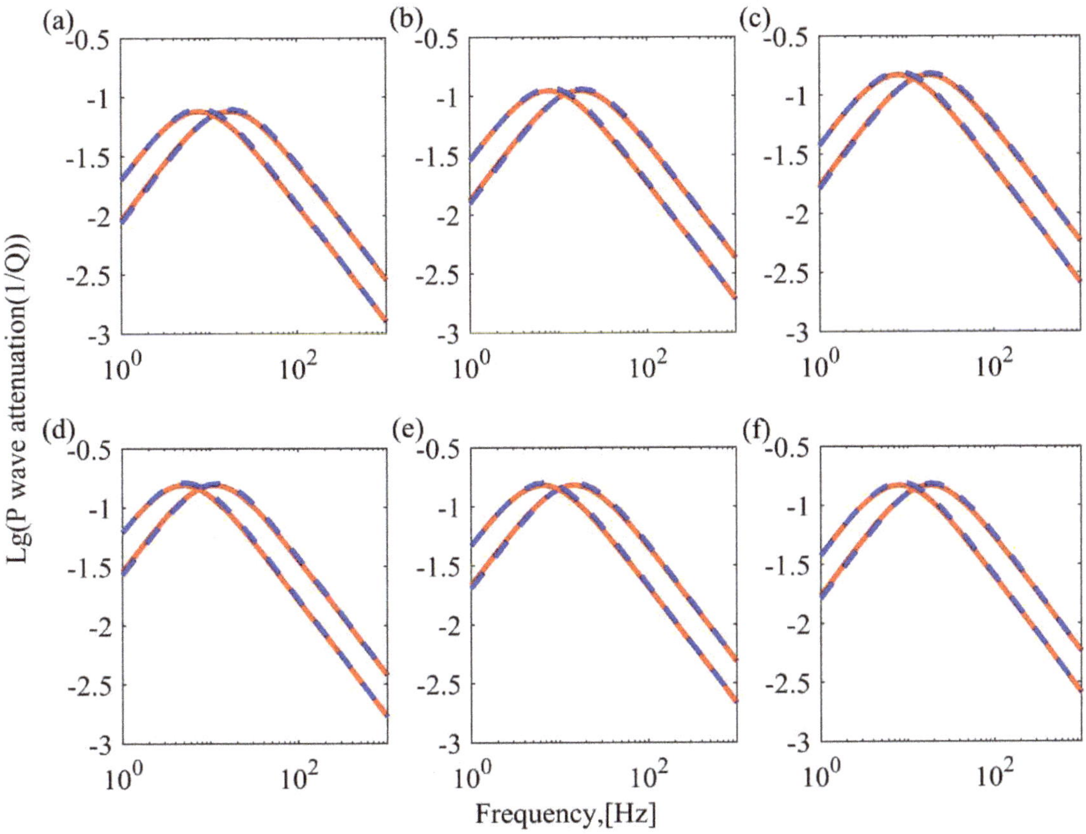

Figure 4. Comparison of 1/Q curves from the pore fluid dissolution model and the SLS model. The red solid curve is the pore fluid dissolution model, and the dash blue curve is the SLS mode. (**a**) Porosity 0.04 and gas saturation 0.25%, (**b**) Porosity 0.06 and gas saturation 0.25%, (**c**) Porosity 0.08 and gas saturation 0.25%, (**d**) Porosity 0.08 and gas saturation 0.15%, (**e**) Porosity 0.08 and gas saturation 0.2%, (**f**) Porosity 0.08 and gas saturation 0.25%.

3.2. Velocity Variation from Logging Data to Seismic Data Based on Rock Physics Model and Backus Averaging Method

In this section, logging data is used for demonstration of velocity difference between upscaling by the pore fluid dissolution model and the Backus averaging method. Firstly, the Backus averaging method is applied, and the velocity variation is shown in Figure 5. Using different sizes of windows, the velocity results from the Backus averaging method show different properties. As the size of the window becomes larger, the velocity curve shows larger divergence from the real data. However, the velocity from all these three cases varies in the same range as that of the real data.

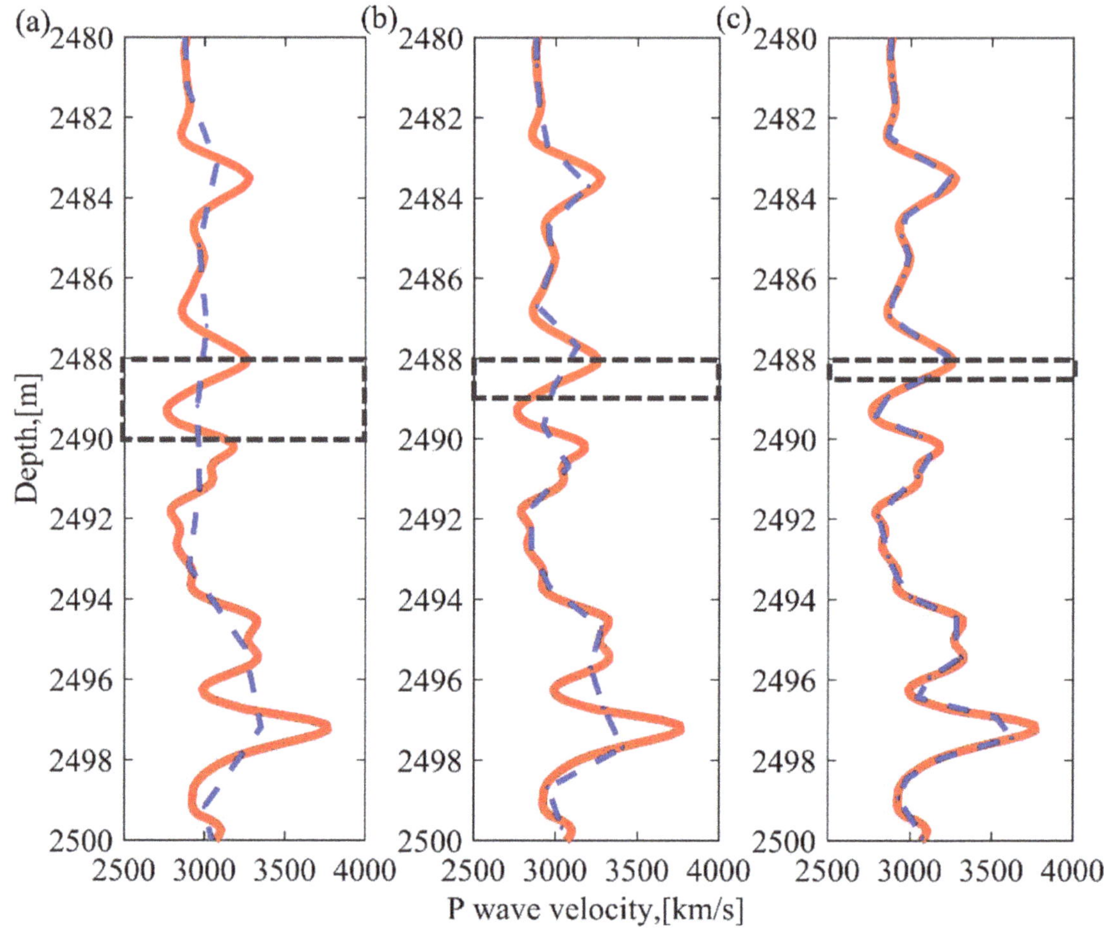

Figure 5. P_wave velocity calculated by using the Backus averaging method. The red solid curve is the real data, and the blue dash curve is the result from the Backus averaging method. (**a**) 16 samples in a window, (**b**) 8 samples in a window, (**c**) 4 sample in a window.

Then, the SLS model and the constant Q model are both applied to the real data for velocity correction. The results are shown in Figure 6. Compared with that of the Backus averaging method, the velocity upscaled by the SLS model and constant Q model contains obvious variation from read data. Both the models result in smaller velocity values than the real data. The SLS model always yields smaller velocity than the constant Q model in these cases.

3.3. Velocity Dispersion Correction on Synthetic Seismic Data

The upscaled velocities from the SLS model, the constant Q model and the Backus averaging method are all applied to the synthetic seismic data for velocity dispersion correction. The synthetic seismic data are generated by convolution of the reflection coefficient with the minimum phase wavelet. The original velocity used is that in the logging band. The amplitude of the data is not changed, as different velocity is used, only the phase. Firstly, the Backus averaging method is applied as shown in Figure 7. For windows with three different sizes, the synthetic data show little variation from the real data. The event at around 0.7 s arrives earlier that of the real data.

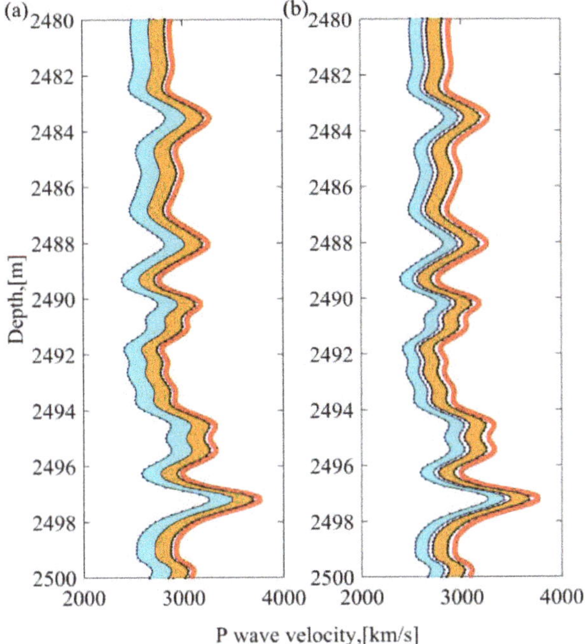

Figure 6. Velocity variation from the pore fluid dissolution model and the constant Q model. The orange area is the velocity range of the constant Q model. The cyan area is the velocity range of the pore fluid dissolution model. (**a**) Porosity ranges from 0.04 to 0.08, gas saturation is 0.25%, (**b**) porosity is 0.08, gas saturation ranges from 0.15% to 0.25%.

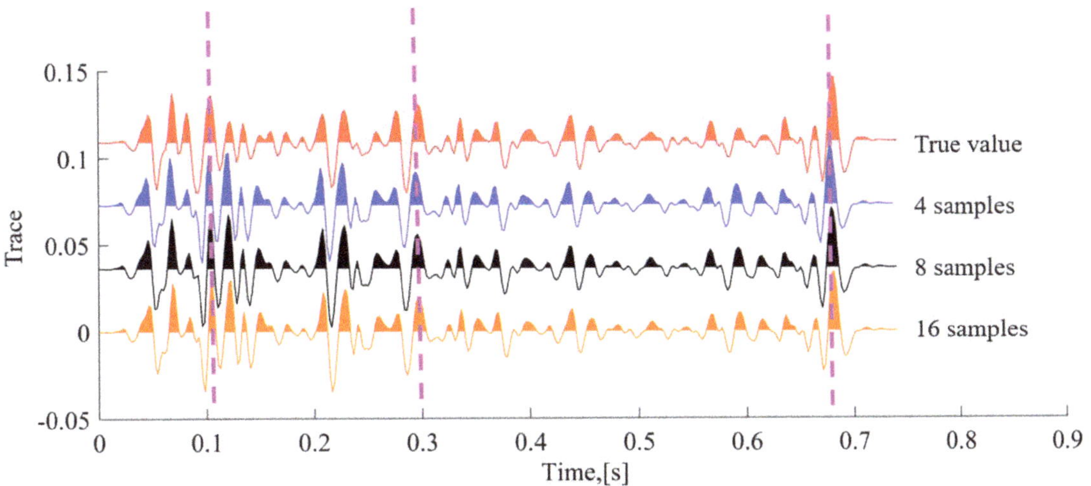

Figure 7. Comparison of synthetic seismic data by using upscaled velocity from three different sizes of sample window using the Backus averaging method.

Then, the constant Q model is applied for the velocity dispersion correction. The results are shown in Figure 8. For the maximum velocity of the constant Q model, the synthetic data are similar to the real data. For the minimum velocity, as the values are smaller than those of logging velocity, the synthetic data stretch obviously, especially for the data after 0.4 s.

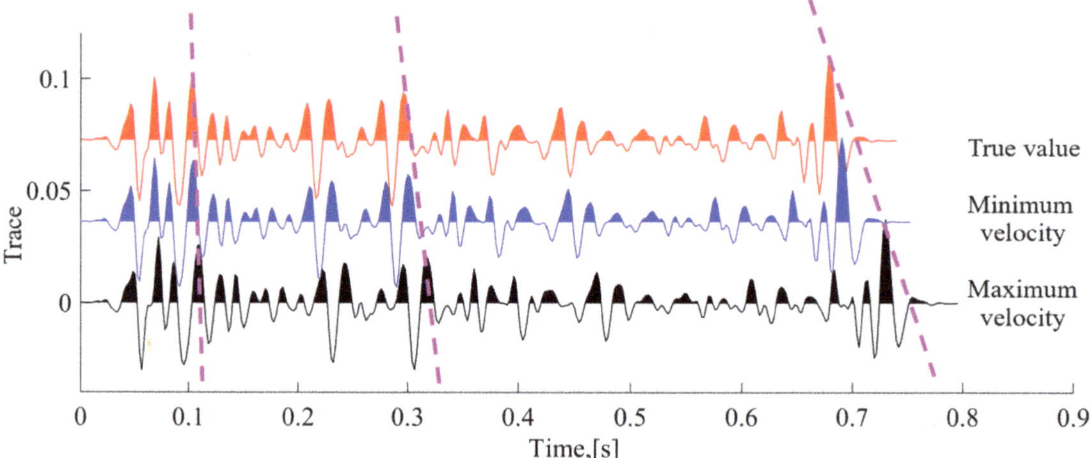

Figure 8. Comparison of synthetic seismic data by using upscaled velocity from the maximum and minimum velocity of constant Q model in Figure 6a.

The SLS model is then applied and shown in Figure 9. As the velocity shows much smaller values in Figure 6a, the synthetic data show much more stretch than both the constant Q model and the Backus averaging method. The stretch can be seen starting from 0.2 s, and it becomes larger at 0.4 s. The stretch is largest at 0.9 s. The whole length of the data is 1.2 s.

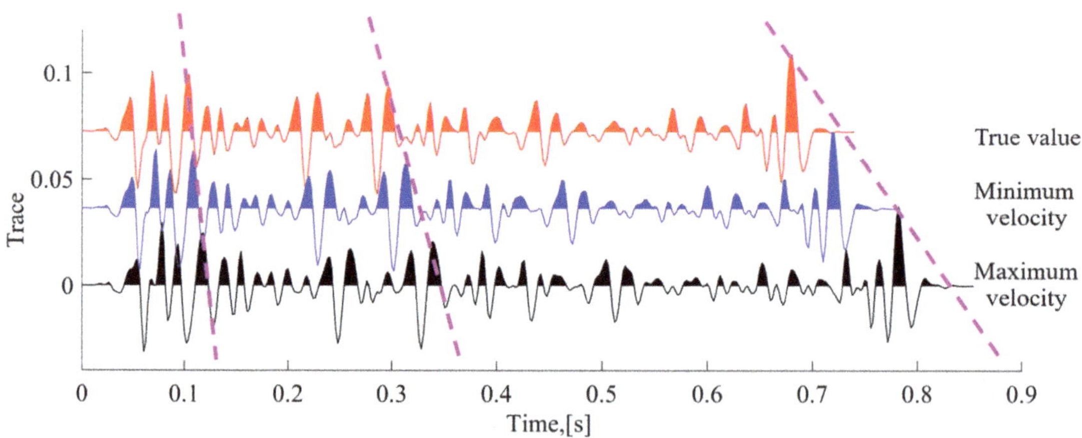

Figure 9. Comparison of synthetic seismic data by using upscaled velocity from the maximum and minimum velocity of the SLS model in Figure 6a.

4. Real Data Application

The proposed method is then applied to field data acquired in northern China. The Q^{-1} results are calculated using logging data by the pore fluid dissolution model. Figure 10a shows the Q^{-1} results at 60 Hz of this model in the depth from 1500 m to 2500 m. The results match well with the Q^{-1} measurements from the core data. The velocity is then upscaled by the SLS model. Figure 10b shows the well-to-seismic calibration before and after using velocity dispersion results from the SLS method as guidance for stretching the events. The event at 1900 ms is fixed, and the event at 1820 ms is stretched 5 ms. The event at 1740 ms is stretched 8 ms with the event at 1820 ms fixed. The corresponding velocity dispersion is calculated as roughly 7% and 4%, which is consistent with that of the SLS model. The upscaled velocity is useful for well-to-seismic calibration and makes the correction more reasonable.

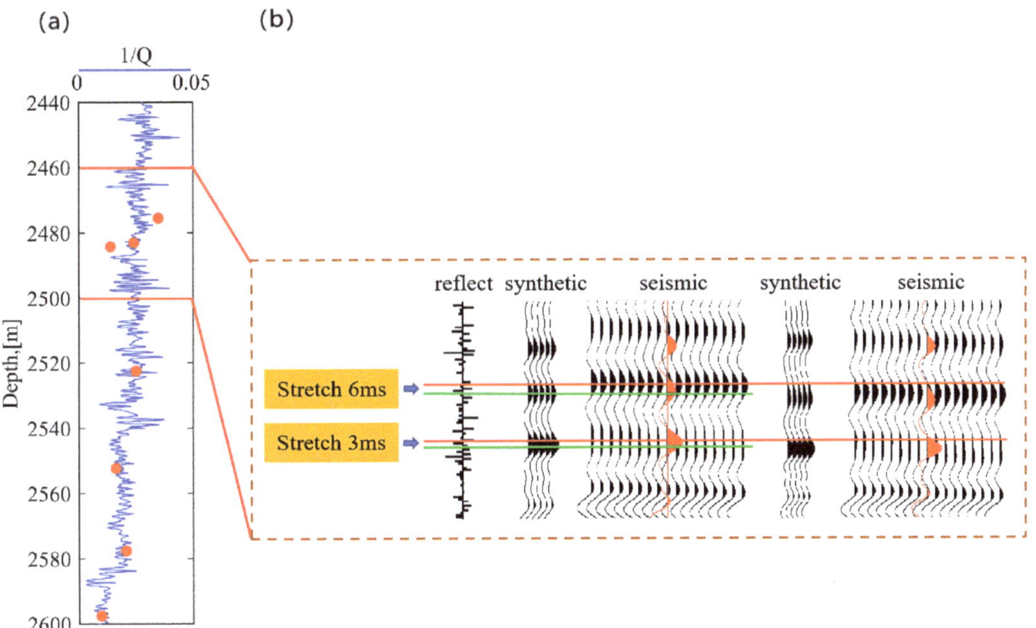

Figure 10. (**a**) Comparison of Q^{-1} results by the WIGED model using logging data and data from core measurements. (**b**) Well-to-seismic calibration before and after using Q^{-1} results from the WIGED method as a guidance.

5. Discussion

The Squirt model and the Pride model are not suitable for predicting the attenuation of rocks containing micro-pores at the seismic frequency band. The energy loss caused by fluid flow is neglectable, and gas dissolution and exsolution become the dominant attenuation mechanisms.

Although the conventional constant Q model is widely accepted, it will not predict accurate velocity dispersion when the corner frequency occurs in the seismic frequency band. For the same rock properties, the velocity range differs from that of the pore fluid dissolution model. The Backus averaging method almost results in a similar velocity range to that of real data. The size of the window chosen will not improve accuracy of velocity. Neither of these two methods are suitable for upscaling. The SLS model is suitable for upscaling the pore fluid dissolution model. The velocity dispersion and attenuation curve can be well modeled by the SLS model for different porosity and gas saturation cases.

Synthetic seismic data generated by using velocities calculated by these methods show obvious difference. Due to similar velocity range from the Backus averaging method, the events in the synthetic seismic data show little stretch. The constant Q model underestimates the velocity dispersion, and thus, the synthetic seismic data show little stretch until 0.4 s. The attenuation becomes observable because of the accumulation of the attenuation effect. The SLS model reasonably predicts the attenuation, and the synthetic seismic data show obvious difference from the traces generated by using logging velocity. When choosing a rock physical model for well-to-seismic calibration, one needs to consider the geology, geochemistry, and core data analysis report, and then select the appropriate rock physics model and update the parameters in the model.

6. Results

In this paper, an appropriate rock physical model is introduced, which can be used in seismic-to-well calibration. For rocks contains micro-pores, the pore fluid dissolution model is suitable for predicting the attenuation in the seismic frequency band. Compared with conventional methods, such as the constant model and the Backus averaging method, the SLS model can be chosen as the appropriate model for upscaling the pore fluid dissolution model. The introduced model can be used for seismic-to-well calibration. Both the synthetic seismic data and the field data test demonstrated the feasibility and accuracy of the proposed method.

Author Contributions: Methodology, Z.J. and X.Z.; Writing—review and editing, W.W.; Supervision, Y.S. All authors have read and agreed to the published version of the manuscript.

Funding: The study was conducted in accordance with the National Natural Science Foundation of China (grant no. 41930431).

Data Availability Statement: Not applicable.

Acknowledgments: The authors are grateful to the National Natural Science Foundation of China (grant no. 41930431).

Conflicts of Interest: The authors declare no conflict of interest.

References

1. Pride, S.R.; Berryman, J.G.; Harris, J.M. Seismic attenuation due to wave-induced flow. *J. Geophys. Res. Solid Earth* **2004**, *109*, 1–19. [CrossRef]
2. Müller, T.M.; Gurevich, B.; Lebedev, M. Seismic wave attenuation and dispersion resulting from wave-induced flow in porous rocks—A review. *Geophysics* **2010**, *75*, A147–A175. [CrossRef]
3. Dvorkin, J.; Mavko, G.; Nur, A. Squirt flow in fully saturated rocks. *Geophysics* **1995**, *60*, 97–107. [CrossRef]
4. White, J.E. Computed seismic speeds and attenuation in rock with partial gas saturation. *Geophysics* **1975**, *40*, 224–232. [CrossRef]
5. White, J.E.; Mikhaylova, N.G.; Lyakhovitskiy, F.M. Low-frequency seismic waves in fluid-saturated layered rocks. *Phys. Solid Earth* **1975**, *11*, 654–659. [CrossRef]
6. Johnson, D.L. Theory of frequency dependent acoustics in patchy-saturated porous media. *Geology* **2001**, *110*, 682–694. [CrossRef]
7. Tisato, N.; Quintal, B.; Chapman, S.; Podladchikov, Y.; Burg, J.P. Bubbles attenuate elastic waves at seismic frequencies: First experimental evidence. *Geophys. Res. Lett.* **2015**, *42*, 3880–3887. [CrossRef]
8. Tisato, N.; Madonna, C. Attenuation at low seismic frequencies in partially saturated rocks: Measurements and descriptions of a new apparatus. *J. Appl. Geophys.* **2012**, *86*, 44–53. [CrossRef]
9. Tisato, N.; Quintal, B. Measurements of seismic attenuation and transient fluid pressure in partially saturated Berea sandstone: Evidence of fluid flow on the mesoscopic scale. *Geophys. J. Int.* **2013**, *195*, 342–351. [CrossRef]
10. Chapman, N.; Quintal, S.; Tisato, B.; Holliger, K. Frequency scaling of seismic attenuation in rocks saturated with two fluid phases. *Geophys. J. Int.* **2017**, *208*, 221–225. [CrossRef]
11. Sams, M.S.; Wolliamson, P.R. Backus averaging, scattering and drift. *Geophys. Prospect.* **1994**, *42*, 541–564. [CrossRef]
12. Cao, D. The upscaling method of the well logging data based on Backus equivalence average method. *Geophys. Prospect. Pet.* **2015**, *54*, 105–111.
13. Dong, E.Q.; Liu, G.Z.; Zhang, Z.P.; Chen, H.Z. Resampling of Logging Data and Matching of Synthetic Seismogram with Seismic Trace by Multiresolution Analysis. *WLT* **1999**, *23*, 264–267.
14. Yang, W. A resonance Q model for viscoelastic rocks. *Acta Geophys. Sin.* **1987**, *30*, 399–411.

15. Ba, J.; Jm, C.; Cao, H.; Du, Q.Z.; Yuan, Z.Y.; Lu, M.H. Velocity dispersion and attenuation of P waves in partially-saturated rocks: Wavepropagation equations in double-porosity medium. *Chin. J. Geophys.* **2012**, *55*, 219–231. (In Chinese)
16. Wang, Z.; Sun, Y.; Xiao, Y. Full-frequency band velocity prediction model and its application on seismic reservoir prediction. *Geology* **2011**, *73*, 23–26.
17. Picotti, S.; Carcione, J.M.; Rubino, J.G.; Santos, J.E.; Cavallini, F. A viscoelastic representation of wave attenuation in porous media. *Comput. Geosci.* **2010**, *36*, 44–53. [CrossRef]
18. Wang, E.; Carcione, J.M.; Ba, J. Wave simulation in double-porosity media based on the Biot-Rayleigh theory. *Geophysics* **2019**, *84*, WA11–WA21. [CrossRef]
19. Zhang, B.; Yang, D.; Cheng, Y.; Zhang, Y. A unified poroviscoelastic model with mesoscopic and microscopic heterogeneities. *Geology* **2019**, *64*, 1246–1254. [CrossRef]
20. Picotti, S.; Carcione, J.M. Numerical simulation of wave-induced fluid flow seismic arrenuation based on the Cole-Cole model. *Geology* **2017**, *142*, 134–145.
21. Lu, J.F.; Hanyga, A. Numerical modelling method for wave peopagation in a linear viscoelastic medium with singulat memory. *Grophysical J. Int.* **2004**, *159*, 688–702. [CrossRef]
22. Bano, M. Modelling of GPR waves for lossy media obeying a complex power law of frequency for dielectric permittivity. *Geophys. Prospect.* **2004**, *52*, 11–26. [CrossRef]
23. Hagin, P.N.; Zoback, M.D. Viscous deformation of unconsolidated reservoir sands—Part 2: Linear viscoelastic models. *Geophysics* **2004**, *69*, 742–751. [CrossRef]
24. Toverud, T.; Ursin, B. Comparison of seismic attenuation models using zero-offset vertical seismic profiling (VSP) data. *Geophisics* **2005**, *70*, F17–F25. [CrossRef]
25. Sacchi, M.D. Reweighting strategies in seismic deconvolution. *Geophys. J. Int.* **1997**, *129*, 651–656. [CrossRef]
26. Rickett, J. Integrated estimation of interval-attenuation distributions. *Geophysics* **2006**, *71*, A19–A23. [CrossRef]

MDPI

Article

Processing the Artificial Edge-Effects for Finite-Difference Frequency-Domain in Viscoelastic Anisotropic Formations

Jixin Yang [1,2,3], Xiao He [1,2,3,*] and Hao Chen [1,2,3]

1 State Key Laboratory of Acoustics, Institute of Acoustics, Chinese Academy of Sciences, Beijing 100190, China; yangjixin@mail.ioa.ac.cn (J.Y.); chh@mail.ioa.ac.cn (H.C.)
2 School of Physical Sciences, University of Chinese Academy of Sciences, Beijing 100049, China
3 Beijing Engineering Research Center of Sea Deep Drilling and Exploration, Beijing 100190, China
* Correspondence: hex@mail.ioa.ac.cn

Abstract: Real sedimentary media can usually be characterized as transverse isotropy. To reveal wave propagation in the true models and improve the accuracy of migrations and evaluations, we investigated the algorithm of wavefield simulations in an anisotropic viscoelastic medium. The finite difference in the frequency domain (FDFD) has several advantages compared with that in the time domain, e.g., implementing multiple sources, multi-scaled inversion, and introducing attenuation. However, medium anisotropy will lead to the complexity of the wavefield in the calculation. The damping profile of the conventional absorption boundary is only defined in one single direction, which produces instability when the wavefields of strong anisotropy are reflected on that truncated boundary. We applied the multi-axis perfectly matched layer (M-PML) to the wavefield simulations in anisotropic viscoelastic media to overcome this issue, which defines the damping profiles along different axes. In the numerical examples, we simulated seismic wave propagation in three viscous anisotropic media and focused on the wave attenuation in the absorbing layers using time domain snapshots. The M-PML was more effective for wave absorption compared to the conventional perfectly matched layer (PML). In strongly anisotropic media, the PML became unstable, and prominent reflections appeared at truncated boundaries. In contrast, the M-PML remained stable and efficient in the same model. Finally, the modeling of the stratified cross-well model showed the applicability of this proposed algorithm to heterogeneous viscous anisotropic media. The numerical algorithm can analyze wave propagation in viscoelastic anisotropic media. It also provides a reliable forward operator for waveform inversion, wave equation travel-time inversion, and seismic migration in anisotropic viscoelastic media.

Keywords: frequency-domain modeling; anisotropy; viscoelastic; edge-effect removing

Citation: Yang, J.; He, X.; Chen, H. Processing the Artificial Edge-Effects for Finite-Difference Frequency-Domain in Viscoelastic Anisotropic Formations. *Appl. Sci.* **2022**, *12*, 4719. https://doi.org/10.3390/app12094719

Academic Editors: Guofeng Liu, Xiaohong Meng and Zhifu Zhang

Received: 22 March 2022
Accepted: 5 May 2022
Published: 7 May 2022

1. Introduction

Regarding the numerical modeling of wave propagation in anisotropic media, numerous studies have been implemented using finite difference and spectral methods in the time domain [1–3] to obtain the numerical solution in discrete geological media. These media may contain rocks with various properties, defined as acoustic, elastic, viscoelastic isotropic, and anisotropic. With the flourishing development of computational technology and numerical algorithms, wavefield simulation methods have played a crucial role in imaging subsurface characteristics. Various imaging methods, such as tomography [4,5], reverse time migration [6–8], and full-waveform inversion [9–11], require an efficient wavefield simulation method to generate synthetic waveform data to match the observed data.

Accurately performing wavefield modeling with high efficiency in complex media, which contains anisotropy and attenuation, is critical. In contrast to time-domain wavefield modeling, although it is challenging to solve the sparse linearized matrix, multi-source modeling can be performed efficiently, and attenuation can be easily introduced [12]. In addition, the physical parameters of the reservoir can be estimated by seismic inversion and imaging methods within a limited number of frequency components [13]. Operto et al. [14] developed frequency-domain finite-difference viscoacoustic wavefield modeling in tilted transversely isotropic (TTI) media. Jeong et al. [15] used the finite element method to simulate the frequency-domain wavefield of vertically transversely isotropic (VTI) media. Zhou et al. [16] implemented seismic wave modeling in elastic anisotropic media using the generalized stiffness reduction method. Yang et al. [17] developed a new version of the generalized stiffness reduction method for anisotropic media and incorporated it into the 2.5D spectral element method. Qiao et al. [3] derived attenuative anisotropic wave equations involving fractional time derivatives and solved them using the pseudospectral method. Reducing artificial reflections is still essential despite numerous frequency-domain modeling methods.

Figure 1 shows a sketch of the absorbing boundary condition (ABC). The middle area is the computational area, and the dashed-dot area is the truncated boundary. When the wavefield propagates to the boundary, the ABC needs to attenuate the reflections inwards. Since Berenger [18] developed a perfectly matched layer (PML) to reduce artificial reflections of the electromagnetic wavefield, many studies have shown success in its applications to acoustic and elastic wave modeling [19,20], and improvements have also been made to the perfectly matched layer method [21]. The PML method has become popular for wave modeling problems because it is applicable to first-order wave equations in the time domain [22] and frequency domain [23]. However, it is not straightforward how to implement it for second-order wave equations [24]. In addition, the most severe problem of PML may exist in absorbing artificial reflections in elastic anisotropic media. This is because PML can only function effectively when all slowness vector components of the waves are parallel to their corresponding group velocity vectors [25]. As is known, an elastic anisotropic wavefield generates multiple modes of waves and becomes more complex when the PML becomes unstable in dealing with the edge-effect problem. Thus, we need to replace PML in wavefield simulations of elastic anisotropic media.

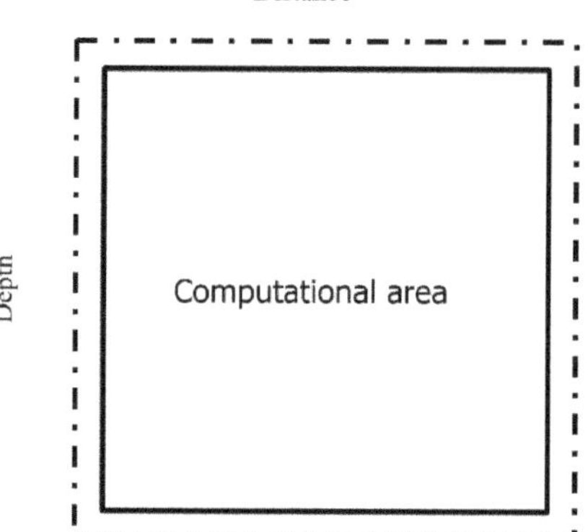

Figure 1. Sketch map of the absorbing boundary condition (ABC). The square area in the middle is the computational area. The thick dashed-dot line represents the ABC applied outside the computational area.

A multi-axial perfectly matched layer (M-PML) boundary condition was developed in the time domain [26], and its stability has been verified in time-domain isotropic elastic modeling [27,28]. In this paper, the M-PML boundary condition is applied to frequency-domain finite-difference modeling in several elastic anisotropic media. The key point of this method is based on a more general version of coordinate stretching for PML, where waves are absorbed in all directions because the damping factor is specified in more than one direction. Then, by adjusting the proportion coefficients of the damping factors according to a specific model, M-PML can be stably performed. Comparing the wavefield generated in two homogenous media shows that PML, frequently used in the first-order wave equation, has the disadvantage of removing artificial reflections. At the same time, the M-PML performs well due to the damping profiles specified in multiple directions. Numerical experiments conducted in a strongly anisotropic medium show that M-PML remains stable in the frequency and time domains for viscoelastic anisotropic wavefield propagation. In addition, the M-PML requires only a modification of the damping factors. Therefore, the additional computational cost remains unchanged for techniques that rely on wavefield simulation, such as wave equation migration and waveform inversion.

2. Theoretical Foundations

2.1. Frequency Domain Elastic VTI Equation

The 2D frequency-domain anisotropic elastic Equation [10] is given as:

$$\begin{aligned} &\rho\omega^2 u_x + C_{11}\frac{\partial^2 u_x}{\partial x^2} + C_{44}\frac{\partial^2 u_x}{\partial z^2} + (C_{13}+C_{44})\frac{\partial^2 u_z}{\partial x \partial z} + f(\omega) = 0, \\ &\rho\omega^2 u_z + C_{44}\frac{\partial^2 u_z}{\partial x^2} + C_{33}\frac{\partial^2 u_z}{\partial z^2} + (C_{13}+C_{44})\frac{\partial^2 u_x}{\partial x \partial z} + g(\omega) = 0, \end{aligned} \tag{1}$$

where ρ is the density, $f(\omega)$ and $g(\omega)$ are Fourier components of horizontal and vertical body forces, respectively; u_x and u_z are the horizontal and vertical displacements, respectively; and ω is the angular frequency. The elastic stiffness is written as:

$$\begin{aligned} &C_{11} = (1+2\varepsilon)\rho v_{pr}^2, \\ &C_{33} = \rho v_{pr}^2, \\ &C_{13} = \sqrt{2\delta C_{33}(C_{33}-C_{44}) + (C_{33}-C_{44})^2} - C_{44}, \\ &C_{44} = C_{55} = \rho v_{sr}^2, \end{aligned} \tag{2}$$

where v_{pr} and v_{sr} are the P-wave and the SV-wave velocity, respectively; ε and δ are Thomsen's anisotropic parameters [29].

The frequency-domain finite-difference method can easily introduce attenuation [30]. The constant-Q model is adopted here, and the complex velocity is given as:

$$\begin{aligned} &\frac{1}{v_p(\omega)} = \frac{1}{v_{pr}}\left(1 - \frac{1}{\pi Q_p}\ln\left|\frac{\omega}{\omega_r}\right|\right)\cdot\left(1 - \frac{i}{2Q_p}\right), \\ &\frac{1}{v_s(\omega)} = \frac{1}{v_{sr}}\left(1 - \frac{1}{\pi Q_s}\ln\left|\frac{\omega}{\omega_r}\right|\right)\cdot\left(1 - \frac{i}{2Q_s}\right), \end{aligned} \tag{3}$$

where v_p, v_s is the complex p-wave and SV-wave velocity, respectively; and Q_p, Q_s is the quality factor of the P-wave and SV-wave; ω_r is the dominant frequency of the source wavelet; *i* represents the imaginary unit. Substituting Equation (3) into Equation (2) yields the viscoelastic anisotropic wave equation.

2.2. Absorbing Boundary Condition

In order to introduce the M-PML absorbing condition to process unwanted reflections, the frequency-domain anisotropic elastic wave equation can be written as:

$$\begin{aligned}
&\rho\omega^2 u_x + \frac{1}{e_x}\left[C_{11}\frac{1}{e_x}\frac{\partial^2 u_x}{\partial x^2} + C_{11}\frac{\partial}{\partial x}\left(\frac{1}{e_x}\right)\frac{\partial u_x}{\partial x} + C_{13}\frac{1}{e_z}\frac{\partial^2 u_z}{\partial x \partial z} + C_{13}\frac{\partial}{\partial x}\left(\frac{1}{e_z}\right)\frac{\partial u_z}{\partial z}\right] \\
&+\frac{1}{e_z}\left[C_{44}\frac{1}{e_x}\frac{\partial^2 u_z}{\partial z \partial x} + C_{44}\frac{\partial}{\partial z}\left(\frac{1}{e_x}\right)\frac{\partial u_z}{\partial x} + C_{44}\frac{1}{e_z}\frac{\partial^2 u_x}{\partial z^2} + C_{44}\frac{\partial}{\partial z}\left(\frac{1}{e_z}\right)\frac{\partial u_x}{\partial z}\right] + f(\omega) = 0, \\
&\rho\omega^2 u_z + \frac{1}{e_x}\left[C_{44}\frac{1}{e_x}\frac{\partial^2 u_z}{\partial x^2} + C_{44}\frac{\partial}{\partial x}\left(\frac{1}{e_x}\right)\frac{\partial u_z}{\partial x} + C_{44}\frac{1}{e_z}\frac{\partial^2 u_x}{\partial x \partial z} + C_{44}\frac{\partial}{\partial x}\left(\frac{1}{e_z}\right)\frac{\partial u_x}{\partial z}\right] \\
&+\frac{1}{e_z}\left[C_{13}\frac{1}{e_x}\frac{\partial^2 u_x}{\partial z \partial x} + C_{13}\frac{\partial}{\partial z}\left(\frac{1}{e_x}\right)\frac{\partial u_x}{\partial x} + C_{33}\frac{1}{e_z}\frac{\partial^2 u_z}{\partial z^2} + C_{33}\frac{\partial}{\partial z}\left(\frac{1}{e_z}\right)\frac{\partial u_z}{\partial z}\right] + g(\omega) = 0,
\end{aligned} \tag{4}$$

where $e_x = 1 - \mathrm{i}\frac{d_x(x)}{\omega}$ and $e_z = 1 - \mathrm{i}\frac{d_z(z)}{\omega}$, x and z are respectively the distance from the inner boundary of the horizontal and vertical direction; $d_x(x)$ is the damping profiles, given as:

$$d_x(x) = 2\pi\alpha_0 f_0\left(\frac{x}{L}\right), \tag{5}$$

where x is the distance from the inner boundary of the horizontal direction; α_0 is the optimized parameter, and here α_0 is given as 1.79 [30]; f_0 is the dominant frequency of the source wavelet and L is the thickness of the PML absorbing boundary condition. The new vertical coordinate has a similar form to the horizontal one:

$$d_z(z) = 2\pi\alpha_0 f_0\left(\frac{z}{L}\right), \tag{6}$$

where z is the distance from the inner boundary of the vertical direction.

Artificial boundary reflection can seriously affect the effect of numerical modeling. Therefore, a frequency-domain multi-axial perfectly matched layer (M-PML) is exploited to absorb the unwanted reflections.

In classical PML, unwanted waves are only attenuated in one direction (uniaxial). Within the left and right boundaries, only the damping function along the x-direction is nonzero and can be formulated as:

$$d_x = d_x^x(x),\ d_z = 0. \tag{7}$$

Similarly, within the bottom and left boundaries, only the damping function along the z-direction takes effect and can be given as:

$$d_x = 0,\ d_z = d_z^z(z). \tag{8}$$

To attenuate unwanted reflections efficiently, M-PML was developed [26]. The basic idea of M-PML is that the damping functions are proportional to each other. The damping function along the x-direction can be defined as:

$$d_x = d_x^x(x),\ d_z = p^{(z/x)} d_x^x(x), \tag{9}$$

where $p^{(z/x)}$ is the proportion coefficient in either the right or left PML boundary.

Similarly, the damping function along the z-direction can be written as:

$$d_x = p^{(x/z)} d_z^z(z),\ d_z = d_z^z(z), \tag{10}$$

where $p^{(x/z)}$ is the proportion coefficient in either the bottom or top PML boundary. During implementation, the proportions $p^{(x/z)}$ are between 0 and 1 [27].

2.3. Numerical Implementation

Introducing the finite differences and derivatives can be summarized as follows:

$$
\begin{aligned}
\frac{\partial u_x}{\partial x} &\approx \frac{1}{2\Delta x}\left[u_x^{i+1,j} - u_x^{i-1,j}\right],\\
\frac{\partial u_x}{\partial z} &\approx \frac{1}{2\Delta z}\left[u_x^{i,j+1} - u_x^{i,j-1}\right],\\
\frac{\partial^2 u_x}{\partial x^2} &\approx \frac{1}{\Delta x^2}\left[u_x^{i+1,j} - 2u_x^{i,j} + u_x^{i-1,j}\right],\\
\frac{\partial^2 u_x}{\partial z^2} &\approx \frac{1}{\Delta z^2}\left[u_x^{i,j+1} - 2u_x^{i,j} + u_x^{i,j-1}\right],\\
\frac{\partial^2 u_x}{\partial x \partial z} &\approx \frac{1}{4\Delta x \Delta z}\left[u_x^{i+1,j+1} - u_x^{i+1,j-1} - u_x^{i-1,j+1} + u_x^{i-1,j-1}\right],
\end{aligned}
\tag{11}
$$

where Δx and Δz are the horizontal and vertical grid sizes, respectively; and the value of grid sizes must be small enough so that there are enough grid points per wavelength [10]; i and j are the indices of the horizontal and vertical grids. The finite differences of u_z are similar to the way of the horizontal one.

Using the finite-difference star stencil [10]:

$$
\begin{matrix}
\mathbf{M}_1 & \mathbf{M}_4 & \mathbf{M}_7\\
\mathbf{M}_2 & \mathbf{M}_5 & \mathbf{M}_8\\
\mathbf{M}_3 & \mathbf{M}_6 & \mathbf{M}_9
\end{matrix}\ .
\tag{12}
$$

Substituting the finite difference of derivatives and each grid point of Equation (12) according to Figure 2 can be collected as follows:

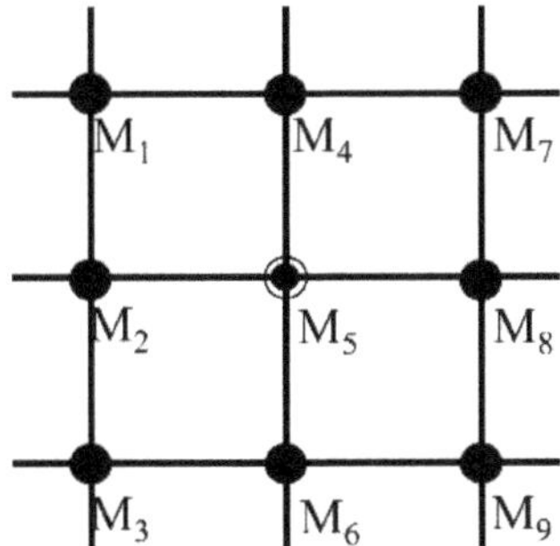

Figure 2. The second-order finite-difference star stencil used to model the viscoelastic anisotropic wave equation.

After using the finite difference method in the frequency domain, the final equation can be written as Equation (14). Thus, for each frequency component, the matrix equation can be given as:

$$
\mathbf{M}(\omega)\mathbf{U}(\omega) = \mathbf{F}(\omega). \tag{14}
$$

The impedance matrix $\mathbf{M}(\omega)$ can be obtained by Equation (13), $\mathbf{U}(\omega)$ is the monochromatic wavefield, and $\mathbf{F}(\omega)$ is the source term. The frequency-domain Ricker wavelet is used in the following numerical examples, and LU decomposition is used to solve the matrix Equation (14) to get the monochromatic wavefield.

$$\mathbf{M}_1 = \frac{(C_{13}+C_{44})_{i,j}}{4\Delta x\Delta z(e_x e_z)_{i,j}}\begin{bmatrix} 0 & 1 \\ 1 & 0 \end{bmatrix},$$

$$\mathbf{M}_2 = \begin{bmatrix} \frac{(C_{11})_{i,j}}{\left(\Delta x(e_x)_{i,j}\right)^2} + \frac{(C_{11})_{i,j}}{2\Delta x(e_x)^3_{i,j}} \cdot \frac{\partial(e_x)_{i,j}}{\partial x} & \frac{(C_{44})_{i,j}}{2\Delta x\left(e_z e_x^2\right)_{i,j}} \cdot \frac{\partial(e_x)_{i,j}}{\partial z} \\ \frac{(C_{13})_{i,j}}{2\Delta x\left(e_z e_x^2\right)_{i,j}} \cdot \frac{\partial(e_x)_{i,j}}{\partial z} & \frac{(C_{44})_{i,j}}{\left((e_x)_{i,j}\Delta x\right)^2} + \frac{(C_{44})_{i,j}}{2\Delta x(e_x)^3_{i,j}} \cdot \frac{\partial(e_x)_{i,j}}{\partial x} \end{bmatrix},$$

$$\mathbf{M}_3 = \frac{-(C_{13}+C_{44})_{i,j}}{4\Delta x\Delta z(e_x e_z)_{i,j}}\begin{bmatrix} 0 & 1 \\ 1 & 0 \end{bmatrix},$$

$$\mathbf{M}_4 = \begin{bmatrix} \frac{(C_{44})_{i,j}}{\left(\Delta z(e_z)_{i,j}\right)^2} + \frac{(C_{44})_{i,j}}{2\Delta z(e_z)^3_{i,j}} \cdot \frac{\partial(e_z)_{i,j}}{\partial z} & \frac{(C_{13})_{i,j}}{2\Delta z\left(e_x e_z^2\right)_{i,j}} \cdot \frac{\partial(e_z)_{i,j}}{\partial x} \\ \frac{(C_{44})_{i,j}}{2\Delta z\left(e_x e_z^2\right)_{i,j}} \cdot \frac{\partial(e_z)_{i,j}}{\partial x} & \frac{(C_{33})_{i,j}}{\left((e_z)_{i,j}\Delta z\right)^2} + \frac{(C_{33})_{i,j}}{2\Delta z(e_z)^3_{i,j}} \cdot \frac{\partial(e_z)_{i,j}}{\partial z} \end{bmatrix},$$

$$\mathbf{M}_5 = \begin{bmatrix} \omega^2\rho_{i,j} - \frac{2(C_{11})_{i,j}}{\left(\Delta x(e_x)_{i,j}\right)^2} - \frac{2(C_{44})_{i,j}}{\left(\Delta z(e_z)_{i,j}\right)^2} & 0 \\ 0 & \omega^2\rho_{i,j} - \frac{2(C_{44})_{i,j}}{\left(\Delta x(e_x)_{i,j}\right)^2} - \frac{2(C_{33})_{i,j}}{\left(\Delta z(e_z)_{i,j}\right)^2} \end{bmatrix}, \tag{13}$$

$$\mathbf{M}_6 = \begin{bmatrix} \frac{(C_{44})_{i,j}}{\left(\Delta z(e_z)_{i,j}\right)^2} - \frac{(C_{44})_{i,j}}{2\Delta z(e_z)^3_{i,j}} \cdot \frac{\partial(e_z)_{i,j}}{\partial z} & \frac{-(C_{13})_{i,j}}{2\Delta z\left(e_x e_z^2\right)_{i,j}} \cdot \frac{\partial(e_z)_{i,j}}{\partial x} \\ \frac{-(C_{44})_{i,j}}{2\Delta z\left(e_x e_z^2\right)_{i,j}} \cdot \frac{\partial(e_z)_{i,j}}{\partial x} & \frac{(C_{33})_{i,j}}{\left((e_z)_{i,j}\Delta z\right)^2} - \frac{(C_{33})_{i,j}}{2\Delta z(e_z)^3_{i,j}} \cdot \frac{\partial(e_z)_{i,j}}{\partial z} \end{bmatrix},$$

$$\mathbf{M}_7 = \frac{-(C_{13}+C_{44})_{i,j}}{4\Delta x\Delta z(e_x e_z)_{i,j}}\begin{bmatrix} 0 & 1 \\ 1 & 0 \end{bmatrix},$$

$$\mathbf{M}_8 = \begin{bmatrix} \frac{(C_{11})_{i,j}}{\left(\Delta x(e_x)_{i,j}\right)^2} - \frac{(C_{11})_{i,j}}{2\Delta x(e_x)^3_{i,j}} \cdot \frac{\partial(e_x)_{i,j}}{\partial x} & \frac{-(C_{44})_{i,j}}{2\Delta x\left(e_z e_x^2\right)_{i,j}} \cdot \frac{\partial(e_x)_{i,j}}{\partial z} \\ \frac{-(C_{13})_{i,j}}{2\Delta x\left(e_z e_x^2\right)_{i,j}} \cdot \frac{\partial(e_x)_{i,j}}{\partial z} & \frac{(C_{44})_{i,j}}{\left((e_x)_{i,j}\Delta x\right)^2} - \frac{(C_{44})_{i,j}}{2\Delta x(e_x)^3_{i,j}} \cdot \frac{\partial(e_x)_{i,j}}{\partial x} \end{bmatrix},$$

$$\mathbf{M}_9 = \frac{(C_{13}+C_{44})_{i,j}}{4\Delta x\Delta z(e_x e_z)_{i,j}}\begin{bmatrix} 0 & 1 \\ 1 & 0 \end{bmatrix}.$$

3. Synthetic Examples

3.1. Comparative Analysis of the Stability of PML and M-PML

In this section, a medium of strong anisotropy was employed to test the stability and performance of PML and M-PML. The material properties of the strongly anisotropic medium are given in Table 1: medium 3. We performed wavefield simulations to test the instability of the boundary conditions and confirmed that M-PML outperformed PML in this strongly anisotropic medium. A square model was considered embedded in an absorbing boundary of four sides. The computational model of medium 3, similar to the model used in Meza-Fajardo [25], has a size of 720 m $\times$ 720 m, and it is discretized by 480 $\times$ 480 grids with a grid size of 1.5 m. A boundary with a thickness of 120 grids is considered to observe the phenomenon of wave propagating inside truncated boundary conditions. A vertical source dominated by 15 Hz is located at the grid point (230 m, 230 m), which is close to the artificial edge. The real parts of the monochromatic wavefields using classical PML and M-PML, respectively, are given in Figure 3. In Figure 3, the monochromatic wavefield at 15 Hz was calculated using classical PML and M-PML, respectively. The asterisk in Figures 3 and 4 is the vertical source. The black dashed line in the figure shows the onset of the absorbing zone. From the results, it can be seen that the PML appears unstable at the boundary intersection in the upper right corner. This phenomenon is due to the complexity of the wavefield in the absorption boundary region, which leads to numerical instability of the PML. Since the PML is a uniaxially defined damping profile, absorption incompleteness is also observed in the other three corners. In the M-PML, however, since the damping profile is multi-axial, the instability that appears in the upper right corner of the PML is well resolved, and the attenuation of the waves is better in the corner regions.

Table 1. Model parameters used for numerical experiments.

Medium	Density (kg/m^3)	Qp	Qs	Elastic Moduli (GPa)
Medium 1	3200	20	10	$C_{11} = 16.70; C_{13} = 6.60;$ $C_{33} = 14.00; C_{44} = 6.63.$
Medium 2	8900	20	10	$C_{11} = 30.70; C_{13} = 10.3;$ $C_{33} = 35.80; C_{44} = 7.55.$
Medium 3	1000	—	—	$C_{11} = 4.000; C_{13} = 7.50;$ $C_{33} = 20.00; C_{44} = 2.00.$

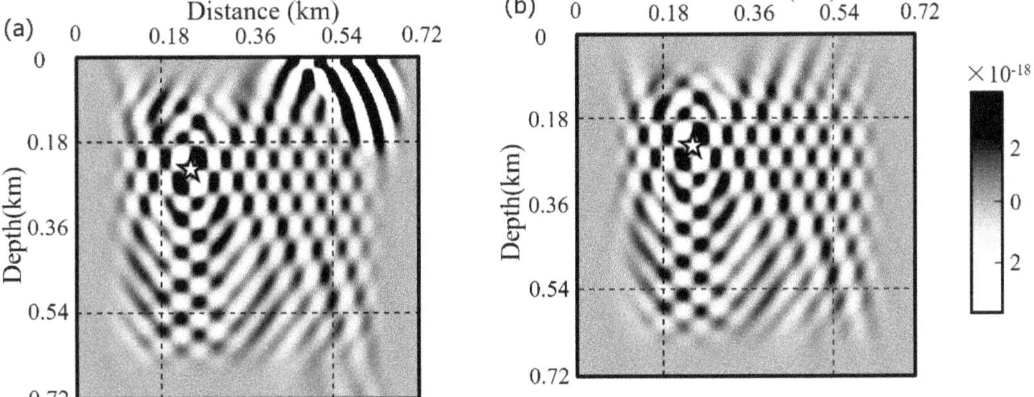

Figure 3. Real part of the horizontal component from the monochromatic wavefield at 15 Hz with (**a**) classical PML and (**b**) M-PML.

Snapshots of the time-domain wavefield corresponding to Figure 3 are given in Figure 4. The black dashed line in Figure 4 is the onset of the absorbing boundary condition, and the source is indicated by an asterisk. Figure 4a,c,e,g are the wavefields calculated by PML, and Figure 4b,d,f,h are the wavefields calculated by M-PML. The results show that the PML-calculated wavefields display a significant instability because at 0.0818 s, the wavefield has not yet propagated to the boundary, but strong energy reflections have already occurred, and these reflections result from the instability remain almost constant as the wavefield propagates. Therefore, PML is unstable in strongly anisotropic media. In contrast, the results of M-PML in the right panel have no unstable reflections, and the waves are successfully absorbed in the truncated boundary regions, so the M-PML with multi-axial damping profiles solves the problem of instability of PML in strongly anisotropic media.

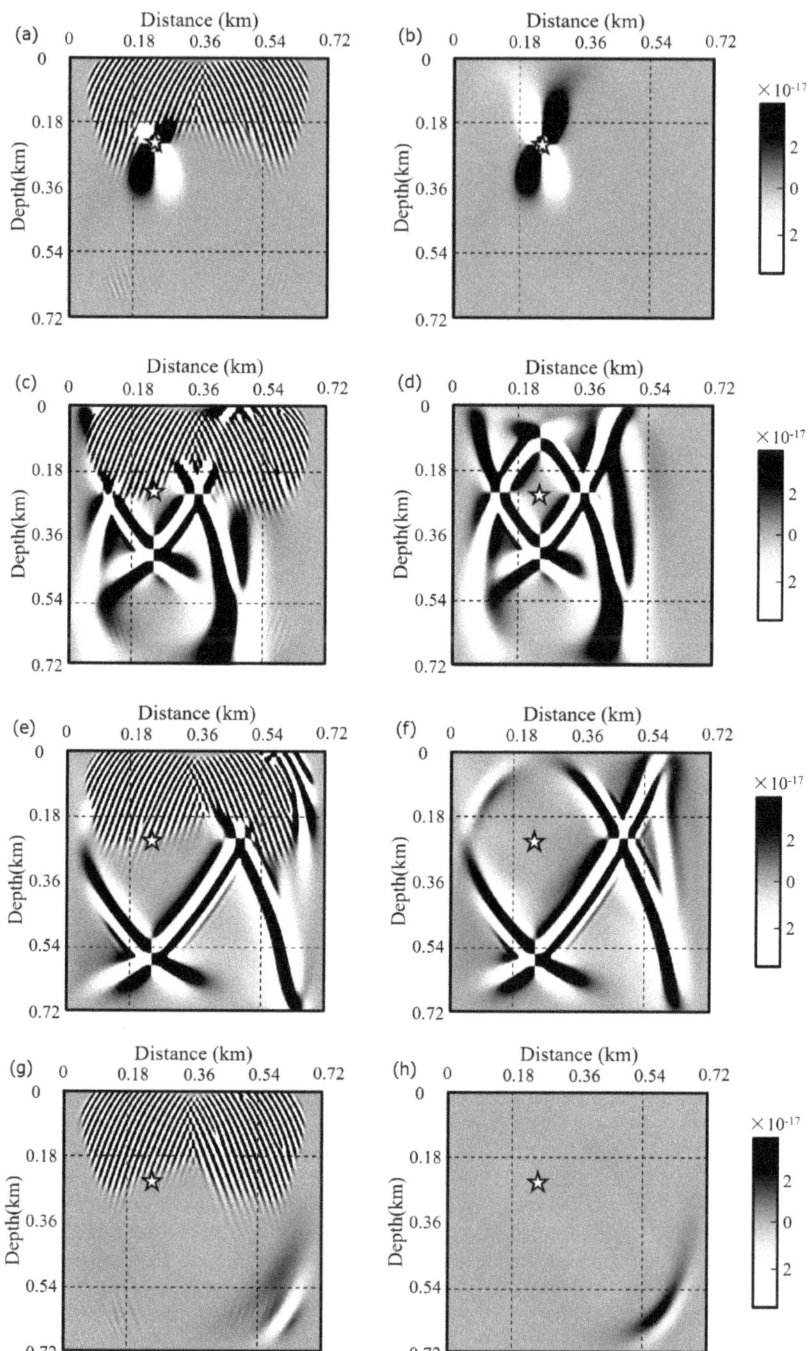

Figure 4. Snapshots of the horizontal component simulated with classical PML (**a**), (**c**), (**e**), (**g**) (left panel) and M-PML (**b**), (**d**), (**f**), (**h**) (right panel) at 0.0818 s, 0.2636 s, 0.4455 s, and 1.3046 s.

3.2. Comparative Analysis in Homogeneous Elastic Anisotropic Media

In the previous section, we verified that M-PML is superior to PML in terms of stability. However, the absorption performance advantages of M-PML and PML were not compared. Therefore, in this section, wavefield simulations are performed using M-PML and PML in different anisotropic media to compare the absorption of the two boundary conditions by observing the time-domain wavefield snapshots and comparing the energy curves. The energy of the wavefield snapshot is the sum of the squares of the amplitude values of each point in the wavefield. Then, the energy decay curve is obtained for each moment of the wavefield snapshot and plotted as a curve with the horizontal axis of time and the vertical axis of amplitude.

In the first test, the physical properties are given in Table 1: medium 1. We set the model size to be 1.0 km $\times$ 1.0 km with a grid space of 2.5 m, and employed an explosive Ricker wavelet source with a dominant frequency of 20 Hz. Figure 5 exhibits snapshots of horizontal displacement and its corresponding analytical group velocity calculated from medium 1. These snapshots were simulated using PML and M-PML, respectively, at different times. Figure 5a,b are the snapshots computed inside the PML and M-PML at 0.1667 s. The consistency between the wavefront calculated from the analytical group velocity and the wavefront of snapshots can be observed in Figure 5c. Similar to the figures above, Figure 5d–f are recorded at 0.2444 s. The wavefront of snapshots is almost the same as the wavefront calculated from the analytical group velocity, and the absorptive capacity of M-PML is superior to that of PML at four corners. Figure 5g,h are recorded at 0.7 s, and the absorptive advantage of M-PML over PML can be observed. Figure 5i shows the wavefield energy decay curves in the computational domain. These results show that the absorption performance of M-PML is better than that of PML when both ABCs remain stable, which can be verified by the comparative observations in Figure 5g,h. Because at 0.7 s, the effective wavefield has all propagated within the boundary, while the boundary reflections observed in PML are stronger at this time, and that of M-PML is much weaker. To further quantitatively compare the performance of the two ABCs, we gave the energy curves of them, where the horizontal axis is time, and the vertical axis is the logarithm of energy, and we can see that the energy of the waves of both ABCs becomes weaker as time passes, but the energy of M-PML decays faster than that of PML. In addition, in order to verify the absorptive capacity and reliability of M-PML in viscous media, the wavefield snapshots of M-PML in viscoelastic anisotropic media are given in Figure 6. From these results, we know that M-PML is still effective in viscous anisotropic media.

Regarding the verification of the results, we calculated the wavefront using the analytical group velocity in anisotropic media [31], and the results are displayed in Figure 5c,f, where the fast propagating wavefront is the P-wave, and the slow one is the S-wave. It can be seen that the wavefield snapshots at the corresponding moment also contain P-waves and S-waves. The wavefront in snapshots can correspond well with the analytical wavefront in Figure 5c,f, which illustrates the reasonableness and applicability of the numerical results.

We further tested our algorithm with medium 2 in this test. The physical properties are given in Table 1. We also set the model size to be 1.0 km $\times$ 1.0 km with a grid space of 2.5 m and employed an explosive Ricker wavelet source with a dominant frequency of 20 Hz. Figure 7 shows snapshots of horizontal displacement and its analytical wavefront calculated from medium 2. These snapshots are calculated within the PML and M-PML boundary conditions, respectively. Figure 7a,b are the snapshots simulated inside the PML and M-PML at 0.1222 s, respectively; Figure 7d,e are simulated inside the PML and M-PML at 0.1889 s, respectively; and the wavefront calculated by the analytical group velocity in Figure 7c,f are consistent with the wavefront of snapshots at 0.1222 s and 0.1889 s, respectively. Figure 7g,h are recorded at 0.6 s, and significant reflections can be observed at the four corners in the snapshot of the PML; however, the M-PML does not. Figure 7i shows the energy decay curves of the wavefield calculated from medium 2. This numerical experiment was conducted to confirm that M-PML has superior absorption performance to PML in arbitrarily different anisotropic media. It can still be seen from the results that the wavefront in the wavefield snapshots Figure 7a,b coincides with the analytical wavefront [31] Figure 7c,f at the corresponding moment, verifying the rationality and applicability of the numerical results. Similar to the previous section, viscosity is still introduced in this medium, and its wavefield snapshots are displayed in Figure 8 to confirm that M-PML can still maintain good absorptive properties and stability in different viscous anisotropic media.

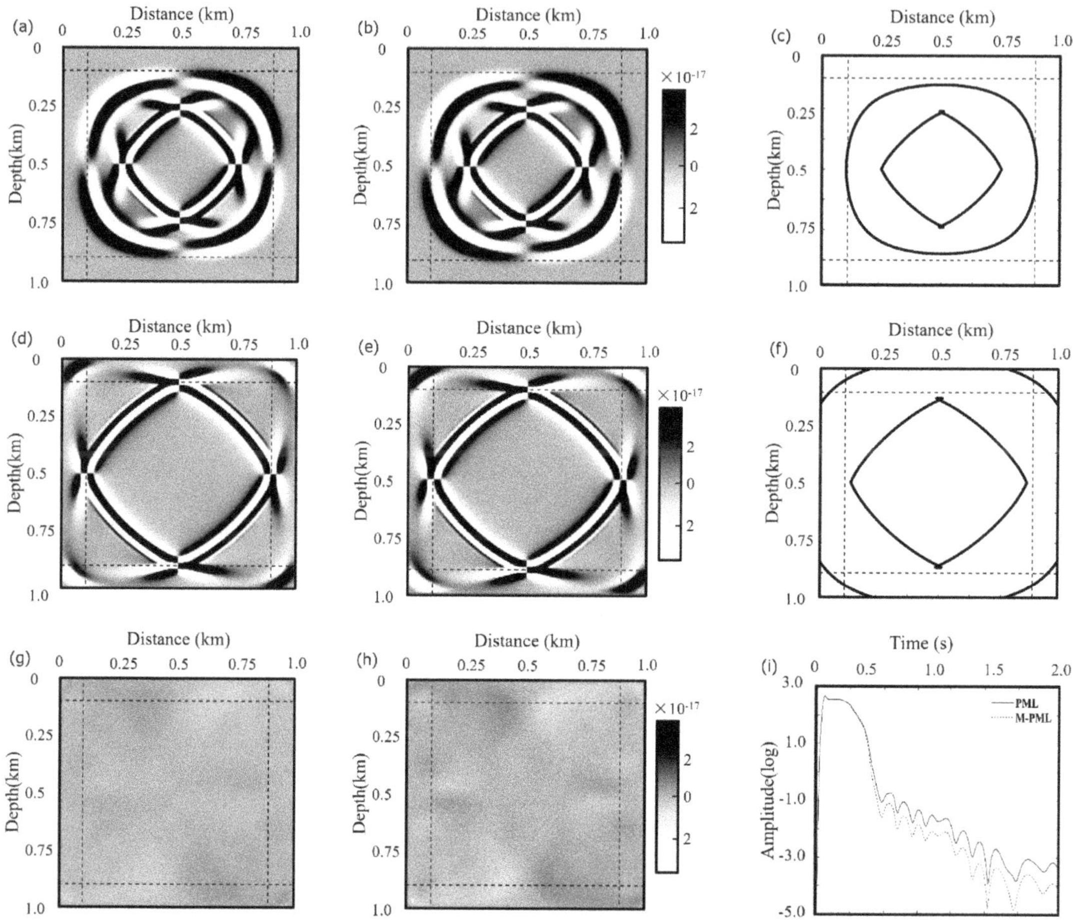

Figure 5. Comparison between snapshots of the horizontal displacement in medium 1 at 0.1667 s, 0.2444 s, 0.7 s (**a**), (**d**), (**g**) calculated by M-PML and (**b**), (**e**), (**h**) PML. Wavefront calculated from analytical group velocity at (**c**) 0.1667 s and (**f**) 0.2444 s. (**i**) is the energy decay curve of PML and M-PML.

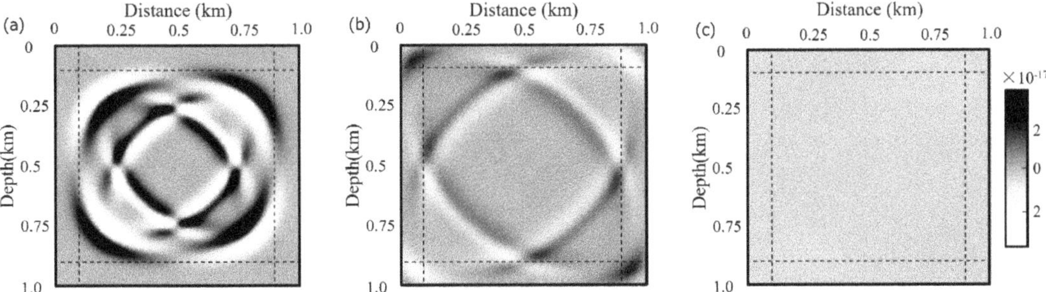

Figure 6. Viscoelastic anisotropic wavefield of medium 1 calculated by M-PML at (**a**) 0.1667 s, (**b**) 0.2444 s, and (**c**) 0.7 s.

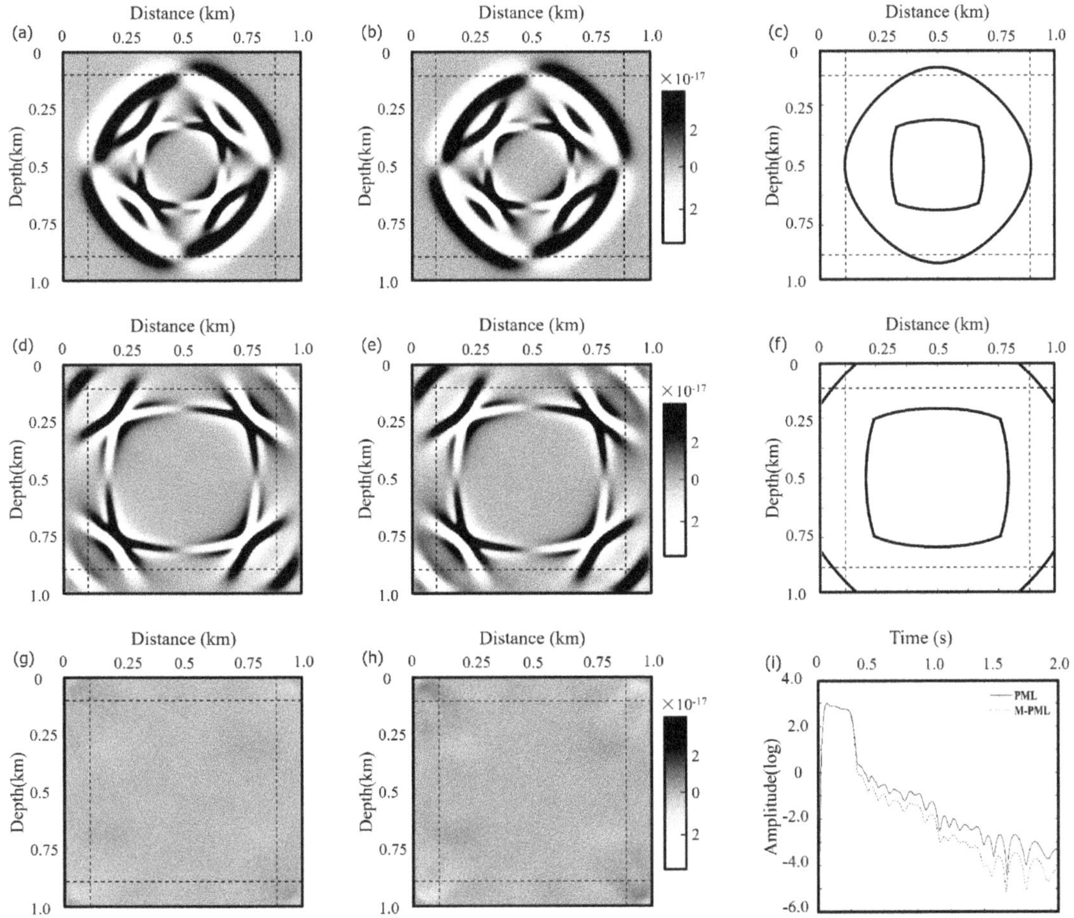

Figure 7. Comparison between snapshots of the horizontal displacement in medium 2 at 0.1222 s, 0.1889 s, 0.6 s. (**a**), (**d**), (**g**) calculated by M-PML, and (**b**), (**e**), (**h**) PML. Wavefront calculated from analytical group velocity at (**c**) 0.1222 s and (**f**) 0.1889 s. (**i**) is the energy decay curve of PML and M-PML.

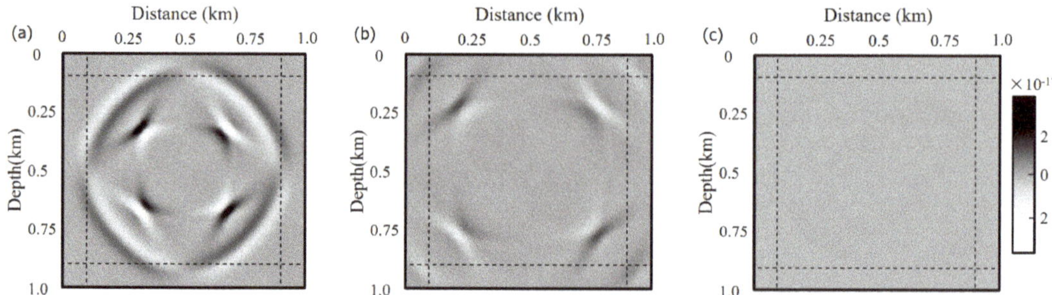

Figure 8. Viscoelastic anisotropic wavefield of medium 2 calculated by M-PML at (**a**) 0.1222 s, (**b**) 0.1889 s, and (**c**) 0.6 s.

Regarding absorption performance, observing the wavefield snapshots of PML and M-PML shows that both ABCs are stable in this homogenous anisotropic medium, so it is possible to compare their absorption performance visually. At 0.6 s, the effective wavefield wholly left the working area, and we can see that the boundary reflections of PML are strong at the four corners in Figure 7h, but those of M-PML are well attenuated. Further, in Figure 7i, it is illustrated that the energy decay curve of M-PML is lower than that of PML, which quantitatively displays the absorption advantage of M-PML over PML.

3.3. Wave Propagation in a Complex Anisotropic Medium

Finally, we used the frequency domain method based on M-PML to perform wavefield simulation for a more complicated cross-well model. The cross-well model is shown in Figure 9. The corresponding parameters selected are as follows: computing area, $0 \leq x \leq 108$ m, $0 \leq z \leq 289$ m; spatial step, $h = \Delta x = \Delta z = 0.5$ m; grid parameter, 578 × 216; the number of M-PML layers, 30; dominant frequency of the source, 130 Hz, the location of the source: (135 m, 0 m); receiver alignment positions: $x = 108$ m, $0.5 \leq z \leq 288.5$ m; receiver interval, 2 m. The cross-well model has eight layers. There are numerical constants in red in Figure 8, which label the layers in the figure. The red constants 1–8 in the figure correspond to layers 1–8 in Table 2, respectively. The asterisk in the figure represents the source. These anisotropic medium parameters are taken from the true sedimentary and are available in the study of Thomsen [28].

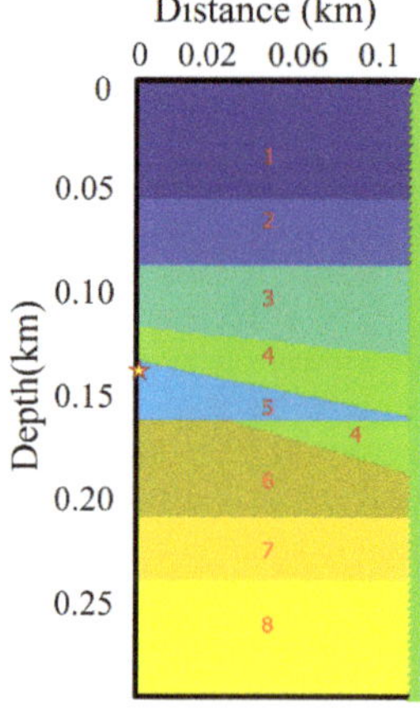

Figure 9. Multi-layer cross-well model with single-excited source well on the left and multiple-received receiver well on the right, with red constants for medium numbers.

Table 2. Physical parameters used for the cross-well model.

Layer Media	Vp (m/s)	Vs (m/s)	Qp	Qs	ε	δ	ρ (kg/m^3)
Layer 1	1875	826	10	10	0.225	0.100	2000
Layer 2	2202	969	10	10	0.015	0.060	2250
Layer 3	2868	1350	15	15	0.970	−0.090	1860
Layer 4	3368	1829	21	18	0.110	−0.035	2500
Layer 5	3688	2774	30	19	0.081	0.057	2730
Layer 6	3901	2682	38	25	0.137	−0.012	2640
Layer 7	4296	2471	42	35	0.081	0.129	2660
Layer 8	4529	2703	50	40	0.034	0.211	2520

Figure 10 shows the mono-frequency wavefield snapshots of elastic and viscoelastic anisotropy obtained by the proposed method at 163.2 Hz, 329.8 Hz, and 429.5 Hz. Furthermore, the time-domain snapshots at the time point of 0.0199 s, 0.0299 s, and 0.0399 s computed by the proposed method are shown in Figure 11, respectively. It is easy to see that there exist no artificial edge effects near the truncated boundary in either the elastic or viscoelastic anisotropic wavefield. Then, we give the single-shot seismograms of the cross-well model, which are shown in Figure 12, and we also see that there exist no numerical dispersion or artificial edge effects in either elastic or viscoelastic anisotropic seismograms. Thus, we may conclude that the M-PML for the finite-difference frequency domain is suitable for wavefield simulation in a viscoelastic anisotropic and complicated medium.

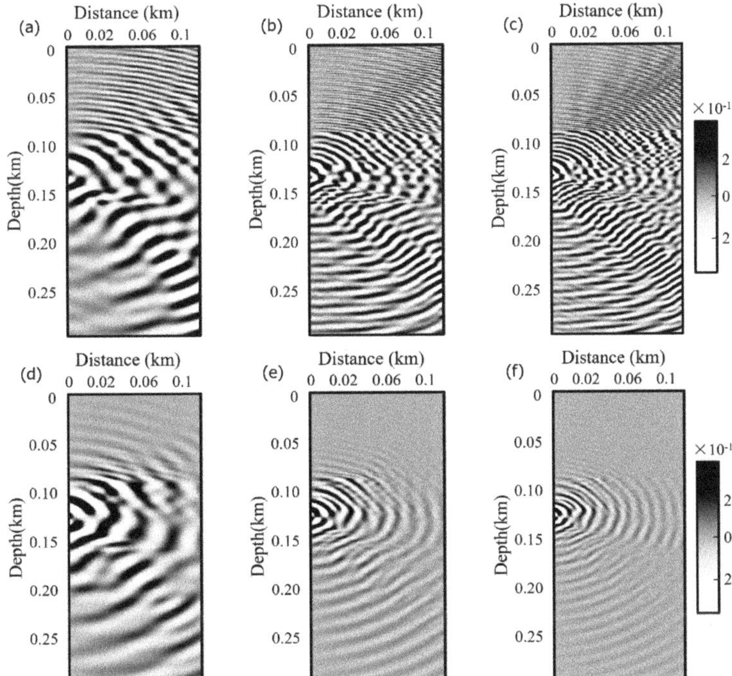

Figure 10. Mono-frequency horizontal elastic anisotropic wavefield snapshots of different frequencies for the cross-well model at (**a**) 163.2 Hz, (**b**) 329.8 Hz, (**c**) 429.5 Hz, and viscoelastic anisotropic wavefield at (**d**) 163.2 Hz, (**e**) 329.8 Hz, and (**f**) 429.5 Hz.

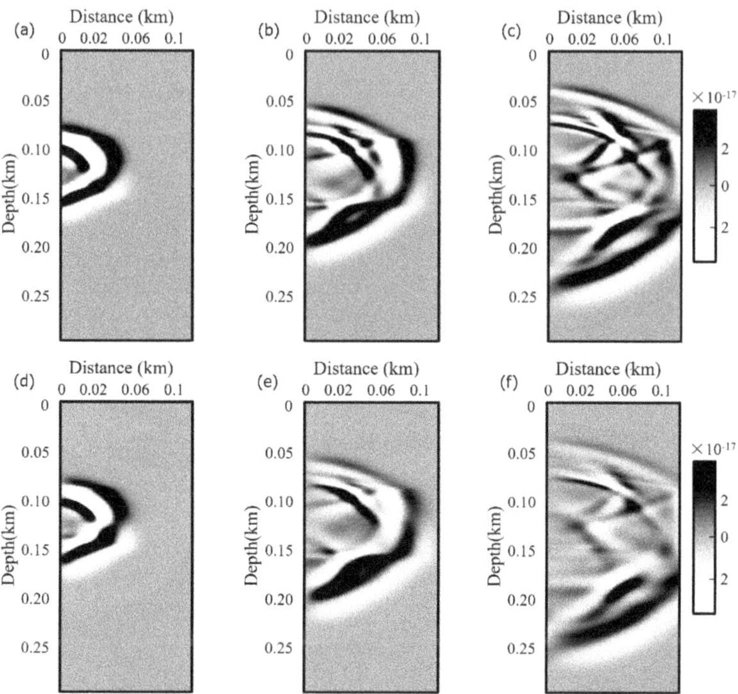

Figure 11. Time-domain horizontal elastic anisotropic wavefield snapshots for the cross-well model at (**a**) 0.02 s, (**b**) 0.03 s, (**c**) 0.04 s, and viscoelastic anisotropic wavefield at (**d**) 0.02 s, (**e**) 0.03 s, and (**f**) 0.04 s.

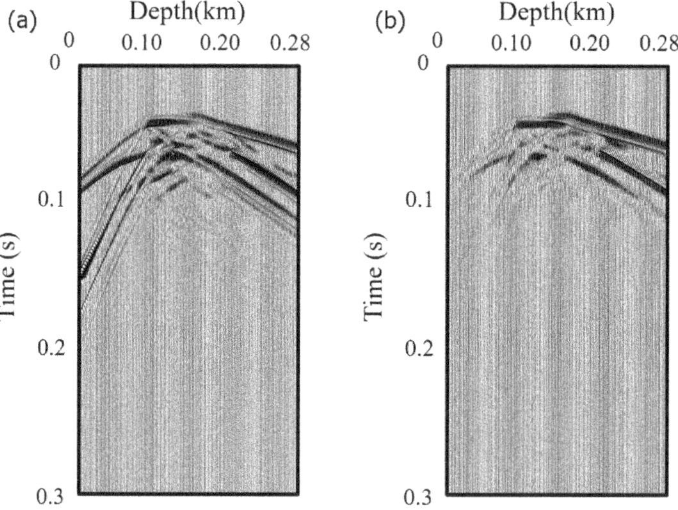

Figure 12. Time-domain horizontal (**a**) elastic anisotropic and (**b**) viscoelastic anisotropic seismogram corresponding to Figure 11.

4. Conclusions

In this study, in order to solve the instability of the absorbing boundary conditions caused by the frequency-domain algorithm for viscoelastic anisotropic wavefields, wave propagation was simulated by the finite-difference frequency-domain, and the reliability of the results was verified by combining the analytical wavefront obtained from the anisotropic analytical group velocity. The multi-axis perfectly matched layer (M-PML) was introduced in the frequency domain. We compared the wavefield snapshots and energy attenuation curves of M-PML and PML in different elastic anisotropic media. Then, we gave the viscoelastic anisotropic wavefield to demonstrate the reliability of M-PML in viscous media. The results indicate that M-PML can still stably and effectively absorb reflections from the truncated boundaries in strongly anisotropic and viscous media. Simulations in a complex cross-well viscoelastic anisotropic model indicate the applicability of this algorithm to a heterogeneous medium.

Moreover, the implementation of M-PML only needs to combine the damping profiles of PML, which improves absorption performance and stability while maintaining computational efficiency. In practice, this algorithm can be used as a forward operator for seismic migration, waveform inversion, and wave equation travel-time tomography. The frequency-domain algorithm can also improve the inversion efficiency of multi-source problems. For complex subsurface media, the proposed method can be applied for inversion of medium anisotropy and attenuation properties. Although our proposed algorithm works well, it has some limitations. Bad choices of $p^{(z/x)}$ and $p^{(x/z)}$ may lead to artificial reflections.

Author Contributions: Conceptualization, J.Y. and X.H.; methodology, J.Y.; investigation, H.C.; writing—original draft preparation, J.Y.; writing—review and editing, J.Y. and X.H.; visualization, J.Y.; supervision, X.H. and H.C. All authors have read and agreed to the published version of the manuscript.

Funding: This research was funded by the National Natural Science Foundation of China under grant numbers 11774373 and 12174421.

Institutional Review Board Statement: Not applicable.

Informed Consent Statement: Not applicable.

Data Availability Statement: Not applicable.

Acknowledgments: We gratefully acknowledge financial support from the National Natural Science Foundation of China (11774373, 12174421).

Conflicts of Interest: The authors declare no conflict of interest.

References

1. Hestholm, S. Acoustic VTI modeling using high-order finite differences. *Geophysics* **2009**, *74*, T67. [CrossRef]
2. Zhu, H.; Zhang, W.; Chen, X. Two-dimensional seismic wave simulation in anisotropic media by non-staggered finite difference method. *Chin. J. Geophys.* **2009**, *52*, 1536–1546.
3. Qiao, Z.; Sun, C.; Wu, D. Theory and modeling of constant-Q viscoelastic anisotropic media using fractional derivative. *Geophys. J. Int.* **2019**, *217*, 798–815. [CrossRef]
4. Devaney, A.J. Geophysical Diffraction Tomography. *IEEE Trans. Geosci. Remote Sens.* **1984**, *GE-22*, 3–13. [CrossRef]
5. Pratt, R.G.; Worthington, M.H. The application of diffraction tomography to cross-hole seismic data. *Geophysics* **1988**, *53*, 1284–1294. [CrossRef]
6. Baysal, E.; Kosloff, D.; Sherwood, J. Reverse time migration. *Geophysics* **1983**, *48*, 1514–1524. [CrossRef]
7. McMechan, G. Migration by extrapolation of time-dependent boundary values. *Geophys. Prospect.* **1983**, *31*, 413–420. [CrossRef]
8. Dai, W.; Wang, X.; Schuster, G. Least-squares migration of multi-source data with a deblurring filter. *Geophysics* **2011**, *76*, R135–R146. [CrossRef]
9. Tarantola, A. Inversion of seismic reflection data in the acoustic approximation. *Geophysics* **1984**, *49*, 1259–1266. [CrossRef]
10. Pratt, R.G.; Worthington, M.H. Acoustic wave equation inverse theory applied to multi-source cross-hole tomography Part I: Acoustic wave-equation method. *Geophys. Prospect.* **1990**, *38*, 287–310. [CrossRef]
11. Virieux, J.; Operto, S. An overview of full-waveform inversion in exploration geophysics. *Geophysics* **2009**, *74*, WCC1–WCC26. [CrossRef]

12. Marfurt, K.J. Accuracy of finite difference and finite element modeling of scalar and elastic wave equations. *Geophysics* **1984**, *49*, 533–549. [CrossRef]
13. Ben-Hadj-Ali, H.; Operto, S.; Virieux, J. An efficient frequency-domain full-waveform inversion method using simultaneous encoded sources. *Geophysics* **2011**, *76*, 109–124. [CrossRef]
14. Operto, S.; Virieux, J.; Ribodetti, A.; Anderson, J.E. Finite-difference frequency-domain modeling of viscoacoustic wave propagation in 2D tilted transversely isotropic (TTI) media. *Geophysics* **2009**, *74*, 75–95. [CrossRef]
15. Jeong, W.; Min, D.J.; Lee, G.H.; Lee, H.Y. 2D frequency-domain elastic full waveform inversion using finite-element method for VTI media. In *SEG Technical Program. Expanded Abstracts 2011*; SEG: San Antonio, TX, USA, 2011; pp. 2654–2658.
16. Zhou, B.; Won, M.; Greenhalgh, S.; Liu, X. Generalizeded stiffness reduction method to remove the artificial edge-effects for seismic wave modeling in elastic anisotropic media. *Geophys. J. Int.* **2020**, *220*, 1394–1408. [CrossRef]
17. Yang, Q.; Zhou, B.; Riahi, M.; Ai-Khaleel, M. A new generalized stiffness reduction method for 2D/2.5D frequency domain seismic wave modeling in viscoelastic anisotropic media. *Geophysics* **2020**, *85*, T315–T329. [CrossRef]
18. Berenger, J.P. A perfectly matched layer for the absorption of electromagnetic waves. *J. Comput. Phys.* **1994**, *114*, 185–200. [CrossRef]
19. Francis, C.; Tsogka, C. Application of the perfectly matched absorbing layer model to the linear elastodynamic problem in anisotropic heterogeneous media. *Geophysics* **2001**, *66*, 294–307.
20. Komatitsch, D.; Tromp, J. A perfectly matched layer absorbing boundary condition for the second-order seismic wave equation. *Geophys. J. Int.* **2003**, *154*, 146–153. [CrossRef]
21. Fang, X.; Niu, F. An unsplit complex frequency-shifted perfectly matched layer for second-order acoustic wave equations. *Sci. China Earth Sci.* **2021**, *64*, 992–1004. [CrossRef]
22. Zhang, W.; Shen, Y. Unsplit complex frequency-shifted PML implementation using auxiliary differential equations for seismic wave modeling. *Geophysics* **2010**, *75*, T141–T154. [CrossRef]
23. Opertp, S.; Virieux, J.; Amestoy, P.; L'Excellent, J.; Giraud, L.; Ali, H. 3D finite-difference frequency-domain modeling of visco-acoustic wave propagation using a massively parallel direct solver: A feasibility study. *Geophysics* **2007**, *72*, 195–211. [CrossRef]
24. Oskooi, A.; Johnson, S.G. Distinguishing correct from incorrect PML proposals and a corrected unsplit PML for anisotropic dispersive media. *J. Comput. Phys.* **2011**, *230*, 2369–2377. [CrossRef]
25. Meza-Fajardo, K.C.; Papageorgiou, A.S. A non-convolutional, split-field, perfectly matched layer for wave propagation in isotropic and anisotropic elastic media: Stability analysis. *Bull. Seism. Soc. Am.* **2008**, *98*, 1811–1836. [CrossRef]
26. Ping, P.; Zhang, Y.; Xu, Y. A multi-axial perfectly matched layer (M-PML) for the long-time simulation of elastic wave propagation in the second-order equations. *J. Appl. Geophys.* **2014**, *101*, 124–135. [CrossRef]
27. Zhang, Z.; Zhang, W.; Chen, X. Complex frequency-shifted multi-axial perfectly matched layer for elastic wave modeling on curvilinear grids. *Geophys. J. Int.* **2014**, *198*, 140–153. [CrossRef]
28. Thomsen, L. Weak elastic anisotropy. *Geophysics* **1986**, *51*, 1954–1966. [CrossRef]
29. Toksoz, M.N.; Johnston, D.H. *Seismic Wave Attenuation*; SEG Geophysical Reprint Series, No. 2; Society of Exploration Geophysicists: Houston, TX, USA, 1981.
30. Zeng, Y.; He, J.; Liu, Q. The application of the perfectly matched layer in numerical modeling of wave propagation in poroelastic media. *Geophysics* **2001**, *66*, 1258–1266. [CrossRef]
31. Daley, P.F.; Hron, F. Reflection and transmission coefficients for transversely isotropic media. *Bull. Seism. Soc. Am.* **1977**, *67*, 661–675. [CrossRef]

Article

The Characteristics of Seismic Rotations in VTI Medium

Lixia Sun [1], Yun Wang [1,*], Wei Li [1] and Yongxiang Wei [2]

1 "MWMC" Group, School of Geophysics and Information Technology, China University of Geosciences, Beijing 100083, China; slx2018@cugb.edu.cn (L.S.); 3010180007@cugb.edu.cn (W.L.)
2 Fujian Earthquake Agency, China Earthquake Administration, Fuzhou 350003, China; wyongx@fjdzj.gov.cn
* Correspondence: yunwang@mail.iggcas.ac.cn

Abstract: Under the assumptions of linear elasticity and small deformation in traditional elastodynamics, the anisotropy of the medium has a significant effect on rotations observed during earthquakes. Based on the basic theory of the first-order velocity-stress elastic wave equation, this paper simulates the seismic wave propagation of the translational and rotational motions in two-dimensional isotropic and VTI (transverse isotropic media with a vertical axis of symmetry) media under different source mechanisms with the staggered-grid finite-difference method with respect to nine different seismological models. Through comparing the similarities and differences between the translational and rotational components of the wave fields, this paper focuses on the influence of anisotropic parameters on the amplitude and phase characteristics of the rotations. We verify that the energy of S waves in the rotational components is significantly stronger than that of P waves, and the response of rotations to the anisotropic parameters is more sensitive. There is more abundant information in the high-frequency band of the rotational components. With the increase of Thomsen anisotropic parameters ε and δ, the energy of the rotations increases gradually, which means that the rotational component observation may be helpful to the study of anisotropic parameters.

Keywords: rotation; translation; isotropic medium; VTI

Citation: Sun, L.; Wang, Y.; Li, W.; Wei, Y. The Characteristics of Seismic Rotations in VTI Medium. *Appl. Sci.* **2021**, *11*, 10845. https://doi.org/10.3390/app112210845

Academic Editor: Jong Wan Hu

Received: 12 October 2021
Accepted: 10 November 2021
Published: 17 November 2021

1. Introduction

It is well known that the six-component records of ground motion include three-component translations and three-component rotations [1]. In the past two decades, with the development of rotational seismometers, the observation of rotations has gradually increased, and rotations caused by earthquakes have attracted increasing attention in the scientific and seismic engineering fields, and great progress has been made in both theoretical research and practical applications [2–6].

In order to study the mechanism of seismological rotations, many meaningful inquiries have been made by means of indirect calculation, direct observation, and numerical simulation. Aiming at the Newport–Inglewood (NI) fault in America, Wang et al. [7] used a finite-difference method to simulate several earthquakes with magnitude 7 in different hypocenters, and found the influence of source and receiving geological conditions on the rotations. They also found that the variations of the source produce a great effect on the rotations. Therefore, rotational observations may contribute to the exploration of the mechanism of NI earthquakes. Ferreira et al. [8] simulated the rotations recorded at the fundamental mode of surface waves with the full ray theory (FRT) for laterally smooth heterogeneous Earth models, and found that the synthetic seismograms showed good agreement with the real ones. With the joint observations of rotation rates and accelerations, the one-dimensional S-wave velocity at the observing locations can be obtained. Pham et al. [9] calculated the amplitudes of rotational rates based on the Kelvin–Christoffel equations and found that the rotational rates caused by anisotropies were large enough in strong earthquakes to predict the underground structures and constrain anisotropic parameters. Tang et al. [10] developed a new method, the generalized reflection and transmission coefficient method, which could obtain the translational and rotational synthetic seismograms in VTI media.

Barak et al. [11] performed a synthetic seismic experiment on a simple two-layer model of a water-covered seabed, where there were isolated high-velocity spheres different from the background medium. They found that the energy of the Scholte waves on rotational components was stronger than that on translational components, which would be helpful to identify different modes of waves and separate the surface waves from the body waves. More importantly, with seven-component seismic data, including the pressure components recorded by hydrophones, displacements recorded by three-component geophones, and rotations recorded by three-component rotational seismometers, their analyses presented the possibility to recover the vector characteristics of wave fields. Additionally, we previously discuss the differences between body waves and surface waves projected on the rotational components in 2D isotropic media [12], and Zhang et al. [13] found that the responses of the anomalies with a lower velocity were more obvious on the rotational components than on the translational components, and the energy of the shear waves on rotational components was stronger than that of P waves.

Although there is much research about rotations at present, the characteristics of rotations in transverse isotropic media with a vertical axis of symmetry (VTI) have not yet been clarified, and the influence of the anisotropic parameters on rotations is still unclear. In order to study the characteristics of the rotations generated from different sources in VTI media and explore the significance of rotations to the research of anisotropic parameters, we simulate the seismic rotations and translations based on the first-order velocity-stress equations using the staggered-grid finite-difference method in 2D VTI medium, analyze the similarities and differences between translations and rotations, and focus on the influence of the Thomsen parameters [14] on rotations so as to promote the relationship of the rotations and the anisotropies of the media, which can be helpful for the wave–field separation, inversion of the anisotropic parameters and the study of the earthquake source mechanism.

2. Theoretical Foundations

Ground motion includes not only three components of linear displacements but also three components of rotation in three-dimensional Cartesian coordinates (Figure 1). The rotation tensor, the curl of the displacement field, is expressed as:

$$\vec{\omega} = \frac{1}{2}\nabla \times \vec{u} \tag{1}$$

where $\vec{\omega}$ is the rotation tensor and $\vec{u}$ is the displacement tensor.

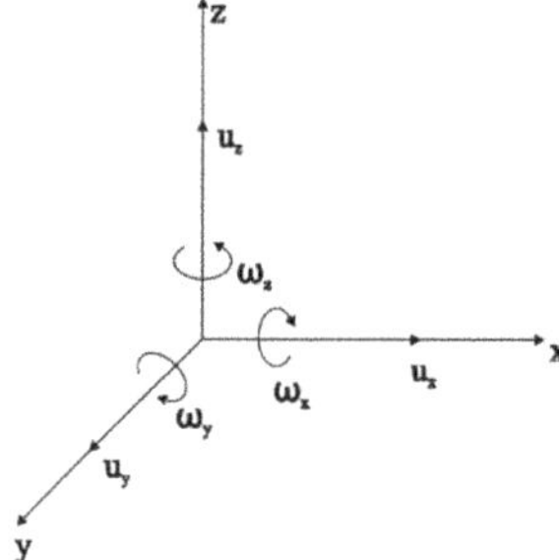

Figure 1. The translations and rotations.

In a two-dimensional plane, the seismic wave fields can be presented as two translational components, X and Z components, and a rotational component, ω_y. In order to correlate with the general seismic velocity observation, the rotation rate vector R_y can be expressed as:

$$R_y = \frac{1}{2}(\frac{\partial v_x}{\partial z} - \frac{\partial v_z}{\partial x}) \tag{2}$$

where v_x and v_z are the X and Z components of the velocity, respectively; x and z are the coordinates on the Cartesian system.

Using the staggered-grid finite-difference method [15], under the assumptions of linear elasticity and small deformation, the one-order velocity-stress equations of the elastic waves in 2D VTI medium can be expressed as:

$$\begin{cases} \frac{\partial \sigma_{xx}}{\partial x} + \frac{\partial \sigma_{xz}}{\partial z} = \rho \frac{\partial v_x}{\partial t} \\ \frac{\partial \sigma_{zx}}{\partial x} + \frac{\partial \sigma_{zz}}{\partial z} = \rho \frac{\partial v_z}{\partial t} \end{cases}$$
$$\begin{cases} \frac{\partial \sigma_{xx}}{\partial t} = c_{11} \frac{\partial v_x}{\partial x} + c_{13} \frac{\partial v_z}{\partial z} \\ \frac{\partial \sigma_{zz}}{\partial t} = c_{13} \frac{\partial v_x}{\partial x} + c_{33} \frac{\partial v_z}{\partial z} \\ \frac{\partial \sigma_{xz}}{\partial t} = c_{44} (\frac{\partial v_z}{\partial x} + \frac{\partial v_x}{\partial z}) \end{cases} \tag{3}$$

where σ_{xx}, σ_{zx} and σ_{zz} are three stress components, t is time, ρ is density; c_{11}, c_{13}, c_{33} and c_{44} are elastic coefficients, which can be calculated with [14]:

$$\begin{aligned} c_{11} &= \rho v_p^2 (1 + 2\varepsilon) \\ c_{33} &= \rho v_p^2 \\ c_{44} &= \rho v_s^2 \\ c_{13} &= \rho \sqrt{\left[(1 + 2\delta) v_p^2 - v_s^2\right] \left(v_p^2 - v_s^2\right)} - \rho v_s^2 \end{aligned} \tag{4}$$

where v_p and v_s are the velocity of the P and S waves, respectively, and ε and δ are the anisotropic parameters. We define different VTI models by giving different Thomsen parameters to study the influence of anisotropic parameters on the translational and rotational components. When $\varepsilon = 0$ and $\delta = 0$, Equation (3) corresponds to an isotropic medium.

Based on Equations (2)–(4), we simulate the seismic waves in discrete models with the grids at second-order time and twelfth-order space differential approximations. In addition, we use the Ricker wavelet with a 60 Hz central frequency to simulate the explosion source, the radial concentrated force source, the vertical concentrated force source, and the shear source, respectively. Furthermore, we utilize the splitting form of perfectly matched absorbing layer boundary condition (SPML) [16] to weaken the boundary reflections.

3. Synthetic Examples

In order to study the influence of the source mechanism on rotations, we defined model 1 as a 2D isotropic full-space homogeneous elastic medium, as illustrated in Table 1. Since P waves theoretically do not generate rotational motion in isotropic media [9], we compared the translational and rotational components generated from the concentrated force and the shear source.

Table 1. Parameters of model 1.

Model	v_p (m/s)	v_s (m/s)	ρ (kg/m^3)	ε	δ
Model 1	2000	1200	2000	0	0

With a size of 200 m $\times$ 200 m, the shot in the center of the model, and the receivers arrayed at a depth of 100 m with 1 m intervals, the translational and rotational components generated from different sources are shown in Figures 2–4, where the sample interval is 0.1 s and the total recording time is 0.1 s.

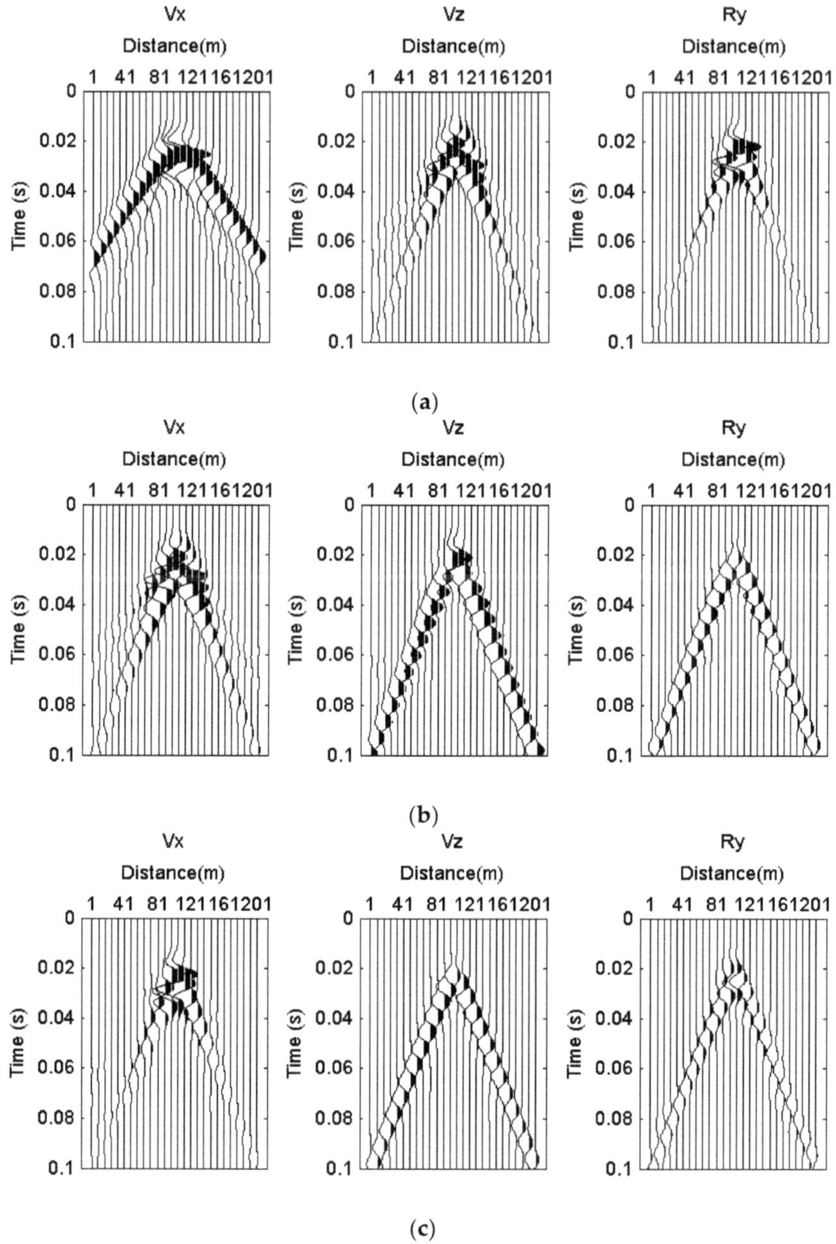

Figure 2. The seismograms. (**a**) The radial concentrated force source. (**b**) The vertical concentrated force source. (**c**) The shear source.

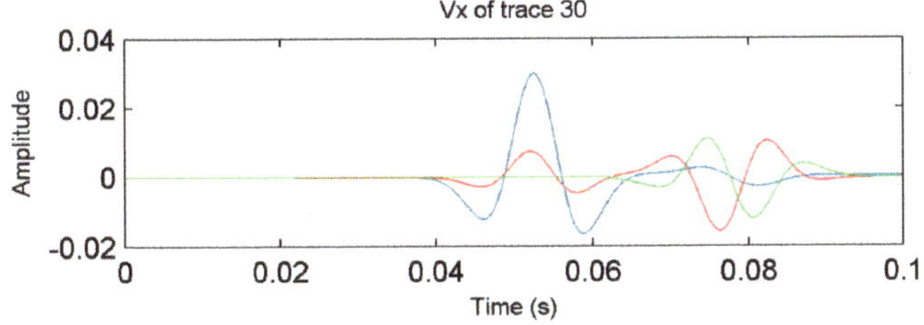

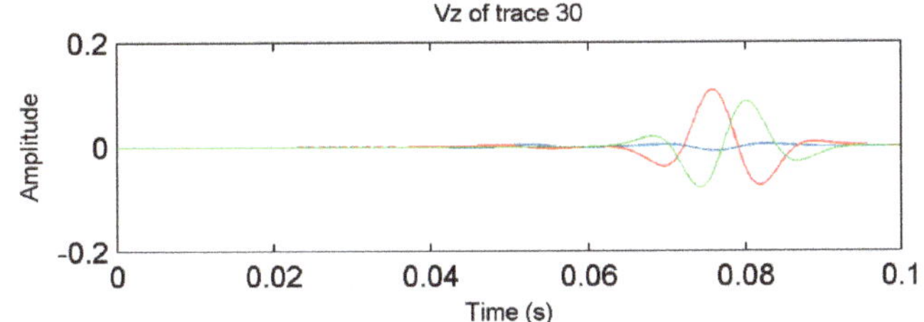

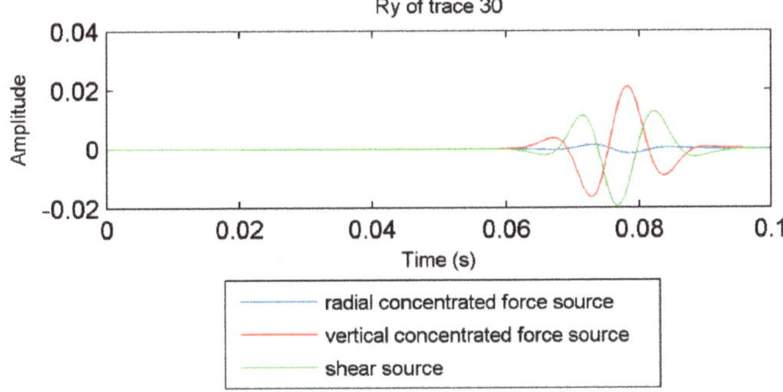

Figure 3. Seismograms of the 30th trace.

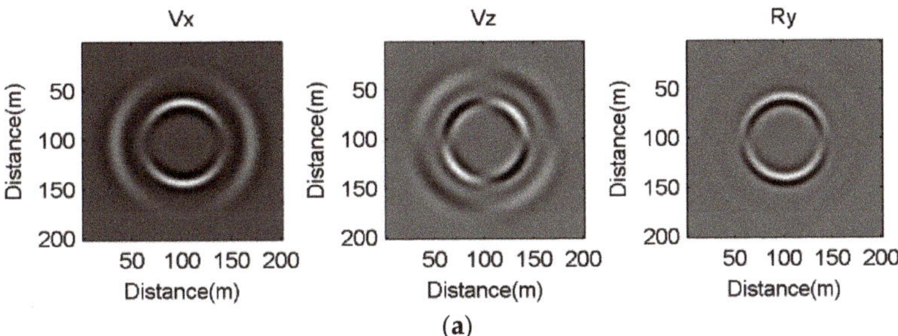

(a)

Figure 4. *Cont.*

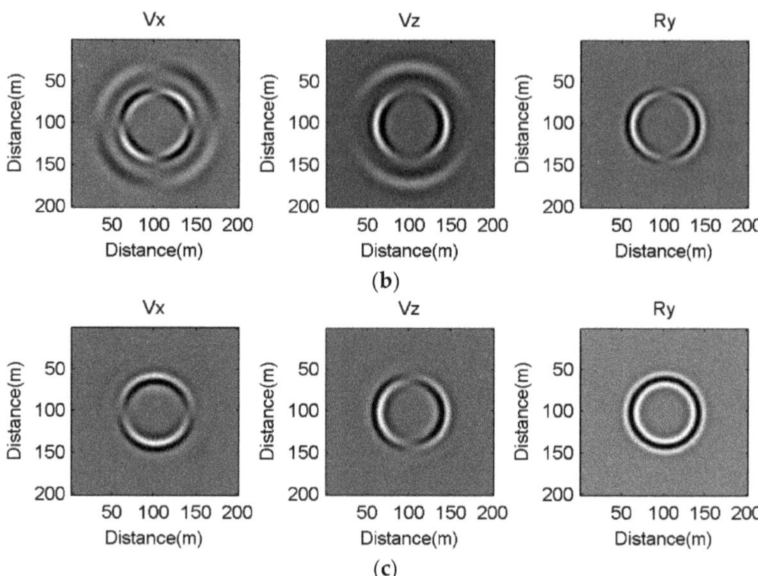

Figure 4. Snapshots of wave fields at 0.05 s. (**a**) Radial concentrated force source. (**b**) Vertical concentrated force source. (**c**) Shear source.

In Figure 2, it is obvious that there are only *S* waves in the three components because the model is isotropic and only *S* waves are produced by the shear source, while it can be clearly seen that there are the first arrivals of *P* and *S* waves generated from the radial and vertical concentrated force sources. The energy of the R_y component generated from the concentrated force source is so weak that it needed to be magnified 10 times to be visible at the same energy level with the other two components on the records. It can be found that *P* waves are much stronger than *S* waves in the *X* component generated from the radial concentrated force source, while *S* waves cause a stronger rotational motion, and *P* waves are hardly visible in R_y components.

We extracted the 30th trace of the seismic records to make a further comparison of translations and rotations, as shown in Figure 3. The influence of sources in the *X* component and R_y component is much stronger than that in the *Z* component in the far offset. The amplitude of *P* waves in the *X* component generated from the radial concentrated force source is much stronger than that from the other two sources. However, the amplitude of the *S* waves generated from the radial concentrated force source is weak in all three components.

All these anomalies can be enhanced in the snapshots of different components generated from different sources (Figure 4), where there are only *S* waves in the R_y components. In all snapshots, the shapes of the wavefronts in the *X* and *Z* components are round, which is consistent with Zhang's research [17]. It can be seen that the wavefront propagates with reverse phase along the horizontal axis in the R_y component generated from the radial concentrated force source, while it propagates with reverse phase along the vertical axis in the R_y component generated from the vertical concentrated force source. However, it propagates with the same phase around the circumference in the R_y component generated from the shear force. The wavefronts in the *X* components and *Z* components are complex and different from the R_y components.

Obviously, the energy of *S* waves is significantly stronger than that of *P* waves on the R_y component, which means the R_y component may be helpful to identify different body waves.

4. Effects of Anisotropic Parameters on Rotations

In order to further explore the effects of anisotropic parameters on rotational motion, we changed the Thomsen parameters δ and ε, based on the same seismic observation system to simulate the three-component velocity fields. Eight different models are defined as illustrated in Table 2: models 2–5 were mainly used to study the effects of ε on rotational components by increasing ε from 0 to 0.3 as δ is constant, and model 4 and models 6–9 were employed to explore the influence of δ by gradually increasing δ from -0.2 to 0.2 as ε is constant. We first analyzed the simulation results under the explosion source in detail, shown in Figures 5–12.

Table 2. Anisotropic parameters of different models.

Parameter	Model 2	Model 3	Model 4	Model 5	Model 6	Model 7	Model 8	Model 9
Ε	0	0.1	0.2	0.3	0.2	0.2	0.2	0.2
Δ	−0.2	−0.2	−0.2	−0.2	−0.1	0	0.1	0.2

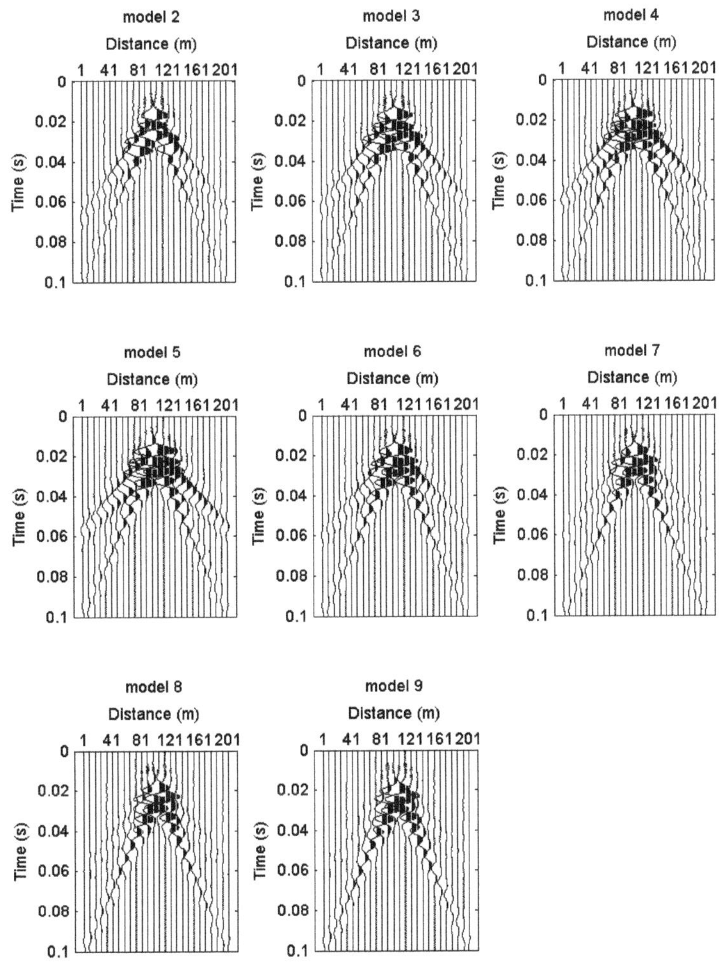

Figure 5. Components generated from the expansion source in different models.

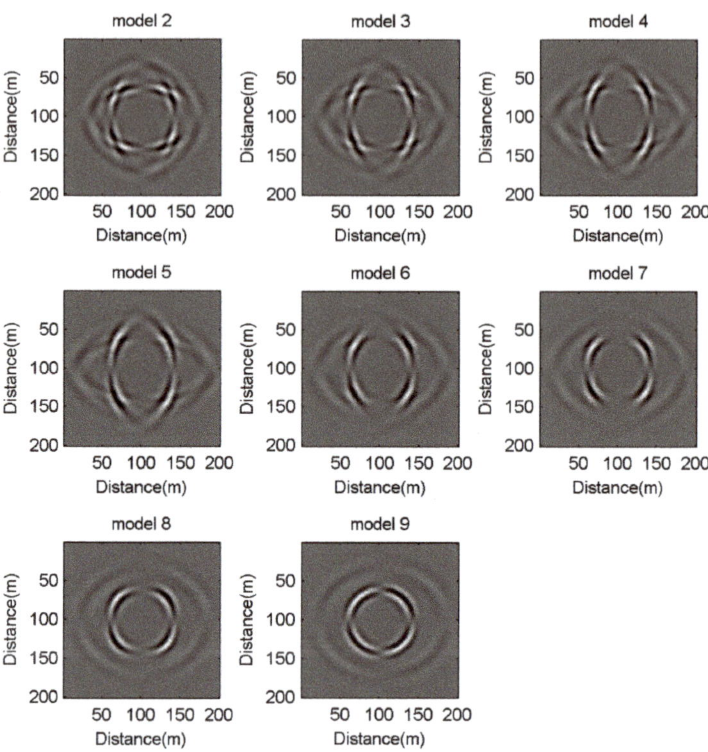

Figure 6. Wave field snapshots of R_y generated from the expansion source at 0.05 s.

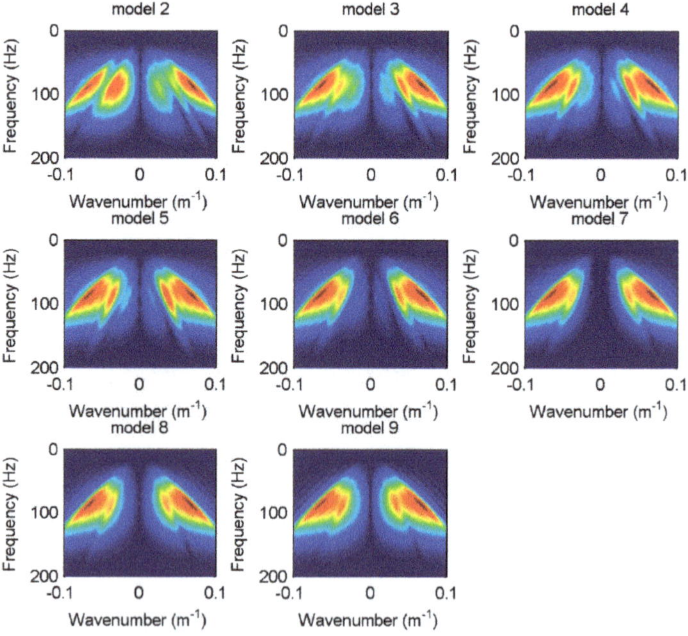

Figure 7. FK spectra of R_y generated from the expansion source in different models.

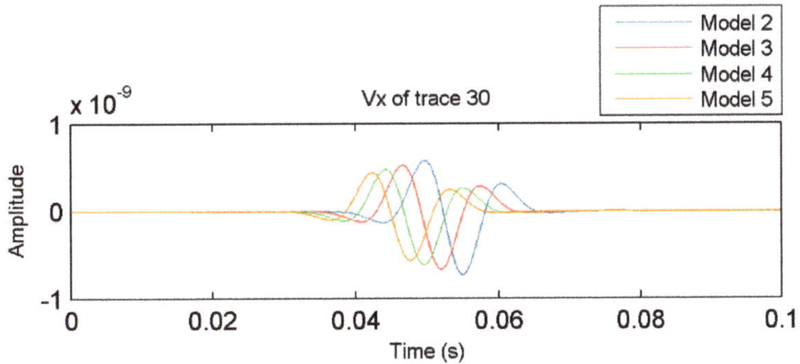

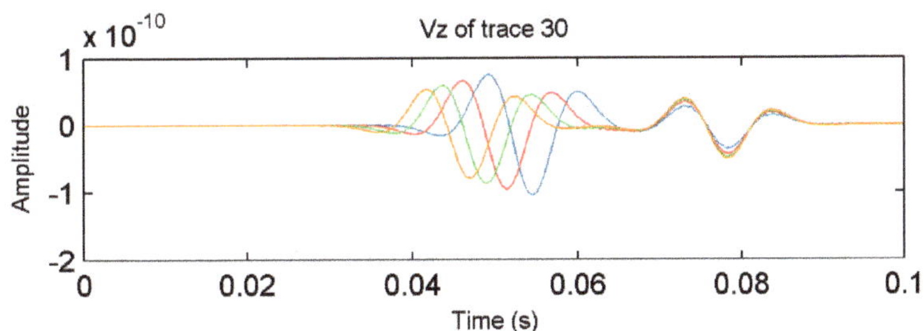

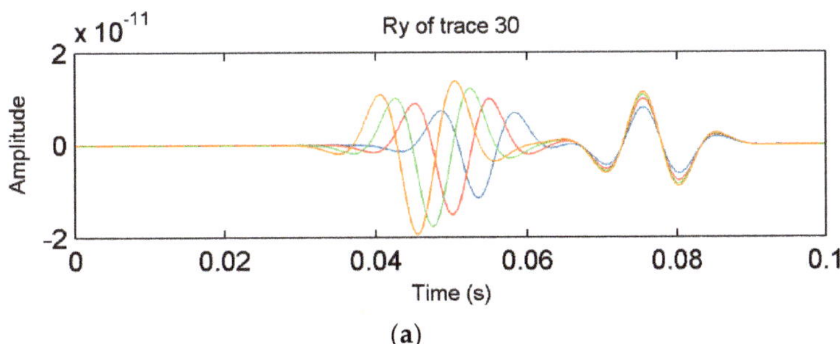

(**a**)

Figure 8. *Cont.*

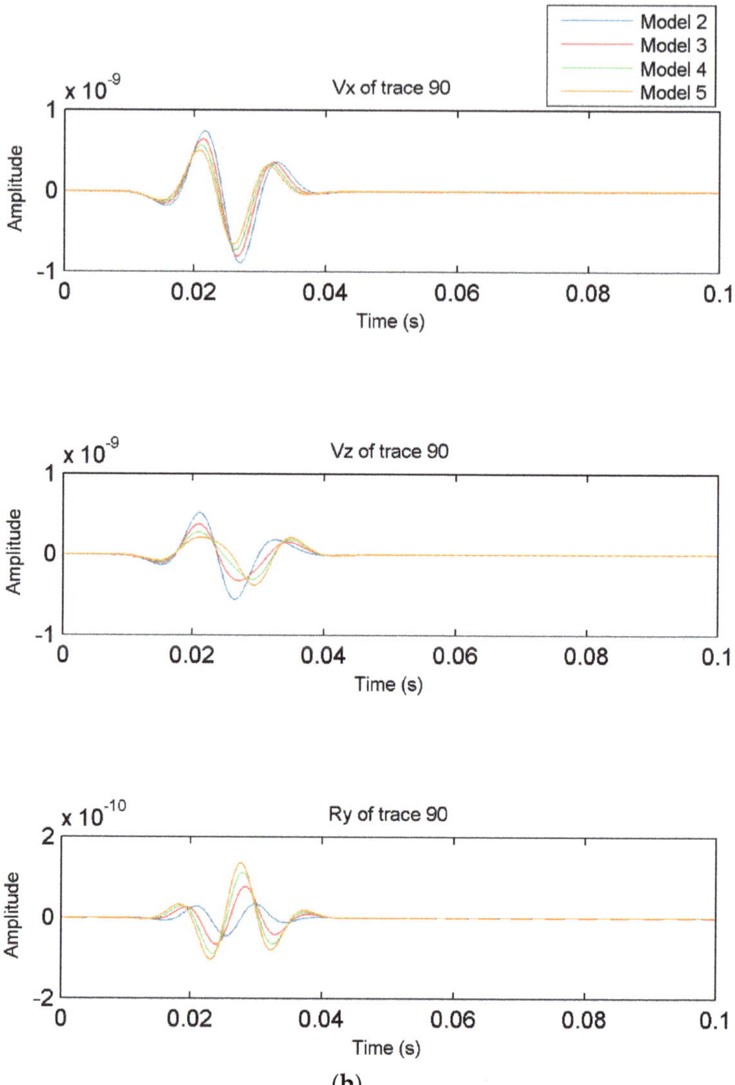

(b)

Figure 8. Influence of ε on seismic records generated from the expansion source: (**a**) the seismic records of the 30th trace; (**b**) the seismic records of the 90th trace.

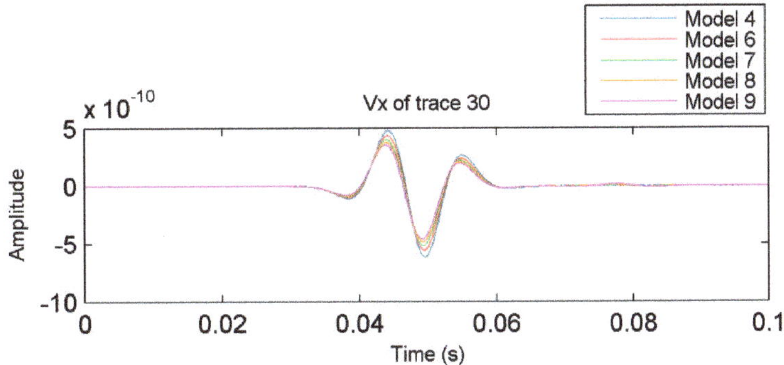

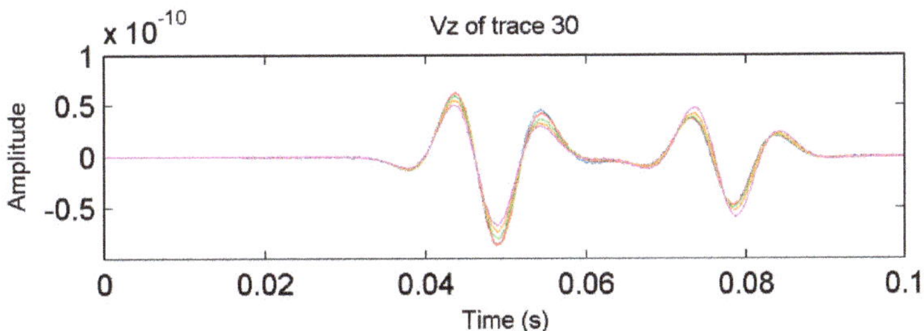

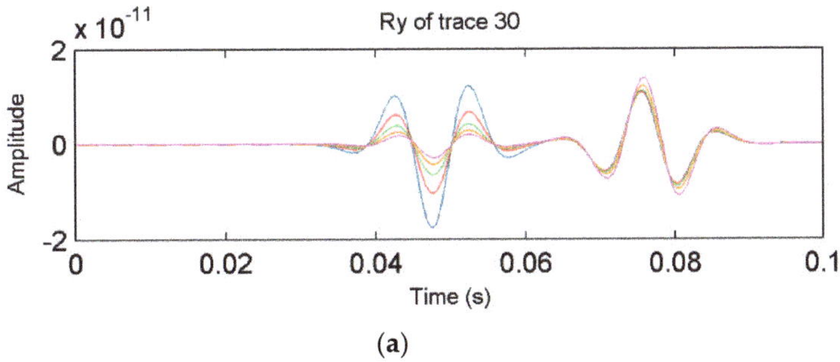

(a)

Figure 9. *Cont.*

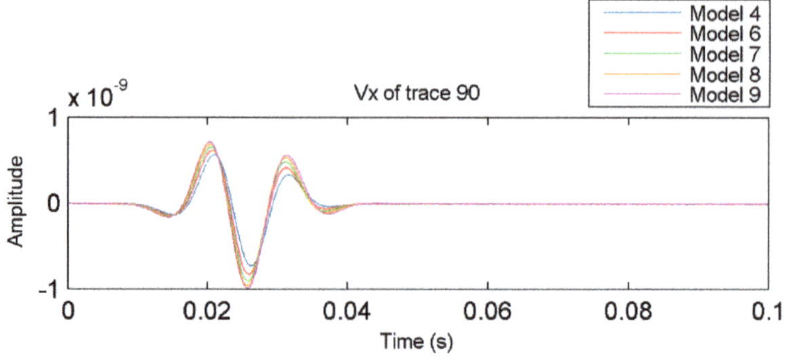

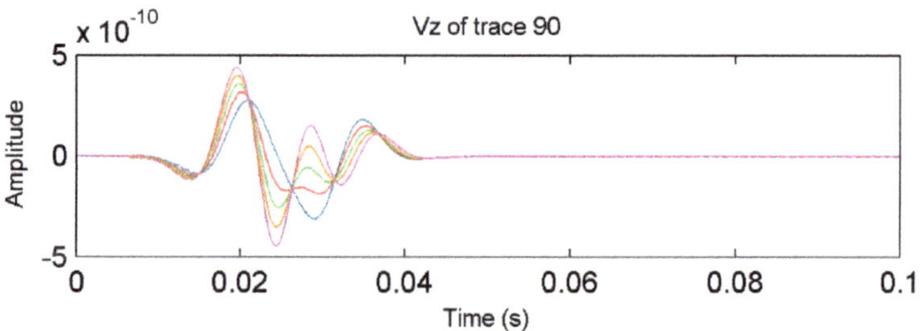

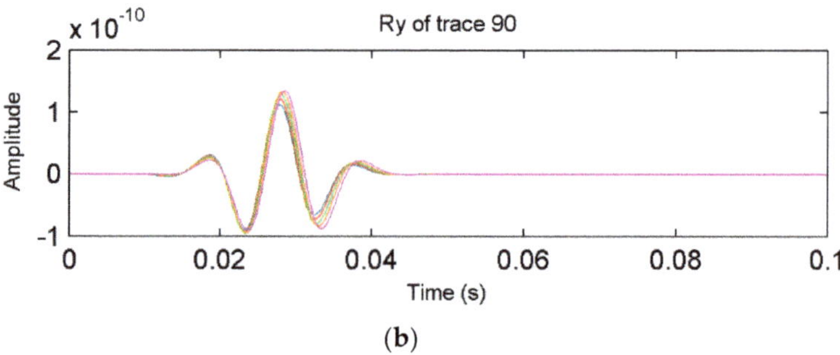

(b)

Figure 9. Influence of δ on seismic records generated from the expansion source: (**a**) the seismic records of the 30th trace; (**b**) the seismic records of the 90th trace.

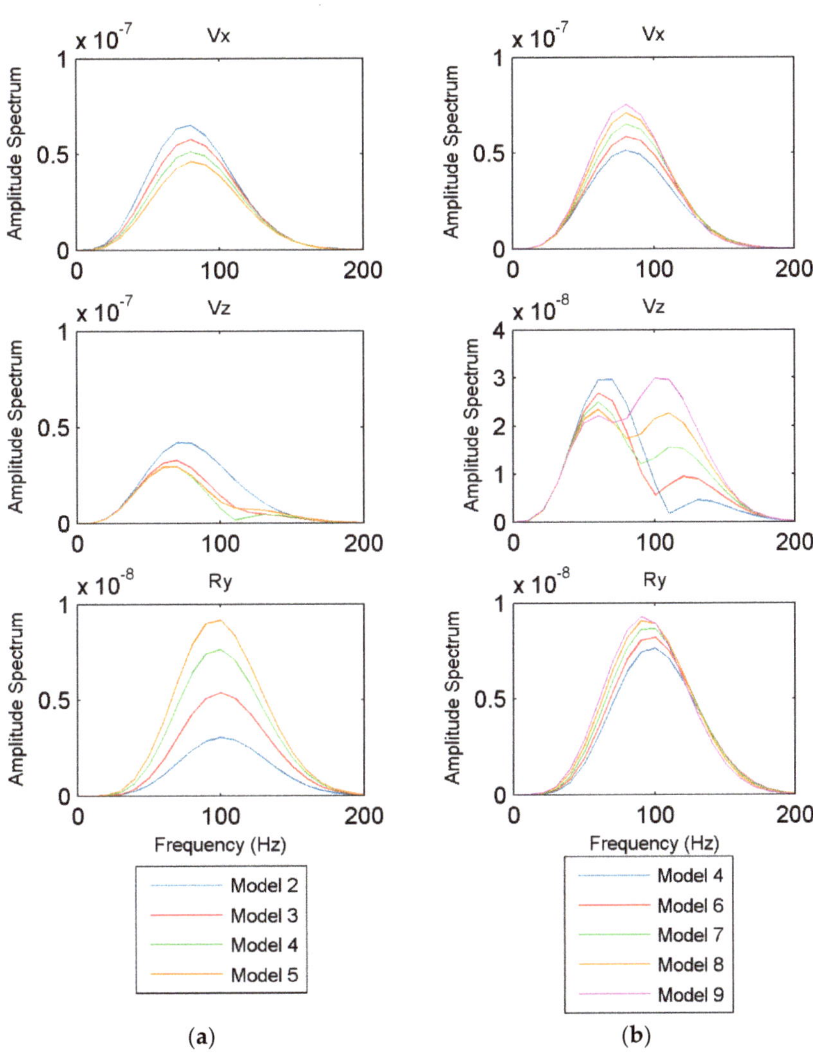

(a) (b)

Figure 10. Influence of anisotropy parameters on amplitude spectra of the 90th trace generated from the expansion source: (**a**) when ε varies; (**b**) when δ varies.

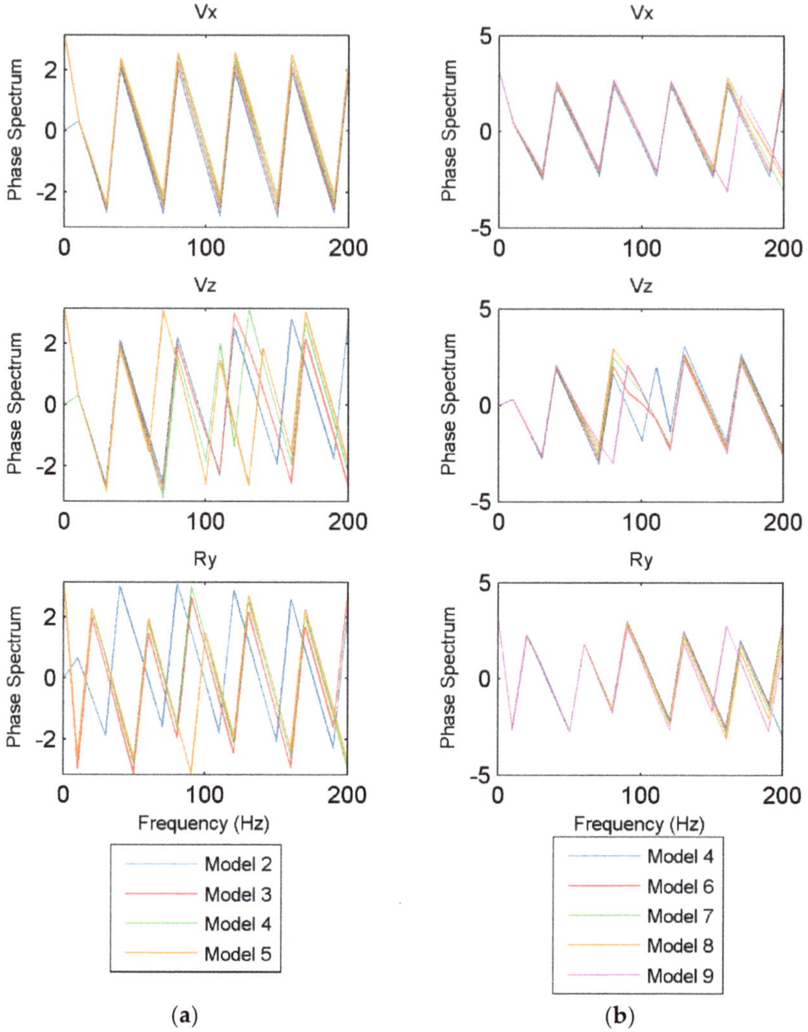

Figure 11. Influence of anisotropy parameters on phase spectra of the 90th trace generated from the expansion source: (**a**) when ε varies; (**b**) when δ varies.

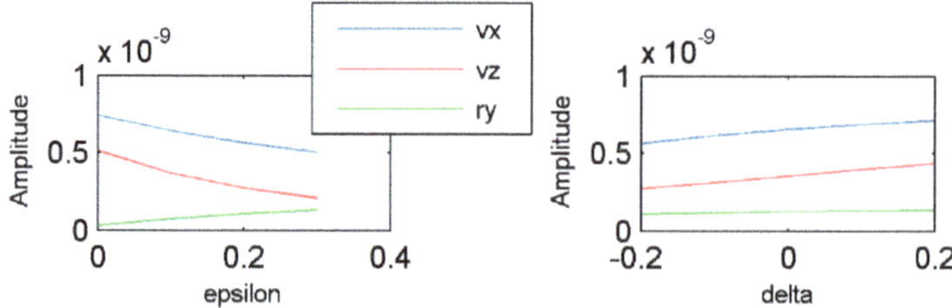

Figure 12. Influence of ε and δ on peak values of the 90th trace generated from the expansion source.

The seismic records, wave field snapshots, and FK spectra of different models are shown in Figures 5–7. Figure 5 demonstrates that, contributed by the velocity anisotropy, both P and S waves existed in the components in the VTI media. The energy of S waves became stronger with increasing ε, while the effects of δ were almost invisible.

The shape of wavefronts in R_y components became elliptical in VTI media, as seen in Figure 6. The eccentricities of the ellipse gradually increased with increasing ε, while the eccentricity of the ellipse gradually decreased with increasing δ.

There are obvious P- and S-wave energy groups in the rotational components in the FK spectra (Figure 7). With the increase of ε, the energy of P waves decreased gradually, but became more concentrated at the same time, while the energy of S waves gradually increased. However, the FK spectra of rotational components were almost unchanged with increasing δ.

We extracted and boosted the 30th and 90th traces' waveforms to study the effects of ε and δ on the amplitude and phase of seismic waves, which represented the far and near offset traces, respectively, to analyze the dynamic characteristics in detail. With the increase of ε, the amplitude and phase of the seismic waveforms showed obvious variations, as illustrated in Figure 8. In the near offset, with the increase of ε, the X components changed slightly. The peak values of Z components decreased gradually, while those of R_y components increased gradually. However, in the far offset, the peak values of R_y components increased slightly, while the phase varied widely with the increase of ε.

The effects of δ on waves' dynamic characteristics are shown in Figure 9. The effects of anisotropy on the X components were nearly negligible in the far offset, while the effects on the R_y components were obvious. The Z components changed substantially with the increase of δ in the near offset. The waves on the rotational components changed more obviously than those on translational components with increasing δ.

Furthermore, the influence of ε and δ on different components is shown in the amplitude and phase spectra (Figures 10 and 11). We can conclude that the rotational components had more high-frequency information than the translational components, since the spectra of X components were mainly in the frequency range 60–100 Hz, Z components are mainly in the frequency range 70–110 Hz, and R_y components in the frequency range 80–120 Hz. The amplitude variation of the R_y components was much greater than that of the X components with the variation of ε. On the Z components, the bandwidths of the wave fields became smaller, and the central frequencies became lower as ε increased. In addition, δ had less of an influence on the amplitude spectra of the three components than ε. With the increase of δ, the amplitudes of the wave fields on the X components and the R_y components increased slightly, while the amplitudes on the Z components showed a greater increase in the high frequencies.

In Figure 11, there are barely visible variations in the phase spectra of the X components with variation of ε, demonstrating that ε had a minor effect on the X components. The Z components of the four models differed mainly in the frequency range 100–200 Hz, while they were almost the same in the low frequencies. With the increase of ε, the phase spectra of R_y components varied more greatly than the translational components. The effects of δ variation on the phase spectra of three components were less pronounced than the effects of ε variation. The phase variation of X components was weak except in the frequency range 180–200 Hz, while it was more substantial in the frequency range 90–130 Hz for the Z components. With the increase of δ, the phase spectra of R_y components differed slightly, but they were significantly different from the isotropic condition. It can be deduced that rotation observations may be preferable to the study of anisotropic parameters.

Peak values of the 90th trace in different models can be seen in Figure 12. We found that the peak values of the rotational components increased gradually with the increase of ε, while they were almost the same with increasing δ.

To demonstrate the influence of the source on rotation, the seismic synthetic data and snapshots of wave fields at 0.05 s generated from different sources for different models are shown in Figures 13 and 14. There are obviously P waves and S waves in the seismic synthetic data generated from the radial concentrated force source, while there are few P waves in the seismograms generated from the other sources. The energy of P waves in the

R_y components generated from the radial concentrated force source was stronger than that generated from the vertical concentrated force source and shear source. With the increase of ε, the energy of P waves generated from the radial concentrated force source was much more enhanced than that of S waves, which is completely opposite to the outcome observed with increasing δ. Since S waves existed in the R_y components generated from the vertical concentrated force and shear source, it can be seen that the energy of S waves gradually increased with the increase of ε and δ.

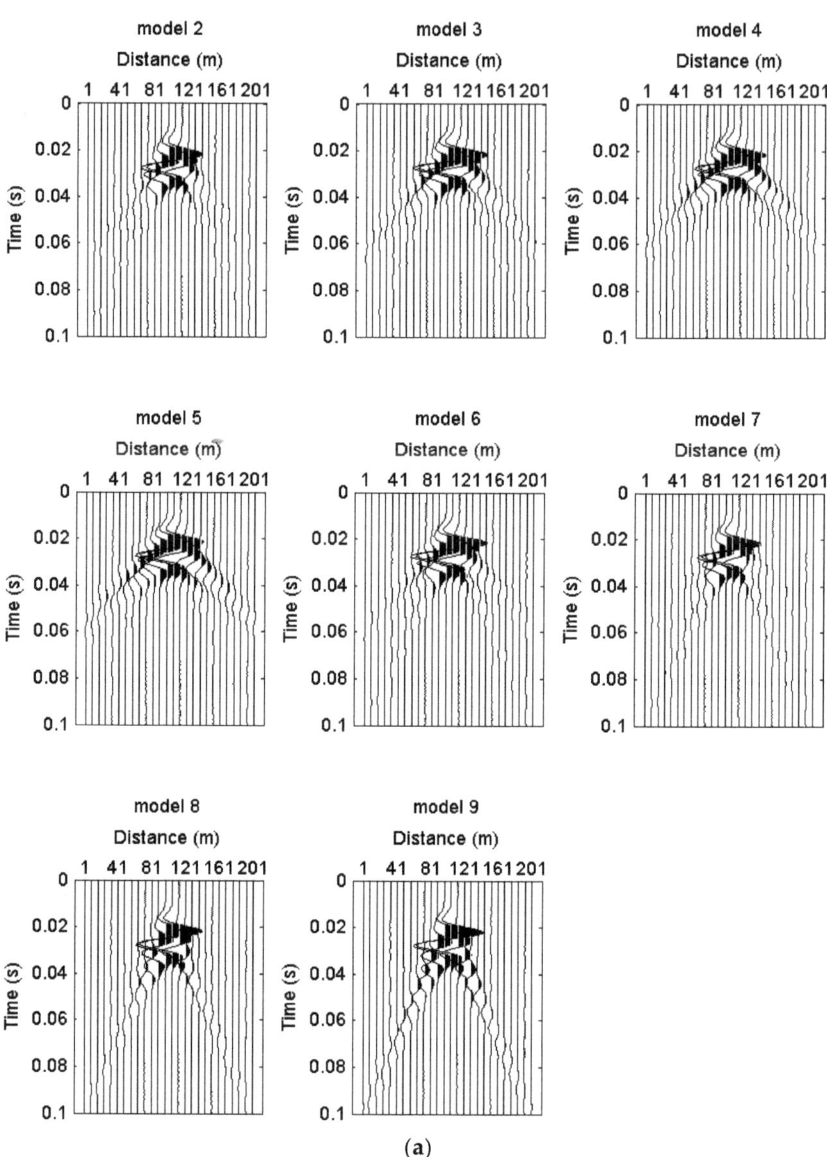

Figure 13. *Cont.*

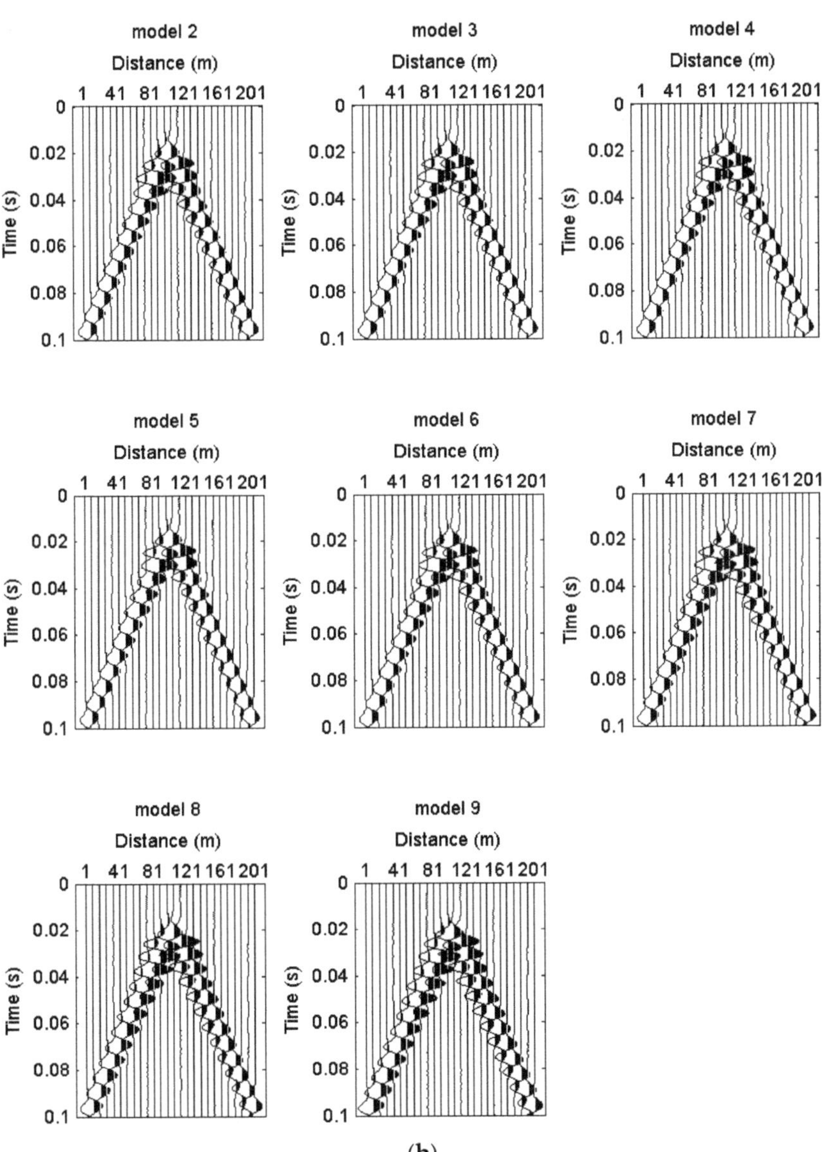

Figure 13. *Cont.*

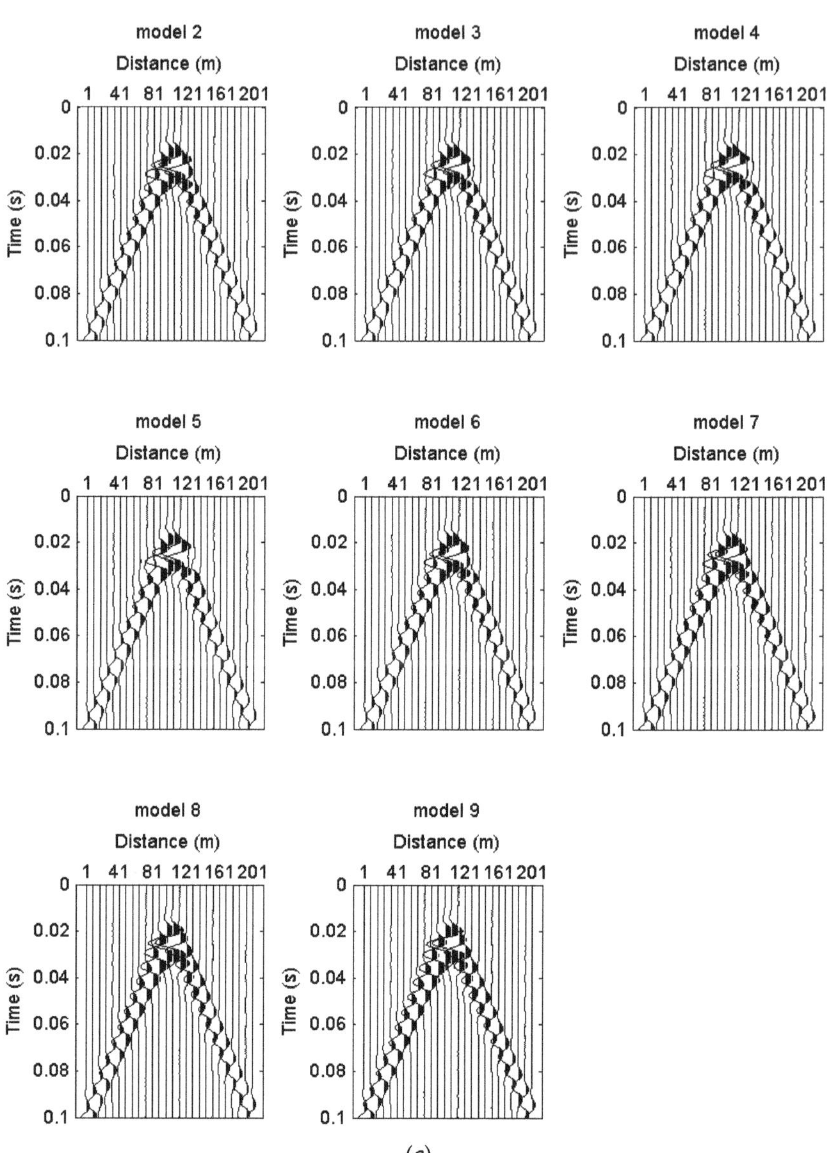

(c)

Figure 13. Seismograms. (**a**) Radial concentrated force source. (**b**) Vertical concentrated force source. (**c**) Shear source.

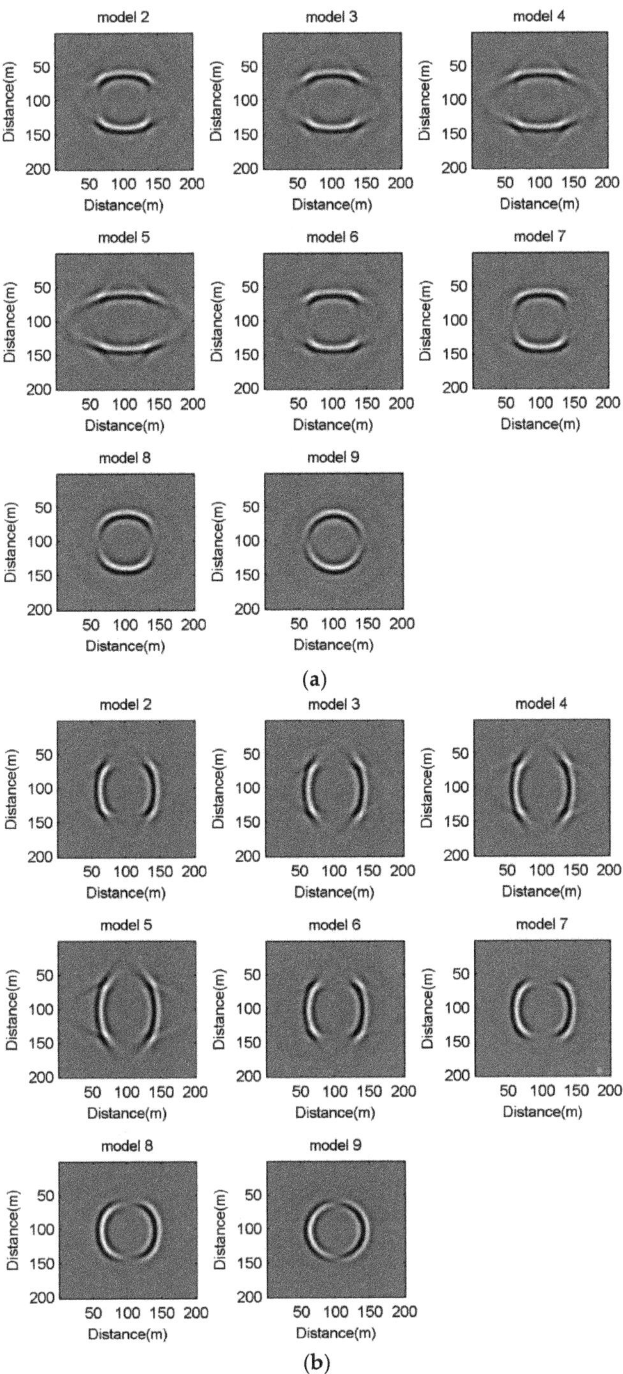

Figure 14. *Cont.*

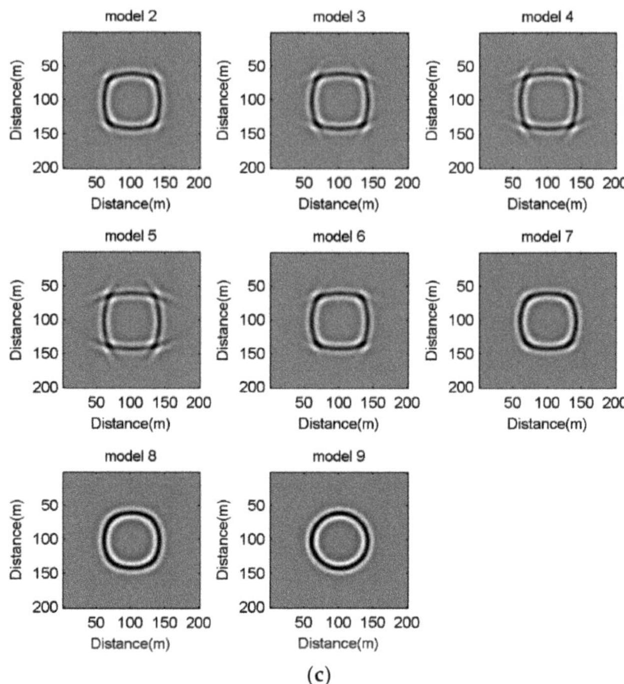

(c)

Figure 14. Snapshots of wave fields at 0.05 s. (**a**) Radial concentrated force source. (**b**) Vertical concentrated force source. (**c**) Shear source.

The shape of the wavefront was an ellipse in VTI media generated from the shear source and concentrated force source in the R_y components, while the wavefront was round in isotropic medium. The energy of P waves was much weaker than that of S waves, although it could barely be seen in the snapshots of wave fields generated from the shear source and vertical concentrated force source. With the increase of ε, the eccentricity of the wavefronts gradually increased. However, with the increase of δ, the eccentricity of the ellipse gradually decreased to near-circular. We conclude that the sources have a great influence on the wave fields of the R_y component.

5. Conclusions

We simulated seismic waves in VTI media with different anisotropic parameters under different sources and analyzed the similarities and differences between the translational components and the rotational components, especially focusing on the effects of anisotropic parameters on the rotational motions. We drew the following conclusions from these synthetic examples:

1. The energy of S waves is significantly stronger than that of P waves in the rotational components, and the rotational components have more high-frequency information.
2. The wave fields of rotations propagate with different phase distributions under different source mechanisms in space, and the vertical concentrated force source and shear source cause much stronger rotations than the others.
3. The amplitude and phase characteristics of the rotations are more complex in VTI media, which is helpful to distinguish VTI media from isotropic media. With the increase of ε and δ, the energy of the rotational components increases. With the increase of ε, the eccentricity of the wavefronts gradually increases, while with the increase of δ the eccentricity of the ellipse gradually decreases to near-circular.

4. Rotations are influenced more significantly by anisotropic parameters than translations, which indicates that the response of the rotations to the anisotropy is more sensitive.

From the simulations and comparisons introduced in this paper, we can deduce that rotation will be beneficial to distinguish VTI media from isotropic media and to predict their anisotropic parameters.

Author Contributions: Conceptualization, L.S. and Y.W. (Yun Wang); investigation, W.L.; writing—original draft preparation, L.S.; writing—review and editing, L.S. and Y.W. (Yun Wang); visualization, Y.W. (Yongxiang Wei). All authors have read and agreed to the published version of the manuscript.

Funding: This research was funded by the National Natural Science Foundation of China, grant number U1839208 and 41425017.

Institutional Review Board Statement: Not applicable.

Informed Consent Statement: Not applicable.

Data Availability Statement: Not applicable.

Acknowledgments: We gratefully acknowledge financial support by the National Natural Science Foundation of China (U1839208, 41425017).

Conflicts of Interest: The authors declare no conflict of interest.

References

1. Trifunac, M.D. Review: Rotations in Structural Response. *Bull. Seismol. Soc. Am.* **2009**, *99*, 968–979. [CrossRef]
2. Jaroszewicz, L.R.; Kurzych, A.; Krajewski, Z.; Marć, P.; Kowalski, J.K.; Bobra, P. Review of the Usefulness of Various Rotational Seismometers with Laboratory Results of Fibre-Optic Ones Tested for Engineering Applications. *Sensors* **2016**, *16*, 2161. [CrossRef] [PubMed]
3. Lee, W.H.K.; Igel, H.; Trifunac, M.D. Recent advances in rotational seismology. *Seismol. Res. Lett.* **2009**, *80*, 479–490. [CrossRef]
4. Kozák, J.T. Tutorial on earthquake rotational effects: Historical examples. *Bull. Seismol. Soc. Am.* **2009**, *99*, 998–1010. [CrossRef]
5. Guidotti, R.; Castellani, A.; Stupazzini, M. Near-field earthquake strong ground motion rotations and their relevance on tall buildings. *Bull. Seismol. Soc. Am.* **2018**, *108*, 1171–1184. [CrossRef]
6. Igel, H.; Cochard, A.; Wassermann, J.; Flaws, A.; Schreiber, U.; Velikoseltsev, A.; Pham Dinh, N. Broad-band observations of earthquake-induced rotational ground motions. *Geophys. J. Int.* **2007**, *168*, 182–196. [CrossRef]
7. Wang, H.; Igel, H.; Gallovič, F.; Cochard, A. Source and basin effects on rotational ground motions: Comparison with translations. *Bull. Seismol. Soc. Am.* **2009**, *99*, 1162–1173. [CrossRef]
8. Ferreira, A.M.G.; Igel, H. Rotational Motions of Seismic Surface Waves in a Laterally Heterogeneous Earth. *Bull. Seismol. Soc. Am.* **2009**, *99*, 1429–1436. [CrossRef]
9. Pham, N.D.; Igel, H.; de la Puente, J.; Käser, M.; Schoenberg, M.A. Rotational motions in homogeneous anisotropic elastic media. *Geophysics* **2010**, *75*, D47–D56. [CrossRef]
10. Tang, L.; Fang, X. Generation of 6-C synthetic seismograms in stratified vertically transversely isotropic media using a generalized reflection and transmission coefficient method. *Geophys. J. Int.* **2021**, *255*, 1554–1585. [CrossRef]
11. Barak, O.; Brune, R.; Herkenhoff, F.; Dash, R.; Rector, J.; Ronen, S. Seven-component seismic data. In *SEG Technical Program Expanded Abstracts 2013*; Society of Exploration Geophysicists: Tulsa, OK, USA, 2013; pp. 5151–5155.
12. Sun, L.; Zhang, Z.; Wang, Y. Six-component elastic-wave simulation and analysis. EGU General Assembly. In *Geophysical Research Abstracts*; EGU2018-14930-1; EGU Press: Munich, Germany, 2018; Volume 20.
13. Zhang, Z.; Sun, L.; Tang, G.; Xu, T.; Wang, Y.; Wang, M.; Guo, X. Numerical simulation of the six-component elastic-wave field. *Chin. J. Geophys.* **2020**, *63*, 2375–2385.
14. Thomsen, L. Weak elastic anisotropy. *Geophysics* **1986**, *51*, 1954–1966. [CrossRef]
15. Madariaga, R. Dynamics of an expanding circular fault. *Bull. Seismol. Soc. Am.* **1976**, *66*, 639–666. [CrossRef]
16. Li, P.; Lin, W.; Zhang, X. Comparisons for regular splitting and non-splitting perfectly matched layer absorbing boundary conditions. *Acta Acust.* **2015**, *40*, 44–53.
17. Zhang, Z. *Geological Seismic Waves in Three-Dimensional Anisotropic Media*; Changchun Institute of Geology Press: Changchun, China, 1989.

Article

Seismic Velocity Anomalies Detection Based on a Modified U-Net Framework

Ziqian Li [1], Jiwei Jia [1,2], Zheng Lu [1], Jian Jiao [3,*] and Ping Yu [3]

1 School of Mathematics, Jilin University, Changchun 130012, China; zqli21@mails.jlu.edu.cn (Z.L.); jiajiwei@jlu.edu.cn (J.J.); luzheng21@mails.jlu.edu.cn (Z.L.)
2 National Applied Mathematical Center (Jilin), Changchun 130012, China
3 College of Geoexploration Science and Technology, Jilin University, Changchun 130061, China; yuping@jlu.edu.cn
* Correspondence: jiaojian001@jlu.edu.cn

Abstract: Accurate and efficient reconstruction of hidden geological structures under the surface is the main task of high-resolution Velocity Model Building (VMB). The most commonly used methods in practice are Tomography and Full Waveform Inversion (FWI), which rely heavily on the initial model. Recently, deep learning types of methods have received widespread attention and have performed well in many tasks such as image segmentation and classification. Therefore, it is of great significance to introduce deep learning algorithms into the VMB procedure to accelerate the production cycle, especially for the velocity anomalies detection, which is crucial for a high-resolution initial model. In this paper, a modified U-Net framework is proposed and applied directly on the seismic shot gathers to identify anomalies in the early stage of VMB, which can provide a suitable initial guess for the following large-scale VMB procedures such as FWI. The numerical examples show the power of the proposed method on synthetic data.

Keywords: velocity anomalies detection; deep learning; seismic data; geological structure; seismic velocity analysis

Citation: Li, Z.; Jia, J.; Lu, Z.; Jiao, J.; Yu, P. Seismic Velocity Anomalies Detection Based on a Modified U-Net Framework. *Appl. Sci.* **2022**, *12*, 7225. https://doi.org/10.3390/app12147225

Academic Editors: Guofeng Liu, Xiaohong Meng and Zhifu Zhang

Received: 9 June 2022
Accepted: 13 July 2022
Published: 18 July 2022

1. Introduction

The main goal of the high-resolution Velocity Model Building process is to reconstruct the subsurface structures, especially to capture potential geological bodies, such as shallow gas clouds, salt bodies and fault plains. On the other hand, these complex geological structures result in certain types of anomalies. Refs. [1,2] proposed different methods to calculate velocity models. Particularly, gas clouds correspond to lower velocity and a smaller Quality Factor (Q) [3], and some quantifiable techniques were proposed to identify shallow gas pockets. A ray-based Q-tomography was developed by [4] and has been applied to field data to estimate the effects of shallow gas by representing the pockets as anomalous Q bodies [5]. One of the most crucial problems in the Q-tomography method is how to predict the Q bodies masks accurately, which is used to indicate the location of strong absorptions. Usually, several iterations were needed to improve the precision of the location information, either by manual editing or introducing some attributes as pilot. Refs. [6,7] show exciting Q-tomography results, employing the FWI model to produce the masks for the subsequent anomalous Q-tomography process to evaluate the shallow gas clouds. However, in the FWI guided Q-tomography flow, the accuracy of the masks is strongly dependent on the FWI process, human intervention, and the initial Q-factor model, which are extremely time-consuming and tedious. The initial Q-factor model can be directly obtained from field data by applying sophisticated methods [8]. Furthermore, the initial velocity model usually starts from a smooth one, which is based on manual velocity analysis or vintage experience from that area only; the local detailed information is excluded. Introducing accurate location information into velocity models automatically in the early stages will help to improve the quality of the VMB results and accelerate the model building process significantly.

Since deep learning was proposed by [9], it has received widespread attention and is widely used in computer vision, voice recognition, natural language processing, etc. Recently, machine-learning based techniques were introduced into the seismic data processing and interpretation community. Ref. [10] presented a supervised-learning-based salt body detection algorithm. Three features—amplitude, second derivative and curve length—were selected to characterize voxels of 3D seismic volume, and the algorithm uses small fraction of the characterized voxels for training to predict the whole volume. Ref. [11] developed a texture classification workflow using seismic attributes, clustering techniques and segmentation by thresholds, followed by second step mathematical morphological and basic operations between volumes to improve the detection. A major strategy of this type of method is to apply data mining algorithms [12] on the post-migration volumes. Ref. [13] developed a novel method based on machine learning techniques to automatically identify and localize faults. The method was introduced in the initial stages of the VMB process, when no seismic data had been migrated, which is different from other types of post-migration methods that use processed seismic data or migrated images [14,15].

In the field of computer vision, it is well known that Fully Convolutional Networks (FCNs) and U-Net perform well on image segmentation tasks. These two frameworks were proposed by [16,17], respectively. U-Net is similar to FCN and has been widely used in medical image segmentation. Compared with FCN, the first feature of U-Net is that it is completely symmetrical, which means the left and right hand side are very similar. However, the decoder of FCN is relatively simple, using only a deconvolution operation. The second difference is skip connection: FCN uses summation, while U-Net uses concatenation. The U-Net model modified and expanded the network on the basis of FCN, so that it can use very few training images to obtain very accurate segmentation results. In addition, an upsampling stage is added, which adds lots of feature channels, allowing more texture information of the original image to spread in high-resolution layers. U-net does not have a fully convolutional layer and uses valid for convolution throughout, which ensures that the results of the segmentation are based on no missing context features.

In this paper, a two-step deep-learning based velocity anomalies detection workflow is established. The workflow starts from the pre-migration shot gathers directly, justifiying the presence of anomalies firstly and then predicting accurate location information of the velocity anomalies prior to VMB process by employing a modified U-Net neural network. A set of two-dimensional synthetic model tests are presented to evaluate the effectiveness of the proposed workflow.

2. Materials and Methods

In this section, the flow chart of the proposed model is presented, including a 2-branch U-Net and a post-processing step.

2.1. Seismic Data Preparation

Firstly, a set of two-dimensional layered velocity models with anomalies are generated; the models include high or low speed anomaly regions. An acoustic wave equation forward modeling scheme is employed for producing seismic data based on the generated velocity models. The detailed settings are listed here.

- Background velocity can be simple layered model or including faults.
- Anomalies are simplified to some elliptical regions, with random center coordinates and major/minor axis length.
- Masks for the designed anomalies are generated simultaneously.
- At most, two anomalies are located in each model, either inside one layer or crossing multiple layers.
- Seismic data generated by different shots are recorded.
- All the data are separated into two parts for training and testing.

2.2. Flow Chart

The flow chart of the proposed framework is presented in Figure 1.

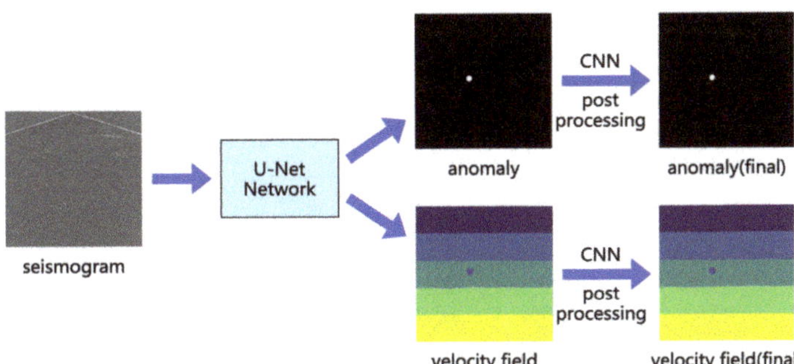

Figure 1. Flow chart.

The pairs of generated seismic and the corresponding mask compose the training dataset and feature label, as the input for the proposed U-Net neural network. The training procedure is shown in Figure 2.

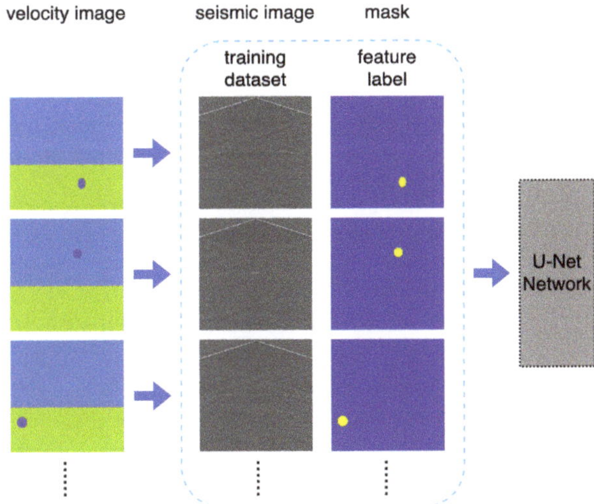

Figure 2. Training procedure.

After the training process, the testing data (not included in training) will be input to the trained U-Net neural network, and the feature label set is predicted for verification purposes, which is shown in Figure 3.

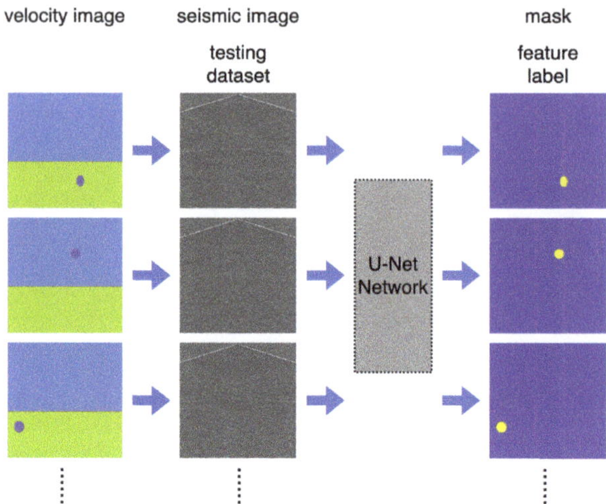

Figure 3. Verification procedure.

2.3. Modified U-Net Neuron Network

U-Net is an efficient deep learning framework for image segmentation task. As Figure 4 shows, it can be divided into the front and latter part. The front is the downsampling part for feature extraction, which is composed of convolution layer and pooling layer. The latter one is the upsampling part for image reconstruction, which is composed of an up-convolution layer and a convolution layer. It should be noted that the net concatenates the front part and the latter part in each parallel layer, so that more information of feature extraction can be reserved.

For the classical U-Net neuron network, suppose that Z_i is the data obtained after upsampling at the i-th layer, Y_i is the data before upsampling, X_i is the data mapped from the left side of the network at the i-th layer and C represents the upsampling process. Then its mathematical formula can be expressed as

$$Z_i = X_i + CY_i. \tag{1}$$

Our task is to invert seismic waveforms to generate velocity field and anomalies locations. A modified U-Net is proposed with a different structure from the classical one. Initially, the convolution operation starts from 1 pixel outside the edge of the seismic image, which guarantees that the generated velocity field does not change the size while acquiring the features of the original data. In addition, since it is found in the classical U-Net experiments that the generated velocity fields are contaminated by the contour of the seismic waveform, and the reason is that the classical U-Net transmits some parts of the data to the output directly, we modify the classical U-Net structure and omit the transmitted part. The structure of the modified U-Net is shown in Figure 4, and the dashed line indicates the transmitting process in traditional U-Net. The corresponding formulation is as follows:

$$\begin{cases} Z_i = CY_i, & if \quad i = 1; \\ Z_i = X_i + CY_i, & if \quad i > 1. \end{cases} \tag{2}$$

In addition, the upsampling process is divided into two branches, which are used to predict the velocity field and the anomaly mask, respectively. Under this setting, which is shown in Figure 5, the velocity and the anomaly mask can be generated from the corresponding seismic waveform simultaneously.

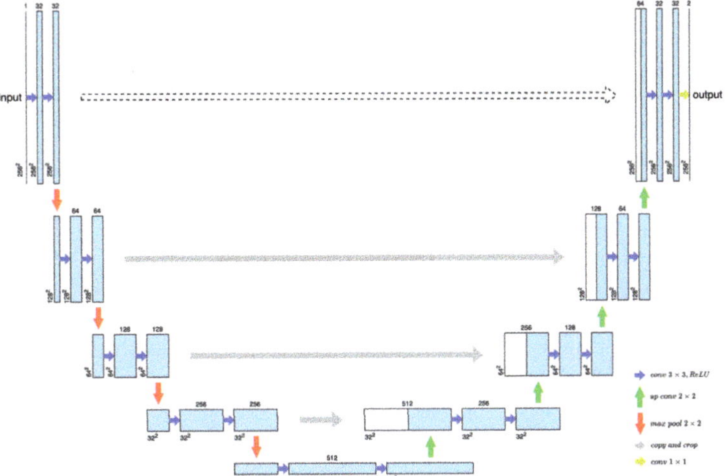

Figure 4. Architecture of the modified U-Net neuron network.

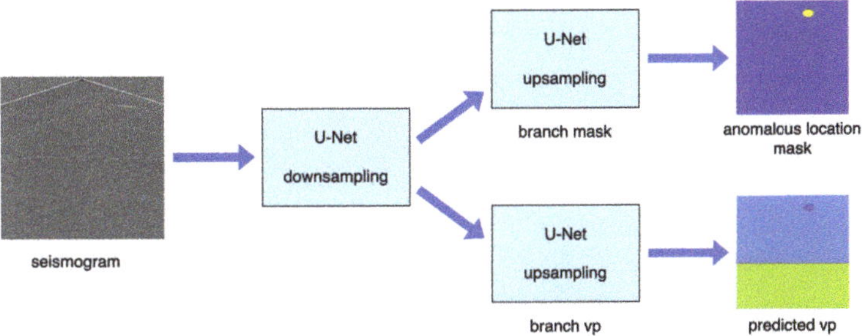

Figure 5. U-Net separated into 2 parts.

Velocity field image generation is a regression problem, and the MSE type of loss function is often used. On the other hand, anomaly image generation is a classification problem, where the BCE type of loss function should be employed. Classification and regression problems cannot use the same loss function as the criterion. Therefore, we define an ensemble loss function of the entire U-Net as a weighted summation of MSE and BCE loss with a fine-tunable parameter α. The accuracy requirement of the model is that the ensemble loss function is less than a threshold, so that the accuracy of two branches can be guaranteed simultaneously. The ensemble loss function is defined as

$$Loss = \alpha \text{BCE} + (1 - \alpha)\text{MSE} \tag{3}$$

2.4. Post-Processing

In practice of the Velocity Model Building, geophysicists usually apply a smoothing post-processing step to obtain a more reliable background velocity field. Following this traditional setting, we added two convolutional layers to the generated velocity field image to achieve a smoothing effect. As shown in Figure 6, the velocity field with a post-processing step has a more reasonable background and without touching the anomalies.

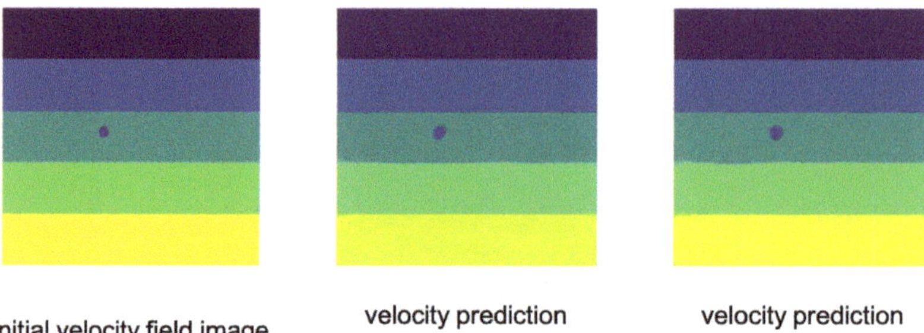

Figure 6. Post-processing step.

3. Results

For an area, the seismic wave propagation speed of different geological structures in the area is different, and there is a cavity in the geological structure (wave velocity is different from other geological structures). In this algorithm, when the data of the velocity field are given, a seismic waveform can be obtained by simulating the ground explosion. The specific method is to input velocity field data and modeling parameters (including shot position, parameters for recording wavefield snapshots, etc.) into the program. Then the corresponding seismic data and seismic waveforms will be generated by using the finite difference method. We made a comparison to the similar machine learning algorithm proposed in [13]. The experiments were performed on a workstation with two 10 Core Intel(R) Xeon(R) Silver 4210R CPU, 2 RTX A5000 GPU, 128GB RAM and an Ubuntu 20.04 operating system that implements Pytorch. The code of our algorithm has been uploaded to GitHub (https://github.com/DavidDeadpool/Unet-seismic/tree/main/Unet-seismic, accessed on 12 July 2022).

First, we generated a set of velocity models with anomalies distributed randomly and also generated the masks to describe the location of the anomalies corresponding to each velocity model. Then we modeled seismic shot gathers based on the generated velocity models. Finally, we paired the shot gathers and the masks together as the input and output of the training pairs of the U-Net network.

3.1. Velocity Models with Anomalies

We have generated various types of velocity models. The basic type is a velocity model with one anomaly and no fault. Through the controlled variable method, the other velocity fields are multiple (two) anomalies with no fault and one anomaly with fault. According to the natural laws of geophysics, the velocity in the upper level (near the surface) is small, and the velocity increases as the depth increases. The location and size of the anomalies are randomly generated, and the default shape of the anomaly is an ellipse. The length of the semi-axis of the ellipse is $[20, 100]$. Figure 7 shows the different types of velocity models.

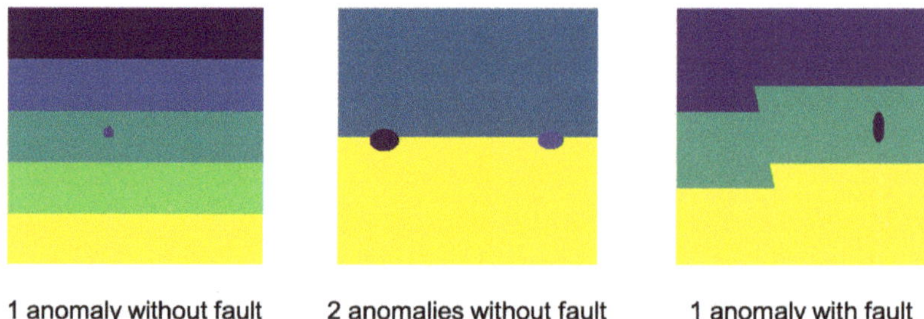

Figure 7. Examples of velocity models.

3.2. Seismic Waveforms

By using a 2D acoustic wave modeling algorithm [18], we calculated the reflection seismic records and wavefield snapshots corresponding to different shot points. We assume that the entire area is 5000 m wide, and the source and receiver position can be in this 5000 m area. The leftmost end of the image is set to $X = 0$ m, and the rightmost end is $X = 5000$ m. We can collect one seismic waveform at each seismic wave launch location. For convenience, for the input of multiple seismic waveforms, the experiment selects the shots at locations of 500 m, 1500 m, 2500 m, 3500 m and 4500 m. Figure 8 shows the seismic waveforms corresponding to different shot positions.

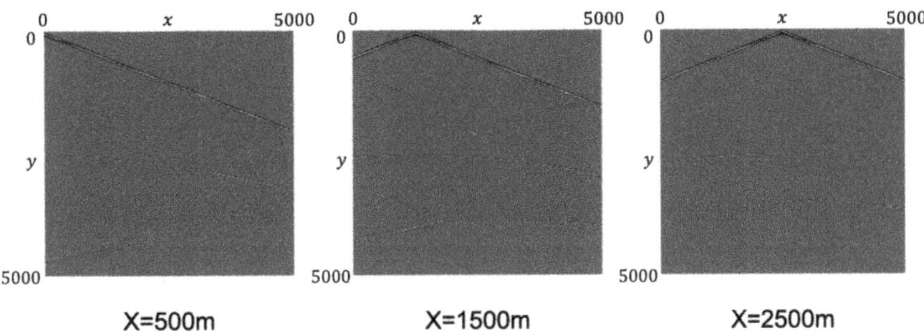

Figure 8. Seismic waveforms at different shot positions.

3.3. Results

In order to understand the influence of fault and the number of anomalies on the experimental results, we conducted multiple sets of comparative experiments. In addition, in order to determine the impact of the number of shot points on the accuracy of the experiment, we conducted single shot point experiments ($X = 2500$ m) and multiple shot point experiments ($X = 500$ m, 1500 m, 2500 m, 3500 m, 4500 m) and completed the comparison in each set of comparison experiments.

3.3.1. One Anomaly without Fault

In this group of experiments, the velocity model is divided into five layers without fault, and there is only one anomaly. We made predictions for a single shot point and multiple shot points. The velocity field and anomaly images obtained through training are shown in Figure 9. There is no obvious difference in the prediction of the velocity field. In order to better view the prediction effect of the velocity field, we have produced vertical velocity profiles, as shown in Figure 10. The prediction error of a single shot point at an anomaly is smaller than that of a multi-shot-point prediction.

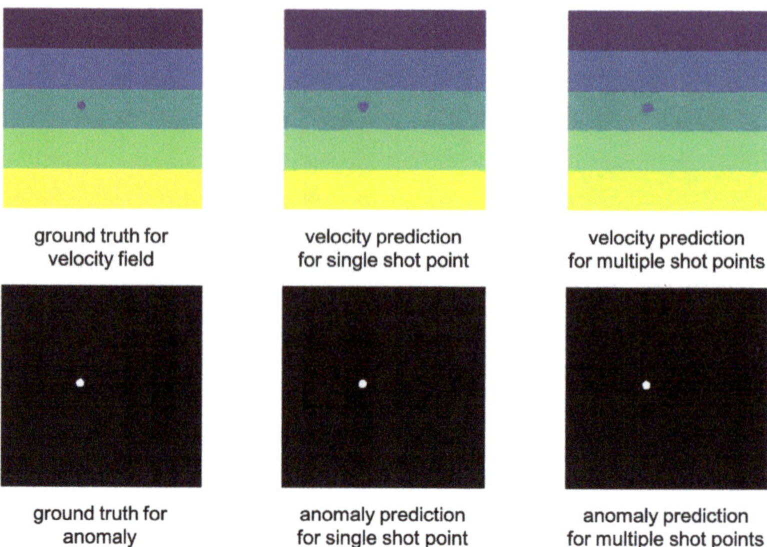

Figure 9. Prediction for velocity field and anomaly in velocity model with one anomaly and no fault.

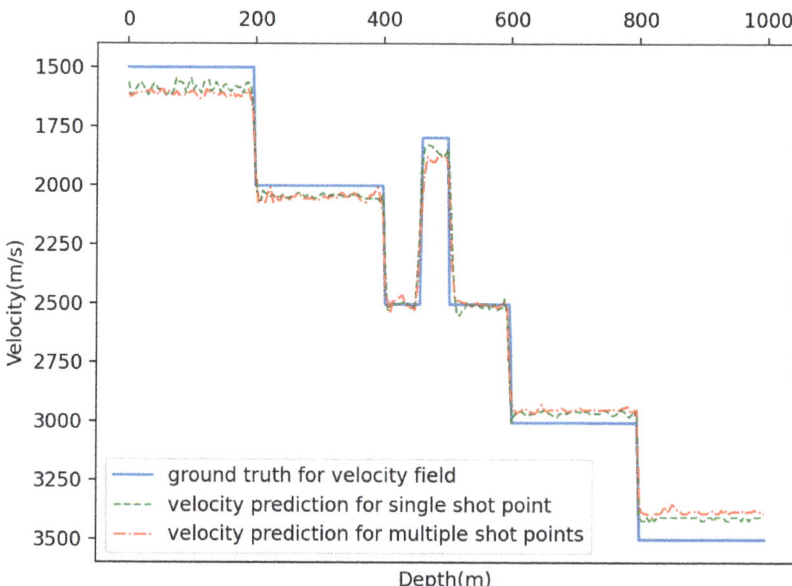

Figure 10. Vertical velocity profiles of velocity field with one anomaly and no fault.

3.3.2. One Anomaly with Fault

In this group of experiments, the velocity field is divided into three layers, with only one abnormal point. We made predictions for a single shot point and multiple shot points. The effect of predicting the velocity field and abnormal points is shown in Figure 11. There is no obvious difference in the prediction of both velocity field and anomaly. In order to better view the prediction effect of the velocity field, we have produced vertical velocity profiles, as shown in Figure 12. The prediction error of a single shot point at an anomaly is smaller than that of a multi-shot-point prediction.

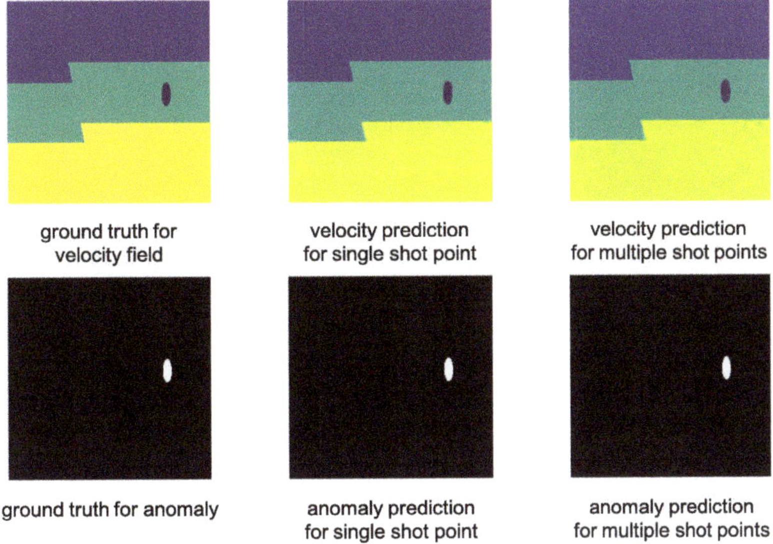

Figure 11. Prediction for velocity field and anomaly with one anomaly and fault.

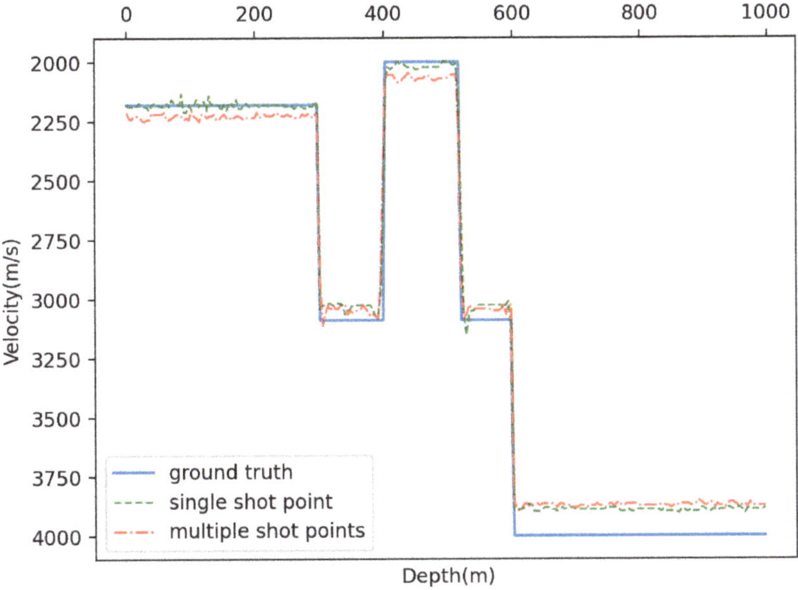

Figure 12. Vertical velocity profiles of velocity field with one anomaly and fault.

3.3.3. Multiple Anomaly without Fault

In this group of experiments, the velocity field is divided into two layers, with two anomalies. We made predictions for a single shot point and multiple shot points. The effect of predicting the velocity field and anomalies is shown in Figure 13. There is no obvious difference in the prediction of the velocity field, but the method of single-shot prediction is more accurate for the prediction of anomalies. In order to better view the prediction effect of the velocity field, we have produced vertical velocity profiles, as shown in Figure 14. The prediction error of a single shot point at an anomaly is smaller than that of a multi-shot-point prediction.

In summary, the neural network algorithm based on U-Net can accurately complete the process of predicting the velocity field and anomalies from the seismic waveform. Both single shot point and multiple shot points have good prediction effects, but the effect of multiple shot points is not necessarily better than single shot point.

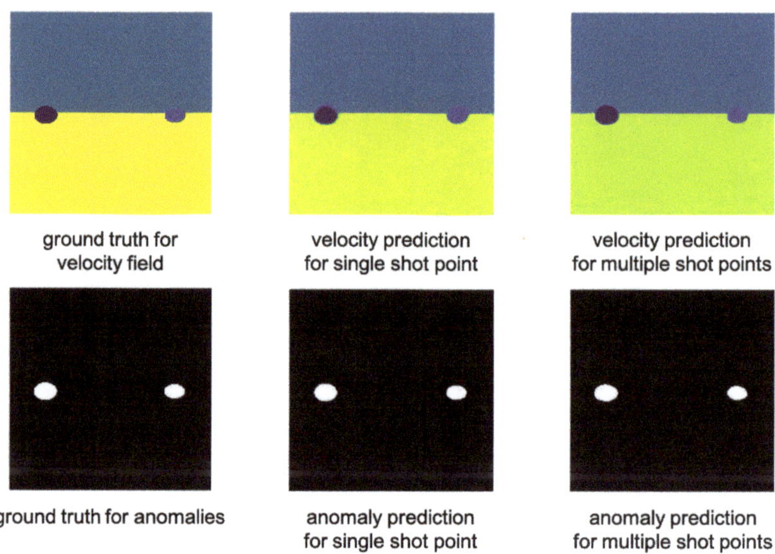

Figure 13. Prediction for velocity field and anomaly with 2 anomalies and no fault.

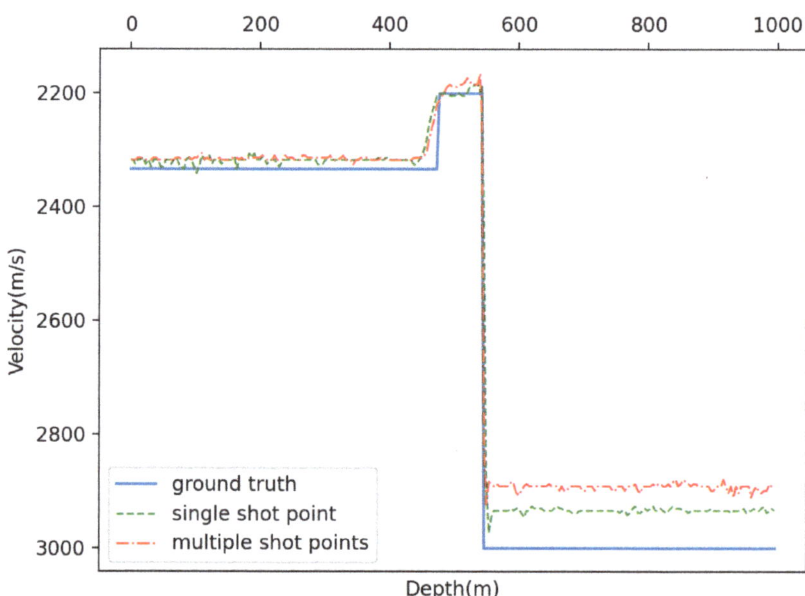

Figure 14. Vertical velocity profiles of velocity field with 2 anomalies and no fault.

4. Discussions and Conclusions

We generated simulated seismic data based on the finite difference method and used the modified U-Net to successfully predict the underground velocity field and the location of anomalies from seismic waveforms, and then used CNN to post-process the generated images. Experimental results show that the effect of multi-shot-point prediction is not necessarily better than that of single shot point. After numerical verification, the predicted velocity field and abnormal point position have very little error with the ground truth. It should be noted that traditional VMB normally needs weeks or months to reconstruct the velocity model; however, our algorithm only needs days, which is shown as Table 1. Next, we will use real seismic data to verify and refine our model.

Table 1. Time of calculation in one anomaly without fault experiment.

Experiment	Time of Training	Time of Prediction
Single shot point	95.3 h	5 min
Multiple shot points	109.3 h	5 min

There are still some problems with our model. First of all, as shown in Figures 10, 12 and 14, as the depth increases, the error of the velocity prediction will gradually increase. In addition, the data used in each of our experiments correspond to 1000 sets of artificially generated information. It is necessary to increase the amount of information and use real seismic data to improve the applicability of the model to the actual situation. Finally, although U-Net can extract image features well and achieve the required training effects, this processing is based on images and not directly obtained from seismic data training. The image resolution will seriously affect the training effect. We will continue to find ways to overcome these problems.

Author Contributions: Methodology, J.J. (Jiwei Jia) and J.J. (Jian Jiao); Resources, P.Y.; Software, Z.L. (Ziqian Li); Validation, Z.L. (Zheng Lu); Writing—Original Draft, Z.L. (Ziqian Li). All authors have read and agreed to the published version of the manuscript.

Funding: The authors are supported by the Natural Science Foundation of Jilin Province (Grant No. 20210101481JC), the Shanghai Municipal Science and Technology Major Project (Grant No. 2021SHZDZX0103), and the Fundamental Research Funds for the Central Universities (Grant No. 93K172020K27).

Institutional Review Board Statement: Not applicable.

Informed Consent Statement: Not applicable.

Data Availability Statement: The data that support the findings of this study are available from the corresponding author upon reasonable request.

Conflicts of Interest: The authors declare no conflict of interest.

References

1. Rointan, A.; Soleimani Monfared, M.; Aghajani, H. Improvement of seismic velocity model by selective removal of irrelevant velocity variations. *Acta Geod. Geophys.* **2021**, *56*, 145–176. [CrossRef]
2. Shahbazi, A.; Ghosh, D.; Soleimani, M.; Gerami, A. Seismic imaging of complex structures with the CO-CDS stack method. *Stud. Geophys. Geod.* **2016**, *60*, 662–678. [CrossRef]
3. Bouchaala, F.; Guennou, C. Estimation of viscoelastic attenuation of real seismic data by use of ray tracing software: Application to the detection of gas hydrates and free gas. *Comptes Rendus Geosci.* **2012**, *344*, 57–66. [CrossRef]
4. Xin, K.; Hung, B. 3-D tomographic Q inversion for compensating frequency dependent attenuation and dispersion. In *SEG Technical Program Expanded Abstracts*; Society of Exploration Geophysicists: Houston, TX, USA, 2009; pp. 4014–4018.
5. Zhou, J.; Birdus, S.; Hung, B.; Teng, K.H.; Xie, Y.; Chagalov, D.; Cheang, A.; Wellen, D.; Garrity, J. Compensating attenuation due to shallow gas through Q tomography and Q-PSDM, a case study in Brazil. In *SEG Technical Program Expanded Abstracts*; Society of Exploration Geophysicists: Houston, TX, USA, 2011; pp. 3332–3336.

6. Zhou, J.; Wu, X.; Teng, K.H.; Xie, Y.; Lefeuvre, F.; Anstey, I.; Sirgue, L. FWI-guided Q tomography and Q-PSDM for imaging in the presence of complex gas clouds, a case study from offshore Malaysia. In *SEG Technical Program Expanded Abstracts 2013*; Society of Exploration Geophysicists: Houston, TX, USA, 2013; pp. 4765–4769.
7. Zhang, Z.; Jia, J.; Fu, G.; Zhang, H.; Chow, D.; Hung, B.; Anstey, I.; Lai, W.L. Fullband Imaging. 2016. Available online: https://archives.datapages.com/data/petroleum-exploration-society-of-australia/news/140/140001/pdfs/52.htm (accessed on 12 July 2022).
8. Matsushima, J.; Ali, M.Y.; Bouchaala, F. A novel method for separating intrinsic and scattering attenuation for zero-offset vertical seismic profiling data. *Geophys. J. Int.* **2017**, *211*, 1655–1668. [CrossRef]
9. Hinton, G.E. Deep belief networks. *Scholarpedia* **2009**, *4*, 5947. [CrossRef]
10. Guillen, P.; Larrazabal, G.; González, G.; Boumber, D.; Vilalta, R. Supervised learning to detect salt body. In *SEG Technical Program Expanded Abstracts*; Society of Exploration Geophysicists: Houston, TX, USA, 2015; pp. 1826–1829.
11. Guillen, P.; Larrazabal, G.; Gonzalez, G.; Sineva, D. Detecting salt body using texture classification. In Proceedings of the 14th International Congress of the Brazilian Geophysical Society & EXPOGEF, Rio de Janeiro, Brazil, 3–6 August 2015; Society of Exploration Geophysicists: Houston, TX, USA, 2015; pp. 1155–1159.
12. Hastie, T.; Tibshirani, R.; Friedman, J. *The Elements of Statistical Learning: Data Mining, Inference, and Prediction*; Springer Science & Business Media: Berlin/Heidelberg, Germany, 2009.
13. Zhang, C.; Frogner, C.; Araya-Polo, M.; Hohl, D. Machine-learning based automated fault detection in seismic traces. In Proceedings of the 76th EAGE Conference and Exhibition, Amsterdam, The Netherlands, 16–19 June 2014; Volume 2014, pp. 1–5.
14. Cohen, I.; Coult, N.; Vassiliou, A.A. Detection and extraction of fault surfaces in 3D seismic data. *Geophysics* **2006**, *71*, P21–P27. [CrossRef]
15. Hale, D. Methods to compute fault images, extract fault surfaces, and estimate fault throws from 3D seismic images. *Geophysics* **2013**, *78*, O33–O43. [CrossRef]
16. Long, J.; Shelhamer, E.; Darrell, T. Fully convolutional networks for semantic segmentation. In Proceedings of the IEEE Conference on Computer Vision and Pattern Recognition, Boston, MA, USA, 7–12 June 2015; pp. 3431–3440.
17. Ronneberger, O.; Fischer, P.; Brox, T. U-net: Convolutional networks for biomedical image segmentation. In *International Conference on Medical Image Computing and Computer-Assisted Intervention*; Springer: Berlin/Heidelberg, Germany, 2015; pp. 234–241.
18. Artru, J.; Farges, T.; Lognonné, P. Acoustic waves generated from seismic surface waves: Propagation properties determined from Doppler sounding observations and normal-mode modelling. *Geophys. J. Int.* **2004**, *158*, 1067–1077. [CrossRef]

MDPI

Article

Fault Imaging of Seismic Data Based on a Modified U-Net with Dilated Convolution

Jizhong Wu [1,*], Ying Shi [2] and Weihong Wang [2]

1 School of Bohai Rim Energy, Northeast Petroleum University, Daqing 163318, China
2 School of Earth Sciences, Northeast Petroleum University, Daqing 163318, China; shiying@nepu.edu.cn (Y.S.); wangweihong@nepu.edu.cn (W.W.)
* Correspondence: wjzkgr@nepu.edu.cn

Abstract: Fault imaging follows the processing and migration imaging of seismic data, which is very important in oil and gas exploration and development. Conventional fault imaging methods are easily influenced by seismic data and interpreters' experience and have limited ability to identify complex fault areas and micro-faults. Conventional convolutional neural network uniformly processes feature maps of the same layer, resulting in the same receptive field of the neural network in the same layer and relatively single local information obtained, which is not conducive to the imaging of multi-scale faults. To solve this problem, our research proposes a modified U-Net architecture. Two functional modules containing dilated convolution are added between the encoder and decoder to enhance the network's ability to select multi-scale information, enhance the consistency between the receptive field and the target region of fault recognition, and finally improve the identification ability of micro-faults. Training on synthetic seismic data and testing on real data were carried out using the modified U-Net. The actual fault imaging shows that the proposed scheme has certain advantages.

Keywords: dilation rate; deep learning; data processing; receptive field

Citation: Wu, J.; Shi, Y.; Wang, W. Fault Imaging of Seismic Data Based on a Modified U-Net with Dilated Convolution. *Appl. Sci.* **2022**, *12*, 2451. https://doi.org/10.3390/app12052451

Academic Editor: Chiara Bedon

Received: 5 January 2022
Accepted: 23 February 2022
Published: 26 February 2022

Publisher's Note: MDPI stays neutral with regard to jurisdictional claims in published maps and institutional affiliations.

1. Introduction

Faults play a major role in lateral sealing of thin reservoirs and accumulation of the remaining oil in conventional and unconventional reservoirs onshore in China [1]. Almost all developed onshore oil and gas fields in China are distributed in rift basins which are rich in oil and gas resources with highly developed and very complex fault systems [2–4]. At present, there are many kinds of fault recognition techniques with different principles, but there are still great difficulties in fine fault imaging. This is because the rift basin experienced a variety of external forces during its growth, and developed a variety of faults, such as normal faults, normal oblique-slip faults, oblique faults, and strike–slip faults, etc. According to their different combinations in plane and section, they also present many forms such as broom shaped, comb shaped, goose row shaped, and parallel interlaced in planes. In rift basins, the filling of sediments, the development and distribution of sedimentary sequences, the formation, distribution and evolution of oil and gas reservoirs (including the formation and effectiveness of traps, and the migration and accumulation of oil and gas) are closely related to the distribution and activity of faults [5]. Therefore, fine detection and characterization of faults in rift basins in China has become a key basic geological problem for oil and gas exploration and development efforts and has become the key topic of basin tectonic research [6].

Continuous and regular event breakpoints constitute faults in seismic imaging data. However, because the accuracy, resolution and signal-to-noise ratio of seismic imaging data cannot reach the theoretical limit and the geological situation is complicated, it is a great challenge for petroleum engineers to describe the spatial distribution of faults from seismic data. In the past, fault characterization has been regarded as an interpretative task, followed by seismic data processing and imaging, because it requires extensive geological knowledge and experience. In recent years, researchers use convolutional neural network to identify faults, focusing on the construction of network architecture, network parameter debugging and optimization and model training. They are less and less constrained by geological knowledge and personal experience, and the processing and mining of seismic data are becoming more and more important. Therefore, it is very reasonable to attribute fault identification via deep learning to the research field of seismic data processing and imaging, and it is also the development trend in the future. Based on this concept, our research employs seismic imaging data to realize the description of fault characteristics through a new neural network model, that is, to realize fault imaging.

In the past 30 years, with the continuous development of computer hardware and software, fault identification has made great progress in efficiency and accuracy. From the perspective of method evolution, fault interpretation has experienced from the initial manual interpretation to the emergence of various identification methods, such as coherence method, curvature attribute method and ant colony algorithm, which describe faults by calculating transverse discontinuities of seismic data. In the past five years, with the rapid development of artificial intelligence [7–9], various fault identification methods based on deep learning have gained remarkable achievements. In 2014, Zheng et al. [10] used deep learning tools to conduct fault identification tests on prestack synthetic seismic records. Araya-Polo et al. [11] applied machine learning and deep neural network algorithms to automate fault recognition, which greatly improved the efficiency and stability of fault interpretation. Waldeland et al. [12] and Guitton et al. [13] successively used a Convolutional Neural networks (CNN) model to make progress in fault description. Xiong et al. [14] used results of the skeletonized seismic coherent self-correction method as training samples to train a CNN model to identify seismic faults. In 2019, Wu et al. [15] realized the identification of small-scale faults by using U-Net network. Wu et al. [6] used a full-convolutional neural network, FCN, to achieve a better characterization of faults. Among these networks, the U-Net architecture is currently very popular, due to its shortcut operation which concatenates attribute maps from the low-level feature (shallow layer) to maps of high-level feature (deep layer), and it can be seen as a special kind of CNN [16–18]. In addition, the U-Net does not have strict requirements on the size of training sets, and smaller training sets can also provide satisfactory results. However, most networks including the U-Net uniformly process all feature maps of the same layer, resulting in the same receptive field of the network in the same layer, thus obtaining relatively single local information. Moreover, with the continuous down-sampling of the network and the convolution operation with step size, the defect that only a single size information can be obtained at the same layer becomes more and more obvious, resulting in the inaccurate identification of faults by the neural network.

To address these issues, this paper introduces a new neural network model, which takes U-Net as the basic network and uses inter-group channel dilated convolution module (GCM) to connect each cross-connection layer between encoding path and decoding path, and uses inter-group space dilated convolution module (GSM) to connect layers after each deconvolution layer in decoding path. Both GCM and GSM use dilated convolution. Dilated Convolution, also known as hole convolution or expanded convolution, is to inject cavities into the standard convolution kernel to increase the reception field of the model. In the CNN structure, convolution and pooling drive most layers, convolution performs feature extraction, while pooling performs feature aggregation. For an image classification task, this network structure has good feature extraction capability, among which the most classical structure is VGGNet (a convolutional neural network was developed by the University of Oxford's Visual Geometry Group and Google DeepMind in 2014). However, this structure has some problems for target detection and image segmentation. For example, the size of receptive field is very important in target detection and image segmentation, and the guarantee of receptive field depends on down-sampling. However, down-sampling will make small targets difficult to detect. If down-sampling is not completed and the number of convolutional layers is only increased, the computation of the network will increase. In addition, if pooling is not carried out for features, the final feature extraction effect will also be affected, and there will be no change in the receptive field. In order to solve these problems in CNN, this paper introduces dilated convolution to increase the receptive field without sacrificing the size of the feature map. Compared with the conventional convolution operation, dilation rate is added to the dilated convolution, which refers to the number of intervals of points in the convolution kernel [19–21]. When dilatation rate is 1, the dilated convolution will degenerate into conventional convolution. The similarity between dilated convolution and conventional convolution lies in that the size of convolution kernel is the same, that is, the number of parameters of neural network remains unchanged [22,23]. The difference lies in that dilated convolution has a larger receptive field and can preserve the structure of internal data [24,25].

2. Illustration of Dilated Convolution

We use a set of pictures to illustrate the principle of dilated convolution. Figure 1a is the conventional convolution kernel, and the dilated convolution is obtained by adding intervals to this basic convolution kernel. Figure 1b corresponds to the convolution of 3×3, with dilation rate of 2 and interval 1, that is, corresponding to 7×7 image blocks. It can be understood that the kernel size becomes 7×7, but only 9 points have parameters, and the rest have parameters of 0. Convolution calculation was performed for the 9 points in Figure 1b and the corresponding pixels in the feature map, and the other positions were skipped. Figure 1b,c are similar, except that the dilation rate is 4, which is equivalent to a 15×15 convolution kernel. As the convolution kernel becomes larger, the receptive field becomes larger naturally.

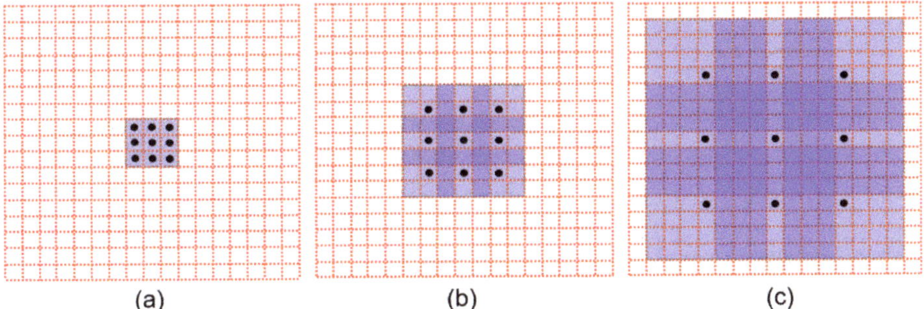

Figure 1. The dilated convolution with dilation rate of 1, 2 and 4, respectively. (**a**) dilation rate = 1; (**b**) dilation rate = 2; (**c**) dilation rate = 4.

In practical application, when the same dilation rate is used for all convolutional layers, a problem called grid effect will appear. Since the convolution calculation points on the feature map are discontinuous, for example, if we repeatedly accumulate 3×3 kernel of dilation rate 2, this problem will occur.

The blue square in Figure 2 is the convolution calculation points participating in the calculation, and the depth of the color represents the calculation times. As can be seen, since the dilation rates of the three convolutions are consistent, the calculation points of the convolution will show a grid expansion outward, while some points will not become calculation points. Such kernel discontinuities, that is, not all pixels are used for calculations, will result in the loss of continuity of information, which is very detrimental for tasks such as pixel-level dense prediction. The solution is to discontinuously use dilated convolution with the same dilation rate, but this is not comprehensive enough. If the dilation rate is multiple, such as 2,4,8, then the problem still exists. Therefore, the best way is to set the dilation rate of continuous dilated convolution as "jagged", such as 1,2,3, respectively, so that the distribution of convolution calculation points will become like Figure 3 without discontinuity.

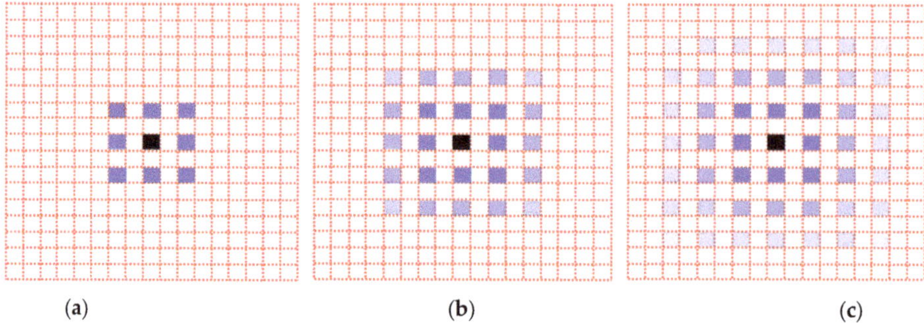

(a) (b) (c)

Figure 2. The dilated convolution with the same dilation rate of 2, respectively. There are grid effects in all three graphs. (**a**) the first convolution with dilation rate of 2; (**b**) the second convolution with dilation rate of 2; (**c**) the third convolution with dilation rate of 2.

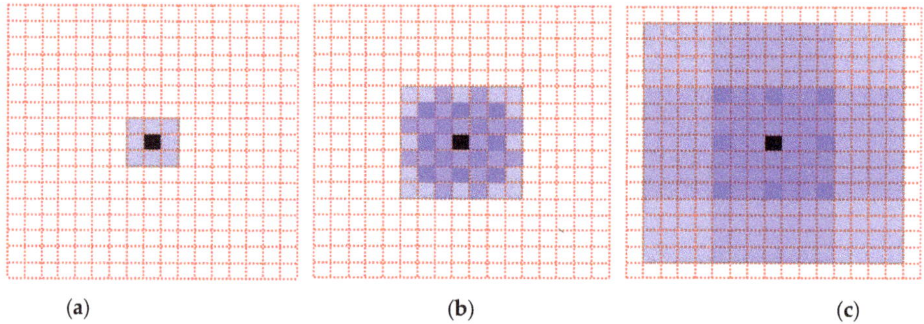

(a) (b) (c)

Figure 3. The dilated convolution with dilation rate of 1, 2 and 3, respectively. (**a**) dilation rate = 1; (**b**) dilation rate = 2; (**c**) dilation rate = 3.

3. The Architecture of the Modified U-Net

The proposed neural network adopts a U-Net of a 4-layer structure as the basic network. In the coding path, feature maps of each layer are connected to the corresponding decoding layer by GCM, in which each layer adopts two 3 × 3 convolution layers and maximum pooling layer for feature extraction. Then, in the decoding path, feature maps of each layer are connected to the corresponding decoding layer by GSM. Each layer adopts a 3 × 3 convolution layer, up-sampling layer and 1 × 1 convolution layer to restore, and the output layer adopts a 3 × 3 convolution layer and 1 × 1 convolution layer to output. In this modified U-Net, the batch regularization (BN) and modified linear units (ReLU) are added to all convolution layers to correct data distribution, except for the output layer. GCM and GSM modules play a key role in the modified U-Net, and their operation mechanism is similar. GCM module can divide the input feature map into four groups on average, and then carry out the dilated convolution operation with dilation rates of 1, 2, 3 and 5, respectively. In addition, the module extracts and outputs the features of the input feature map through pooling, convolution, batch regularization, activation, softmax and other conventional operations, and finally obtains the channel information of all groups. For the GSM module, it can divide the input feature map into three groups on average, and each group carries out the dilated convolution operation with dilation rates of 1, 2 and 4, respectively. This module carries out feature extraction and output in sequence by down-sampling, convolution, batch regularization, activation, up-sampling, convolution, batch regularization, activation and softmax, and finally obtains the spatial information of all groups.

Figure 4 shows the structure of GCM in the modified U-Net. This module divides the input feature map into four groups evenly, and each group performs dilated convolution operations with dilation rates of 1, 2, 3, and 5. The size of the target area of fault identification determines the value of the dilation rate. After dilated convolution, four groups of feature maps with different scales are obtained. Besides, four groups of channel information are returned by softmax, which were taken as weights and multiplied by four groups of feature maps with different scales obtained via dilated convolutions to acquire new feature maps. The receptive field corresponding to the group with the largest weight contributed the most to the final network prediction. Finally, the four groups of new feature maps are spliced together and then a residual operation is performed with the input feature map to obtain the final prediction result. The GCM module uses the idea of grouping and realizes the automatic selection of inter-group multi-scale information under the guidance of channel information.

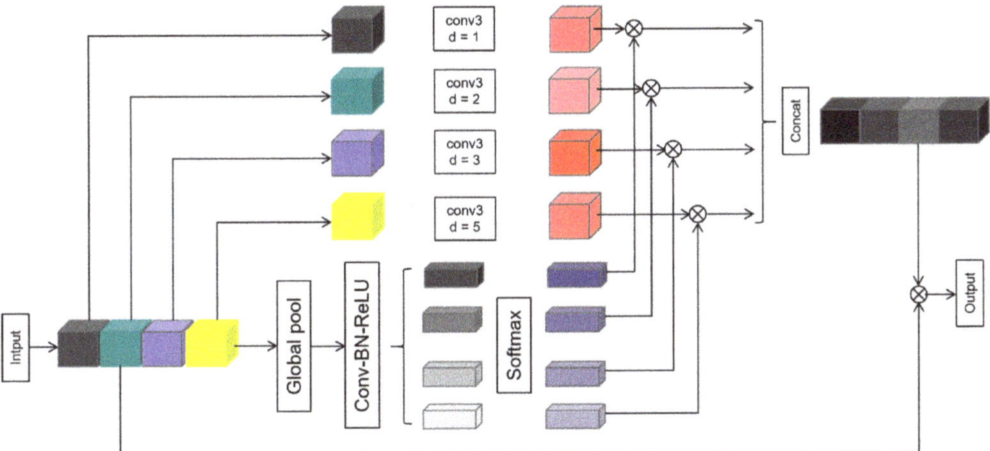

Figure 4. The architecture of GCM.

Figure 5 shows the structure of GSM, which realizes the selection of multi-scale information between groups in another way and enhances the consistency of receptive field and target region recognition. In this module, the input feature map was divided into three groups, and then three groups of feature maps with different scales are obtained by the dilated convolution with dilation rates of 1, 2 and 4. At the same time, three feature maps with spatial weights are cropped from the input feature map through a series of conventional operations. In these operations, the purpose of down-sampling is to obtain more global information, the purpose of up-sampling is to restore the size of feature maps, and the purpose of softmax is to enable the module to automatically select multi-scale information. Three feature maps with spatial weights are multiplied by three feature maps of different scales obtained by dilated convolution to get three new feature maps. Finally, after splicing the three groups of new feature maps, a residual operation is performed with the input feature map to acquire the final prediction results. In summary, under the guidance of spatial information, the GSM module can select multi-scale information among a group of feature maps.

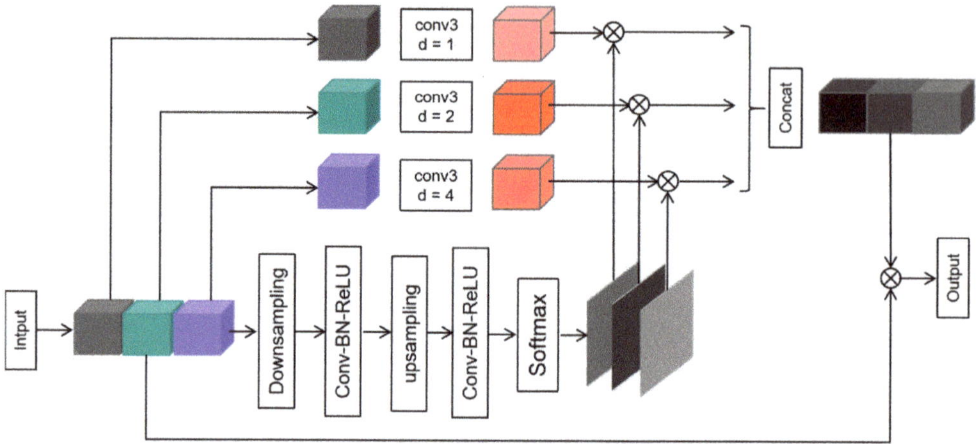

Figure 5. The architecture of GSM.

The proposed neural network is based on the U-Net network and has two functional modules, the GCM and GSM, which can finely describe faults of different scales. Its architecture is shown in Figure 6. Due to the powerful multi-scale information selection ability of GCM and GSM modules, this paper only uses a 4-layer U-Net based on a coding–decoding structure as the basic network. In the coding path, only two 3 × 3 convolution and maximum pooling are used to quickly obtain feature maps with different resolutions. In the decoding path, multiple simple decoding blocks are used to quickly and effectively recover feature maps with high resolution. In this neural network, the data distribution after convolution is corrected by BN and ReLU, and the GCM module is placed at the connection layer of the network to automatically select multi-scale information, which makes up for the lack of transmitting single information to the decoder by the encoder of conventional U-Net. At the same time, the GSM module between groups is placed in the path of decoder to realize the function of multi-scale information selection, which makes up for the disadvantage of losing global information in up-sampling process.

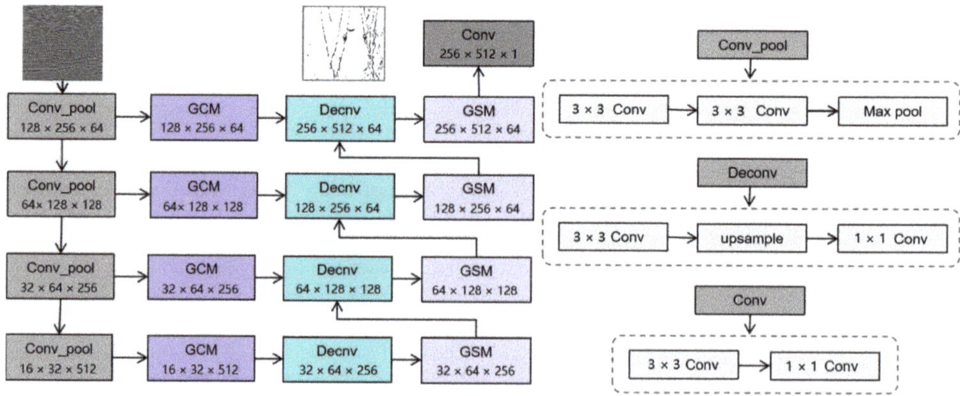

Figure 6. The architecture of the modified U-Net.

4. Loss Function

The neural network will produce deviation between prediction and reality during training, and the deviation value is represented by loss function. During the training, the stochastic gradient descent (SGD) algorithm is used to update the network parameters and reduce the value of the loss function, so that the prediction and the actual convergence gradually, tend to be consistent [26]. The final result of neural network output is fault probability body, where 1 represents fault and 0 represents non-fault. In this study, fault recognition is regarded as a binary segmentation task. In the fault probability body, the most part is non-fault, and the least part (less than 10%) has a value of 1. There are strong data imbalance and uneven fault distribution area. In this case, the binary cross entropy loss (BCE) function is most often used [6,27]. Dice loss function is commonly used to serve the segmentation and recognition tasks of small-scale targets in medical research [28]. In this research, BCE and Dice are combined to solve problems such as data imbalance, uneven fault distribution area and insufficient accuracy in fault identification. The expression of the combined loss function is as follows:

$$\begin{cases} L_{BCE} = -\frac{1}{N}\sum\limits_{i}^{N}(g_i\log(p_i) + (1-g_i)\log(1-p_i)) \\ L_{Dice} = 1 - \frac{2\sum\limits_{i}^{N} p_i g_i + \varepsilon}{\sum\limits_{i}^{N} p_i + \sum\limits_{i}^{N} g_i + \varepsilon} \\ L = \lambda L_{BCE} + L_{Dice} \end{cases}$$

where N is the total number of pixels in the input image. $p_i \in [0,1]$ and $g_i \in [0,1]$ represent the prediction probability and label value of pixel, respectively, ε is the smoothing factor, whose value range is (0.1,1). λ is the balance coefficient of Dice loss and BCE loss.

5. Training and Validation

In the neural network training, we randomly selected 1000 seismic images from an open-source dataset [15] for training, and the corresponding label data were also completed by manual marking in advance, marked as 1 in places with faults and 0 in places without faults. The purpose of network training is to optimize the parameters of the whole network. With each training, the deviation between the prediction and the actual represented by the loss function will decrease until the prediction and the actual tend to be consistent.

Figure 7 shows randomly seismic data sets with their corresponding labels. We used another 200 images as test and validation data, which were not included in the training. In the process of training, SGD is used to optimize the network, and the number of images sent into the network is 10 each time. The network model can be trained when the number of epochs reaches 30 times. Figure 8a shows the change of training accuracy and validation accuracy of the modified U-Net with the number of epochs. After 30 epochs, the accuracy rate tends to be above 0.9. Figure 8b shows the changes of training loss and validation loss of the modified U-Net with the number of epochs. After 30 epochs, the loss value tends to 0.01. After training, save the network parameters. In the process of training and validation, in order to increase the diversity of training data sets and make the trained neural network have better classification or recognition performance, data enhancement is used to improve the diversity of training data sets. The data enhancement operation mainly includes data reversal and rotation around the time axis.

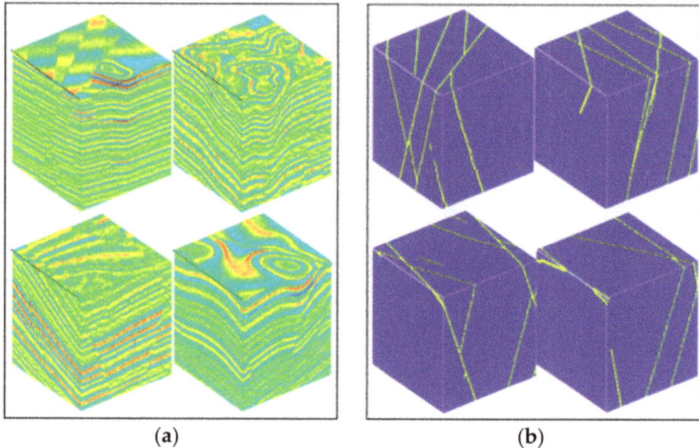

Figure 7. (a) Seismic data sets and (b) their fault labels.

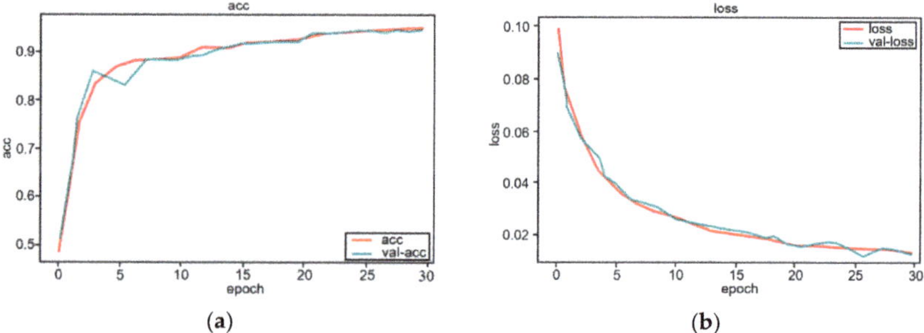

(a) (b)

Figure 8. (a) The training and validation accuracy both will increase with epochs, whereas (b) the training and validation loss decreases with epochs.

6. Application

This paper used field data to verify the effectiveness of the trained network, and the fault probability volume is shown in Figure 9. In order to facilitate the interpretation of results, the opacity of the fault probability cube was adjusted and superimposed on the original seismic image. At the same time, we used a trained U-Net to identify faults of this data, and other parameters were completely the same except for GCM and GSM modules. The study area is located in a sandstone oil field in China, and the faults are mainly Y-shaped throughout the section and occur in almost every formation [29]. In the 700 ms–1500 ms time window, the number of faults is the largest, and the characteristics of faults are the most complex [30]. As depth increases, the imaging accuracy of seismic data decreases and the difficulty of fault imaging becomes more and more. On the plane, the fault is affected by the tension and strike-slip stress mechanism, and the fault direction is mainly NE and NW. This data set consists of 495 (lines) $\times$ 580 (CDPs) with a CDP spacing of 25 m and a line spacing of 25 m. The data are sampled at 1 ms with a length of 2 s. The inline number and xline number are in the range of (2410,2905) and (3600,4180), respectively. By using an NVIDIA TITAN Xp GPU, it takes about 110 min to calculate the fault probability volume. The randomly selected vertical profiles and time slice are inline 2510, xline 4025 and time slice at 1540 ms, respectively. The fault imaging results of the modified U-Net and U-Net are shown in Figures 10–12.

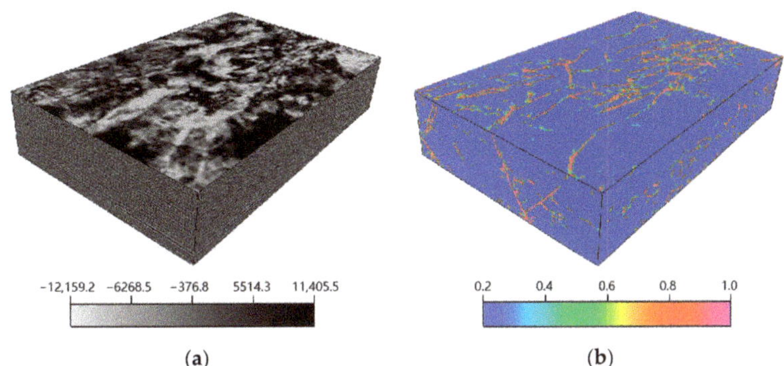

Figure 9. (**a**) Original 3D seismic data volume and (**b**) its fault probability volume.

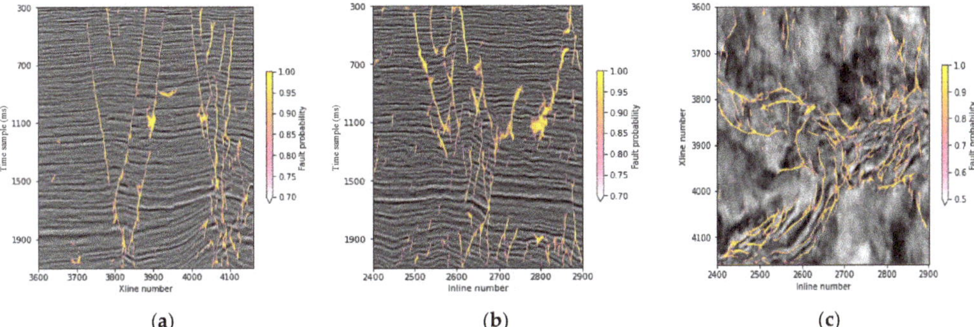

Figure 10. Three seismic images are displayed with faults that are imaged using the trained modified U-Net model. (**a**) Inline 2510; (**b**) Xline 4025; (**c**) a time slice at 1540 ms.

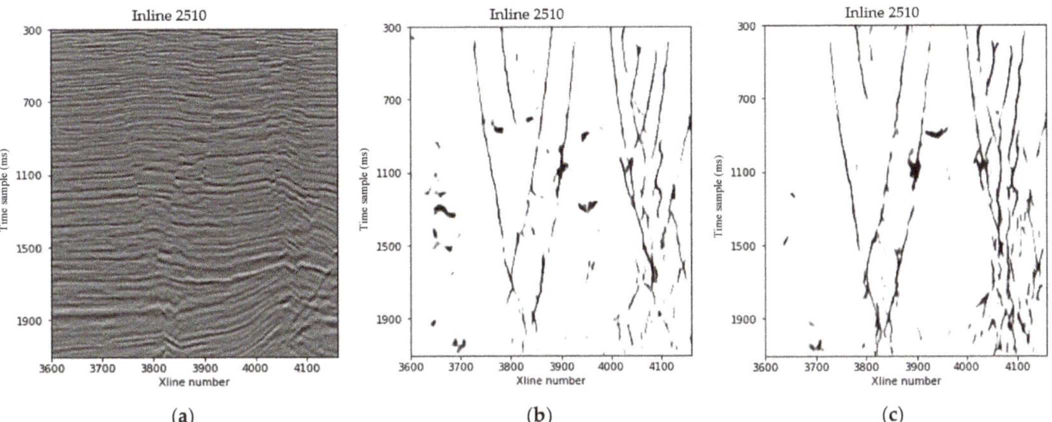

Figure 11. (**a**) A seismic image is displayed with faults that are imaged using (**b**) the trained U-Net model and (**c**) the trained modified U-Net model.

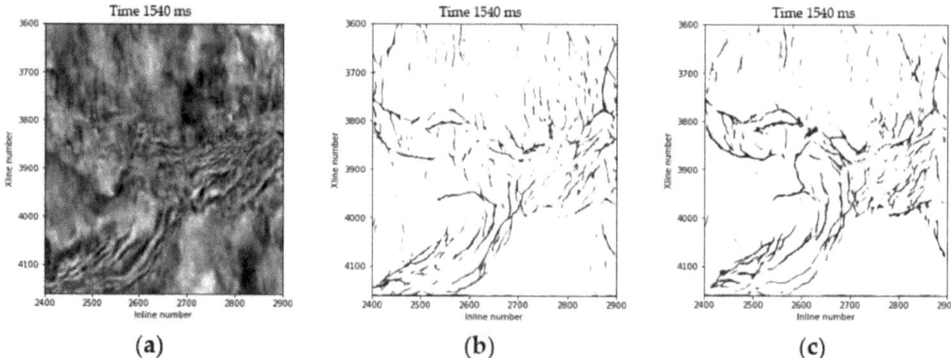

Figure 12. (**a**) A time slice is displayed with faults that are imaged via (**b**) the trained U-Net model and (**c**) the trained modified U-Net model.

Figure 10 represents three seismic images in different directions with different scales faults that are imaged using the trained modified U-Net model. Figure 11b shows the fault image predicted by the trained U-Net model and Figure 11c shows the modified U-Net prediction results. The U-Net result (Figure 11b) is reliable enough to depict faults in this seismic image, however, much of the detail is still missing compared to features predicted by the modified U-Net (Figure 11c). Figure 12b,c illustrate fault imaging results of different slices. We observe that most faults can be clearly imaged under the trained modified U-Net model, and multiple groups of faults in different directions can be distinguished on horizontal slices. Figure 12b is the result of U-Net prediction, some small fracture information has not been portrayed. In summary, the field data example shows that the proposed method based on the modified U-Net has superior performance in detecting faults of multiple scales, and provides relatively high sensitivity and continuity.

7. Conclusions

We developed a modified U-Net-based method to automatically image faults in the sandstone reservoirs in China. The proposed network containing GCM and GSM modules is designed and trained to enhance the ability of the network to select multi-scale information. The GCM and GSM module can select multi-scale information obtained by convolution of different dilation rates between groups, enhance the consistency of receptive field and fault recognition target region, and jointly improve the recognition ability of micro-faults. The field data applications demonstrate the effectiveness of this approach. For sandstone oil and gas reservoirs in China with abundant faults, this method has great advantages in improving fault imaging accuracy, but further research is needed in improving computational efficiency and optimizing network architectures.

Author Contributions: Conceptualization, J.W. and Y.S.; methodology, W.W.; software, Y.S.; validation, J.W.; formal analysis, Y.S.; investigation, W.W.; resources, J.W.; data curation, Y.S.; writing—original draft preparation, J.W.; writing—review and editing, J.W.; visualization, W.W.; project administration, W.W. All authors have read and agreed to the published version of the manuscript.

Funding: This research was funded by the Key Project of National Natural Science Foundation of China (41930431), China Postdoctoral Science Foundation (2020M680840) and Northeast Petroleum University's special fund (1305021889).

Institutional Review Board Statement: Not applicable.

Informed Consent Statement: Not applicable.

Data Availability Statement: Not applicable.

Conflicts of Interest: The authors declare no conflict of interest.

References

1. Liu, B.; Sun, J.; Zhang, Y.; He, J.; Fu, X.; Yang, L.; Xing, J.; Zhao, X. Reservoir space and enrichment model of shale oil in the first member of Cretaceous Qingshankou Formation in the Changling sag, southern Songliao Basin, NE China. *Pet. Explor. Dev.* **2021**, *48*, 608–624. [CrossRef]
2. Xu, J.; Zhang, L. Genesis of Cenozoic basins in Northwest Pacific Ocean margin (1): Comments on basin-forming mechanism. *Oil Gas Geol.* **2000**, *21*, 93–98.
3. Chen, W.-C.; Yan, J.-J. On the Evolutional Characteristics of Cenozoic Episodic rifting of Nanpu Sag. *J. Jining Norm. Coll.* **2020**, *3*, 115–119.
4. Peacock, D.C.P.; Sanderson, D.J.; Rotevatn, A. Relationships between fractures. *J. Struct. Geol.* **2018**, *106*, 41–53. [CrossRef]
5. Tong, H.; Zhao, B.; Cao, Z.; Liu, G.; Dun, X.M.; Zhao, D. Structural analysis of faulting system origin in the Nanpu sag, Bohai Bay basin. *Acta Geol. Sin.* **2013**, *87*, 1647–1661.
6. Wu, J.; Liu, B.; Zhang, H.; He, S.; Yang, Q. Fault Detection Based on Fully Convolutional networks (FCN). *J. Mar. Sci. Eng.* **2021**, *9*, 259. [CrossRef]
7. LeCun, Y.; Bengio, Y.; Hinton, G. Deep learning. *Nature* **2015**, *521*, 436–444. [CrossRef] [PubMed]
8. Smith, J.A. LAI Inversion Using a Back-propagation Neural Network Trained with a Multiple Scattering Model. *IEEE Trans. Geosci. Remote Sens.* **1993**, *31*, 1102–1106. [CrossRef]
9. Krizhevsky, A.; Sutskever, I.; Hinton, G.E. ImageNet classification with deep convolutional neural networks. *Commun. ACM* **2017**, *60*, 84–90. [CrossRef]
10. Zheng, Z.H.; Kavousi, P.; Di, H.B. Multi-Attributes and Neural network-Based Fault Detection in 3D Seismic Interpretation. *Adv. Mater. Res.* **2014**, *838–841*, 1497–1502. [CrossRef]
11. Araya-Polo, M.; Dahlke, T.; Frogner, C.; Zhang, C.; Poggio, T.; Hohl, D. Automated fault detection without seismic processing. *Lead. Edge* **2017**, *36*, 208–214. [CrossRef]
12. Waldeland, A.; Solberg, A. Salt classification using deep learning. In Proceedings of the 79th EAGE Conference and Exhibition 2017, Paris, France, 12–15 June 2017.
13. Guitton, A.; Wang, H.; Trainor-Guitton, W. Statistical imaging of faults in 3D seismic volumes using a machine learning approach. In *SEG Technical Program Expanded Abstracts*; Society of Exploration Geophysicists: Tulsa, OK, USA, 2017; pp. 2045–2049.
14. Xiong, W.; Ji, X.; Ma, Y.; Wang, Y.; AlBinHassan, N.M.; Ali, M.N.; Luo, Y. Seismic fault detection with convolutional neural network. *Geophysics* **2018**, *83*, 97–103. [CrossRef]
15. Wu, X.; Liang, L.; Shi, Y.; Fomel, S. FaultSeg3D: Using synthetic data sets to train an endto-end convolutional neural network for 3D seismic fault segmentation. *Geophysics* **2019**, *84*, IM35–IM45. [CrossRef]
16. Ronneberger, O.; Fischer, P.; Brox, T. U-Net: Convolutional networks for biomedical image segmentation. In *Medical Image Computing and Computer-Assisted Intervention, Proceedings of the MICCAI 2015, Munich, Germany, 5–9 October 2015*; Lecture Notes in Computer Science; Navab, N., Hornegger, J., Wells, W.M., Frangi, A.F., Eds.; Springer: Cham, Switzerland, 2015; Volume 9351, pp. 234–241.
17. Sevastopolsky, A. Optic disc and cup segmentation methods for glaucoma detection with modification of U-Net convolutional neural network. *Pattern Recognit. Image Anal.* **2017**, *27*, 618–624. [CrossRef]
18. Zhang, H.; Han, J.; Li, Z.; Zhang, H. Extracting Q Anomalies from Marine Reflection Seismic Data Using Deep Learning. *IEEE Geosci. Remote Sens. Lett.* **2021**, *19*, 7501205. [CrossRef]
19. Liang, C.; Wang, N.; Zhu, M.; Yang, X.; Li, J.; Gao, X. Based on Multi-Scale Feature Fusion Faces—Sketch Synthesis. China Science, Information Science, 1–14. Available online: http://kns.cnki.net/kcms/detail/11.5846.TP.20220126.1627.002.html (accessed on 4 January 2022).
20. Ma, L.; Liu, X.; Li, H.; Duan, J.; Niu, B. Neural Network Lightweight Method Using Cavity Convolution. Computer Engineering and Applications: 1–14. Available online: http://kns.cnki.net/kcms/detail/11.2127.TP.20210419.1339.035.html (accessed on 4 January 2022).
21. Yu, Q.; Zhang, J.; Wei, X.; Zhang, Q. Segmentation of liver tumors based on cascated separable cavity residual U-NET. *Chin. J. Appl. Sci.* **2021**, *39*, 378–386.
22. Fisher, Y.; Koltun, V. Multi-scale context aggregation by dilated convolutions. *arXiv* **2015**, arXiv:1511.07122.
23. Zhe, Z.; Bilin, W.; Zhezhou, Y.; Zhiyuan, L. Dilated Convolutional Pixels Affinity network for Weakly Supervised Semantic Segmentation. *Chin. J. Electron.* **2021**, *30*, 1120–1130. [CrossRef]
24. Gao, H.; Cao, L.; Yu, D.; Xiong, X.; Cao, M. Semantic Segmentation of Marine Remote Sensing Based on a Cross Direction Attention Mechanism. *IEEE Access* **2020**, *8*, 142483–142494. [CrossRef]
25. Meng, D.; Sun, L. Some New Trends of Deep Learning Research. *Chin. J. Electron.* **2019**, *28*, 1087–1091. [CrossRef]
26. Rumelhart, D.E.; Hinton, G.E.; Williams, R.J. Learning representations by back-propagating errors. *Nature* **1986**, *323*, 533–536. [CrossRef]
27. Xie, S.N.; Tu, Z.W. Holistically-Nested Edge Detection. In Proceedings of the International Conference on Computer Vision, Santiago, Chile, 7–13 December 2015; pp. 1395–1403. [CrossRef]
28. Milletari, F.; Navab, N.; Ahmadi, S.A. V-Net: Fully Convolutional Neural networks for Volumetric Medical Image Segmentation. In Proceedings of the 2016 Fourth International Conference on 3D Vision (3DV), Stanford, CA, USA, 25–28 October 2016.

29. Liu, B.; Zhao, X.; Fu, X.; Yuan, B.; Bai, L.; Zhang, Y.; Ostadhassan, M. Petrophysical characteristics and log identification of lacustrine shale lithofacies: A case study of the first member of Qingshankou Formation in the Songliao Basin, Northeast China. *Interpretation* **2020**, *8*, SL45–SL57. [CrossRef]
30. Gao, D. Integrating 3D seismic curvature and curvature gradient attributes for fracture characterization:Methodologies and interpretational implications. *Geophysics* **2013**, *78*, O21–O31. [CrossRef]

 applied sciences

Article

Multi-Task Deep Learning Seismic Impedance Inversion Optimization Based on Homoscedastic Uncertainty

Xiu Zheng [1], Bangyu Wu [1,*], Xiaosan Zhu [2] and Xu Zhu [1]

[1] School of Mathematics and Statistics, Xi'an Jiaotong University, Xi'an 710049, China; zxxxjtu@stu.xjtu.edu.cn (X.Z.); zhuxu@mail.xjtu.edu.cn (X.Z.)
[2] Institute of Geology, Chinese Academy of Geological Sciences, Beijing 100037, China; zhuxiaosan@yahoo.com
* Correspondence: bangyuwu@xjtu.edu.cn; Tel.: +86-02982667846

Abstract: Seismic inversion is a process to obtain the spatial structure and physical properties of underground rock formations using surface acquired seismic data, constrained by known geological laws and drilling and logging data. The principle of seismic inversion based on deep learning is to learn the mapping between seismic data and rock properties by training a neural network using logging data as labels. However, due to high cost, the number of logging curves is often limited, leading to a trained model with poor generalization. Multi-task learning (MTL) provides an effective way to mitigate this problem. Learning multiple related tasks at the same time can improve the generalization ability of the model, thereby improving the performance of the main task on the same amount of labeled data. However, the performance of multi-task learning is highly dependent on the relative weights for the loss of each task, and manual tuning of the weights is often time-consuming and laborious. In this paper, a Fully Convolutional Residual Network (FCRN) is proposed to achieve seismic impedance inversion and seismic data reconstruction simultaneously, and a method based on the homoscedastic uncertainty of the Bayesian model is used to balance the weights of the loss function for the two tasks. The test results on the synthetic datasets of Marmousi2, Overthrust, and Volve field data show that the proposed method can automatically determine the optimal weight of the two tasks, and predicts impedance with higher accuracy than single-task FCRN model.

Keywords: seismic impedance inversion; fully convolutional residual network; multi-task learning; homoscedastic uncertainty

Citation: Zheng, X.; Wu, B.; Zhu, X.; Zhu, X. Multi-Task Deep Learning Seismic Impedance Inversion Optimization Based on Homoscedastic Uncertainty. *Appl. Sci.* **2022**, *12*, 1200. https://doi.org/10.3390/app12031200

Academic Editor: Andrea Paglietti

Received: 20 December 2021
Accepted: 19 January 2022
Published: 24 January 2022

1. Introduction

Reflection seismic exploration is used to detect changes in impedance in the subsurface through an active seismic source. Seismic inversion refers to the process of estimating the properties of underground rocks using surface acquired seismic data. Classical seismic inversion methods usually start with a smooth model of underground properties, and then perform forward simulation to generate synthetic seismic data. The differences between the synthetic and actual seismic data can be used to update the model parameters [1]. Traditional inversion methods are usually physics-driven, which are limited by expensive computational costs and physical theory/assumptions. Due to the increased complexity of the subsurface structures, and the difficulty in obtaining a good initial model to converge to the high-resolution target model for conventional methods, advanced techniques are required for effective and efficient seismic inversion. More recently, with the successes of deep learning in the computer vision community, time series forecasting [2], and natural language processing, researchers have developed various data-driven seismic inversion techniques. The amount of available seismic data is growing exponentially and the deep learning methods are becoming integral components of geophysical exploration workflows [3], such as P-wave detection [4], seismic fault detection [5–8], seismic data noise attenuation [9,10], seismic data interpolation [11–15], and seismic slope estimation [16].

Deep neural networks are built by a composition of hierarchical linear and non-linear functions (layers). The high-capacity networks trained using large datasets enable tasks beyond traditional methods, such as high-resolution velocity model building [17]. At the same time, seismic impedance inversion has also made many contributions using deep learning methods. In 2019, Biswas et al. used Convolutional Neural Networks (CNNs) to estimate acoustic impedance and elastic impedance from seismic data [18]. Das et al. used a Fully Convolutional Neural Network (FCN) to invert P-wave impedance [19]. Alfarraj et al. [20] introduced a Recurrent Neural Network (RNN) based on serial modeling to estimate petrophysical properties and Mustafa et al. [21] introduced a Temporal Convolutional Network (TCN) to estimate various rock properties from seismic data. Li et al. [22] used geological and geophysical model-driven CNNs (GGCNNs) to estimate elastic properties from pre-stack seismic data. Wu et al. proposed a Fully Convolutional Residual Network (FCRN) combined with transfer learning for seismic impedance inversion [23], and then extended their work to semi-supervised learning seismic impedance inversion based on a Generative Adversarial Network (GAN) [24,25]. Wang et al. [26] proposed a novel seismic impedance inversion method based on a Cycle-consistent Generative Adversarial Network (Cycle-GAN). To fully explore the multichannel characteristics of the seismic data [27], Wu et al. proposed a deep learning method for multidimensional seismic impedance inversion [28].

These studies show that neural networks have great potential for seismic inversion. All of these methods are based on learning the mapping from seismic data to well logging data, and then using the learned mapping to estimate properties for off-well locations. This approach usually requires a large amount of labeled training data to improve generalization performance. However, due to the high drilling cost, the number of wells in most exploration operations is limited, leading to a trained model with poor generalization. MTL provides an effective way to mitigate this problem [29,30]. Compared with single task learning, multi-task learning network is in the fashion of "single input and multi output", with an incomparable advantage over a single task network [31]. Simultaneously learning multiple related tasks can improve the model generalization ability, thus improving the performance of the main task on the same amount of labeled data [32]. Meanwhile, multi-task learning can help extract multi-scale texture information from datasets [33]. There are two main ways to implement MTL: hard parameter sharing and soft parameter sharing. Hard parameter sharing involves the hidden layer of a two-task sharing network, and uses different output layers to complete different tasks. The soft parameter sharing method means that each task has its own model and parameters, and the distance between model parameters is then regularized to increase the similarity of models [34]. When there is a high correlation between tasks, the hard parameter sharing method is more suitable, and the higher the correlation between tasks, the greater the proportion of sharing layers in hard parameter sharing [35]. Mustafa et al. proposed an example of multi-task learning via representation sharing where multiple tasks (seismic impedance inversion and data reconstruction) are learnt simultaneously in the hope that the network can learn more generalizable feature representations leading to better performance in all tasks [1]. This is especially the case when the tasks are highly related to each other. In this paper, we propose a multi-task FCRN for simultaneous seismic impedance inversion and seismic data reconstruction. The performance of hard parameter sharing deep learning is highly dependent on the loss weight of each task, and the weighted linear sum of the loss for each individual task is usually used for training. However, it is tedious to manually adjust the weights for the loss of different tasks. It was found that the optimal weights for each task are closely related to the task magnitude and ultimately depend on the task's noise level [36]. Therefore, in this paper, we propose an automatic weight adjustment method for a multi-task loss function based on homoscedastic uncertainty for seismic impedance inversion. The test results from the two synthetic datasets of the Marmousi2 and Overthrust model, and one field dataset of the Volve model, show that the proposed method can automatically determine the optimal weight of the two tasks, and generate higher accurate impedance results than single-task model.

The remainder of this paper is organized as follows. Section 2 describes the methodology, which includes the network architecture and the theory of multi-task loss function based on homoscedastic uncertainty. In Section 3, the three datasets and experimental results are presented. We discuss the limitations of the proposed method in Section 4. Conclusions are given in Section 5.

2. Methodology

2.1. Network Architecture

The single-task network used in this paper is a FCRN from Wu et al. [23], and the multi-task network is a hard parameter sharing network developed from the single-task FCRN. The multi-task network structure is shown in Figure 1. In order to better capture the low-frequency characteristics of seismic data, the first convolution layer of FCRN has 16 kernels of size 299 × 1. After the first convolution layer, three residual blocks are stacked, and each residual block is composed of two convolution layers. The first layer and the second layer, respectively, have 16 convolution kernels of size 299 × 1 and 3 × 1. To enable the network to complete multi-task inversion, two output channels are set after the residual block. Each output channel contains two one-dimensional convolution layers, and each one-dimensional convolution layer has a kernel of 3 × 1. The convolution step size is 1, and zero-padding is used to all convolution layers to ensure the same input and output sizes. Rectified linear unit (ReLU) and batch normalization (BN) are introduced into the network to accelerate network training and convergence.

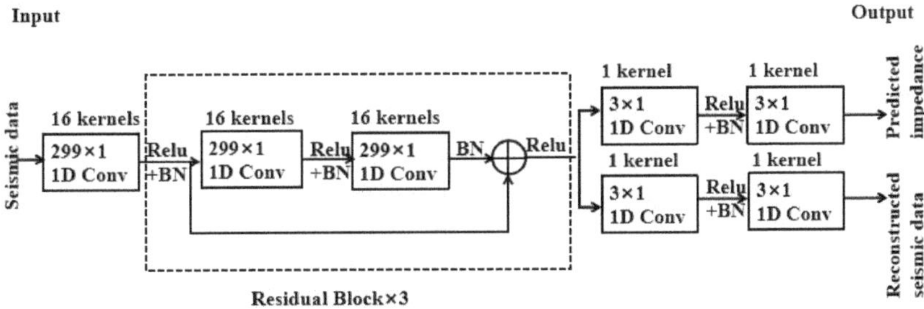

Figure 1. Architecture of the multi-task FCRN.

2.2. Multi-Task Loss Function Based on Homoscedastic Uncertainty

The performance of hard parameter sharing is highly dependent on the loss weight of each task, and simply performing a weighted linear sum of the loss for each individual task is usually undertaken to carry out training. Manual tuning of the weights is often troublesome. Thus, a method based on homoscedastic uncertainty of the Bayesian model is used to balance the weights of the loss function of multiple tasks.

In Bayesian modeling, there are two main types of uncertainty, i.e., epistemic and aleatoric uncertainty. In a model, the aleatoric uncertainty captures the randomness of the model prediction, which depends on the noise inherent in input observations, and the epistemic uncertainty captures what a model does not know due to lack of training data [37]. Aleatoric uncertainty can again be divided into two subcategories, heteroscedastic uncertainty and homoscedastic uncertainty. Heteroscedastic uncertainty depends on the inputs to the model, with some inputs potentially having more noisy outputs than others. Homoscedastic uncertainty can be described as task dependent, which stays constant for all input data and varies between different tasks [36]. In this paper, we derive a multi-task loss function based on maximizing the Gaussian likelihood with homoscedastic uncertainty. The derivation is as follows:

(1) Given a dataset $X = \{x_1, \ldots, x_N\}, Y = \{y_1, \ldots, y_N\}$, we define $f^w(x^*)$ as the output of a neural network with weights w on input x^*. For regression tasks, we define the likelihood as a Gaussian distribution that takes the model output as the mean:

$$p(y^*|f^w(x^*)) = N\left(f^w(x^*), \sigma^2\right) \tag{1}$$

where σ is an observation noise scalar capturing how much noise is in the outputs. When w and x^* are determined, the establishment of a probabilistic model of observation is equivalent to the determination of epistemic uncertainty and heteroscedastic uncertainty. Only homoscedastic uncertainty is considered in this paper. Different tasks have different homoscedastic uncertainties.

(2) In maximum likelihood inference, we want to maximize the logarithmic likelihood of the model, that is, to maximize the following equation:

$$log\ p(y^*|f^w(x^*)) \propto -\frac{1}{2\sigma^2}||y^* - f^w(x^*)||^2 - log\sigma \tag{2}$$

(3) Construct the maximum likelihood function for multi-task. There are two tasks in our model: impedance prediction and data reconstruction. We use y_{pre} and y_{rec} to denote the outputs of the two tasks and assume that y_{pre} and y_{rec} following a Gaussian distribution:

$$p(y_{pre}, y_{rec}|f^w(x)) = p(y_{pre}|f^w(x))p(y_{rec}|f^w(x)) = N\left(y_{pre}, f^w(x), \sigma_{pre}^2\right)N\left(y_{rec}, f^w(x), \sigma_{rec}^2\right) \tag{3}$$

(4) To maximize the logarithmic likelihood function, that is, to minimize the negative logarithmic likelihood function, then the multi-task loss function is:

$$\begin{aligned} L(w, \sigma_{pre}, \sigma_{rec}) &= -log\ p(y_{pre}, y_{rec}|f^w(x)) \propto \tfrac{1}{2\sigma_{pre}^2}||y_{pre} - f^w(x)||^2 + \tfrac{1}{2\sigma_{rec}^2}||y_{rec} - f^w(x)||^2 + log\sigma_{pre}\sigma_{rec} \\ &= \tfrac{1}{2\sigma_{pre}^2}L_{pre}(w) + \tfrac{1}{2\sigma_{rec}^2}L_{rec}(w) + log\sigma_{pre}\sigma_{rec} \end{aligned} \tag{4}$$

the process of minimizing the loss function is to learn the optimal weight of $L_{pre}(w)$ and $L_{rec}(w)$ automatically according to the data. As σ_{pre} increases, the weight of task impedance prediction decreases. However, with too great an increase in noise, the data will be ignored, so the last term of the objective function $log\sigma_{pre}\sigma_{rec}$ is the noise term regularizer. When the loss function reaches its minimum, we can obtain the corresponding σ_{pre} and σ_{rec} . We use w_{pre} and w_{rec} to denote the weights of the two tasks; then, from Equation (4) we can obtain the optimal weight between the two tasks as:

$$w_{pre} : w_{rec} = \frac{1}{2\sigma_{pre}^2} : \frac{1}{2\sigma_{rec}^2} = \sigma_{rec}^2 : \sigma_{pre}^2 \tag{5}$$

Further processing makes the two weights add to 1; then, the final optimal weight is:

$$w_{pre} : w_{rec} = \frac{\sigma_{rec}^2}{\sigma_{pre}^2 + \sigma_{rec}^2} : \frac{\sigma_{pre}^2}{\sigma_{pre}^2 + \sigma_{rec}^2} \tag{6}$$

3. Experiments

Two synthetic datasets and one field dataset were used in this study, which are denoted as the Marmousi2 model, the Overthrust model, and the Volve model, respectively. The three datasets were used to train the single-task network and the hard parameter sharing multi-task network respectively, and then the trained networks were used for impedance inversion.

3.1. Experiment on the Marmousi2 Model

The impedance of the Marmousi2 model and its corresponding synthetic data are the same as in [23], and are shown in Figure 2. For this model, 101 traces were selected from the synthetic seismic data and impedance data as the training set through isometric sampling, and 1350 traces were randomly selected from the remaining 13,500 traces as the validation set and the whole 13,601 traces comprised the test set. The Adam optimization method was adopted in this paper. The weight decay rate was set to 1×10^{-7}. The learning rate was set to 0.001. The number of epochs was set to 50, and the batch size was set to 10. The above hyperparameters and network structure were decided from an ablation study by adjusting the convergence of the training set and the validation set. The training of the network was implemented under the PyTorch framework, and a GPU was applied to accelerate the calculation.

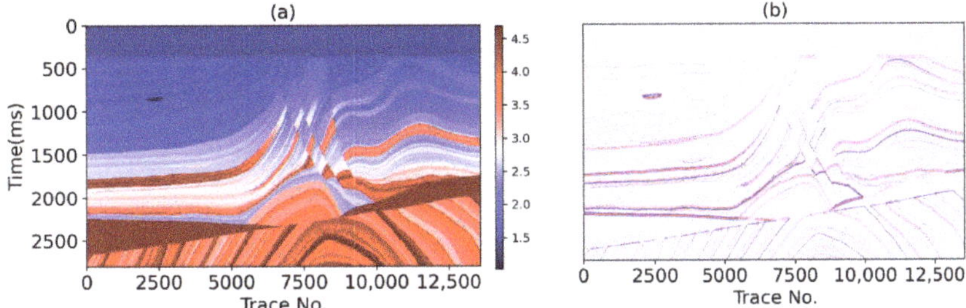

Figure 2. Marmousi2 model dataset: (**a**) impedance; (**b**) synthetic seismic data generated by 30 Hz 0° phase Ricker wavelet.

The validation Mean Squared Error (MSE) of the two tasks with different weights is shown in Table 1. The training/validation procedure was executed five times and the average results were output for comparison. The performance of the model for a single task is shown in the first and last rows. When the validation loss is minimal during the 50 epochs, the corresponding σ_{pre} = 0.890 and σ_{rec} = 0.759, then we can calculate that the optimal weight for impedance prediction and seismic data reconstruction is 0.421:0.579 by Equations (5) and (6). Figure 3 shows the variation at different epochs of σ_{pre}, σ_{rec} and the corresponding w_{pre}, w_{rec}. Under this optimal weight, the validation MSE of the two task is 0.0365 and 0.00054. Compared with results in Table 1, it can be seen that the MSEs of the multi-task network of the two tasks under different weights are smaller than that of the single-task network, and the optimal weight determined by the proposed method makes the MSE smaller for both tasks, which proves the effectiveness of the method. The training and validation loss curves of different tasks under the optimal weight are shown in Figure 4. Figure 5 illustrates the profiles predicted by the single-task network (Figure 5a) and the hard parameter sharing multi-task network (Figure 5b) with the optimal weight for visual comparison. The MSEs of the profiles predicted by the single-task network and the hard parameter sharing multi-task network are 0.0478 and 0.0353, respectively, which proves the advantage of the hard parameter sharing multi-task network from a quantitative perspective.

Table 1. The validation MSEs of the two tasks with different weights on the Marmousi2 model.

Task Weights		MSE of Validation	
w_{pre} (Impedance prediction)	w_{rec} (Data reconstruction)	Impedance prediction	Data reconstruction
0	1	/	0.00080
0.1	0.9	0.0411	0.00028
0.2	0.8	0.0412	0.00040
0.3	0.7	0.0407	0.00045
0.4	0.6	0.0397	0.00056
0.5	0.5	0.0372	0.00065
0.6	0.4	0.0374	0.00066
0.7	0.3	0.0397	0.00060
0.8	0.2	0.0385	0.00072
0.9	0.1	0.0375	0.00077
1	0	0.0446	/

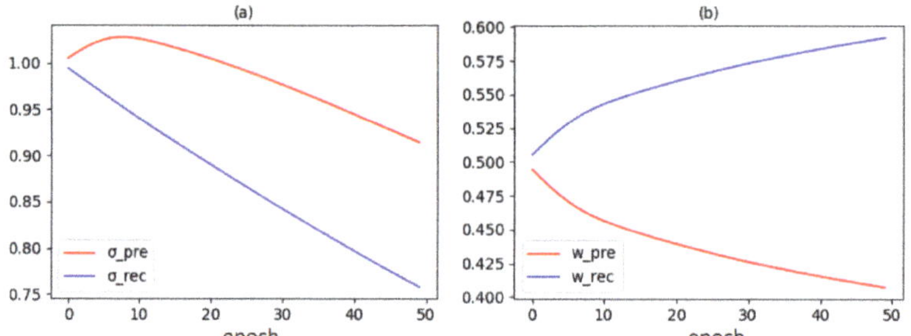

Figure 3. The variation of (**a**) σ_{pre}, σ_{rec} and (**b**) w_{pre}, w_{rec} with epochs.

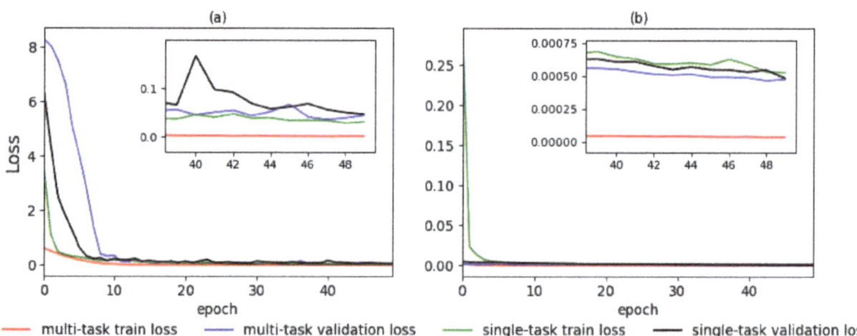

Figure 4. The training and validation loss curves of the Marmousi2 model: (**a**) impedance; (**b**) data reconstruction.

To further demonstrate the effectiveness of the proposed method, Figure 6 shows the results of impedance prediction for the 731st (Figure 6a) and 8991st (Figure 6b) trace data points. We can see that the impedance predicted by the hard parameter sharing multi-task network (green) matches the true impedance (red) better than that of the single-task network (blue). The blue and green dotted lines in Figure 6 represent residuals between the true impedance and the impedance predicted by the two networks. Table 2 shows the Pearson Correlation Coefficient (PCC) between the predicted value and the truth. Under the optimal weights, the PCC between the predicted value and the ground truth of the 731st and 8991st traces of the multi-task network (third column) are both higher. We also tested the tolerance of the two networks to noise. We added six levels of Gaussian noise with a signal-to-noise ratio (SNR) of 0, 5, 15, 25, 35, and 45 dB to the test dataset. The MSEs between the prediction of different SNR data and the true impedance are presented in

Table 3, and shows that the accuracy of the multi-task network is higher than that of the single-task network under all six test datasets with different SNRs. This further proves the superiority of the method in this paper.

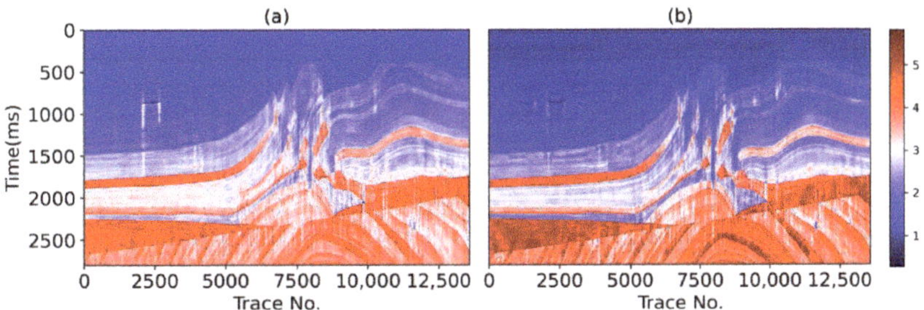

Figure 5. Profiles of the Marmousi2 model predicted by (**a**) the single-task network and (**b**) the hard parameter sharing multi-task network under the optimal weights.

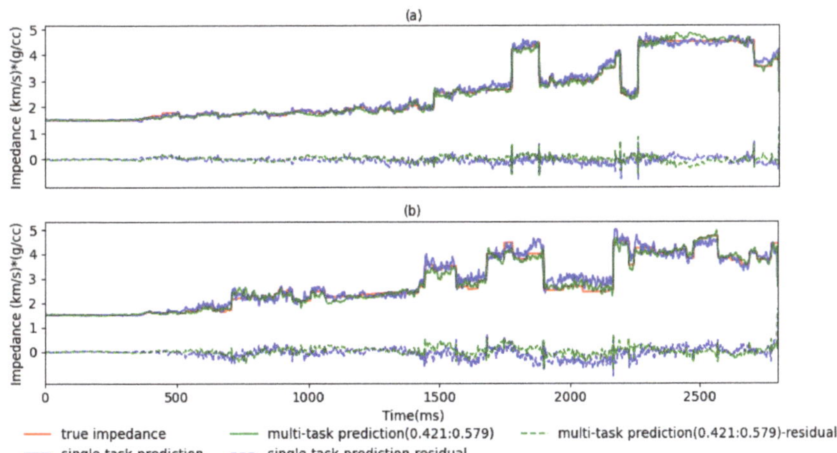

Figure 6. Impedance traces' prediction of the Marmousi2 model by the two networks: (**a**) trace 731; (**b**) trace 8991.

Table 2. PCC between the traces' predicted value and the ground truth of the Marmousi2 model.

Trace No.	Single-Task	Multi-Task (0.421:0.579)
731	0.9952	0.9956
8991	0.9805	0.9890

Table 3. The MSEs between the prediction of different SNR data and the true impedance of the Marmousi2 model

SNR (dB)	Single-Task	Multi-Task (0.421:0.579)
0	1.1270	0.5948
5	0.2020	0.1260
15	0.0546	0.0379
25	0.0484	0.0355
35	0.0478	0.0353
45	0.0478	0.0353

3.2. Experiment on the Overthrust Model

The impedance of the Overthrust model and its corresponding synthetic data are also that same as in [23] and are shown in Figure 7. For this model, five traces were selected as the training set from the synthetic seismic data and impedance data by isometric sampling, 39 traces were randomly selected from the remaining 392 traces as the validation set, and the whole 401 traces were the test set. Due to the small amount of sample data, there may be overfitting. Data enhancement methods are usually used to avoid overfitting. In contrast to the Marmousi2 model test, we used cubic spline interpolation to generate 20 new impedance and seismic data traces between every two selected traces in this model. Ultimately, we used 85 traces to train the network. The interpolation data are shown in Figure 8. The frequency-wavenumber spectra of the original and interpolated seismic data are shown in Figure 9 for comparison. The number of epochs was set to 1000 and the other hyperparameters and network structure were same as those of the Marmousi2 model. The validation MSEs of the two tasks with different weights are shown in Table 4. The performance of the model for a single task is shown in the first and the last rows. When the validation loss is minimal during the 1000 epochs, the corresponding σ_{pre} = 0.192 and σ_{rec} = 0.120, then we can calculate that the optimal weight is 0.28: 0.72, and the MSE of the two tasks under this weight is 0.0079 and 0.00001. Figure 10 shows the variation with epochs of σ_{pre}, σ_{rec} and the corresponding w_{pre}, w_{rec}. It can be seen that the MSEs obtained by the multi-task network under different weights for both tasks are all smaller than that obtained by the single-task network, and the optimal weight minimizes the MSEs. The training and validation loss curves of different tasks under the optimal weight are shown in Figure 11. Figure 12 illustrates the profiles predicted by the single-task network (Figure 12a) and the hard parameter sharing multi-task network (Figure 12b) with the optimal weight. The MSEs of the predicted profiles are 0.0119 and 0.0063, respectively, which also shows that the hard parameter sharing multi-task network outperforms the single-task network. Similarly, we chose two traces to further demonstrate the performance of the proposed method. Figure 13 shows the predicted results of impedance prediction for the 99th (Figure 13a) and 316th (Figure 13b) traces, and their corresponding PCC between the predicted value and the ground truth are shown in Table 5. It is clear that the impedance predicted by the hard parameter sharing multi-task network (green) matches the true impedance (red) better than that by the single-task network (blue). The blue and green dotted lines in Figure 13 represent residuals between the true impedance and the impedance predicted by the two networks. In addition, the MSEs between the prediction of different SNR data and the true impedance are presented in Table 6. It shows that, with the exception of 0 dB noise, the accuracy of the multi-task network is higher than that of the single-task network under the other five testing data with different SNRs, which further proves the advantage of the proposed method.

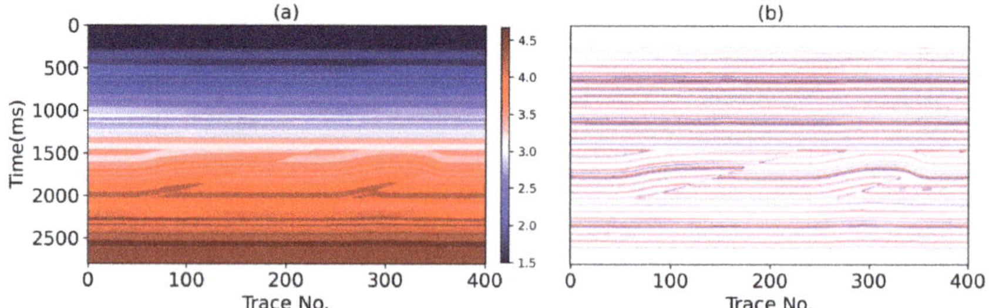

Figure 7. Overthrust model dataset: (**a**) impedance; (**b**) synthetic seismic data generated with 30 Hz 0° phase Ricker wavelet.

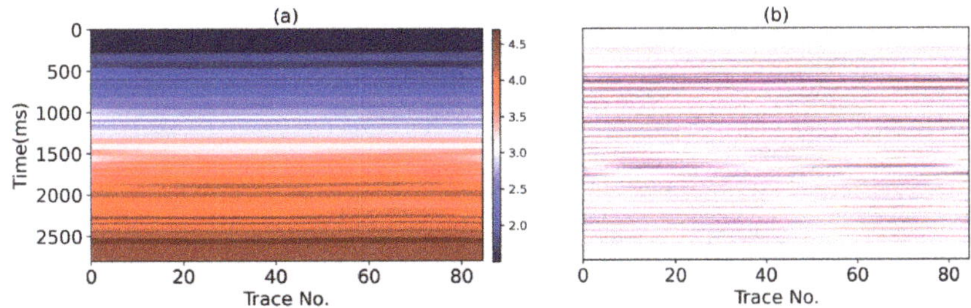

Figure 8. Interpolated data of the Overthrust model. (**a**) impedance; (**b**) synthetic seismic data.

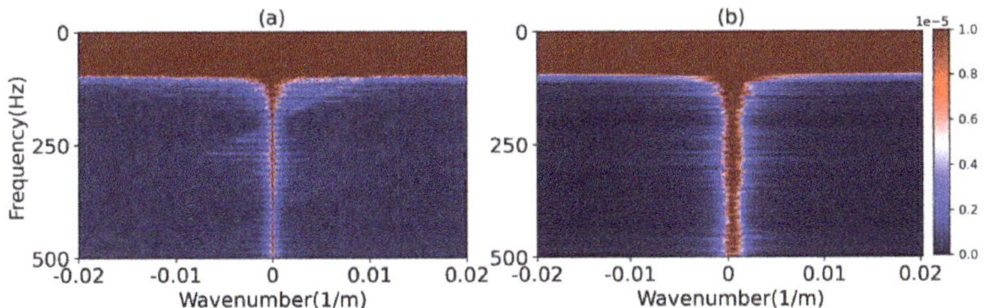

Figure 9. The frequency-wavenumber spectra of the original (**a**) and interpolated seismic data (**b**).

Table 4. The validation MSEs of the two tasks with different weights of the Overthrust model.

Task Weights		**MSE of Validation**	
w_{pre} (Impedance prediction)	w_{rec} (Data reconstruction)	Impedance prediction	Data reconstruction
0	1	/	0.00005
0.1	0.9	0.0081	0.00002
0.2	0.8	0.0088	0.00003
0.3	0.7	0.0085	0.00001
0.4	0.6	0.0083	0.00002
0.5	0.5	0.0111	0.00004
0.6	0.4	0.0084	0.00003
0.7	0.3	0.0096	0.00001
0.8	0.2	0.0093	0.00006
0.9	0.1	0.0090	0.00007
1	0	0.0120	/

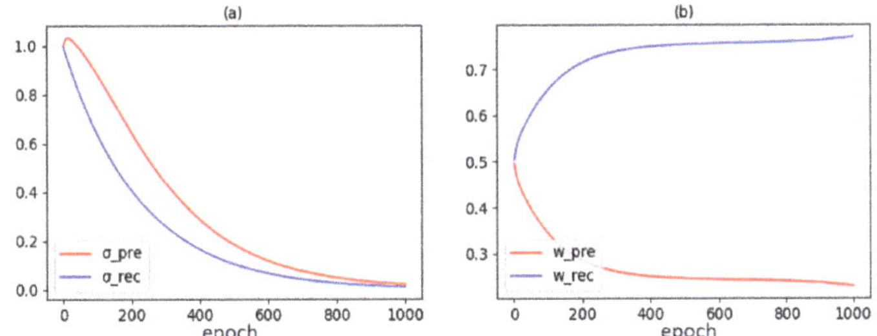

Figure 10. The variation of (**a**) σ_{pre}, σ_{rec} and (**b**) w_{pre}, w_{rec} with epochs.

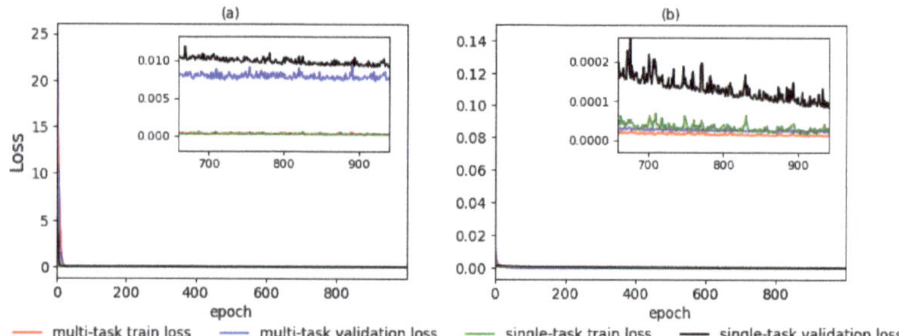

Figure 11. The training and validation loss curves of the Overthrust model: (**a**) impedance; (**b**) data reconstruction.

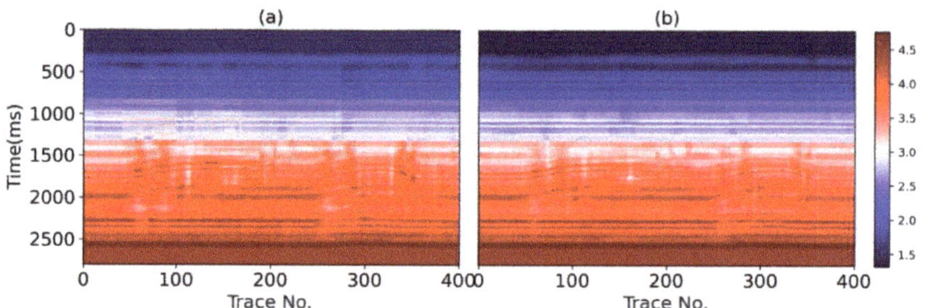

Figure 12. Profiles of the Overthrust model predicted by (**a**) the single-task network and (**b**) the hard parameter sharing multi-task network under the optimal weight.

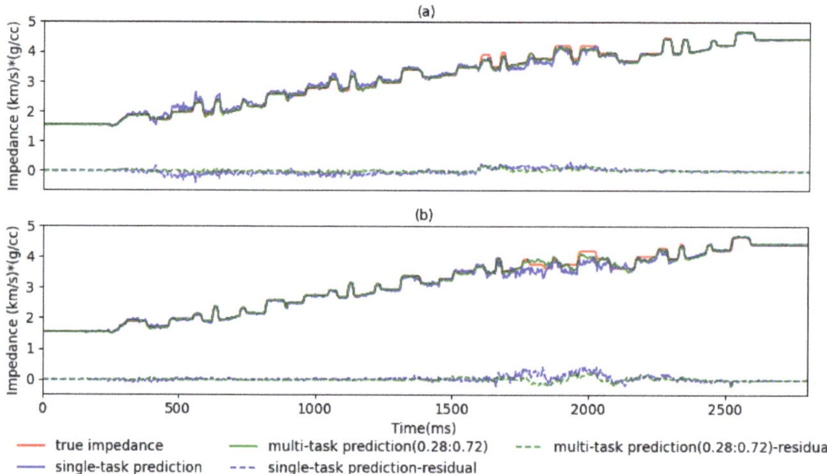

Figure 13. Impedance traces prediction of the Overthrust model by the two networks: (**a**) trace 99; (**b**) trace 316.

Table 5. PCC between the traces' predicted value and the ground truth of the Overthrust model.

Trace No.	Single-Task	Multi-Task (0.28:0.72)
99	0.9964	0.9993
316	0.9960	0.9990

Table 6. The MSEs between the prediction of different SNR data and the true impedance of the Overthrust model

SNR (dB)	Single-Task	Multi-Task (0.28:0.72)
0	0.1300	0.2163
5	0.0295	0.0199
15	0.0128	0.0068
25	0.0120	0.0063
35	0.0119	0.0063
45	0.0119	0.0063

3.3. Experiment on the Volve Model

The field Volve data used in this paper are from the open-source code of Das et al. [19]. The Volve field shown in Figure 14a is located in offshore Norway and is a clastic reservoir. There are 1300 labeled traces in this dataset, which were generated using the data augmentation method in [19] based on the statistical characteristics of the well position log data shown in Figure 14b. We randomly selected 750 traces as the training set and the remaining 550 traces were set to be the validation set. We used the single true well log data to test the performance of the two networks. Unlike the previous two datasets, the kernel size was set as 80 to adapt the 160 time sampling points of the Volve model. The number of epochs was set to 500, and the other hyperparameters and network structure were set to be the same as in the former two models. Similarly, when the validation loss is minimal during the 500 epochs, the corresponding σ_{pre} = 0.829 and σ_{rec} = 0.828, then we can calculate that the optimal weight for impedance prediction and seismic data reconstruction as 0.499:0.501. Figure 15 shows the predicted results of impedance prediction by the single-task model (blue) and the multi-task model (green) under the optimal weight. We can see that the impedance predicted by the multi-task model under the optimal weight matches the true impedance (red) better than the single-task model, especially in the time range between 40 and 50 ms. The blue and green dotted lines in Figure 15 represent residuals between the true impedance and the impedance predicted by the two networks. The MSE of the impedance predicted by the multi-task model under the approximate optimal weight is 0.0239, whereas the MSE predicted by the single-task model is 0.0352. In addition, the PCC between the ground truth and the values predicted by the multi-task model and the single-task model are 0.872 and 0.832, respectively. The above two metrics prove the superiority of the multi-task model under the optimal weight. It is worth noting that the test MSE and PCC in reference [19] are 0.0255 and 0.82, which is comparable with our single-task model, but is slightly inferior to our multi-task model. The Volve field data test shows that the proposed method in this paper improves the accuracy of impedance prediction on real seismic data.

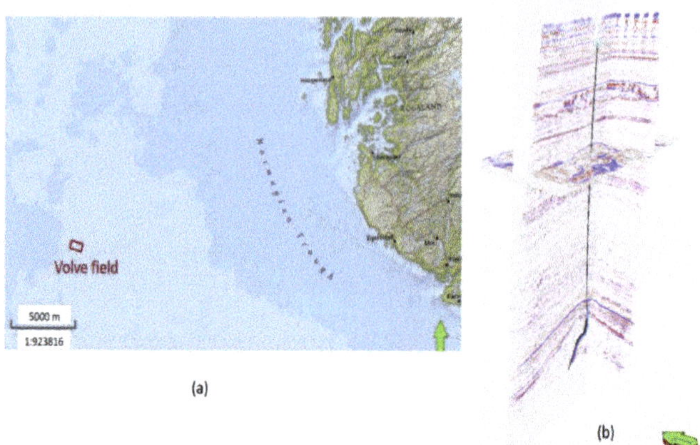

Figure 14. (**a**) The Volve field, as shown on a map, is located in the offshore North Sea area and (**b**) seismic data with a well trajectory from the Volve field (figure is from reference [19]).

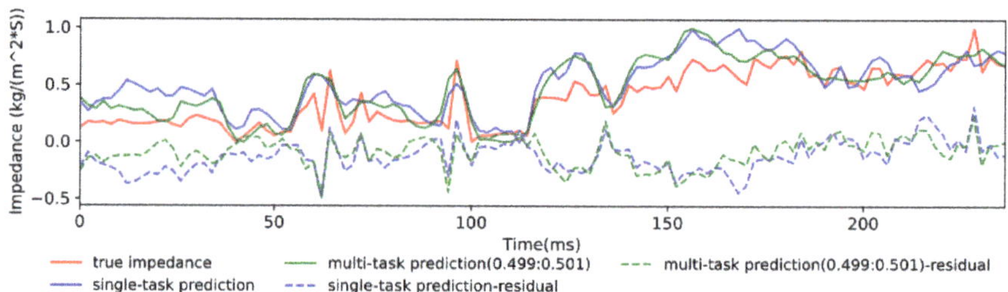

Figure 15. Impedance predicted for the well location of the Volve model by the two networks.

4. Discussion

The proposed multi-task FCRN model was directly developed from the single-task FCRN of Wu et al. [23]. However, this is an open framework, enabling many other networks for seismic impedance prediction to also be explored. We believe seismic reconstruction is a task that is not limited to impedance inversion, and other deep learning tasks such as seismic fault interpretation can also benefit in similar way. Close inspection of Figures 3 and 4 indicates that the almost flat total loss variation range corresponds to a relatively large range of $(\sigma_{pre}, \sigma_{rec})$. This means that there maybe not a single optimal weighting for all tasks, which was also an observation of Kendall et al. [36].

For the first two experiments, we attempted to add six levels of Gaussian noise, having an SNR of 0, 5, 15, 25, 35, and 45 dB, to the test set to test the tolerance of the two networks to noise. The test results of the Overthrust model show that the accuracy of the impedance prediction of the multi-task network is lower than that of the single-task network under the 0 dB noise. A possible reason for this is that when we add noise to the test dataset, we should train the model again to obtain a new optimal weight for the two tasks.

5. Conclusions

In this paper, we propose a multi-task network for both impedance inversion and seismic data reconstruction, and use a loss function based on the homoscedastic uncertainty of a Bayesian model to determine the optimal weight of the two tasks in the loss function. Test results from the Marmousi2, Overthrust, and Volve models show that: (1) the multi-task model has better generalization performance than the single-task model when the amount of labeled data is the same; and (2) the method proposed in this paper can be used to automatically determine the optimal weight of two tasks, and generates more accurate impedance than the single-task model. The proposed method can be extended to other multi-task learning approaches in a similar fashion.

In the future, we may apply the proposed method to high-dimensional seismic inversion, pre-stack inversion, and additional kinds of neural networks to verify its application value.

Author Contributions: Conceptualization, X.Z. (Xiu Zheng) and B.W.; Data curation, X.Z. (Xiu Zheng); Formal analysis, X.Z. (Xiu Zheng) and B.W.; Funding acquisition, X.Z. (Xiaosan Zhu); Investigation, X.Z. (Xiu Zheng); Methodology, X.Z. (Xiu Zheng) and B.W.; Supervision, B.W., X.Z. (Xiaosan Zhu) and X.Z. (Xu Zhu); Visualization, B.W.; Writing—original draft, X.Z. (Xiu Zheng). All authors have read and agreed to the published version of the manuscript.

Funding: This work was supported by the National Natural Science Foundation of China under Grant 41974122.

Institutional Review Board Statement: Not applicable.

Informed Consent Statement: Not applicable.

Data Availability Statement: The Volve field data is obtained from https://github.com/vishaldas/CNN_based_impedance_inversion/tree/master/Volve_field_example (accessed on 30 December 2020).

Acknowledgments: We thank two anonymous reviewers for their constructive comments on this paper. We also would like to thank Vishal Das et al., for the open-source code and prediction results of Volve field data, and thank Delin Meng, Jiaxu Yu and Zhenhui Jin for their programming help.

Conflicts of Interest: The authors declare no conflict of interest.

Abbreviations

The following abbreviations are used in this manuscript:

MTL	Multi-Task Learning
FCRN	Fully Convolutional Residual Network
CNN	Convolutional Neural Networks
FCN	Fully Convolutional Neural Network
TCN	Temporal Convolutional Network
GGCNNs	Geological and Geophysical Model Driven CNNs
GAN	Generative Adversarial Network
Cycle-GAN	Cycle-consistent Generative Adversarial Network
ReLU	Rectified Linear Unit
BN	Batch Normalization
MSE	Mean Squared Error
PCC	Pearson Correlation Coefficient
SNR	Signal-to-Noise Ratio

References

1. Mustafa, A.; AlRegib, G. Joint learning for seismic inversion: An acoustic impedance estimation case study. In *SEG Technical Program. Expanded Abstracts 2020*; SEG Library: Houstin, TX, USA, 2020; pp. 1686–1690.
2. Spadon, G.; Hong, S.; Brandoli, B.; Matwin, S.; Rodrigues, J.F., Jr.; Sun, J. Pay Attention to Evolution: Time Series Forecasting with Deep Graph-Evolution Learning. *arXiv preprint* **2020**, arXiv:2008.12833. [CrossRef]
3. Adler, A.; Araya-Polo, M.; Poggio, T. Deep learning for seismic inverse problems: Toward the acceleration of geophysical analysis workflows. *IEEE Signal. Proc. Mag.* **2021**, *38*, 89–119. [CrossRef]
4. Hu, Y.; Zhang, Q.; Zhao, W.; Wang, H. TransQuake: A transformer-based deep learning approach for seismic P-wave detection. *Earthq. Res. Adv.* **2021**, *1*, 100004. [CrossRef]
5. Wu, X.; Liang, L.; Shi, Y.; Fomel, S. FaultSeg3D: Using synthetic data sets to train an end-to-end convolutional neural network for 3D seismic fault segmentation. *Geophysics* **2019**, *84*, IM35–IM45. [CrossRef]
6. Gao, K.; Huang, L.; Zheng, Y.; Lin, R.; Hu, H.; Cladohous, T. Automatic fault detection on seismic images using a multiscale attention convolutional neural network. *Geophysics* **2022**, *87*, N13–N29. [CrossRef]

7. Wang, Z.; Li, B.; Liu, N.; Wu, B.; Zhu, X. Distilling knowledge from an ensemble of convolutional neural networks for seismic fault detection. *IEEE Geosci. Remote Sens. Lett.* **2020**, *19*. [CrossRef]
8. Zhou, R.; Yao, X.; Wang, Y.; Hu, G.; Yu, F. Seismic fault detection with progressive transfer learning. *Acta Geophys.* **2021**, *69*, 2187–2203. [CrossRef]
9. Jiang, J.; Ren, H.; Zhang, M. A Convolutional Autoencoder Method for Simultaneous Seismic Data Reconstruction and Denoising. *IEEE Geosci. Remote Sens. Lett.* **2021**, *19*. [CrossRef]
10. Qiu, C.; Wu, B.; Liu, N.; Zhu, X.; Ren, H. Deep learning prior model for unsupervised seismic data random noise attenuation. *IEEE Geosci. Remote Sens. Lett.* **2021**, *99*, 1–5. [CrossRef]
11. Wang, B.; Zhang, N.; Lu, W.; Wang, J. Deep-learning-based seismic data interpolation: A preliminary result. *Geophysics* **2019**, *84*, V11–V20. [CrossRef]
12. He, T.; Wu, B.; Zhu, X. Seismic Data Consecutively Missing Trace Interpolation Based on Multistage Neural Network Training Process. *IEEE Geosci. Remote Sens. Lett.* **2021**, *19*, 1–5. [CrossRef]
13. Li, X.; Wu, B.; Zhu, X.; Yang, H. Consecutively Missing Seismic Data Interpolation based on Coordinate Attention Unet. *IEEE Geosci. Remote Sens. Lett.* **2021**, *19*. [CrossRef]
14. Yu, J.; Wu, B. Attention and Hybrid Loss Guided Deep Learning for Consecutively Missing Seismic Data Reconstruction. *IEEE Trans. Geosci. Remote Sens.* **2021**, *60*. [CrossRef]
15. Greiner, T.A.L.; Lie, J.E.; Kolbjørnsen, O.; Evensen, A.K.; Nilsen, E.H.; Zhao, H.; Gelius, L.J. Unsupervised deep learning with higher-order total-variation regularization for multidimensional seismic data reconstruction. *Geophysics* **2021**, *87*, 1–62.
16. Huang, W.L.; Gao, F.; Liao, J.P.; Chuai, X.Y. A deep learning network for estimation of seismic local slopes. *Petroleum Sci.* **2021**, *18*, 92–105. [CrossRef]
17. Zhang, Z.; Lin, Y. Data-driven seismic waveform inversion: A study on the robustness and generalization. *IEEE Trans. Geosci. Remote Sens.* **2020**, *58*, 6900–6913. [CrossRef]
18. Biswas, R.; Sen, M.K.; Das, V.; Mukerji, T. Prestack and poststack inversion using a physics-guided convolutional neural network. *Interpretation* **2019**, *7*, SE161–SE174. [CrossRef]
19. Das, V.; Pollack, A.; Wollner, U.; Mukerji, T. Convolutional neural network for seismic impedance inversion. *Geophysics* **2019**, *84*, R869–R880. [CrossRef]
20. Alfarraj, M.; AlRegib, G. Petrophysical property estimation from seismic data using recurrent neural networks. In Proceedings of the 2018 SEG International Exposition and Annual Meeting, Anaheim, CA, USA, 16 October 2018.
21. Mustafa, A.; Alfarraj, M.; AlRegib, G. Estimation of acoustic impedance from seismic data using temporal convolutional network. In Proceedings of the SEG Technical Program, San Antonio, TX, USA, 15–20 September 2019; pp. 2554–2558.
22. Li, H.; Lin, J.; Wu, B.; Gao, J.; Liu, N. Elastic Properties Estimation From Prestack Seismic Data Using GGCNNs and Application on Tight Sandstone Reservoir Characterization. *IEEE Trans. Geosci. Remote Sens.* **2021**, *60*. [CrossRef]
23. Wu, B.; Meng, D.; Wang, L.; Liu, N.; Wang, Y. Seismic impedance inversion using fully convolutional residual network and transfer learning. *IEEE Geosci. Remote Sens. Lett.* **2020**, *17*, 2140–2144. [CrossRef]
24. Wu, B.; Meng, D.; Zhao, H. Semi-supervised learning for seismic impedance inversion using generative adversarial networks. *Remote Sens.* **2021**, *13*, 909.
25. Meng, D.; Wu, B.; Wang, Z.; Zhu, Z. Seismic Impedance Inversion Using Conditional Generative Adversarial Network. *IEEE Geosci. Remote Sens. Lett.* **2021**, *19*, 1–5. [CrossRef]
26. Wang, Y.P.; Wang, Q.; Lu, W.; Ge, Q.; Yan, X. Seismic impedance inversion based on cycle-consistent generative adversarial network. *Petroleum Sci.* **2021**. [CrossRef]
27. Ghaderpour, E. Multichannel antileakage least-squares spectral analysis for seismic data regularization beyond aliasing. *Acta Geophys.* **2019**, *67*, 1349–1363. [CrossRef]
28. Wu, X.; Yan, S.; Bi, Z.; Zhang, S.; Si, H. Deep learning for multidimensional seismic impedance inversion. *Geophysics* **2021**, *86*, R735–R745. [CrossRef]
29. Ruder, S. An overview of multi-task learning in deep neural networks. *arXiv preprint* **2017**, arXiv:1706.05098.
30. Zhang, Y.; Yang, Q. A survey on multi-task learning. *arXiv preprint* **2017**, arXiv:1707.08114. [CrossRef]
31. Cao, D.; Ji, S.; Cui, R.; Liu, Q. Multi-task learning for digital rock segmentation and characteristic parameters computation. *J. Petroleum Sci. Eng.* **2022**, *208*, 109202. [CrossRef]
32. Abu-Mostafa, Y.S. Learning from hints in neural networks. *J. Complex.* **1990**, *6*, 192–198. [CrossRef]
33. Wang, Q.; Wang, Y.; Ao, Y.; Lu, W. Seismic inversion based on 2D-CNN and multi-task learning. In Proceedings of the 82nd EAGE Annual Conference & Exhibition, Amsterdam, Netherlands, Online, October 2021; pp. 1–5.
34. Ku, B.; Min, J.; Ahn, J.K.; Lee, J.; Ko, H. Earthquake event classification using multitasking deep learning. *IEEE Geosci. Remote Sens. Lett.* **2020**, *18*, 1149–1153. [CrossRef]
35. Duong, L.; Cohn, T.; Bird, S.; Cook, P. Low resource dependency parsing: Cross-lingual parameter sharing in a neural network parser. In Proceedings of the 53rd Annual Meeting of the Association for Computational Linguistics and the 7th International Joint Conference on Natural Language Processing, Beijing, China, 26–31 July 2015; Volume 2, pp. 845–850.
36. Kendall, A.; Gal, Y.; Cipolla, R. Multi-task learning using uncertainty to weigh losses for scene geometry and semantics. In Proceedings of the IEEE Conference on Computer Vision and Pattern Recognition, Salt Lake City, UT, USA, 18–22 June 2018; pp. 7482–7491.

37. Feng, R.; Grana, D.; Balling, N. Uncertainty quantification in fault detection using convolutional neural networks. *Geophysics* **2021**, *86*, M41–M48. [CrossRef]

applied sciences

Article

Near-Surface Geological Structure Seismic Wave Imaging Using the Minimum Variance Spatial Smoothing Beamforming Method

Ming Peng [1,2], Dengyi Wang [1,2], Liu Liu [3,*], Chengcheng Liu [4,*], Zhenming Shi [1,2], Fuan Ma [5] and Jian Shen [1,2]

[1] Key Laboratory of Geotechnical and Underground Engineering of Ministry of Education, Department of Geotechnical Engineering, Tongji University, Shanghai 200092, China; pengming@tongji.edu.cn (M.P.); 1832670@tongji.edu.cn (D.W.); 94026@tongji.edu.cn (Z.S.); 2130172@tongji.edu.cn (J.S.)
[2] Department of Geotechnical Engineering, College of Civil Engineering, Tongji University, Shanghai 200092, China
[3] State Key Laboratory of Geomechanics and Geotechnical Engineering, Institute of Rock and Soil Mechanics, Chinese Academy of Sciences, Wuhan 430071, China
[4] Academy for Engineering and Technology, Fudan University, Shanghai 200433, China
[5] Guangxi Nonferrous Survey & Design Institute, Nanning 530031, China; mafuan_mfa@126.com
* Correspondence: liuliu@mail.whrsm.ac.cn (L.L.); chengchengliu@fudan.edu.cn (C.L.)

Abstract: Erecting underground structures in regions with unidentified weak layers, cavities, and faults is highly dangerous and potentially disastrous. An efficient and accurate near-surface exploration method is thus of great significance for guiding construction. In near-surface detection, imaging methods suffer from artifacts that the complex structure caused and a lack of efficiency. In order to realize a rapid, accurate, robust near-surface seismic imaging, a minimum variance spatial smoothing (MVSS) beamforming method is proposed for the seismic detection and imaging of underground geological structures under a homogeneous assumption. Algorithms such as minimum variance (MV) and spatial smoothing (SS), the coherence factor (CF) matrix, and the diagonal loading (DL) methods were used to improve imaging quality. Furthermore, it was found that a signal advance correction helped improve the focusing effect in near-surface situations. The feasibility and imaging quality of MVSS beamforming are verified in cave models, layer models, and cave-layer models by numerical simulations, confirming that the MVSS beamforming method can be adapted for seismic imaging. The performance of MVSS beamforming is evaluated in the comparison with Kirchhoff migration, the DAS beamforming method, and reverse time migration. MVSS beamforming has a high computational efficiency and a higher imaging resolution. MVSS beamforming also significantly suppresses the unnecessary components in seismic signals such as S-waves, surface waves, and white noise. Moreover, compared with basic delay and sum (DAS) beamforming, MVSS beamforming has a higher vertical resolution and adaptively suppresses interferences. The results show that the MVSS beamforming imaging method might be helpful for detecting near-surface underground structures and for guiding engineering construction.

Keywords: near-surface; seismic imaging; beamforming; underground structure; reflection seismic

Citation: Peng, M.; Wang, D.; Liu, L.; Liu, C.; Shi, Z.; Ma, F.; Shen, J. Near-Surface Geological Structure Seismic Wave Imaging Using the Minimum Variance Spatial Smoothing Beamforming Method. *Appl. Sci.* **2021**, *11*, 10827. https://doi.org/10.3390/app112210827

Academic Editors: Guofeng Liu, Zhifu Zhang, Xiaohong Meng and Jong Wan Hu

Received: 29 September 2021
Accepted: 12 November 2021
Published: 16 November 2021

1. Introduction

Most underground geotechnical engineering is carried out near the surface. Unfortunately, the near-surface geological conditions are highly complex due to the abundance of joints, faults, boulders, caves, and so on. Encountering unexplored weak formations and caves can cause a variety of geological disasters, such as uneven settlement or collapse, which can inflict immense economic losses and even death [1,2]. According to the report [3] of the current status of sinkhole collapses in the karst area in China, more than 1500 karst collapsing events have been recorded and these events formed more than 45,000 sinkholes. More than 75% sinkholes were triggered by human activities. These activities are mainly around the near-surface, including mine drainage, foundation engineering, and tunnel constructions. Subsidence sinkholes result from both subsurface dissolution and the downward gravitational movement of the undermined overlying material. These sinkholes, which are invisible from the surface, are the most important from a hazard and engineering perspective [4]. Hence, to provide guidance for construction and design efforts, geophysical prospecting methods are often used in engineering to detect underground structures as an important supplement to drilling and excavation. However, the near-surface is complex and traditional migration imaging methods may suffer from types of artifacts and interferences. High-precision prospecting in near-surface detection is quite challenging since surface waves and S-waves are powerful near the surface. There is an urgent need to deal with the artifacts and interferences. Furthermore, the detection capability is limited by the construction site and time frame [5]. Therefore, the development of accurate geophysical prospecting methods for near-surface detection is of great significance to geotechnical engineering.

Seismic methods are among the most commonly used geophysical techniques for near-surface detection with the advantages of a high efficiency, a high accuracy, and a low cost; moreover, seismic techniques are nondestructive [6]. Seismic reflection is extensively used in both academic and practical engineering [7]. One of the main goals of seismic data processing is seismic imaging. For instance, seismic migration imaging [8] is used to map underground structures. Although advanced imaging methods have emerged in recent decades, such as reverse time migration [9,10] and Gaussian beam migration [11–13], Kirchhoff migration [14] is still the most popular approach due to its ability to provide an image of sufficient quality and at an affordable computational cost. Zhang et al. [15] processed a model example and seismic field data to demonstrate the validity of prestack Kirchhoff time migration. Yuan et al. [16] applied Kirchhoff prestack time migration to seismic data of coal seam reflections and obtained better images than with poststack time migration. Wang et al. [17] used tomographic travel-time inversion and prestack Kirchhoff depth migration-based migration velocity analysis (MVA) and obtained a precise, high-resolution migration velocity model. Liu and Zhang [18] proposed a novel approach that attached a prediction shaping filter to Kirchhoff prestack time migration to mitigate the stretching effect and demonstrated their method with a numerical example and field data. Zhang et al. [7] applied Kirchhoff poststack migration to a case of seismic scattered wave imaging and obtained reliable imaging results in both the synthetic data and field data. Nevertheless, in near-surface seismic exploration, where the depth of the target area is usually less than a hundred meters, Kirchhoff migration still suffers from the following problems: (1) Kirchhoff migration relies on the preprocessing of signals to deal with noise and interference such as white noise, S-wave signals, and multiples; (2) Kirchhoff migration is affected by high-energy surface waves, which appear as massive artifacts in the imaging results. The surface wave can be muted from the observed data, but it requires a manual cost; and (3) Kirchhoff migration is limited by the degradation in the imaging resolution that occurs when the wavelength is not much smaller than the size of the imaging area in near-surface detection.

In the medical field, the widely applied ultrasound beamforming method, which detects human body structures using acoustic properties, gives us another possible way to acquire seismic signals and solve the abovementioned problems. The delay and sum (DAS) beamforming algorithm is the most popular beamforming technique for imaging human body structures because of its real-time imaging capability [19,20]. In recent decades, various adaptive beamforming methods have been proposed to improve the resolution and stability of beamformers in medical ultrasound imaging. The most common approaches are based on the minimum variance (MV) beamformer devised by Capon [21], and numerous MV beamforming algorithms have been developed to continue improving the imaging quality [22–25]. Asl [26] proposed an MV beamforming method combined with adaptive coherence weighting and achieved an excellent performance in an application to medical ultrasound imaging. Ma et al. [27] introduced the multiple delay and sum with enveloping method to efficiently suppress sidelobes and artifacts. Most recently, Chen et al. [28] proposed the multioperator MV adaptive beamformer to promote real-time imaging. Accordingly, because the transmission and receiver concepts in medical imaging and seismic reflection are similar, the medical beamforming method can be adapted for seismic imaging.

In this study, we propose a minimum variance spatial smoothing (MVSS) beamforming method for near-surface reflection seismic exploration in a homogeneous assumption. Synthetic near-surface geological models are established to carry out a numerical simulation, and the image quality of the proposed MVSS beamforming method is compared with that of Kirchhoff migration and basic DAS beamforming. Moreover, the robustness to interferences and noise, robustness to other wave components and a delay time correction to enhance the focus effect are presented. Finally, the computational efficiency, a comparison between MVSS beamforming and RTM (one of the best imaging methods for seismic data), and the potential application of the CF matrix in time domain prestack migration methods are discussed.

2. Methods

The MVSS method was proposed in this paper to improve the signal-to-noise ratio (SNR), to enhance the focusing effect, and to improve the resolution. The core concept of MVSS beamforming is to assign a weight to each receiver. When higher weights are assigned to receivers with higher SNRs, the imaging quality is improved. The receiver array is divided into several subarrays to improve the robustness when obtaining the weights. In each subarray, to improve the resolution of the image and increase the SNR, we designed a weight for each receiver according to the covariance matrix of signals. After the imaging results of each subarray are superimposed, the obtained image is multiplied by the signal coherence matrix to reduce the influences of sidelobe interference and focusing errors. The proposed method is based on a homogeneous subsurface assumption, and it is a time domain raytracing method, which means the propagation path of waves is static. Notably, MVSS beamforming reduces to basic DAS beamforming without spatial smoothing, neglecting to calculate the weight of each receiver, and excluding the coherence factor (CF) matrix. Though DAS beamforming and Kirchhoff migration are coded in different ideas, they share similar principles, such as they are both prestack time migration and raytracing methods. Background velocity plays an important role in Kirchhoff migration and other migration methods, but it does not in DAS beamforming. DAS beamforming can be understood as an extremely simplified Kirchhoff migration, in which we assume that the wave propagates along a straight line and the subsurface is homogeneous.

2.1. MVSS Beamforming Imaging

Figure 1 shows the workflow of MVSS beamforming, which is divided into three parts: (1) signal delay; (2) superposition with calculated weights from minimum variance matrix in subarrays; and (3) imaging and processing, including time-depth conversion and shot stacking. The detailed mathematics can be found in Appendix A.

First, the signal delay times are calculated according to the distance from the target point to each receiver and the background velocity. At the top of Figure 1 is a target point in the detection area. The waves reflected from the target point are received by the array, which is arranged in a straight line. The signals recorded by the receivers have different arrival times because the receivers are situated at different distances from the target point. Thus, delay processing ($\tau_1 \ldots \tau_5$) is applied to these signals so that the fluctuations from this target point are aligned on the time axis.

Second, the signals from different receivers are superimposed with an MVSS beamformer. Calculating the superposition ($\sum$) with weights ($\omega_1 \ldots \omega_5$) of the delayed signals amplifies the signals from the target point while suppressing the reflections from other scattering points in the imaging area. The weights are calculated from a minimum variance matrix of subarrays with diagonal loading to enhance robustness. The weights are designed to minimize the output interference-plus-noise power while maintaining a distortionless response to the target signal. Then, the CF is used as a weight matrix to enhance the image, which can obviously amplify the reflection signals.

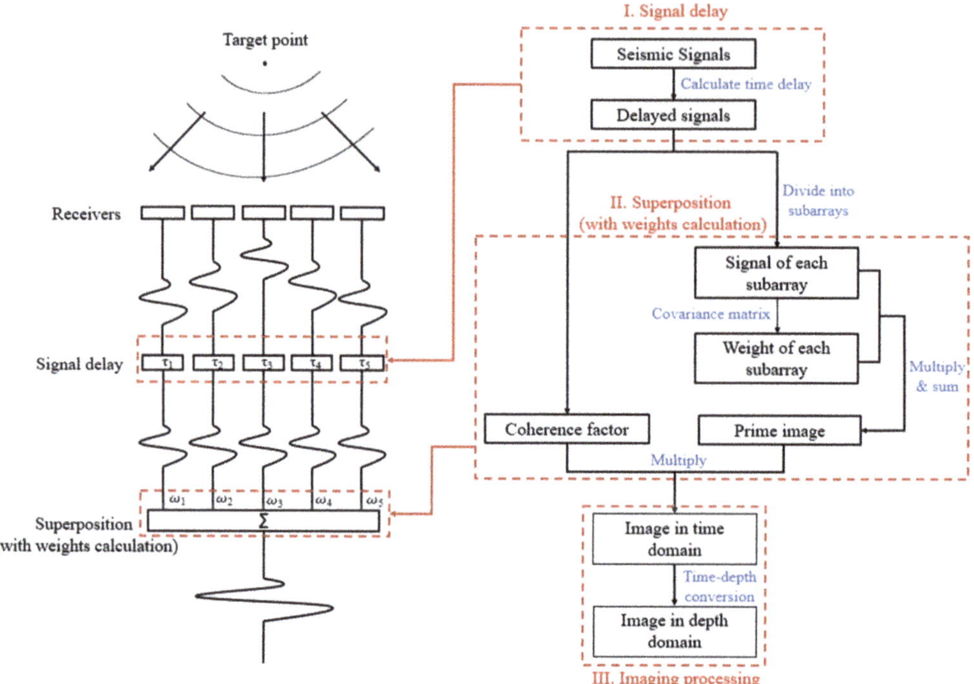

Figure 1. Workflow of MVSS beamforming imaging.

Finally, after performing time-depth conversion and stacking the images of all shots, we obtained the beamforming result. Underground structures can be determined by the amplitudes in the image, where a higher amplitude corresponds to a greater possibility of a reflection interface or a stronger reflection.

2.2. A Simple Example

We took a circular cave model (Figure 2a) as an example to illustrate the workflow of MVSS beamforming. The seismic data of the 10th shot obtained by finite-different time-domain (FDTD) numerical simulations (Liu et al., 2018 [29]) are shown in Figure 2b. The receivers were arranged from 0 m to 30 m with an interval of 0.2 m. There were 151 receivers in total.

For example, we took point A (16 m, 7 m) as the target point (Figure 2). The delay time was calculated as Equation (A1), and the results are shown in Figure 3. The 151 black dots represent the delay time of each receiver, and the 20 red circles represent the delay time of each shot. The first source shared the same spatial position and delay time with the 36th receiver. We moved the target point and repeated the delay time calculation using Equation (A1).

As per the definition of a subarray in Equation (A11), we divided the array of 151 receivers into 76 subarrays, and each subarray was composed of 75 receivers. Taking the first subarray as an example, we calculated the weight of each receiver in the subarray (Figure 4b) according to the DL and MV method using Equation (A13). We multiplied the 75 delayed signals (Figure 4a) with the weight of each receiver (Figure 4b) and then added them together to obtain the amplitudes on the axis line $x = x_A$ (Figure 4c). Two reflection interfaces were observed from the curve. The delay time was reduced by $t = t_m + t_n = 0.0186$ s to exclude the delay time of point A itself. Before the 0.0186 s was excluded, the delay time is not absolutely correct. If we see point A as the target point, the delay time of the receiver which is above point A should be 0. In delay time calculations according to Equation (A1), the reference point is not point A (16,7) but it is the point on x axis (16,0). This results in the 0.0186 s of delay time of point A itself. So, the 0.0186 s should be excluded from the delay time and the correct time-axis position of point A can be obtained. This can also be found in Equation (A3) as $(t - \tau_m)$. Then, we repeated the process above with each subarray to obtain the amplitude curve on the axis line $x = x_A$ (Figure 4d). In this certain case, the amplitude of a reflected wave was 0.005 times that of a surface wave in seismic signal. Eventually, the amplitude was 0.05 times that of a surface wave on the superposed curve for all subarrays. Then, we picked the peak amplitude and the corresponding estimated time (0.0093 s) as the amplitude result of point A.

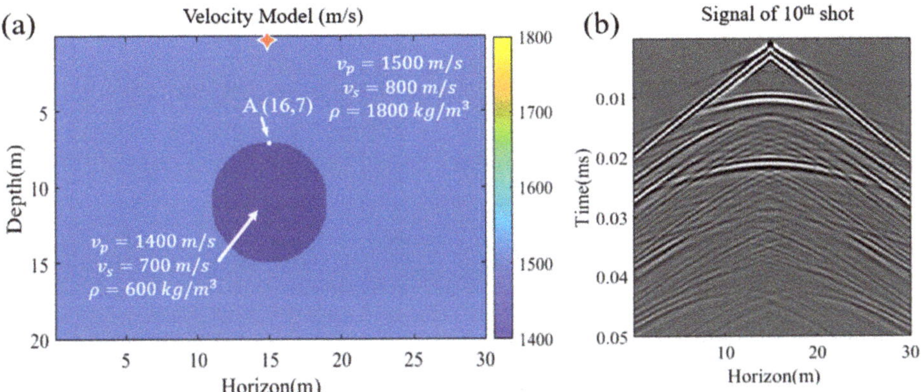

Figure 2. (**a**) Circular cave model. (**b**) Signal of the 10th shot with 20 stacked traces.

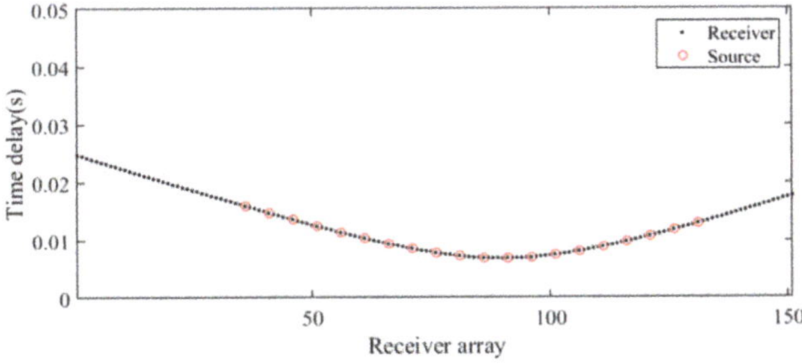

Figure 3. Time delay of each receiver and source.

We repeated the delaying, weighting, and superposition processes with the subarrays at every point in the imaging area to obtain the preliminary image for this shot (Figure 5a). Then, the CF (Figure 5b) was obtained according to Equation (A15). The CF matrix was used as another weight to enhance the SNR by multiplying it with the preliminary image. In this step, surface wave artifacts were suppressed because the adopted methods were developed for reflected waves to only amplify the reflected signals with focus. Other interferences were suppressed for the same reason. Then, we obtained the final image of the 10th shot.

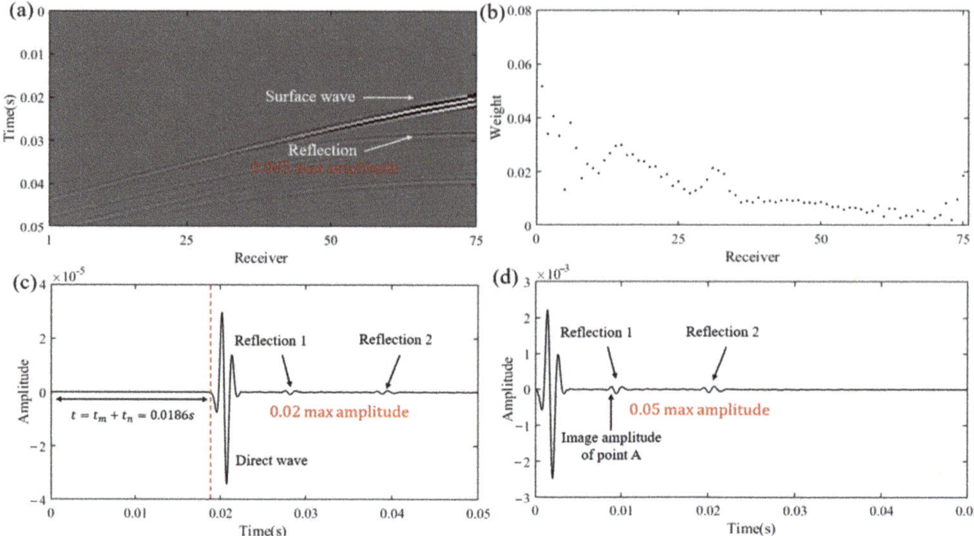

Figure 4. Delayed signal and weighted superposition of point A. (**a**) Delayed signal of the 1st subarray. (**b**) Weight of each receiver in the 1st subarray according to the MV method. (**c**) Superposed curve of weighted signals. (**d**) Amplitude curve on axis line $x = x_A$.

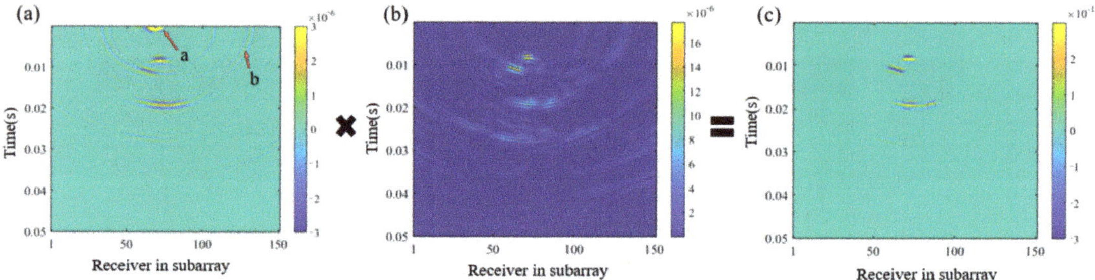

Figure 5. Preliminary image multiplied with the CF matrix to obtain the time-domain image of one shot. (**a**) Preliminary image. (**b**) CF matrix. (**c**) Image result of the 10th shot.

The time-domain image was converted into a depth-domain image according to the background velocity. The same processing scheme, including beamforming imaging and time-depth conversion, was applied to the remaining shots. The final image was a stacked result of all shots (Figure 6).

2.3. Related Works

2.3.1. Kirchhoff Migration

Kirchhoff migration was adopted as a related work in this paper. The methodology is mainly from the authors of [30]. The involved work migrates a single shot record using prestack time migration. The algorithm is a simple travel time path summation with few options. Travel time from source to scatter point (also known as a target point in beamforming methods) is approximated by a Dix equation using the RMS velocity from the model at the lateral position halfway between the source and receiver and at the vertical travel time of the scatter point. Similarly, from the scatter point to a receiver, a Dix equation using the RMS velocity at halfway between the scatter point and the receiver is used. There is no topographic compensation. The source and receivers are assumed to be on the same horizontal plane.

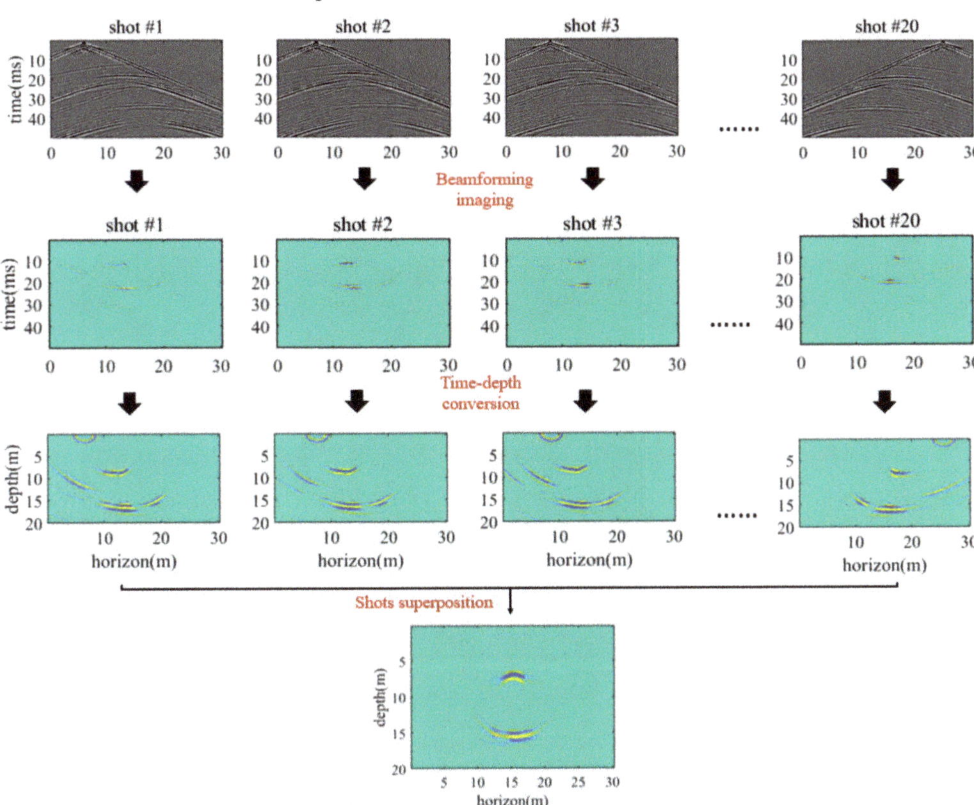

Figure 6. Stacked MVSS beamforming image of 20 shots.

2.3.2. Reverse Time Migration

Reverse time migration (RTM) is one of the best imaging methods for seismic data. RTM is a depth migration method. It takes the seismic records as input and reconstructs the underground wave field using the inverse time wave field continuation, then obtains the image amplitude according to the cross-correlation with the source wave field. There is no high-frequency approximation assumption of ray tracing methods or the angle limitation of one-way wave migration. Thus, RTM has a high imaging quality but also costs more in terms of computational resources.

RTM was adopted as a related work in this paper. The adopted reverse time migration program involved was from the authors of [30]. The algorithm dose is a prestack reverse time migration of a shot record. The migration computes the forward-in-time propagation of the source field and the reverse-time propagation of the input shot (receiver field). Thus, two wavefields (source field and receiver field) are simulated by finite-difference time stepping. The migrated shot comes from the cross correlation of these two fields at equal times. This requires that the source wavefield must be available at each time step of the receiver field and this is a major complexity.

3. Numerical Experiments

3.1. Numerical Experiments

Six synthetic models were used to validate the proposed imaging method. For this purpose, we constructed an area with a height of 20 m and a width of 30 m (Figure 7) to simulate near-surface seismic exploration. The seismic data came from numerical simulations based on staggered-grid finite differences in the time domain [29,31]. A perfectly matched layer (PML) was used to absorb reflections on the left, right, and bottom boundaries of the model. The upper boundary of the model was a free boundary.

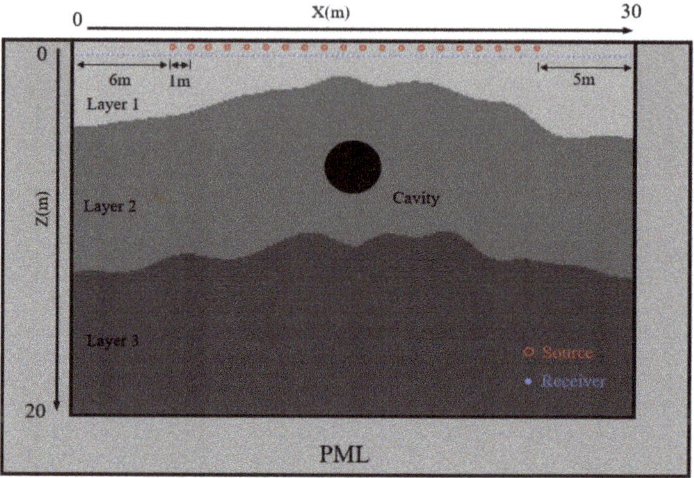

Figure 7. Sketch map of the geological model.

According to the near-surface geological conditions, the background P-wave velocity was set as 1500 m/s, the S-wave velocity was set as 700 m/s, and the density was set as 1.8 g/cm^3. A 600 Hz Ricker wavelet was used as the source at the ground surface. Twenty shots were deployed with an interval of 1 m, and 151 receivers were evenly arranged at a 0.2 m intervals across the top of the model.

The signal time length was 0.05 s. According to the size of the area and background velocity, the estimated P-wave travel time of a vertically down and up path was 0.026 s and the travel time associated with the far-right source, reflection at bottom left, and far-right receiver pair was 0.045 s (Figure 8). Therefore, the target P-wave signal integrity was guaranteed, which meant that a 0.05 s signal would be long enough to cover all the reflected objectives and the reflection of most part of the area could be received by multiple receivers.

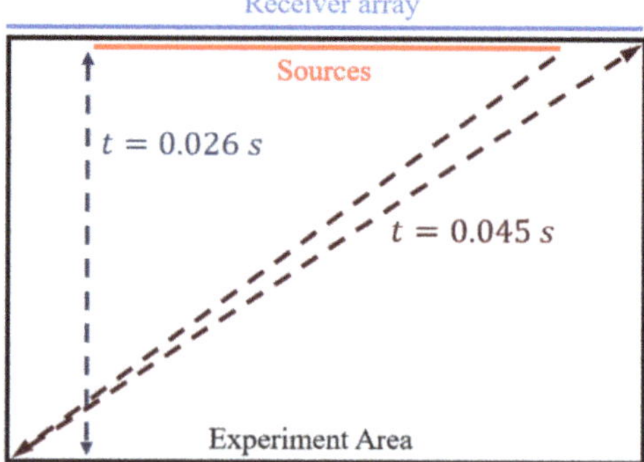

Figure 8. Illustration of travel time.

All parameters in the numerical experiments are shown in Table 1. Note that the sampling frequency was 2×10^5 which is consistent with the time step of the numerical simulation. In the numerical experiments, adapting a shorter recorded signal can help save computational resources. However, in real cases, the recorded signal can be longer in order to obtain more information.

Table 1. Numerical experiment parameters.

Parameter	Value
Receiver spacing interval	0.2 m
Source frequency	600 Hz
Sampling frequency	2×10^5 Hz
Background P-wave velocity	1500 m/s
Receiver array length	30 m
Number of receivers	151
Signal length	0.05 s

The receivers are velocity sensors that collect vibrations perpendicular to the ground surface. The geological models are produced by a random near-surface model generation algorithm (Figure 9). For example, horizontal parallel layers (1), folds (2), random fluctuations (3), faults (4) and caves (5) were incorporated into the models. More details on the model generation methods can be found in Appendix B.

Figure 9. Geological model generation workflow.

The structural details of each model are shown in Table 2. The models in group A contained cave-type geological anomalies. The models in group B were composed of multiple strata. The models in group C were combinations of cavities and multiple strata. The medium parameters are shown in Table 3.

3.2. *Imaging Results*

3.2.1. Cave Models

The imaging results of model A-1 are shown in Figure 10. In the MVSS beamforming imaging (Figure 10b), the cave roof and floor can be observed, but the right and left sides of the cave are missing. A slight artifact caused by surface waves is observed at the top of the image (Mark 1). The ceiling and floor can be observed, but the walls on both sides are missing. The reason is that the roof is convex and reflects the waves to both sides. Some of these reflected waves exceed the range of the receiver array, resulting in a lack of information from the roof. In contrast, the concave floor can better reflect all arriving waves toward the receiver array, forming a better image. The imaging results of the three methods show that the imaged cave is slightly larger in the vertical direction than the cave actually is.

Figure 10c shows the imaging result via Kirchhoff migration. There is a massive artifact (Mark 1) at the top of the image caused by surface waves. Some minor artifacts can be observed between the top of the image and the cave (Mark 2). Arcuate noise signals caused by S-waves are present on both sides (Mark 4). The upper roof is formed as a dot (Mark 3).

Table 2. Model list.

Model	Geological Structure Type	Caves	Cave Center Location (m) (Horizontal, Vertical)	Cave Radius (m)	Map	Description
A-1	-	1	15, 11	1.8		A circular homogeneous cavity.
A-2	-	2	13, 11 and 17, 11	1.8	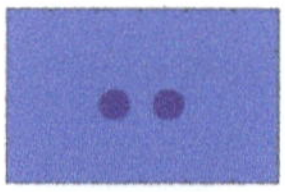	A model with two horizontally aligned caves.
A-3	-	2	15, 11 and 15, 16	1.8		A model with two vertically aligned caves.
B-1	Syncline fold	-	-	-	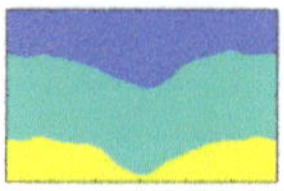	Syncline model with a fold depth of 4 m.
B-2	Fault	-	-	-	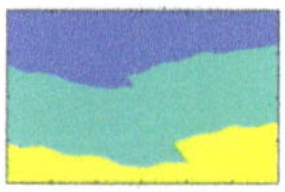	Fault model with a dislocation of 2 m and a great angle (*arctan*1.5).
B-3	Fault	-	-	-		Fault model with a great angle weak fault layer.
B-4	Fault	-	-	-		Fault model with a small angle weak fault layer (*arctan*0.5).
C-1	Syncline fold	1	15, 11	1.8	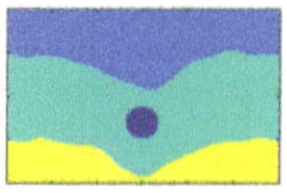	Syncline model composed of three layers with a cavity.
C-2	Fault	1	15, 11	1.8		Fault model composed of three layers with a cavity.

Table 3. Elastic parameters of the model media.

Sequence	P-Wave Velocity (m/s)	S-Wave Velocity (m/s)	Density (kg/m^3)
Cave 1	1400	700	1600
Layer 1	1500	800	1800
Layer 2	1600	900	2000
Layer 3	1800	1000	2200

Figure 10d shows the DAS beamforming imaging result. The artifacts caused by surface waves are still present at the top of the image (Mark 1). The roof of the cave is also formed as a dot shape (Mark 3). The artifact caused by S-waves is again observed (Mark 4) and is stronger than that in the Kirchhoff migration results.

Compared with RTM (Figure 10e), the advantage for suppressing the artifact caused by the surface wave and S-waves remains in MVSS beamforming. The cave location in MVSS beamforming is more accurate than that in RTM.

The imaging results of model A-2 are shown in Figure 11. In the MVSS beamforming imaging result shown in Figure 11b, the locations of the two caves are well determined, and the roofs and floors of both caves can be observed. In addition to the same surface wave artifacts detected in model A-1, some interferences were found beneath the floors of the caves (Mark 1) due to the superposition of different S-waves.

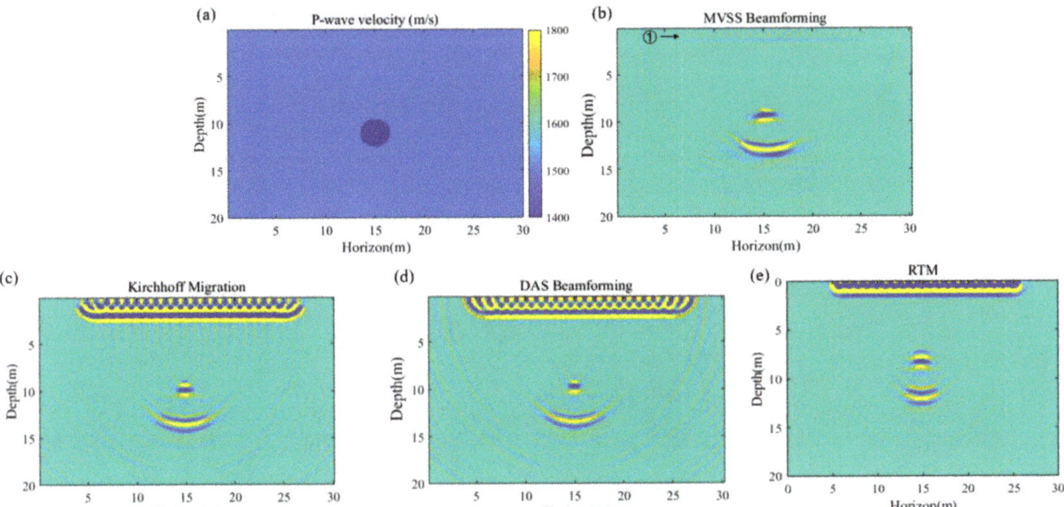

Figure 10. Model A-1. (**a**) True velocity model. (**b**) Imaging result of MVSS beamforming. (**c**) Imaging result of Kirchhoff migration. (**d**) Imaging result of DAS beamforming. (**e**) Imaging result of RTM.

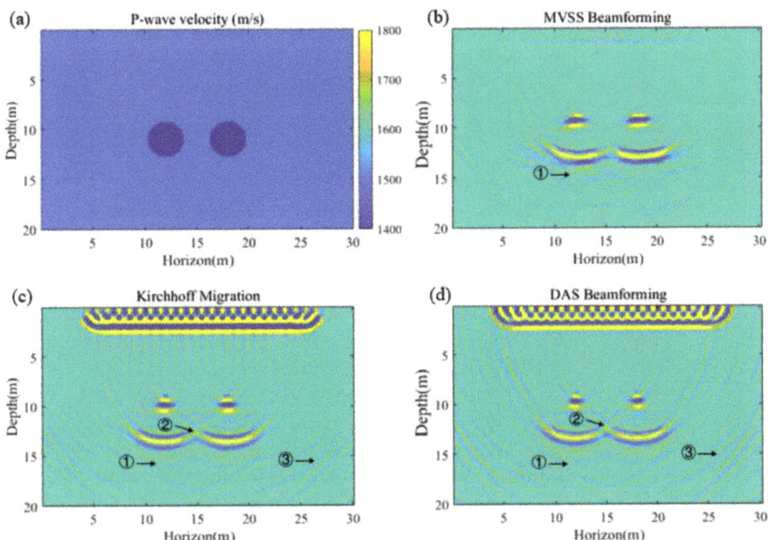

Figure 11. Model A-2. (**a**) True velocity model. (**b**) Imaging result of MVSS beamforming. (**c**) Imaging result of Kirchhoff migration. (**d**) Imaging result of DAS beamforming.

Compared with MVSS beamforming, in addition to the same interferences and artifacts in model A-1, the two floors overlap (Mark 1) in the Kirchhoff migration result (Figure 11c). This overlap appears again in the DAS beamforming result. This finding indicates that MVSS beamforming has a higher horizontal resolution than the other two methods. These imaging results further demonstrate that the imaging of these caves did not suffer from signal interference because the propagation paths of the waves reflected from the cave boundaries do not overlap with each other. Besides, the S-wave artifacts appears in Kirchhoff migration and DAS beamforming (Mark 2).

Compared with those in model A-1, the interferences in the Kirchhoff migration and DAS beamforming results in model A-2 are more intensive due to the more complicated structure (Mark 3). However, these interferences are still much weaker in the MVSS beamforming result, which means that MVSS beamforming exhibits better S-wave suppression.

The imaging results of model A-3 are shown in Figure 12, in which two caves are aligned vertically in the middle of the area (a beaded cave model). The surface wave artifacts and S-wave interferences are similar to those in model A-1 and model A-2. In addition, there is some overlap between the roof of the shallower cave and the floor of the deeper cave (Mark 1), which means that the three methods have similar vertical resolutions in this cave model.

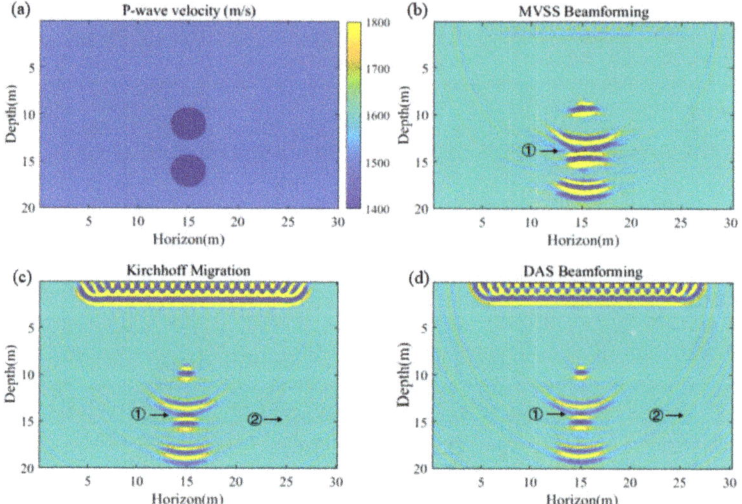

Figure 12. Model A-3. (**a**) True velocity model. (**b**) Imaging result of MVSS beamforming. (**c**) Imaging result of Kirchhoff migration. (**d**) Imaging result of DAS beamforming.

Compared with the MVSS beamforming imaging result, the Kirchhoff migration imaging result suffers from more intensive and complicated interferences on both sides of the image caused by S-waves. These S-wave interferences are also relatively intensive in the DAS beamforming result.

3.2.2. Layer Models

The imaging results of model B-1 are shown in Figure 13. In the MVSS beamforming imaging result (Figure 13b), two interfaces can be clearly observed. There is some light interference on both sides of the image (Mark 2) and between the interfaces (Mark 1). Furthermore, some parts of the deeper interface are missing (Mark 3).

In the Kirchhoff migration imaging result (Figure 13c), the interfaces between the layers can be observed clearly, but artifacts appear between the two layers and on both sides of the image (Mark 1), and some arc-shaped interferences can be observed on both sides of the image (Mark 2). The DAS beamforming imaging result is shown in Figure 13d; the interfaces can be observed, but arc-shaped interferences and a massive surface wave artifact at the top of the image can also be observed.

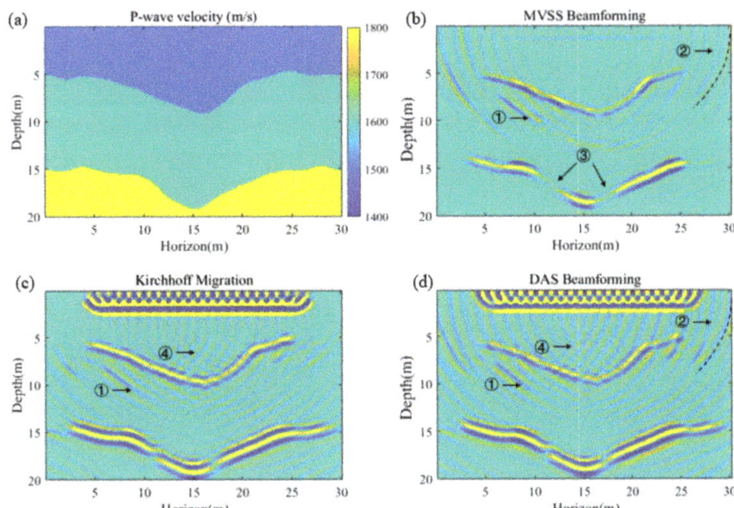

Figure 13. Model B-1. (**a**) True velocity model. (**b**) Imaging result of MVSS beamforming. (**c**) Imaging result of Kirchhoff migration. (**d**) Imaging result of DAS beamforming.

Comparing the three imaging methods reveals interferences between the different interfaces in the Kirchhoff migration and DAS beamforming imaging results. In the MVSS beamforming image, the amplitude of the surface wave artifact is weak, and the interferences between the interfaces are suppressed. The thickness of the interface between the two layers is approximately 0.4 m. The best resolution given the wavelength of the wavelet is 0.24 m. In contrast, the interfaces are thicker (approximately 1.6 m on the deeper interface) in the Kirchhoff migration and DAS beamforming images, which means that the vertical resolution of MVSS is higher in this layer model. However, artifacts caused by the superposition of S-wave energy still exist and caused interference between the interfaces. Kirchhoff migration and DAS beamforming suffer from similar intersecting interferences (Mark 4) caused by S-waves. However, at shallow depths, these intersecting interferences are replaced by vertical bars in the Kirchhoff migration image.

Compared with the MVSS beamforming imaging results obtained for the cave models, the interfaces can be readily determined, but the amplitudes of the interfaces in this layer model seem weaker than the amplitudes of those in the cave models (using the same imaging and display parameters in every model). In addition, the layers are missing on two sides of the image because the shots range horizontally from 5 m to 25 m.

The imaging results of model B-2 are shown in Figure 14. Figure 14b shows the imaging result of MVSS beamforming. The interfaces between the layers can be observed, and the location of a fault can be inferred from the shapes of the interfaces, but the details of the fault are difficult to discern. Lightly surface wave artifact can be observed (Mark 1).Dislocations in the shallow interface can be observed. For the deeper interface, although the upper and lower stratigraphic boundaries can be seen, the dislocations cannot. Slight interferences can be observed on the sides of the image and beneath the interface (Mark 2).

Figure 14c shows the imaging result of Kirchhoff migration. The artifact caused by surface waves can be observed at the top of the image. At the same time, interferences can be observed above the shallower interface and beneath the deeper interface.

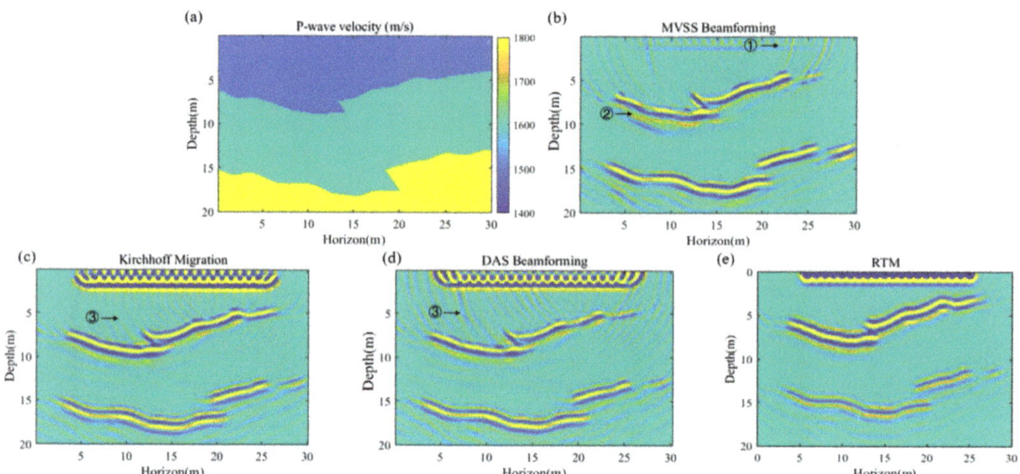

Figure 14. Model B-2. (**a**) True velocity model. (**b**) Imaging result of MVSS beamforming. (**c**) Imaging result of Kirchhoff migration. (**d**) Imaging result of DAS beamforming. (**e**) Imaging result of RTM.

Figure 14d shows the imaging result of DAS beamforming. Likewise, the artifact caused by surface waves can be observed at the top of the image (Mark 1). In addition, artifacts caused by S-waves are present on both sides of the image. We also detected interferences between the two interfaces and beneath the deeper interface. The interferences above the shallower interface appear with different shapes in the Kirchhoff migration and DAS beamforming imaging results (Mark 3).

Compared with the RTM (Figure 14e), there were less interferences beneath the interfaces in RTM but MVSS beamforming showed a higher vertical resolution. Still, the surface wave caused an intensive artifact at the top in the RTM result.

Compared the imaging results of the three methods, the artifacts caused by direct waves and surface waves are weaker in the MVSS beamforming result. Similarly, the artifacts on both sides of the image and beneath the deeper interface are also weaker. The dislocations across the deeper interface are missing because the reflection angles generated at such dislocations are large, and thus, the signal could not be recorded by the receiver array.

In summary, compared with the imaging results in model B-1, the shallower interface is complete, and the fault dislocation is well determined. However, the dislocation of the deeper interface is missing because the reflected waves were not received by the receiver array.

Both model B-3 (Figure 15) and model B-4 (Figure 16) contained low-velocity areas near the dislocation. The imaging results of model B-3 are not very different from those of model B-2. The weak layer results in more S-wave interferences above the shallower interface (Mark 1). The reason is that when seismic waves arrive at the weak layer, the reflected waves are not recorded by the receivers. Thus, because of the shadow created by the weak fault layer, the layer beneath the shallower interface is distorted (Mark 2).

Compared with Kirchhoff migration and DAS beamforming, MVSS beamforming still suppressed the S-wave interferences more effectively and provided a better contrast.

In model B-4, the fault angle is shallower than that in model B-3. The shallow part of the weak layer can be observed, but the deep part could not because the imaging results were disturbed by the weak layer. Arcuate interferences (Mark 1) and artifacts along the fault (Mark 2) can be observed. Furthermore, the imaging results were disturbed by the complicated structure due to the fault dislocation (Mark 2). However, the position of the deeper interface was not influenced by the weak fault layer.

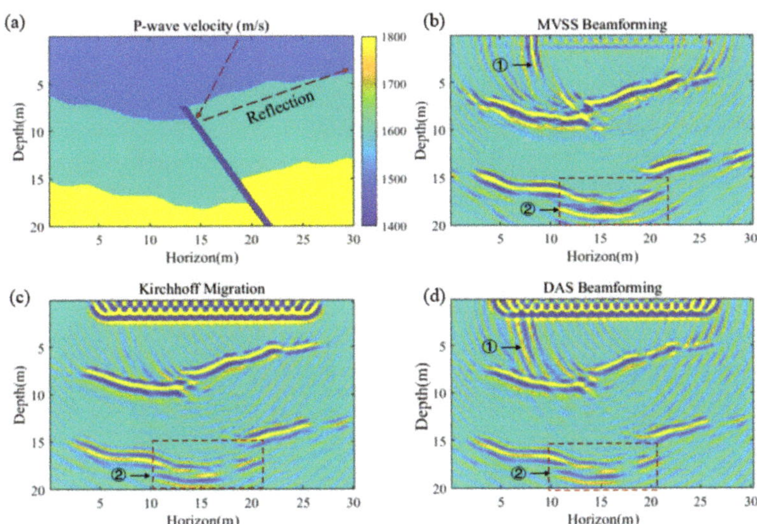

Figure 15. Model B-3. (**a**) True velocity model. (**b**) Imaging result of MVSS beamforming. (**c**) Imaging result of Kirchhoff migration. (**d**) Imaging result of DAS beamforming.

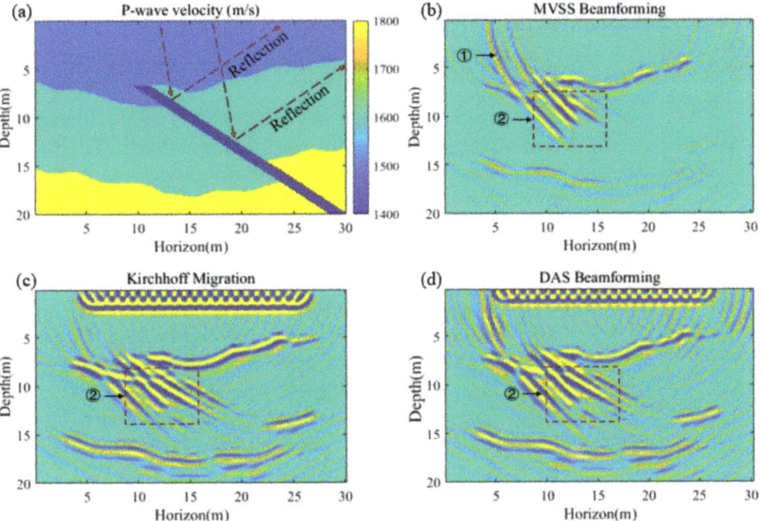

Figure 16. Model B-4b. (**a**) True velocity model. (**b**) Imaging result of MVSS beamforming. (**c**) Imaging result of Kirchhoff migration. (**d**) Imaging result of DAS beamforming.

Comparing MVSS beamforming with Kirchhoff migration and DAS beamforming reveals that the artifacts (Mark 2) in the latter two methods were more intensive than those in the former. For the deeper interface, Kirchhoff migration provided a better contrast than MVSS beamforming.

3.2.3. Cave-Layer Hybrid Models

The imaging results of model C-1 are shown in Figure 17. The imaging result of MVSS beamforming is shown in Figure 17b. The shapes and locations of the interfaces and caves can be clearly observed. There were some interferences between the layer interfaces and the cave boundaries. On both sides of the layer beneath the shallower interface, some artifacts caused by S-waves can be observed.

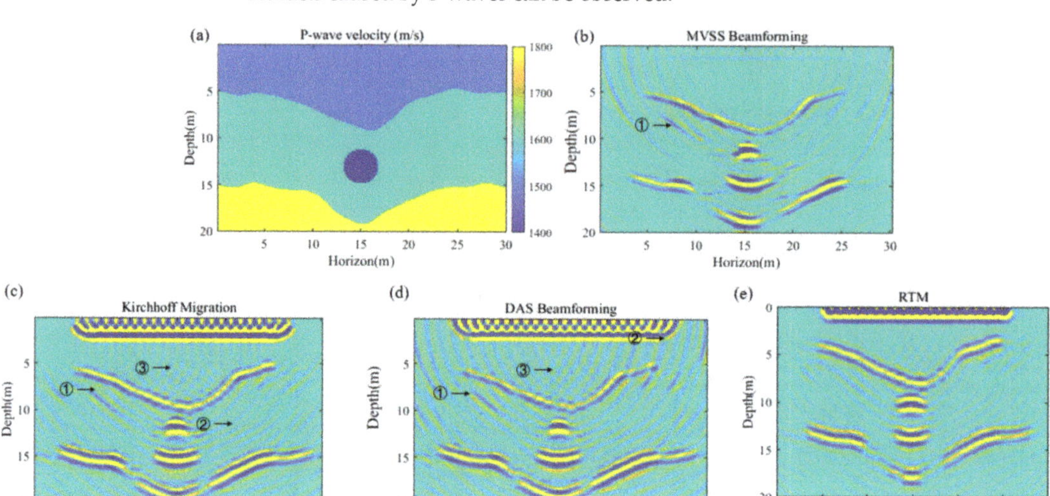

Figure 17. Model C-1. (**a**) True velocity model. (**b**) Imaging result of MVSS beamforming. (**c**) Imaging result of Kirchhoff migration. (**d**) Imaging result of DAS beamforming. (**e**) Imaging result of RTM.

The imaging result of Kirchhoff migration is shown in Figure 17c, indicating that some interferences appeared near (Mark 1) and between (Mark 2) the layer interfaces. The DAS beamforming imaging result is shown in Figure 17d, revealing multiple arcuate interferences between the two interfaces (Mark 2) and above the shallow interface (Mark 3).

When the three imaging methods are compared, the imaging results of Kirchhoff migration and DAS beamforming still exhibit artifacts of direct waves and surface waves, and there are more interferences between the layers and near the interfaces. Nevertheless, the MVSS method achieved the best resolution and most effectively suppressed the direct and surface waves.

Compared with those for model B-1, the imaging results for model C-1 display more interferences between the interfaces and on both sides of the caves because of the increased stacking of S-waves from different reflection interfaces. In addition, compared with model A-1, it can be noted that the image of the cave in model C-1 is more precise.

Compared with the RTM (Figure 17e), in the composed model of cave and fold, MVSS beamforming still made a better vertical resolution. There was a strong artifact caused by a surface wave in the RTM result.

Model C-2 comprises a combination of a fault, multiple layers, and a cave (model A-1 and model B-2). The imaging results are shown in Figure 18, where Figure 18b shows the imaging result of the MVSS beamforming method. The shallower interface of the layer can be clearly observed, and the cave between the layers can be detected as well. However, the deeper interface is not complete at widths of 10 m to 20 m (Mark 4).

Comparing the three methods makes it evident that the artifacts caused by surface waves are still noticeable in the Kirchhoff migration and DAS beamforming imaging results. Furthermore, the incompleteness of the deeper interface is as notable as it is in the MVSS beamforming imaging result, while the interferences between the layers and on both sides of the image (Mark 2) are weaker in the MVSS beamforming imaging results. However, the disturbances on the deeper interface (Mark 3) are similar among all three methods.

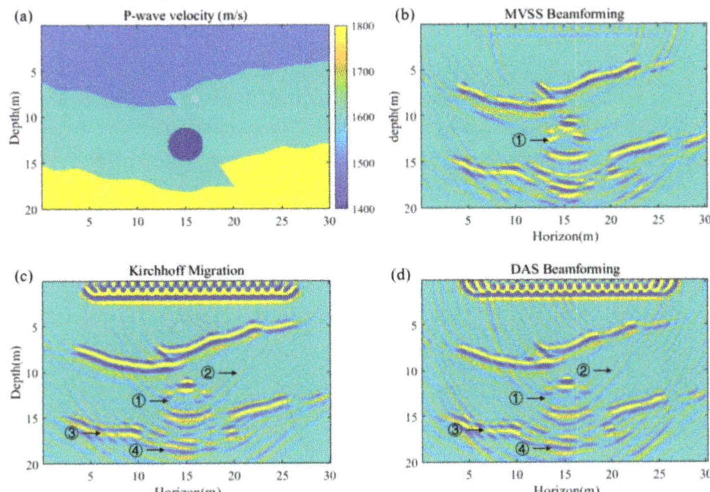

Figure 18. Model C-2. (**a**) True velocity model. (**b**) Imaging result of MVSS beamforming. (**c**) Imaging result of Kirchhoff migration. (**d**) Imaging result of DAS beamforming.

Compared with model B-2, a part of the deeper interface of the layer is blurred in model C-2. There are two main reasons for this blurriness. The structure is particularly complicated in the middle of the area due to the coexistence of the fault and cave, which causes the wave to reflect back and forth between those structures; this produces many multiples. Thus, the reflected waves from the deeper interface are seriously disturbed by multiples from the complex structure in the middle of the model. Moreover, compared with the features of model A-1, the cave's roof and floor are not clear (Mark 1) because the waves are disturbed by the fault.

3.3. Robustness to Other Wave Components

This numerical experiment was carried out based on elastic wave propagation theory, in which the signals received contain various components, such as P-P, P-S, and S-S waves, surface waves, and direct waves (Figure 19). For these imaging methods, we employed the P-P signal. Other components in the signal were considered to be noise or interference in imaging, and S-waves and surface waves were significant causes of imaging artifacts.

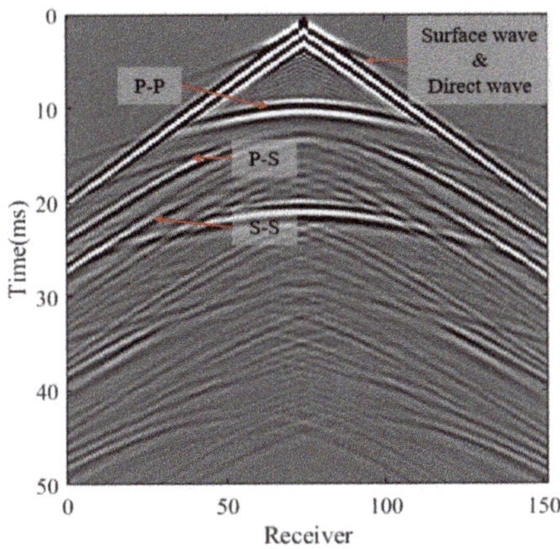

Figure 19. Components in a seismic signal.

Figure 20 shows the normalized amplitude curves and Hilbert transforms at $x = 15$ m in model B-2 (Figure 20a,b) and model A-3 (Figure 20c,d). The curves indicate the following: (1) The amplitude in the shallow parts of the Kirchhoff curve and DAS curve is high (greater than the reflection amplitude). This is the result of a surface wave. However, in the MVSS curve, the surface wave amplitude is suppressed. The surface wave is suppressed because the MVSS method amplifies the reflected wave by using the CF matrix. (2) From the Hilbert transform, it can be noted that the MVSS beamforming method can recognize two close interfaces, while the Kirchhoff and DAS methods might combine these interfaces as one. Thus, MVSS beamforming has a better vertical resolution than the other two methods. (3) In the MVSS curve, the reflection amplitudes are greater than those in the Kirchhoff migration and DAS beamforming curves; this is also partly due to the suppression of surface waves. This means that MVSS beamforming offers a better contrast than the other two imaging methods.

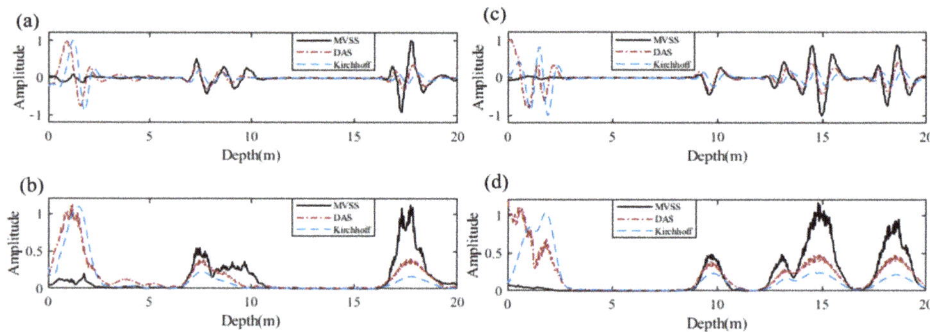

Figure 20. Imaging amplitude curves at $x = 15$ m and the Hilbert transform of each. (**a**,**b**) model B-2; (**c**,**d**) model A-3.

Figure 21a shows the imaging result of Kirchhoff migration with all signal components. The interfaces between the layers at depths of 5 m and 15 m are clear. However, massive horizontal artifacts can be detected at the top of the image; in addition, artifacts obliquely penetrate the shallower interface, and noise exists beneath the interface at a depth of 15 m. Likewise, Figure 21d shows the imaging result of DAS beamforming with all signal components. The two interfaces are clear, but horizontal artifacts can again be observed at the top of the image, and there are slight artifacts on both sides of the interfaces. Noise also appears beneath the interface at a depth of 15 m. Figure 21g shows the imaging result of MVSS beamforming with all wave components. The two interfaces are clear at depths of 5 m and 15 m. However, horizontal artifacts remain near the top of the image, and noise can be observed beneath the deeper interfaces. Nevertheless, compared with the imaging results of Kirchhoff migration and DAS beamforming, MVSS beamforming can suppress both the horizontal artifacts caused by surface waves and the artifacts on both sides of the image caused by S-waves.

These results confirm that the MVSS beamforming method can significantly suppress artifacts caused by direct waves and surface waves. The suppression of surface waves in reflection seismic exploration has long been a topic of interest [32]. Strong surface waves are significant causes of artifacts in imaging. Hence, to determine how surface waves impact the imaging results, we performed image processing by excluding direct waves and surface waves and then compared the results with the original imaging results. Figure 21b shows the Kirchhoff migration imaging results excluding direct waves and surface waves. Compared with Figure 21a, the surface wave artifact at the top of the image disappears, and the interference across the shallower interface is weaker. However, the artifacts on both sides of the interfaces and beneath the deeper interface still exist. Figure 21e shows the DAS imaging result excluding direct waves and surface waves. Compared with Figure 21d, the surface wave artifact at the top of the image disappears, but the S-wave artifacts across the interfaces still exist. Figure 21h shows the MVSS beamforming result excluding direct waves and surface waves. Compared with Figure 21g, the artifact at the top of the image disappears, but the interferences beneath the deeper interface still exist. These comparisons suggest that the main consequence of direct waves and surface waves is the presence of artifacts at the top of the image.

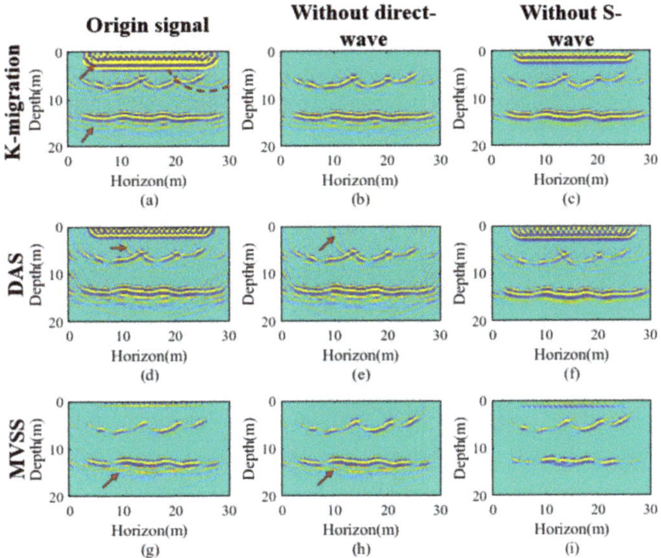

Figure 21. Imaging results after eliminating either surface waves or S-waves. (**a**) Kirchhoff migration result using the original signals. (**b**) Kirchhoff migration result using signals without surface waves. (**c**) Kirchhoff migration result using signals with only P-waves. (**d**) DAS beamforming result using the original signals. (**e**) DAS beamforming result using signals without surface waves. (**f**) DAS beamforming result using signals with only P-waves. (**g**) MVSS beamforming result using the original signals. (**h**) MVSS beamforming result using signals without surface waves. (**i**) MVSS beamforming result using signals with only P-waves.

S-waves (or converted waves) are another main factor causing artifacts and interference in seismic imaging. S-wave signals and converted waves (reflected S-waves converted from P-waves at interfaces) have similar in-phase axes as P-wave signals. However, the velocities of S-waves and converted waves do not match the estimated velocity (P-wave velocity), so the use of S-waves (or converted waves) can cause errors. Figure 21c shows the Kirchhoff migration imaging result ignoring S-waves. Compared with Figure 21a, artifacts at the top of the image still exist, but the artifacts are thinner and weaker. Moreover, the interferences across the shallower interface and beneath the deeper interface both disappear. Figure 21f shows the DAS beamforming result ignoring S-waves. Compared with Figure 21d, the artifacts at the top of the image still exist but are weaker, and the interferences beneath the deeper interface and on both sides of the shallower interface disappear. Finally, Figure 21i shows the MVSS beamforming result ignoring S-waves. Compared with Figure 21g, the artifacts at the top of the image still exist, albeit with weaker energy, and the interference beneath the deeper interface disappears. In addition, compared with Figure 21f, the artifacts on both sides of the interfaces disappear.

From this discussion, we have found the following:

1. The horizontal artifacts at the top of the images are caused by surface waves, while the artifacts beneath the interfaces are caused by S-waves, probably P-S waves;
2. When S-waves are eliminated, the direct wave still exists, but the surface wave does not. This greatly weakens the artifacts at the top of the image;

3. The ability of MVSS beamforming to suppress surface wave artifacts at the top of the image and the S-wave artifacts on both sides of the interfaces is superior. However, the S-wave artifacts beneath the interfaces still affect MVSS beamforming as much as they do the other methods.

3.4. Robustness to Random Noise

The signals in the numerical experiments were clear synthetic signals (Figure 22a). The SNR is defined as follows, where P_{signal} and P_{noise} are the power levels of the signal and noise, respectively, and A_{signal} and A_{noise} are the amplitudes of the signal and noise, respectively:

$$SNR(\text{dB}) = 10\ \log_{10} \frac{P_{signal}}{P_{noise}} = 20\ \log_{10} \frac{{A_{signal}}^2}{{A_{noise}}^2} \quad (1)$$

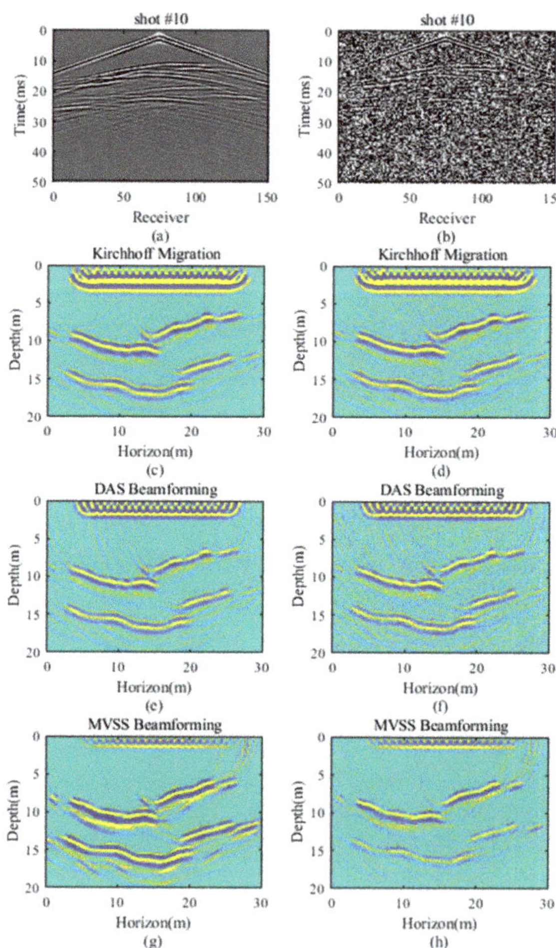

Figure 22. Imaging result using the clear signal (**a**) and noisy signal (**b**). (**c**) Kirchhoff migration imaging result using the clear signal. (**d**) DAS beamforming result using the clear signal. (**e**) MVSS beamforming result using the clear signal. (**f**) Kirchhoff migration imaging result using the noisy signal. (**g**) DAS beamforming result using the noisy signal. (**h**) MVSS beamforming result using the noisy signal.

We added Gaussian noise to the signal (Figure 22b) to make the SNR 0 dB (the power of noise equals the power of the target reflection signals) and then generated images using Kirchhoff migration, DAS beamforming, and MVSS beamforming.

In the Kirchhoff migration result with the noisy signal (Figure 22d), the interference caused by white noise can be observed as dots in the image. In the imaging result of DAS beamforming for the signal with Gaussian noise (Figure 22f), compared with the imaging result containing clear signals (Figure 22e), the white noise in the signal is detectable in the imaging result as well.

Figure 22h shows the MVSS beamforming result for the signal with Gaussian noise. Compared with the Kirchhoff migration and DAS beamforming results, the influence of white noise is much less apparent for the MVSS beamforming results. However, relative to the imaging result with a clear signal (Figure 22h), although the noise is well suppressed, the brightness of the reflection interface decreases. The reason is that the energy of all signals increases when white noise is added, which narrows the energy gap separating the reflection signal and other signals from empty space. As a consequence, the energy of the reflective interface is relatively reduced. This phenomenon corresponds to the principle of keeping the effective signal gain unchanged while minimizing the energy of the whole image in the MVSS beamforming method.

3.5. Focus Enhancing by a Signal Advance Correction

In real-world applications, the velocity assigned to the background is inaccurate. Although the phenomenon we discuss in this section was not apparent in our numerical experiments, it does exist in practice. Therefore, another model was designed at a smaller scale with a higher velocity to observe more obvious phenomena and to verify how MVSS beamforming works when an inaccurate background velocity is provided. Figure 23 displays the imaging results when different background velocities were provided while the true background velocity was 3000 m/s. When the background velocity reaches 2250 m/s, MVSS beamforming can barely differentiate three different targets (dots) in the region and loses focus. Interestingly, MVSS beamforming does not appear to work optimally under an accurate velocity (3000 m/s), whereas it performs successfully under a velocity of 2750 m/s. The reason is that unlike in oil and gas exploration, the shot wavelet length in near-surface exploration cannot be ignored because the scale of the model is small.

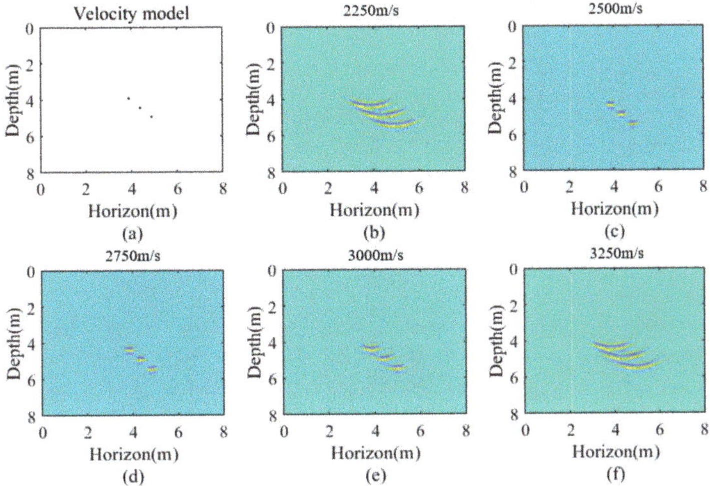

Figure 23. Imaging results using different background velocities while the true velocity is 3000 m/s. (**a**) True velocity model; (**b**) 2250 m/s; (**c**) 2500 m/s; (**d**) 2750 m/s; (**e**) 3000 m/s; (**f**) 3250 m/s.

In the signals received in a near-surface situation, the time from zero to the first arrival is t. However, t is not the most appropriate value to use because the time employed in the algorithm is the peak-to-peak time or the time from first arrival to first arrival (Figure 24). Thus, we fixed all signals by half a cycle to correct this phenomenon, called 'defocusing', to match the wave peak times, and the signal was advanced by a quarter of a cycle, half a cycle, and three-quarters of a cycle of the source wavelet, producing a set of images using MVSS beamforming. The results show that the imaging algorithm works best when the signal is advanced by half a cycle. However, the location of the objective becomes shallower than the real location as a result.

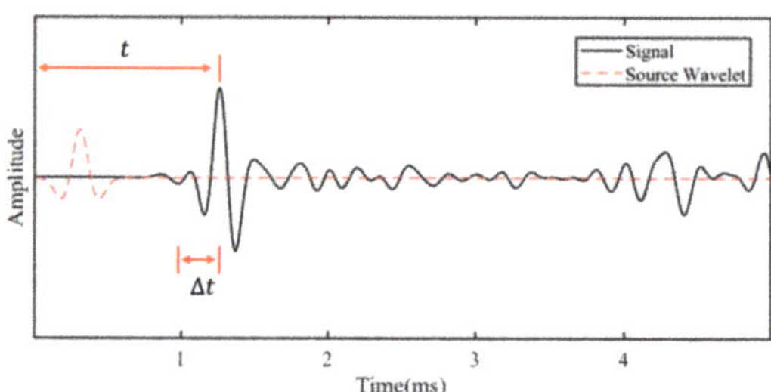

Figure 24. Signal advance correction to match the peak time to the arrival time.

To better observe the signal correction described above, we implemented the imaging algorithm in a model with smaller point anomalies and a high background velocity and excluded both surface waves (direct waves) and S-waves. The results with no processing, a 1/4-cycle advance, a 1/2-cycle advance, and a 3/4-cycle advance are shown in Figure 25. Upward-concave artifacts can be detected near the point target in the image without processing; this is the phenomenon known as defocusing. The reason for this defocusing is the mismatch between the delay information and the reflection signals. With an increase in the signal correction amount, the best focusing effect was achieved when the correction value was 1/2 of a cycle. When the correction value exceeds this value, the imaging defocuses in the other direction (concave-downward); the reason for this is again the mismatch between the delay information and the reflection signals.

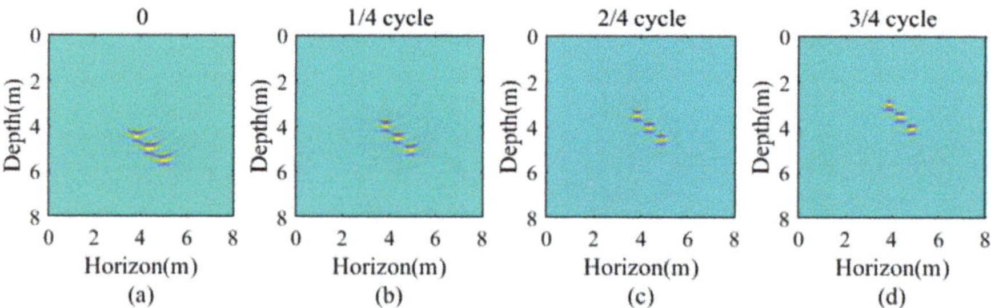

Figure 25. Imaging after time advance corrections: (**a**) no processing; (**b**) 1/4-cycle advance; (**c**) 1/2-cycle advance; (**d**) 3/4-cycle advance.

This phenomenon also appears in the 30 m × 20 m model, although it is not particularly obvious due to the lower frequency and slower wave velocity. The red line in Figure 26 is a reference line at a depth of 5 m (Figure 26a), and the red box marks the comparison of different time advance corrections. After the signal advance correction was applied, the focusing effect improved, and the focusing effect was best when the correction value was 1/2 of a cycle (Figure 26d). When the correction value increased to 3/4 of a cycle (Figure 26e), the image began to defocus to the other side (shallower side). It is worth noting that due to the signal advance correction, all the imaging targets moved to a shallower position. Therefore, this processing scheme will enhance the focusing effect while causing the target position to be slightly shifted.

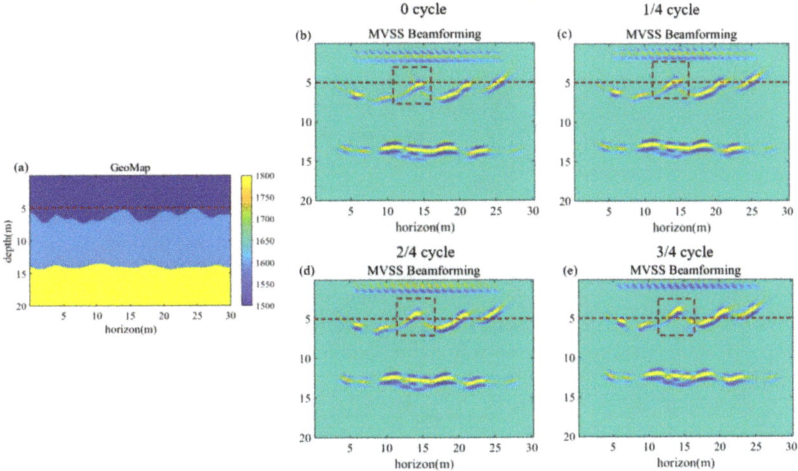

Figure 26. Imaging after applying a time correction in model B-2: (**a**) true velocity model; (**b**) no processing; (**c**) 1/4-cycle advance; (**d**) 1/2-cycle advance; (**e**) 3/4-cycle advance.

4. Discussion

4.1. Computational Efficiency

In terms of the size of our numerical experiment (20 shots, 151 receivers, 1001 time grids per signal), the calculation time needed to implement each imaging method ranged from several seconds to minutes, as summarized in Table 4. DAS beamforming was the most efficient method, with a fast imaging time of 3.9 s, followed by MVSS beamforming at 201.1 s, and Kirchhoff migration took the longest at 520.1 s. If the time delay was calculated according to the background velocity and stored prior to imaging, the calculation times of DAS and MVSS beamforming were further shortened. In the Kirchhoff migration adopted in this paper, the travel times from source to scatter point and from scatter point to receiver were approximated by a Dix equation using the RMS velocity. On the other hand, the travel times were calculated based on the geometry relationship between the source, scatter point, and receiver. To some extent, DAS beamforming is a simplified option of Kirchhoff migration, which results in the computational efficiency difference. Conceivably, if the condition of the medium at each point is known, the Kirchhoff migration method should theoretically have the best imaging results. In this experiment, the geological information was known approximately (background velocity), so the accuracy advantage of Kirchhoff migration was almost negligible, but the efficiency advantage of beamforming persisted. Beamforming methods assume that the wave propagates in a straight line and that the seismic velocity in the area is the same. This assumption is also available for Kirchhoff migration. However, in the adopted algorithm it still took time to estimate the travel time by a Dix equation using the RMS velocity desired from a homogeneous background model. This meant that the travel time was almost same as that in DAS beamforming, in which the travel time was simply calculated according to geometry and background velocity, but the calculation time was quite different. Therefore, the efficiency of DAS beamforming was as high as 130 times that of Kirchhoff migration. For this reason, we would like to apply beamforming imaging to detect underground anomalies in a scene with a single background velocity. The computational efficiency and accuracy of an algorithm have always been contradictory, but in near-surface exploration, the prior information is simple, which is suitable for beamforming methods to improve efficiency.

Table 4. Calculation time of each method.

Method	Kirchhoff Migration	DAS Beamforming	MVSS Beamforming
Calculation Time (s)	520.1	3.9	201.1

4.2. Coherence Factor Matrix

As stated in 2.5, the coherence factor (CF) matrix is an effective method to reduce artifacts and interferences in the image. The CF matrix is obtained in the delayed signal. Since the CF matrix is used as a weight to amplify the reflection wave and suppress the artifacts for each shot, it could be theoretically applied into other prestack time domain migration methods. Figure 27 here presents the results when the CF matrix was applied into Kirchhoff migration and DAS beamforming. The results shows that the CF matrix can obviously reduce the artifact caused by surface wave and interference on two sides of the images. However, there was also a contrast reduction at the lower floor in Kirchhoff. This might be a result of different travel time (delay time) calculation algorithm between MVSS beamforming and Kirchhoff migration.

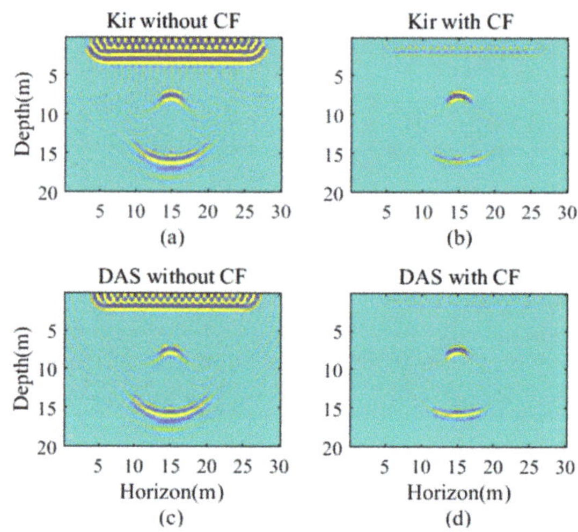

Figure 27. Results when the CF matrix was applied into Kirchhoff migration and DAS beamforming. Kirchhoff migration (**a**) without and (**b**) with CF matrix applied. DAS beamforming (**c**) without and (**d**) with CF matrix applied.

5. Conclusions

This research introduced a medical imaging method, the minimum variance spatial smoothing (MVSS) beamforming imaging method, for near-surface seismic detection to achieve rapid and accurate imaging. To increase the robustness of the algorithm, the DL method was employed to add spatial white noise to the recorded wavefield and to adjust the weight calculation in the proposed method. The CF method was used to improve the image quality by amplifying the amplitudes of points with a strong focusing effect and depressing those with a weak focusing effect. We observed that reducing the delay time by one-half of a wavelet cycle provided the best image focusing effect when the wavelet length was not short enough for the imaging area in a near-surface situation. The method was tested on multiple synthetic near-surface geological models and compared with DAS beamforming, Kirchhoff migration, and reverse time migration. The results indicate that the MVSS beamforming method has a high computational efficiency while maintaining a high imaging quality, a high imaging resolution and a high contrast. The MVSS beamforming method provides the superior suppression of artifacts caused by surface waves as well as other imaging artifacts or interferences caused by S-waves, surface waves, and white noise. Meanwhile, we suggest that future works should focus on adapting inhomogeneous velocity models into beamforming imaging.

Author Contributions: Conceptualization, C.L.; Formal analysis, D.W.; Funding acquisition, Z.S.; Methodology, D.W.; Project administration, M.P. and Z.S.; Resources, M.P., L.L., C.L., Z.S. and F.M.; Software, L.L. and J.S.; Writing—original draft, D.W.; Writing—review and editing, M.P., D.W. and L.L. All authors have read and agreed to the published version of the manuscript.

Funding: This research was funded by the National Natural Science Foundation of China, grant numbers 42172296, 42071010 and 42061160480, and the Shanghai Natural Science Foundation, grant number 20ZR1461300.

Institutional Review Board Statement: Not applicable.

Informed Consent Statement: Not applicable.

Data Availability Statement: The data presented in this study are available on request from the corresponding author.

Conflicts of Interest: The authors declare no conflict of interest.

Appendix A. MVSS Beamforming Methodology

Appendix A.1. Signal Time Delay Calculation

The time delay τ_i of each receiver can be calculated by Equation (A1) in a 2D imaging problem where c is the background velocity. By adding delayed signals together, the amplitude at the target point can reveal whether the waves were reflected from the target point. As shown in Figure A1, the waves produced by the shot source (x_s, z_s) arrive at the scatter point $\left(x_f, z_f\right)$ and are reflected; then, the reflected wave signals are received by a receiver (x_r, z_r). Here, the time difference among the signals at different receivers is caused by the position differences among the receivers and sources. The delay time of each shot-receiver pair includes two parts: τ_n, the time from the source point (x_s, z_s) to the target point $\left(x_f, z_f\right)$, and τ_m, the time from the target point $\left(x_f, z_f\right)$ to the receiver (x_r, z_r):

$$\tau_i = \tau_s + \tau_r = \frac{\sqrt{\left(x_s - x_f\right)^2 + \left(z_s - z_f\right)^2}}{c} + \frac{\sqrt{\left(x_r - x_f\right)^2 + \left(z_r - z_f\right)^2}}{c} \tag{A1}$$

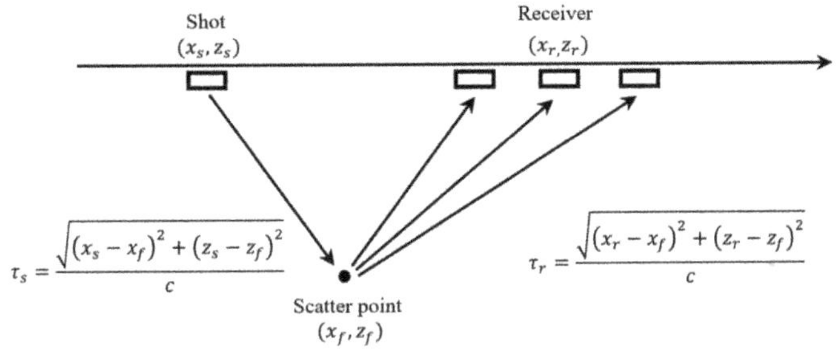

Shot (x_s, z_s)
Receiver (x_r, z_r)
$\tau_s = \frac{\sqrt{(x_s - x_f)^2 + (z_s - z_f)^2}}{c}$
$\tau_r = \frac{\sqrt{(x_r - x_f)^2 + (z_r - z_f)^2}}{c}$
Scatter point (x_f, z_f)

Figure A1. Time delay calculation.

Appendix A.2. Superposition with Weight Calculation

Superposition is a weighted averaging process that filters the receiver array signal in the spatial domain. In this paper, instead of preset and constant weighting coefficients, the adaptive weighting method was applied to obtain weights that cause different receivers to have different influences on the imaging result.

In signal superposition, we suppose that the number of receivers is M and that the signal of the receiver array is $x(t)$, which includes the signal from target point $s(t)$ as well as noise $i(t)$ and interference $v(t)$:

$$x(t) = s(t)a + i(t) + v(t) \tag{A2}$$

where a is the direction vector of the beam and t is the time sequence of the signal. The beamformer $b(t)$ is defined as:

$$b(t) = w^H y(t) = \sum_{m=1}^{M} w_m{}^* x_m(t - \tau_m) \tag{A3}$$

where w is the beamforming weight vector, H represents the conjugate transpose, and $w_m{}^*$ represents a conjugate matrix of w_m.

$$y(t) = [x_1(t-\tau_1)\ x_2(t-\tau_2)\ \ldots\ x_3(t-\tau_3)] \tag{A4}$$

is the signal of each receiver after a time delay. τ_m is the delay value of each receiver. The signal is similar to a vertically incident plane wave at each receiver after time delay processing. In this case, the direction vector a is a vector composed entirely of 1:

$$a = [1\,1\ldots 1]^{\mathrm{T}} \tag{A5}$$

where T represents matrix transposition. When w changes with the received signal to improve the image quality, $b(t)$ becomes an adaptive beamformer. The weight vector w (Capon, 1969) [21] can be found from the maximum of the signal-to-interference-plus-noise ratio (SINR) [33]:

$$\mathrm{SINR} = \frac{\sigma_s^2 \left|w^H a\right|^2}{w^H R_{i+n} w} \tag{A6}$$

where σ_s^2 is the signal power, $R_{i+n}(t)$ is the interference-plus-noise covariance, and a is the direction vector in Equation (A5). The Maximizing Equation (A6) is equivalent to minimizing the output interference-plus-noise power while maintaining a distortionless response to the desired signal. In practice, the exact interference-plus-noise $R_{i+n}(t)$ is unavailable, so the sample covariance matrix $R(t)$ is used in N receivers:

$$R(\mathrm{t}) = \frac{1}{N} \sum_{n=0}^{N} y(t) y^H(t) \tag{A7}$$

Then, this issue can be equivalent to:

$$\min_{w} w^H R w \text{ , subject to } w^H a = 1 \tag{A8}$$

By using the Lagrange multiplier method, the analytic formula is:

$$w_{\mathrm{MV}} = \frac{R^{-1} a}{a^H R^{-1} a} \tag{A9}$$

where w_{MV} is the MV beamforming weight vector and a is a known vector according to Equation (A5). The next step is to obtain the estimated covariance matrix R of the signal data. The spatial smoothing method is used to obtain a good estimation. The receiver array is divided into several overlapping subarrays, and the covariance matrix of these subarrays is averaged to obtain a robust covariance matrix estimation. This is also defined as spatial smoothing processing [22]:

$$\hat{R}(\mathrm{t}) = \frac{1}{M-L+1}\sum_{l=0}^{M-L}\overline{y_l}(t)\overline{y_l}^H(t) \quad \text{(A10)}$$

where M is the total number of receivers and L is the length of the subarray. As shown in Figure A2, there are M receivers in total, and they are divided into $M-L+1$ subarrays. Each of the subarrays is L receivers long. From left to right, subarray No. 1 to subarray No. $M-L+1$ covers all receivers.

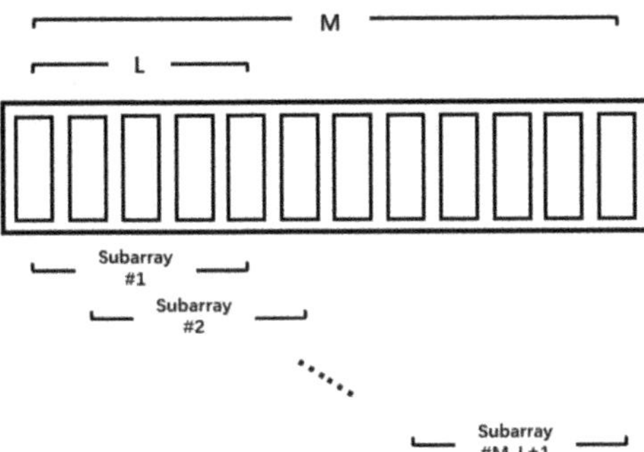

Figure A2. Illustration of subarrays.

$\overline{y_l}(t)$ is the delayed signal of No. l subarray:

$$\overline{y_l}(t) = \begin{bmatrix} y_l(t) \\ y_{l+1}(t) \\ \vdots \\ y_{l+L-1}(t) \end{bmatrix} \quad \text{(A11)}$$

The size of each subarray must satisfy $L \leq M/2$ (Asl, 2009 [26]) to ensure that the estimated covariance matrix is invertible (full rank). However, reducing L increases the robustness but reduces the resolution, and setting $L = 1$ makes the beamformer a DAS beamformer with uniform weights. Therefore, there is a trade-off between the robustness and resolution. The critical value $L = M/2$ is taken to achieve the highest possible imaging resolution.

Diagonal loading (DL) [34] is adopted to improve the robustness of the algorithm by adding spatial white noise to the recorded wavefield and to adjust the weight calculation [25] using:

$$R_{dl} = \hat{R}(t) + \delta \cdot \mathrm{trace}\{\hat{R}(t)\} \cdot \mathrm{I} \quad \text{(A12)}$$

to replace $\hat{R}(t)$, where δ represents the DL factor and I represents the identity matrix. The DL factor should satisfy $\delta \ll 1/L$. Then, we replace R in (8) with R_{dl}, and the weight of the beamformer becomes:

$$w_{MV} = \frac{{R_{dl}}^{-1}a}{a^H {R_{dl}}^{-1}a} \quad \text{(A13)}$$

The beamformer can be written as:

$$b(t) = \frac{1}{M-L+1}\sum_{l=0}^{M-L} w_{MV}^H \overline{y_l}(t) \quad \text{(A14)}$$

A weighting factor modified from the CF map is presented as:

$$\mathrm{CF}(t) = \frac{\left|\sum_{l=0}^{M-1} y_l(t)\right|^2}{M\sum_{l=0}^{M-1}|y_l(t)|^2} \tag{A15}$$

CF ranges from 0 to 1: when $\mathrm{CF}(t) = 1$, it is a perfect focused point at time t, and when $\mathrm{CF}(t) = 0$, the signals of different receivers are not coherent. After multiplication by the amplitude beamformer $b(t)$, the final image of the Nth shot $I_t(N)$ is obtained as:

$$I_t(N) = b(t)\cdot\mathrm{CF}(t) \tag{A16}$$

Appendix A.3. Image Processing and Stacking

The result obtained by the beamformer is the imaging result in the time domain. We estimated the depth of the reflection interface based on the background velocity and the wave travel time, which is a conversion from time domain to depth domain. For each shot, an image under a particular illumination angle is obtained. The final image is obtained by stacking the amplitudes of each shot, where N is the total number of shots, I_d is the final image with all shots stacked together, and c is the estimated background velocity:

$$I_d = c\sum_{N=1}^{N_{max}} I_t(N) \tag{A17}$$

Appendix B. Near-Surface Model Generation Method

For an elastic wave numerical simulation, the model is established mainly to determine the spatial distribution of the elastic parameters. It includes five parameters of the medium involved: the P-wave velocity (v_p), S-wave velocity (v_s), density (ρ), Lame constant μ and λ. The relation among these parameters can be given by:

$$v_p = \sqrt{\frac{\lambda + 2\mu}{\rho}},\ v_s = \sqrt{\frac{\mu}{\rho}} \tag{A18}$$

in which three of the five parameters must be chosen to determine the other two parameters. Here, we control v_p, v_s, and ρ and determine the Lame constant according to these parameters.

Appendix B.1. Horizontal Layer Model

We start with a simple layered model with N layers and $N-1$ interfaces. The P-wave velocity map is used to describe the model. For each column in the area, the points are filled with preset mediums using:

$$M(X,Y) = m_u\ ,\ y_n < Y \le y_{n+1}\ ,\ n = 0,1,2\ldots,N \tag{A19}$$

where $M(X,Y)$ is the elastic parameter of the point (X,Y), m_k is the preset parameter of medium u, y_n and y_{n+1} are respectively the depth of the nth and $(n+1)$th interface. The three-layer (two-interface) model is shown in Figure A3.

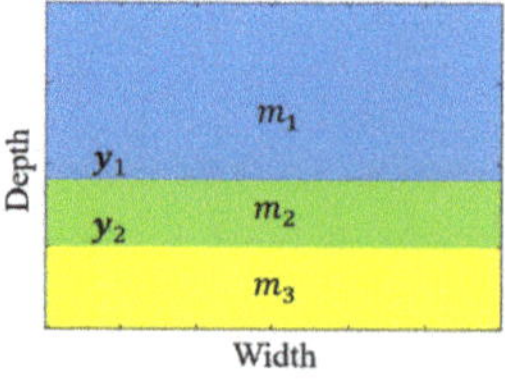

Figure A3. Horizontal layered model.

Appendix B.2. Fold

A Gaussian curve is used to form folds in the model because a Gaussian curve can satisfactorily present folds when the PML layer and fault dislocation are cut (Figure A4).

$$y_g(x) = ae^{-(x-b)^2/2c^2} \tag{A20}$$

where a is the fold depth, b is the horizontal location of the fold and c is a parameter controlling the width of the fold.

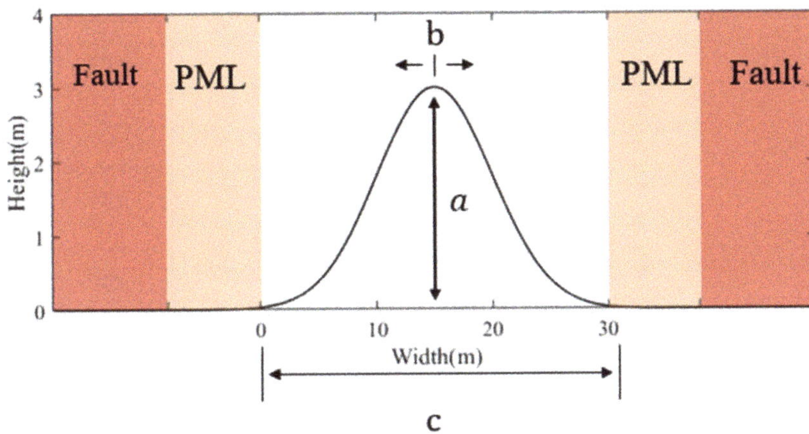

Figure A4. Gaussian curve used to control the fold shape.

The interfaces between the layers are often not totally smooth. The curve $y_g(x)$ is used as a baseline to create fluctuations. The fluctuation y_f is composed of random floating values with 0 as the mean. Now, the interfaces y_i between layers (Figure A5) can be obtained by Equation (A20). Since the Gaussian curve and fluctuations are randomly generated, the interfaces can be either parallel or nonparallel (Figure A5).

$$y_i(x) = y_n + y_g(x) + y_f(x) \tag{A21}$$

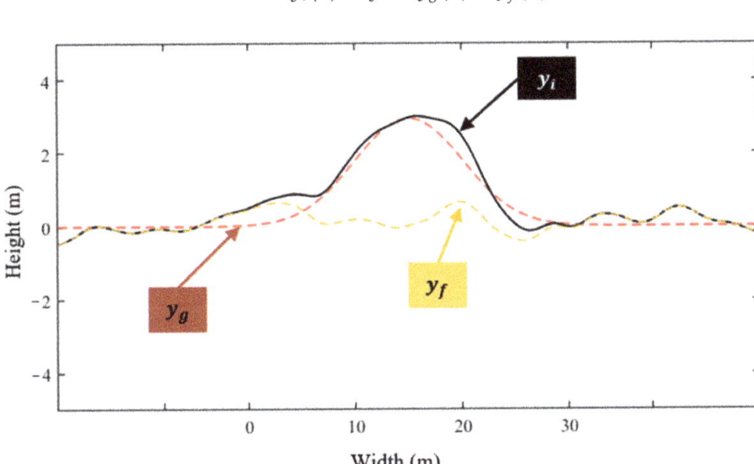

Figure A5. Interface with fluctuations.

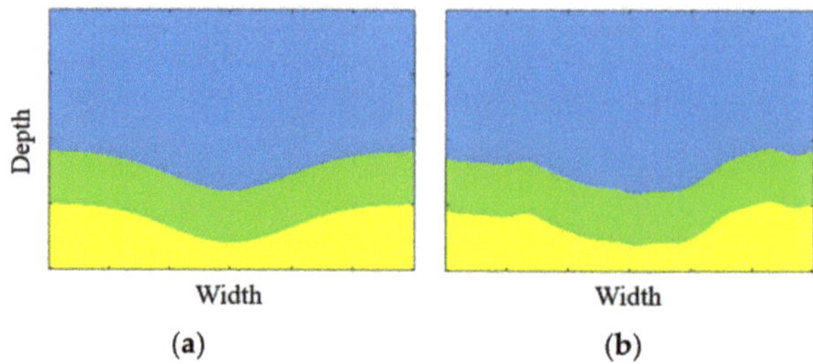

Figure A6. Fold model (**a**) without and (**b**) with random fluctuations.

Appendix B.3. Fault

The fault is determined by the fault line and dislocation distance. The fault line controls the position and inclination angle of the fault. The dislocation distance controls the amount of fault displacement. We select a point (x_0, y_0) in the area as the reference point and use a point-slope linear equation with a slope of k.

$$\frac{Y - y_0}{X - x_0} = k \tag{A22}$$

When the horizontal displacement Δx is determined, the fault can be generated as:

$$M(X, Y) = M(X + \Delta x, Y + k\Delta x),\ X < (Y - y_0)/k + x_0, (X, Y) \in \Omega \tag{A23}$$

Equation (A23) shows that the points on the left side of the fault line move along the fault line by Δx in the horizontal direction and $k\Delta x$ in the vertical direction. Fault generation is also a random procedure. To ensure the stability of random processing and the existence of sufficient space for the PML boundary after the fault slips, a 10 m-thick layer is added around the model. As shown in Figure A7, blue represents the model area, orange shows the PML boundary, red indicates the layer reserved for fault dislocation, and light blue shows the simulation area after fault generation.

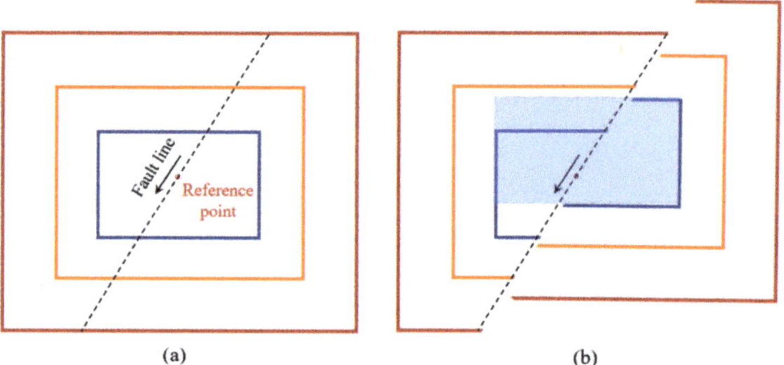

Figure A7. Fault generation illustration. (**a**) Model before fault displacement with the fault line and reference point. (**b**) Generated fault model.

Appendix B.4. Cave

Caves filled with low-velocity materials (m_c) are added into the model as anomalies by simply controlling the cave center location (x_c, y_c) and radius r_c:

$$M(X, Y) = m_c, (X - x_c)^2 + (Y - y_c)^2 \le {r_c}^2, (X, Y) \in \Omega \tag{A24}$$

Now, we can obtain a model with horizontal parallel layers, folds, random fluctuations, faults, and caves incorporated (Figure A8).

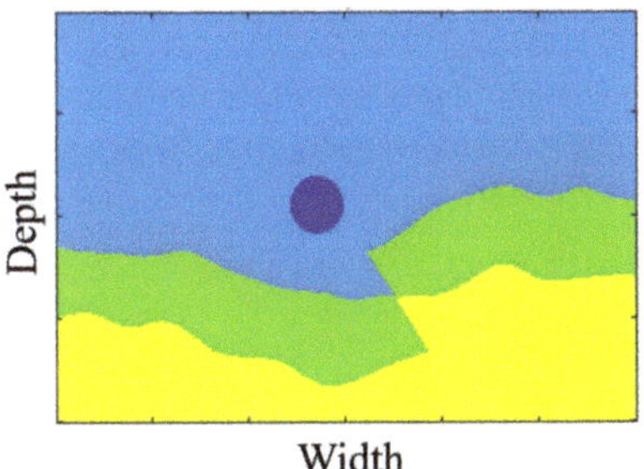

Figure A8. Model with folds, random fluctuations, faults, and caves.

References

1. Li, L.; Lei, T.; Li, S.; Zhang, Q.; Xu, Z.; Shi, S.; Zhou, Z. Risk assessment of water inrush in karst tunnels and software development. *Arab. J. Geosci.* **2014**, *8*, 1843–1854. [CrossRef]
2. Xue, Y.; Li, S.; Ding, W.; Fang, C. Risk Evaluation System for the Impacts of a Concealed Karst Cave on Tunnel Construction. *Mod. Tunn. Technol.* **2017**, *54*, 41–47.
3. Lei, M.; Gao, Y.; Jiang, X. Current Status and Strategic Planning of Sinkhole Collapses in China. In *Engineering Geology for Society and Territory*; Springer: Berlin/Heidelberg, Germany, 2015; Volume 5, pp. 529–533.
4. Gutiérrez, F.; Parise, M.; de Waele, J.; Jourde, H. A review on natural and human-induced geohazards and impacts in karst. *Earth Sci. Rev.* **2014**, *138*, 61–88. [CrossRef]
5. Li, X.; Dou, S.E.; Qu, H. A View on Application and Development of Engineering Geophysical Prospecting and Testing in City. *Chin. J. Eng. Geophys.* **2008**, *05*, 564–573.
6. Bakulin, A.; Silvestrov, I.; Dmitriev, M.; Neklyudov, D.; Protasov, M.; Gadylshin, K.; Dolgov, V. Nonlinear Beamforming for Enhancement of 3D Prestack Land Seismic Data. *Geophysics* **2020**, *85*, 283–296. [CrossRef]
7. Zhang, J.; Liu, S.; Yang, C.; Liu, X.; Wang, B. Detection of urban underground cavities using seismic scattered waves: A case study along the Xuzhou Metro Line 1 in China. *Near Surf. Geophys.* **2020**, *19*, 95–107. [CrossRef]
8. Bednar, J.B. A brief history of seismic migration. *Geophysics* **2005**, *70*, 3MJ–20MJ. [CrossRef]
9. Baysal, E.; Kosloff, D.D.; Sherwood, J.W.C. Reverse time migration. *Geophysics* **1983**, *48*, 1514–1524. [CrossRef]
10. Xue, Z.G.; Chen, Y.K.; Fomel, S.; Sun, J.Z. Seismic imaging of incomplete data and simultaneous-source data using least-squares reverse time migration with shaping regularization. *Geophysics* **2016**, *81*, S11–S20. [CrossRef]
11. Hill, N.R. Gaussian-Beam Migration. *Geophysics* **1990**, *55*, 1416–1428. [CrossRef]
12. Hill, N.R. Prestack Gaussian-beam depth migration. *Geophysics* **2001**, *66*, 1240–1250. [CrossRef]
13. Yang, F.; Sun, H. Application of Gaussian beam migration to VSP imaging. *Acta Geophys.* **2019**, *67*, 1579–1586. [CrossRef]
14. Vidale, J. Finite-difference calculation of travel times. *Bull. Seismol. Soc. Am.* **1988**, *78*, 2062–2076.
15. Zhang, K.; Cheng, J.B.; Ma, Z.T.; Zhang, W. Pre-stack time migration and velocity analysis methods with common scatter-point gathers. *J. Geophys. Eng.* **2006**, *3*, 283–289. [CrossRef]
16. Yuan, Y.; Gao, Y.; Bai, L.; Liu, Z. Prestack Kirchhoff time migration of 3D coal seismic data from mining zones. *Geophys. Prospect.* **2011**, *59*, 455–463. [CrossRef]
17. Wang, Y.; Xu, J.; Xie, S.; Zhang, L.; Du, X.; Chang, X. Seismic imaging of subsurface structure using tomographic migration velocity analysis: A case study of South China Sea data. *Mar. Geophys. Res.* **2015**, *36*, 127–137. [CrossRef]
18. Liu, Q.; Zhang, J. Trace-imposed stretch correction in Kirchhoff prestack time migration. *Geophys. Prospect.* **2018**, *66*, 1643–1652. [CrossRef]
19. Havlice, J.F.; Taenzer, J.C. Medical ultrasonic imaging: An overview of principles and instrumentation. *Proc. IEEE* **1979**, *67*, 620–641. [CrossRef]
20. Kollmann, C. Diagnostic Ultrasound Imaging: Inside Out (Second Edition). *Ultrasound Med. Biol.* **2015**, *41*, 622. [CrossRef]
21. Capon, J. High-resolution frequency-wavenumber spectrum analysis. *Proc. IEEE* **1969**, *57*, 1408–1418. [CrossRef]
22. Tie-Jun, S.; Kailath, T. Adaptive beamforming for coherent signals and interference. *IEEE Trans. Acoust. Speech Signal Process.* **1985**, *33*, 527–536. [CrossRef]
23. Mann, J.A.; Walker, W.F. A constrained adaptive beamformer for medical ultrasound: Initial results. In Proceedings of the IEEE International Ultrasonic Symposium, Munich, Germany, 8–11 October 2002; pp. 1807–1810.
24. Sasso, M.; Cohen-Bacrie, C. Medical ultrasound imaging using the fully adaptive beamformer. In Proceedings of the IEEE International Conference on Acoustics, Speech, and Signal, Philadelphia, PA, USA, 18–23 March 2005.

25. Synnevag, J.F.; Austeng, A.; Holm, S. Adaptive Beamforming Applied to Medical Ultrasound Imaging. *IEEE Trans. Ultrason. Ferroelectr. Freq. Control.* **2007**, *54*, 1606–1613. [CrossRef]
26. Asl, B.M.; Mahloojifar, A. Minimum Variance Beamforming Combined with Adaptive Coherence Weighting Applied to Medical Ultrasound Imaging. *IEEE Trans. Ultrason. Ferroelectr. Freq. Control.* **2009**, *56*, 1923–1931. [CrossRef]
27. Ma, X.; Peng, C.; Yuan, J.; Cheng, Q.; Xu, G.; Wang, X.; Carson, P.L. Multiple Delay and Sum with Enveloping Beamforming Algorithm for Photoacoustic Imaging. *IEEE Trans. Med. Imag.* **2020**, *39*, 1812–1821. [CrossRef] [PubMed]
28. Chen, J.; Chen, J.; Zhuang, R.; Min, H. Multi-Operator Minimum Variance Adaptive Beamforming Algorithms Accelerated with GPU. *IEEE Trans. Med. Imag.* **2020**, *39*, 2941–2953. [CrossRef]
29. Liu, L.; Shi, Z.; Peng, M.; Liu, C.; Tao, F.; Liu, C. Numerical modeling for karst cavity sonar detection beneath bored cast in situ pile using 3D staggered grid finite difference method. *Tunn. Undergr. Space Technol.* **2018**, *82*, 50–65. [CrossRef]
30. Margrave, G.F.; Lamoureux, M.P. *Numerical Methods of Exploration Seismology: With Algorithms in MATLAB®*; Cambridge University Press: Cambridge, UK, 2019.
31. Graves, R.W. Simulating seismic wave propagation in 3D elastic media using staggered-grid finite differences. *Bull. Ssmol. Soc. Am.* **1996**, *86*, 1091–1106.
32. Wang, D.-Y.; Ling, Y. Phase-shift- and phase-filtering-based surface-wave suppression method. *Appl. Geophys.* **2016**, *13*, 614–620. [CrossRef]
33. Hudson, J.E. Introduction to Adaptive Arrays. *Electron. Power* **1981**, *27*, 491. [CrossRef]
34. Ning, M.; Goh, J.T. Efficient method to determine diagonal loading value. In Proceedings of the 2003 IEEE International Conference on Acoustics, Speech, and Signal Processing, Hong Kong, China, 6–10 April 2003.

MDPI
St. Alban-Anlage 66
4052 Basel
Switzerland
www.mdpi.com

Applied Sciences Editorial Office
E-mail: applsci@mdpi.com
www.mdpi.com/journal/applsci

www.ingramcontent.com/pod-product-compliance
Lightning Source LLC
LaVergne TN
LVHW071613170726
843515LV00009B/2386

* 9 7 8 3 0 3 6 5 9 7 8 8 1 *